Wiltshire Record Society

(formerly the Records Branch of the Wiltshire Archaeological and Natural History Society)

VOLUME 74

1841 June 30th — **Swede's**, finish'd sow'g with Machine
= upper part of Sandhill Piece
Drying day some little rain

" **Vetch's**, the 2 Acres of Winter **Finish'd**
= full old at bottom. began 4th June

" **Vetch's** Spring Sheep began, in full bloom
= sow'd 24th March

Died at Calne at 12 o'clock at night, after
a short illness, very much respected
Mr. Thos. Bethell
Aged 43 Years 14th Novr last
his Father Saml. died the 11th May 1820 aged 51

" **Railroad**, opened from London to Bristol
July 1st — Plough'g a Vetch drift for Turnips
" **Rain**, nearly all day
" **Poney**, returned from Kennett
2 — Dryer day some little rain
" **Turnips**, sow'd the drift winter Vetches fed
Horses Thrash'g Wheat
3 — Do Do finish'd Mow
" **Poney**, to Cheverell present'd to Mr. L
" Very fine day the first since 18th June
4 Sunday, some storms in morn'g, very warm
5 — Foggy morn'g fine drying day
" **Hay**, began carry'g the Field, the first
= cutting quite **Spoil'd**, being cut
= 21 days & **rain** on it 14 days

A typical page of Thomas Pinniger's diary, entries for June and July 1841

THE FARMING DIARIES OF THOMAS PINNIGER 1813–1847

edited by

ALAN WADSWORTH

CHIPPENHAM

2021

Published on behalf of the Wiltshire Record Society
by The Hobnob Press,
8 Lock Warehouse, Severn Road,
Gloucester GL1 2GA
www.hobnobpress.co.uk

c/o Wiltshire and Swindon History Centre,
Cocklebury Road, Chippenham SN15 3QN

www.wiltshirerecordsociety.org.uk

ISBN 978-0-901333-51-3

Typeset by John Chandler

CONTENTS

Foreword vi
Preface vii
Lists of Tables and Illustrations viii

INTRODUCTION ix
The Bedwyn Years xvi
From Bedwyn to Beckhampton via Chippenham lviii
The Beckhampton Years lxvii
Bibliography cxl
Editorial Note cxlix
List of Abbreviations clii
Glossary cliii
Genealogical Tables clxi
Illustrations clxiv

THE DIARIES 1823-1838
1823 1
1824 22
1825 45
1826 56
1827 67
1828 80
1829 96
1830 121
1831 146
1832 167
1833 188
1834 212
1835 238
1836 262
1837 286
1838 308

APPENDICES (with list) 333
INDEX OF PERSONS AND PLACES 387
INDEX OF SUBJECTS 404

List of officers 411
List of members 411
List of publications 414

FOREWORD

The diaries of Thomas Pinniger provide an opportunity to understand the pressures and influences on the agricultural world in the first half of the nineteenth century. The effects of the weather on the work of the farm, the fluctuations in livestock and grain prices, along with the interventions by the government of the day, are all reflected in his diary entries, and these are no different to the modern-day constraints which I encounter as a tenant farmer in the south of the county of Wiltshire.

Whilst focussed on the management and operation of his farms, he makes reference to the wider community in which he operated. In particular, during his time at Beckhampton, he purchased an inn and constructed the first racing stables there in 1835. The fact that these are the origins of an establishment which still operates today is heartening to discover, involved as I was in the racing industry in my earlier years.

I am sure that he would have been very proud to know that his descendants are still farming in the county today, continuing the family's agricultural heritage, albeit using different technologies at a different scale. Whilst farming is still about looking after the health of the soil to produce food, the heart of agriculture lies in the people involved, about whom Pinniger provided glimpses into their characters and the conditions in which they lived and worked. Overall, this book enables a better understanding of the operation of the farming world in one part of Wiltshire from two centuries past, and for this to be more widely known and appreciated.

Minette Batters
President, National Farmers Union
Downton, Salisbury

PREFACE

I would like to thank the Wiltshire Record Society for the opportunity to prepare this volume, giving me a foothold back into the county of Wiltshire where I lived for 27 years.

In preparing this volume, the guidance and process knowledge provided by Steven Hobbs has made a major contribution to the end result: in addition, Claire Skinner has provided encouragement along the way and comments on an early draft, thereby ensuring that this volume made it into print. I am also grateful to John Chandler for ensuring that the volume has been produced to high professional standards, in line with other volumes in this series. The staff at the Wiltshire and Swindon History Centre have been as helpful as ever, as have the staff at the National Archives at Kew, enabling me to consult a wider range of records than simply the diaries themselves, thereby unlocking and explaining some of the conundrums within. I am also grateful to the Pinniger family for their permission to use the diaries of Thomas Pinniger as the basis for this volume.

Finally, I would like to thank my wife Sally for proof-reading the early drafts of my manuscripts and for her encouragement throughout this project.

Alan M Wadsworth (Dr)
Badsey, April 2021

LIST OF TABLES

Table 1: Land at Mansion House, 1809
Table 2: Land at Home Farm, 1809
Table 3: Land at Jockey Farm, 1809
Table 4: Arable crops – sowing rates
Table 5: Wheat yields at Little Bedwyn, 1814
Table 6: Wool weights, 1815
Table 7: Lot 2 – Sambourne Farm
Table 8: Summary of Terrier of Beckhampton Farm, 1788
Table 9: Land at Beckhampton Farm, 1794
Table 10: Land at Beckhampton Farm, 1794, as recorded by Pinniger
Table 11: Land at Beckhampton Farm, 1846
Table 12: Beckhampton Inn, 1846
Table 13: Days of rain at Beckhampton Farm by month, 1829
Table 14: Days with and without rain at Beckhampton Farm, 1839
Table 15: Settings for seed drill, 1846
Table 16: Crop rotation on the down at Beckhampton Farm, 1829 to 1836
Table 17: Calculation of fine at Melksham Bridge toll gate, 1840
Table 18: Heights of Pinniger family members
Table 19: Weights of Pinniger family members as at 14 December 1837
Table 20: Road measurement, Avebury, 1839

LIST OF ILLUSTRATIONS

Figure 1: The farming diaries of Thomas Pinniger - eight volumes
Figure 2: Extract from Volume 7 of the diaries
Figure 3: Dial from a barometer made c. 1815
Figure 4: Lester's cultivator or seven-share plough
Figure 5: Extract from diary for 7 March 1820
Figure 6: Extract from 1784 Powell's map of Chippenham
Figure 7: Hurdle stack with thatched roof at Priddy, Somerset

INTRODUCTION

THE DIARIES

Thomas Pinniger kept a diary from 1813 to 1847 in eight volumes, written in a neat consistent hand.[1] During this period, he lived at Little Bedwyn (1813 to 1825), Chippenham (1825 to 1829) and Beckhampton (1829 to 1847). Although he was not farming whilst residing in Chippenham, his interest in agricultural activities is clear from the diaries. The majority of the text contained in the diaries are daily entries and most agricultural activities are covered including livestock husbandry, arable husbandry, labour, farm buildings and equipment, as well as the weather. Whilst the over-riding content of the diaries is related to farming, there are occasional entries relating to external events and family, as well as the recording of the deaths of people close to, or known by, Pinniger himself. This became more frequent in the diaries from 1824 onwards, reflecting the impact of the passage of time on their mortality: initially these related to his own family members and then widened to include friends and acquaintances or worthy locals. With the exception of the first couple of years, the eight diaries present a continuous daily record of farming activities, although the records for Sundays tend to be limited to the weather. A further volume was kept by his son, Thomas Large Pinniger from 1 January 1844 until 31 December 1857.[2]

There are few records in the public domain describing detailed daily agricultural activity on individual Wiltshire farms for the first half of the nineteenth century. Whilst no other farming diaries are known to

---

1 WSA: 4381/1/1, *Thomas Pinniger's farming diary, Little Bedwyn, 1813-1816;* WSA: 4381/1/2, *Thomas Pinniger's farming diary, Little Bedwyn, 1817-1819;* WSA: 4381/1/3, *Thomas Pinniger's farming diary, Little Bedwyn 1820-1823;* WSA: 4381/1/4, *Thomas Pinniger's farming diary, Little Bedwyn and Chippenham, 1823-1828;* WSA: 4381/1/5, *Thomas Pinniger's farming diary, Beckhampton farm, Avebury, 1828-1832*; WSA: 4381/1/6, *Thomas Pinniger's farming diary, Beckhampton farm, Avebury, 1832-1838*; WSA: 4381/1/7, *Thomas Pinniger's farming diary, Beckhampton farm, Avebury, 1839-1846*; WSA: 4381/1/8, *Thomas Pinniger's farming diary, Beckhampton farm, Avebury, 1847 Jan-Mar.*

2 WSA: 4381/1/9, *Thomas Pinniger's farming diary, Beckhampton farm, Avebury, 1844-1857.*

exist in public archives for this period covering Wiltshire, it is not known how many others are still to be discovered in dusty lofts or lodged on the top of dressers in the parlours of Wiltshire houses. There are however farm account books to be found in various archives.[3] Further, only a small number of farming diaries have been published by record societies in other counties, including Hertfordshire, Norfolk and the West Riding of Yorkshire.[4]

The value of these diaries is considerable. The detailed accounts of farming activities, on a daily basis, provide evidence of agricultural husbandry in part of Wiltshire: at both Little Bedwyn and Beckhampton, Pinniger was practising sheep-and-corn husbandry.[5] Texts exist which

---

3 For example, at the Wiltshire and Swindon History Centre (WSHC): WSA: 939/1-9, *Farm account books: Wyndham family estates in Dinton, Teffont Magna and Sutton Mandeville, 1821-1850;* at the Museum of English Rural Life (MERL): MERL: FR WIL 11/2/4-6, *Farm account books of Aldbourne Farm, Baydon Farm and Warren Farm, Aldbourne, Co. Wilts, 1836-1856*; MERL: FR WIL 11/2/18-20, *Account books of corn sold from Aldbourne Farm, Baydon Farm and Warren Farm, Aldbourne, Co. Wilts, 1831-1859*; MERL: FR WIL 11/2/23, *Account book of sheep bought and sold for Aldbourne Farm and Warren Farm, Aldbourne, Co. Wilts, 1831-1879.*

4 Susan Flood (ed.), *The diary of John Carrington, Farmer of Bramfield, 1798-1810: Volume 1, 1798-1804* (Rickmansworth: Hertfordshire Record Society, 2015); Susan Flood (ed.), *The diaries of John Carrington, Farmer of Bramfield, 1798-1810: Volume 2, 1805-1810 and John Carrington junior, 1810-1812* (Rickmansworth: Hertfordshire Record Society, forthcoming); Susanna Wade Martins and Tom Williamson (eds), *The farming journal of Randall Burroughes (1794-1799)* (Norwich: Norfolk Record Society, 1995); Richard Davies, Alan Petford and Janet Senior (eds), *The diaries of Cornelius Ashworth 1782-1816* (Hebden Bridge: Hebden Bridge Local History Society, 2011).

5 Sheep-and-corn husbandry was the practice of keeping sheep to provide the necessary nutrients for corn crops. The associated practice of 'folding' or 'sheepfold' involved grazing the sheep during the day and then walking them to, and folding (enclosing using hurdles) them at night on the land to be used for arable crops. The daytime pasture comprised the water-meadows, the downs and root crops grown for winter feed at different times of the year. Stocking was intensive with around 2,000 sheep per acre: the treading of the sheep's feet consolidated the seedbed and their dung provided nitrogen for the next arable crop. The sheepfold was used in the summer and autumn for wheat and in the winter and spring for turnips and barley. In this way, the feed consumed during the day was converted into fertiliser at night. G. G. S. Bowie, 'Northern Wolds and Wessex Downlands: Contrasts in Sheep Husbandry and Farming Practice, 1770-1850', *Agricultural History Review*, 38.2 (1990), 121. Edward Little, 'Farming of Wiltshire', *Journal of the Royal Agricultural Society of England*, V (1845), 168.

describe the general pattern of such practices, but here we have descriptions specific to two farms in different locations within the county.[6]

Agricultural diaries and journals such as this are an important source of information for the study of agricultural history, in that they are written by practising farmers, rather than by interested observers of the time.[7] They can also be an important source of information on the weather, which played a crucial part in the success or failure of the farming year, and Pinniger is particularly helpful in this. Each volume is entitled Farming Memorandums which suggests that it was his intention that these notebooks should be a source of reference for his farming activities, rather than simply a diary. Indeed, at certain points along the way, he makes notes of rules to be followed in the future, as well as year-on-year comparisons of yields. They are also important in that they allow a local view of agriculture to be compared with printed agricultural material from the time, as well as published summaries.[8] These diaries also present a different view, in that neither farm was part of a large estate with the resources to introduce improvements in livestock breeding, use of machinery and new crop varieties which were being practised across the broader farming community at this time. Notwithstanding this, it will be seen that he does make improvements in his own way demonstrating that he was aware of wider developments in agriculture at the time.

It is important to bear in mind that Pinniger wrote about what interested him. Given this, the diaries lack the statistical detail to allow a full analysis of agricultural productivity and economics at the time. Whilst there are some details of crop yields and animal growth rates (transcribed in the Appendices), there is little to enable an understanding of wage rates or standards of living for example. However, what they do provide is a record of daily activity for two specific farms located in the centre of Wiltshire over a period of thirty-four years. Whilst these diaries can also provide useful information for social historians, the introduction which follows will examine the content from an agricultural viewpoint, given that they are after all primarily concerned with farming activity.

---

6 For example: G. E. Mingay (ed.), *The agrarian history of England and Wales, Volume VI, 1750-1850* (Cambridge: Cambridge University Press, 1989); Thomas Davis, *General View of the Agriculture of Wiltshire* (London: Board of Agriculture, 1811).

7 For example: Arthur Young, *A six week tour through the southern counties of England and Wales, 2nd edition* (Dublin, Ireland: J. Milliken, 1771); William Cobbett, *Rural Rides* (London: William Cobbett, 1830).

8 For example: John Sinclair, *The Code of Agriculture* (London: Sherwood, Neely and Jones, 1817); *Farmer's Magazine*, published quarterly from 1800.

The diaries are bound in leather and measure 19cm (7½ inches) x 12cm (4¾ inches) x 3cm (1¼ inches), with slight variations in thickness: each volume has between 360 and 420 pages, although only nine pages are used in Volume 8. The spines are embossed with lettering and a variety of designs in gold leaf. As well as the title Farming Memorandums, each volume is numbered in sequence in Roman numerals. The exact covering dates are given for the first three volumes: see Figure 1. A double-page spread in shown in Figure 2 which gives an indication of the appearance of the diary entries.

Enclosed in the final volume is a loose tri-folded sheet of paper: inside this sheet, there are two columns of notes, with the centre left blank.[9] The notes on the left are clearly draft diary entries and cover the period of 13 days from 20 March to 1 April 1847, following the last entry for 20 March in Volume 8.[10] From this, it appears that Pinniger compiled his diary by making initial notes which were written up soon after: this approach probably explains the neatness of the completed diaries. However, there was some sudden disruption to his diary keeping, in that he stopped writing up his diary in the middle of 20 March, with some entries for that date in the diary and others on the loose sheet. Whilst the notes covering the previous years would have been destroyed, the survival of this single sheet could be attributed to the fact that Pinniger was unable to write it up, due to illness or incapacity, after 34 years of meticulous and methodical diary-keeping.

The editor first became aware of the diaries in March 2018, when he was invited to examine them at the farmhouse of Mr Peter Pinniger,

---

9 WSA: 4381/1/8, *Thomas Pinniger's farming diary, Beckhampton farm, Avebury, 1847 Jan-Mar.*

10 The year of the entries on this sheet has been determined by cross-referencing in two ways. Firstly, on 24 March, Pinniger notes *General Fast and Humiliation. All Labour suspended*: on that date in 1847, there was *a public fast and humiliation to obtain pardon of our sins and to send up prayers and supplications to the Divine Majesty for the removal of [a scarcity and dearth of divers articles of sustenance and necessaries of life]*, Philip Williamson, 'State Prayers, Fasts and Thanksgivings: Public Worship in Britain 1830-1897', *Past & Present,* 200 (2008), 172. There was concern at the time about fast days leading to the loss of a day's labour. Secondly, Pinniger notes *W Tanners Funeral at Avebury*: William Tanner of Blackland House buried on 25 March 1847 at Avebury, aged 69, *Wiltshire, England, Church of England Deaths and Burials, 1813-1916 for William Tanner* [accessed via Ancestry, 30 July 2020]. This also links to the diary entry on 18 March 1847: *Died suddenly, last night, being at Calne Market in his usual health in the morn'g Mr William Tanner of Blackland and Kennett Farm Aged […].*

a descendant of Thomas Pinniger. Following that visit and discussions with staff at the Wiltshire and Swindon History Centre (hereafter WSHC), the diaries were then archived in the Wiltshire and Swindon Archives (hereafter WSA) and permission was given by Peter Pinniger for the diaries to be used as the basis for this volume. Apart from a brief mention in a book published some twenty years ago, without reference or acknowledgement, there is no other known scholarship relating to these diaries.[11]

The Pinniger genealogical memoranda book has also been used to prepare this volume: it was compiled initially by Thomas Pinniger probably around the time of his father's death, in that the following inscription appears at the front: *Thomas Pinnigers Memorials of his Forefathers, as given him by his Father, a short time before his Death.*[12] With subsequent material added by members of the family, it covers a period from the early-eighteenth century to the mid-twentieth century, and contains descriptions of various Pinniger families over this period, as well as other Wiltshire families including Large, Spackman, Alexander and Henly. This has been used as the basis of four genealogical charts included in this volume:

The children of Christopher and Susannah Pinniger (Thomas's parents);
The parents of Thomas Pinniger and Mary Pinniger (neé Large);
The children of Thomas Pinniger and Mary Pinniger (neé Large).

Thomas was born on 6 November 1779 and baptised on 15 November that year at Compton Bassett.[13] His parents Christopher and Susannah had married in the same church on 12 May 1774, Susannah Alexander's parish being stated as Hilmarton: Christopher was recorded as a yeoman in the marriage register, whilst in his will, he described himself as a gentleman.[14] His father was born on 10 March 1748 and baptised

---

11 Patricia Parslew, *Beckhampton: Time present and time past* (Salisbury: The Hobnob Press, 2004).

12 WSA: 4381/2/1, *Pinniger genealogical memoranda book, c.1820-1943*, p.5.

13 WSA: 4381/2/1, *Pinniger genealogical memoranda book, c.1820-1943*, p.8. *Wiltshire, England, Church of England Baptisms, Marriages and Burials, 1538-1812 for Thomas Pinegar* [accessed via Ancestry, 8 July 2020].

14 *Wiltshire, Church of England, Marriages and Banns, 1754-1916 for Christopher Pinnigar* [accessed via Ancestry, 10 August 2020]; TNA: PROB11/1738/65, *Will of Christopher Pinniger, Gentleman of Bremhill, Wiltshire, 7 March 1828;* WSA: 4381/2/1, *Pinniger genealogical memoranda book, c.1820-1943*, p.8. Susannah died on 27 October 1789, WSA: 4381/2/1, *Pinniger genealogical*

at Compton Bassett on 14 March 1748, the son of John Pinniger.[15] His mother was also baptised at Compton Bassett on 26 November 1749, the daughter of William Alexander.[16]

On 28 December 1820, Thomas Pinniger married Mary Large, born 6 April 1794, the daughter of Abbot Large and Mary Large (neé Henly), at Lyneham church.[17] As an adult, we know from entries in his diaries that he was three-quarters of an inch short of six feet tall (1.81m) and that he weighed around 11¾ stone (75kg). He died at Beckhampton aged 67 on 5 April 1847 and was buried a week later at Lyneham church.[18]

APPROACH TAKEN

With regard to the transcription of these diaries, which forms the main part of this volume, an initial assessment was made which determined that it would not be possible to transcribe all of the diaries into a single volume. The entries in earlier diaries tend to be somewhat sparse and repetitive from one year to the next and on that basis, transcription started with the entry for 1 January 1823: this provides two full farming years at Little Bedwyn as well as the move away from there in 1825. During his time in Chippenham, the entries are often single lines, not amounting to a large extent, but for continuity are included in full.

---

memoranda book, c.1820-1943, p.7. Christopher married his second wife, Mary White, at Bremhill on 28 June 1804, WSA: 4381/2/1, *Pinniger genealogical memoranda book, c.1820-1943,* p.8; *Wiltshire, Church of England, Marriages and Banns, 1754-1916 for Christopher Pinneger* [accessed via Ancestry, 8 July 2020]. Similarly, Thomas Pinniger was described as a yeoman in his marriage licence bond, WSA: D1/62/4/1820/Pinniger, *Marriage Licence Bonds, 1820.* In his will, he was described as a gentleman, TNA: PROB11/2063/389, *Will of Thomas Pinniger, Gentlemen of Avebury, Wiltshire, 5 October 1847.*

15 WSA: 4381/2/1, *Pinniger genealogical memoranda book, c.1820-1943,* p.6; *Wiltshire, England, Church of England Baptisms, Marriages and Burials, 1538-1812 for Christopher Pinniger* [accessed via Ancestry, 8 July 2020]. NOTE: New Style date given for baptism.

16 *Wiltshire, England, Church of England Baptisms, Marriages and Burials, 1538-1812 for Susannah Alexander* [accessed via Ancestry, 8 July 2020]; WSA: 4381/2/1, *Pinniger genealogical memoranda book, c.1820-1943,* pp.73-74.

17 WSA: 4381/2/1, *Pinniger genealogical memoranda book, c.1820-1943,* p.18; *Wiltshire, Church of England, Marriages and Banns, 1754-1916 for Thomas Pinniger* [accessed via Ancestry, 8 July 2020].

18 WSA: 4381/2/1, *Pinniger genealogical memoranda book, c.1820-1943,* p.7; *Wiltshire, England, Church of England Deaths and Burials, 1813-1916 for Thomas Pinniger* [accessed via Ancestry, 8 July 2020].

Finally, the diaries for his time in Beckhampton are once again extensive and contain information not only about the farm, but also the impact of external events such as the Swing Riots and the Corn Laws. After 1840, the entries return to recording the routine activities of the farm and from 1844 tend to become less extensive, sometimes being reduced to a single line per day with the weather described in a few words and the farming activities similarly summarised. This brevity may reflect the fact that Pinniger was by now in his sixties, possibly lacking the energy and drive of his earlier years: further, his son was also maintaining a diary in parallel from 1844. Therefore, the decision was made to transcribe in full the diary entries from 1 January 1823 to 31 December 1838: that is, sixteen out of the total of thirty-four years.

As for the analysis contained in this Introduction, this is based on the whole period covered by the diaries from 1813 to 1847. The content of the diaries has been examined using a number of themes: the ownership and occupation of the property, crops grown, livestock kept, animal and field husbandry practices, buildings, farm equipment and labour relations. The pervasive theme of the weather and climate are considered, along with the affairs of family, friends and neighbours, and the church, as well as the local and national events which impinged on Pinniger's life and work, to a greater or lesser extent. Throughout the diaries, the regular beat marking his life was the routine and rhythm of the agricultural year: managing the sheep fold; haymaking and harvest; the preparation and marketing of crops; and noting the first cuckoo in Spring.

OTHER SOURCES USED

In addition to the diaries themselves, a number of other sources have been used to provide background to various aspects of the diaries. With respect to the farm at Little Bedwyn, the papers of the Kent family estate, especially the particulars from the sale of the estate in 1809, have provided information about the extent of the farm at the time that Thomas Pinniger took it over in 1813.[19] Similarly, a combination of Powell's 1784 map of Chippenham and the 1830 particulars for the sale of Anthony Guy's property have enabled the farmhouse occupied by Thomas Pinniger between 1825 to 1829 to be identified.[20] Moving forward to his time at

19 WSA: 79b/9-12, *Title deeds, Kent family, 1575-1810*; WSA: 79b/26, *Estate accounts, Kent family, 1792-1808;* WSA: 79b/40-41, *Papers relating to the Kent family estate, 1790-1810.*

20 WSA: G19/1/53L, *Map of the parish of Chippenham by John Powell, 1784*;

Beckhampton, the tithe map and award for the parish of Avebury have been used to understand the extent of his farm there, along with an earlier terrier of the farm and the enclosure award.[21]

Other sources referred to have included the Land Tax assessments for Little Bedwyn, Chippenham and Beckhampton, as well as Land Tax Redemption Certificates.[22] Extracts from local newspapers and census records have been used, along with other records in both the National Archives (TNA) and the Wiltshire and Swindon Archives (WSA).

THE BEDWYN YEARS (1813 TO 1825)

Writing the diary

During this period, many of the diary entries only run to two or three lines, usually summarising the work done that day or from time to time a count of sheep stocks or crops harvested. On occasions, Pinniger would take the opportunity to summarise the position on the farm, as this example from 27 July 1815 shows:

> Ploughing Fold drift in the Breach & put both Ewes & Lambs fold together on it, intendd to do for sowing when 7 Shear'd at sowing time | Rolling & harrowing Marls & burning Couch | WHEAT the straw geting Foxey with the Blight | EWE DIED, Blow'd in NF in the evening, the Clover fed off but a few days before & just shot again with an hollow East wind, the cause | Vetches in the Jockey ground now generally blooming but very weak & spiry, the greater part sow'd 18th May | Sweeds, Parlourfield was sown 9th June, for want of rain not grown. Pulling Charlick were the dung was put, being

---

WSA: 415/53/1, *Sale particulars relating to property in Chippenham, of Anthony Guy, bankrupt, with valuation, 1830*; WSA: 4381/1/4, *Thomas Pinniger's farming diary, Little Bedwyn and Chippenham, 1823-1828.*

21 WSA: 1126/65, *Avebury tithe award and map (parish copy), 1845*; WSA: 248/195, *Terrier of Beckhampton Farm in Avebury, 1788;* WSA: EA95, *Avebury Enclosure Map, 1794.*

22 WSA: A1/345/27, *Land Tax Assessments, Little Bedwyn, 1780-1831, 1869-1881*; WSA: A1/345/99C, *Land Tax Assessments, Chippenham, 1816-1831*; WSA: A1/345/22A, *Land Tax Assessments, Beckhampton, 1782-1841*; WSA: A1/345/22B, *Land Tax Assessments, Beckhampton, 1842-1884*; TNA: IR24/15, *Land Tax Redemption Office: Registers of Redemption Certificates. Registered numbers: Nos 6000 – 6999, 1799*; TNA: IR24/16, *Land Tax Redemption Office: Registers of Redemption Certificates. Registered numbers: Nos 7000 – 7999, 1799*; TNA: IR24/43, *Land Tax Redemption Office: Registers of Redemption Certificates. Registered numbers: Nos 31000-31999, 1799.*

> Sweeds enough, - the remainer Ploughing up, being quite gone. | Sweeds in lane piece sow'd the 23rd June finish'd cuting the Thistles out of them, and began hoeing them being good in places. | Vetches in Pearces sow'd 24th June, the weather being very dry but a part came up. – the Rain yesterday & storms before brot up the remainer do not grow for want of Rain.[23]

In the early years of the diaries, there are very few entries for Sundays, one in 1813 and three in 1814: these dates are either missing or they simply state *Sunday* followed by a description of the weather and nothing more. This is a clear reflection of the fact that Sunday was not a working day at that time.

The farm

The first entry in Volume 1 of the diaries is dated 10 February 1813, in which it was announced that *Thos Pinniger commenced residing at BEDWIN Little*: however, it is important to note that this does not mean that he was only occupying Little Bedwyn Farm. At the time, he was aged 33 and a single man: to understand exactly what he had taken on, it is necessary to look back some twenty to thirty years before.[24]

When William Kent made his will in November 1785, his wife Elizabeth was to receive the rents, issues and profits of all of his freehold and leasehold estate in Little Bedwyn and elsewhere, holding them until the age(s) of their son(s) reached 21 years.[25] At this time, the couple had a son, William Streat Kent (baptised on 13 February 1785) and Elizabeth was pregnant with their second child, a daughter, also named Elizabeth (baptised on 11 February 1786). William Kent was buried on 22 February 1786.[26]

An early view of the estate comes from an indenture drawn up in October 1788 between the Trustees of the late William Kent and Joseph Hawkins for Little Bedwyn Farm and Jockey Farm.[27] The lease was to be for twenty years from 29 September 1786 (when son William Streat

---

23 As explained in the editing conventions, the original layout of the diaries has not been preserved, given the narrow width of the pages. Entries within a single day are separated with a vertical stroke as in this example. Vetches – also known as tares.

24 WSA: 4381/2/1, *Pinniger genealogical memoranda book, c.1820-1943*, p.8.

25 WSA: 315/27/5, *Copy will of William Kent of Little Bedwyn, 1785;* TNA: PROB11/1140/178, *Will of William Kent, Gentleman of Little Bedwin, Wiltshire, 21 March 1786.*

26 WSA: 1955/25, *Parish records, Little Bedwyn, 1730-1813.*

27 WSA: 79b/11, *Title deeds, Kent family, 1575-1810.*

would have reached the age of 21) at an annual rent of £201 1s 0d, with a rebate of £20 for the first four years, suggesting that the land and/or the buildings needed some improvement. Hawkins remained at the farm up to 1790: in 1792, Mr Whit[e]lock[e], who appears to have been acting as agent for the guardians of the estate of William Kent junior, had to settle with Hawkins for two-and-a half years rent which was owing, albeit that he was allowed half-year land tax amounting to £9 15s 4d.[28] Thereafter a number of tenants occupied the farm.[29]

In March 1804, William Streat Kent was in Bengal, India where he had joined the Artillery (or Engineers).[30] His situation as an absentee landlord with inadequate supervision of his tenants appears to have led to the estate declining. In a letter dated 7 October 1805, W. Haynes of Shalbourne [who appears to be addressing either the trustees or Kent himself], makes a report on the poor state of Little Bedwyn Farm which he attributes to the bad management of Thomas Baker (tenant from 1794 to 1805). With Kent approaching the age of 21, Haynes notes that the intention was to sell the estate.[31] Further, he notes that:

> Nothing but good management aided by the power of a good purse can save the farm or the farmer from damnation[32]

From 1806, the land tax assessments have been realigned in respect of the estate to correspond with the three parts, namely, The Mansion House (Land Tax £4 2s 8d), Little Bedwyn or Home Farm (£12 9s 0d) and Jockey Farm (£4 4s 0d) and Mr William Kent was named as the proprietor of all three.[33] The three parts were occupied respectively by James Whitelock, Edward Horn and Giles Shepherd: at the time, Little Bedwyn Farm extended to 300 acres, whilst Jockey Farm was 140 acres.

---

28 WSA: A1/345/27, *Land Tax Assessments: Little Bedwyn, 1780-1831, 1869-1881;* WSA: 79b/26, *Estate accounts, Kent family, 1792-1808.* This land tax value equates to one half of £16 12s 4d and £2 18s 4d, being the two entries in the land tax assessment.

29 WSA: 1033/89, *Deed of the manor house in Little Bedwyn and Jockey Farm and Little Bedwyn Farm, 1792*; WSA: A1/345/27, *Land Tax Assessments: Little Bedwyn, 1780-1831, 1869-1881.*

30 The Asiatic Annual Register, Volume 7, 1807, p.121 (via Google Books); The Asiatic Annual Register, Volume 8, 1809, p.164 (via Google Books).

31 WSA: 79b/12, *Title deeds, Kent family, 1575-1810.*

32 WSA: 79b/12, *Title deeds, Kent family, 1575-1810.*

33 WSA: A1/345/27, *Land Tax Assessments: Little Bedwyn, 1780-1831, 1869-1881.*

William Streat Kent died intestate on 30 March 1808 at Berhampore, India.[34] As a result, the Little Bedwyn estate descended to his sister, Elizabeth, through the rules of intestacy.[35]

On 22 November 1809, a sale by auction of the reputed manor of Little Bedwyn was held in London: it is from these sales particulars that a detailed description of the estate can be derived.[36] It was noted that the buildings were in a good state in that a large sum had been spent recently on repairs and improvements (*the power of a good purse* – see above). Lot 1 comprised the Mansion House, described as a very comfortable family house with three reception rooms and six bedrooms along with the usual domestic offices. In a detached courtyard, there were two coach houses, stabling for six and a granary, all of which were brick-built and tiled, with a farmyard contiguous to the courtyard with a large brick & tile barn, cowhouse and stable. Table 1 presents a summary of the land sold with the Mansion House.

Table 1: Land at Mansion House, 1809

	No	*Description*	*Usage*	*a*	*r*	*p*
	12, 13	Garden, Nursery, Orchard, Woodyard and close		6	1	10
		Cottage & Garden in Cowleaze Close Cottage & garden in Farm Close		0	1	26
	3, 4	Sprout's Godding's & Robin's Meadow		3	1	2
*	22	Cowleaze	Pasture	13	0	23
*	34, 36	Brick Kiln Dell (now in one)	Pasture	12	3	36
*	28	Bonning's	Meadow	2	0	0
*	15, 16	Pedlar's Pieces (now in one)	Arable	8	0	23
*	17	Gate Close	Arable	3	2	10
*	20	Mooshing's Ground	Arable	4	1	39

34 Berhampore is a city in the state of West Bengal, India. Calcutta Gazette, 12 May 1808. The London Gazette, part 1, 1811, p.625 (via Google Books): William Streat Kent's name (now captain of artillery) appeared here in a list of intestates in Bengal. That he died intestate is contrary to a schedule of deeds and papers relating to the general title to all parts of the Little Bedwyn estate, which lists the will of Wm. Streat Kent, dated 20 March 1804, probably written before he departed for India. WSA: 79b/40, *Papers relating to the Kent family estate, 1790-1810.*

35 Later in 1811, the Salisbury and Winchester Journal (8 April 1811) reported "Lately was married John Farmer Newton Esq. of Jesus College, Cambridge, to Miss Kent, daughter and heiress of the late William Kent Esq of Little Bedwin".

36 WSA: 315/28, *Printed sale particulars relating to properties, including the manor of Little Bedwyn, 1809.*

	No	*Description*	*Usage*	*a*	*r*	*p*
★	23	Sheppard's Little Ground	Arable	2	3	4
★	25	Sheppard's Great Ground	Arable	6	2	24
★	29	Bonning's Four Acres	Arable	4	1	26
★	32	Six Acres	Arable	5	2	32
★	33	Eighteen Acres	Arable	18	1	8
		An allotment in East Field, adjoining the Orchard (14)	Arable Pt Clover and Santfoin	21	3	37
★ - These lands in the occupation of Mr Edward Horne, tenant of Lot 2			Sub-total	114	0	20
		Woodland in hand		20	0	34
			Total	134	1	14

Lot 2 was Home Farm, described as an excellent, highly improveable farm, adjoining Lot 1. As well as a very comfortable farmhouse, described as substantially brick-built and recently repaired at great expense, the farm buildings comprised two large barns, stables, cart and cow houses, granary, rick house, store-staddles, and convenient outbuildings with rick barton, yard, garden and orchard. This lot also included a pew in the church and the particulars confirm that the farm was let to Edward Horne for fourteen years from Michaelmas 1805 at £1 per acre, determinable by the landlord after seven years.

Table 2: Land at Home Farm, 1809

No	*Description*	*Usage*	*a*	*r*	*p*
7, 8, 9	Cottage called Greenways with barn, outhouse, yard and garden		2	0	15
5	Court	Meadow	2	1	37
2	East	Water Meadow	3	3	34
39, 40	The Moors (now in one)	Pasture	5	2	12
42	Picked Ground	Meadow	3	1	4
48	Hitching's Ground	Arable	6	0	9
41	Bartlett's Five Acres	Arable	5	1	6
	A border of wood and hedge in the land adjoining No. 41		0	1	13
1	North Field Allotment, including the river and the greenway	Arable	121	1	39
49	South Field Allotment (part) including the Meadow cut off by the canal (late Potter's)	Arable	110	1	15
	A cottage in two tenements in Bartlett's Mead (No. 43)		0	0	30

No	*Description*	*Usage*	*a*	*r*	*p*
44	Bartlett's	Old Meadow	4	2	16
45	Bartlett's Ground, with a small border of Wood	Arable	5	1	29
		Sub-total	271	0	19
	Freehold woodlands		21	1	16
		Total	292	1	35

Finally, Jockey Farm, a very desirable and improvable farm, was sold as Lot 3, adjoining Lots 1 and 2. In addition to the neat and comfortable farmhouse (brick built in very good repair) there were two barns, a large stable and carthouse along with newly-built buildings with a yard: again, a pew in the church was included in this lot. The farm was let to Giles Sheppard at £130 per annum on a lease which ended at Michaelmas 1812. Table 3 describes the land at Jockey Farm at the time of the sale.

Table 3: Land at Jockey Farm, 1809

No	*Description*	*Usage*	*a*	*r*	*p*
59	Buildings and yard		0	2	0
51	Paradise	Water Meadow	0	1	31
52	Spain, including the river	Water Meadow	3	1	10
70	Merril Down Allotment	Pasture	24	2	26
69	Bushell's Great Ground and borders	Arable	5	0	16
67	Bushell's Ground	Arable	2	3	33
66	Bushell's Hill Ground	Arable	3	3	17
71	Madam's Ground	Arable	2	1	26
73	Great West Leaze	Arable	7	1	36
42	Three Corners' Ground	Arable	3	0	19
64	West Borough and border	Arable	8	0	6
56	Parlour Field	Arable	21	1	26
53	Three Acre	Arable	2	1	1
57 61	Hither Cope's Ground Further Cope's Ground (now in one)	Arable	12	1	34
50	Marles	Arable	18	1	16
58 60	Hither Jockey Ground Further Jockey Ground (now in one)	Arable	10	2	29
49	Wansdike Furlong (an allotment in South Field)	Arable	14	1	36
		Sub-total	141	2	2
	Freehold woodlands		20	3	29
		Total	162	1	31

The outcome of the 1809 sale is unclear. A copy of an agreement of sale held at the Wiltshire and Swindon Archives (written on a further printed copy of the sales particulars) shows that Thomas Scott of London as agent for James Whitelock sold the manor of Little Bedwyn, along with the farms as described in the sales particulars, to Richard Debary, agent for Richard Randall of Masons Hall, Basinghall Street, London for £25,500.[37] Therefore, it would appear that a sale was completed: however, the land tax assessment for 1810 still named Miss Kent as proprietor of the estate, but by 1811 Anthony Guy was the new proprietor. Given that the evidence does not confirm Guy as the purchaser, there are three possible explanations: firstly, that Randall was acting for Anthony Guy (in the same way as Whitelock was acting for Miss Kent); secondly, that Randall subsequently sold the estate on to Guy; or thirdly, that Guy was leasing the estate from Randall.[38]

According to the land tax records for 1811, the Mansion part of the estate was occupied by Anthony Guy himself whilst Edward Horn occupied Home Farm and Giles Shepherd was at Jockey Farm up to the end of their leases at Michaelmas 1812: thereafter, Thomas Pinniger occupied all three parts of the estate. For the period of his occupation of the farm up to 1825, Anthony Guy was listed as the proprietor at which point he sold the estate: that Guy owned, rather than leased, the estate is confirmed on the third page of Volume 5 of the diaries.[39]

---

37 WSA: 79b/40, *Papers relating to the Kent family estate, 1790-1810.* Worth £1.817m at 2018 prices. Measuring Worth, https://www.measuringworth.com/calculators/ukcompare/relativevalue.php [accessed 23 March 2020]. This price agrees with that quoted in an undated note estimating the value of Miss Kent's property, wherein it states that Little Bedwyn estate was sold, exclusive of timber and underwood, for £25,500, WSA: 79b/12, *Title deeds, Kent family, 1575-1810.*

38 WSA: A1/345/27, *Land Tax Assessments: Little Bedwyn, 1780-1831, 1869-1881.* For the first two explanations, it is not unknown for these assessments to contain outdated information. For the third possibility, it should be noted that the proprietor is not necessarily the owner: hence Guy could have been leasing the estate from the owner.

39 WSA: A1/345/27, *Land Tax Assessments: Little Bedwyn, 1780-1831, 1869-1881.* In WSA: 4381/1/5, *Thomas Pinniger's farming diary, Beckhampton, 1828-1832*, under the heading: *1829. Questions which Miss Forty of Chippenham propose puting to Mr Guy, on his becoming Insolvent* is written the following: *What have you done with the money arising from … Little Bedwin £30000 .. 0 .. 0.* The value of £30,000 for Little Bedwyn is somewhat lower than the values recorded in the diary entry of 15 January 1825 (see Appendix F) which states that the estate sold for £32,800 with a further £10,000 for the timber, live and dead stock, fixtures and makeout (see Glossary).

As well as the Little Bedwyn estate, Anthony Guy also acquired two further parcels of land in the parish. In 1799, when Mary Blandy of Reading decided to redeem the land tax (£8 10s 6d annually) on her land at Little Bedwyn, the farm consisted of one messuage with appurtenances and 170 acres of arable, meadow and pasture land, at that time in the occupation of Stephen Wentworth.[40] Eleven years later, this freehold farm in Little Bedwyn extending to 180 acres was to be sold on 20 June 1810, by order of the Executors of Mrs Blandy. The farm was still let to Mr Wentworth for an unexpired term of two years from Michaelmas 1811 and comprised arable, pasture, water meadow and woodland with a substantial farmhouse, abundant offices and labourers' cottages.[41]

Similarly, when John Pearce of Standen redeemed the land tax on his property in 1799, the annual land tax due on a messuage, malthouse and 32 acres of land in Little Bedwyn was £1 11s 8d: this land was also in the occupation of Stephen Wentworth.[42] This property was advertised for sale by auction in 1808 and was described as two freehold tenements now occupied as one dwellinghouse with malthouse, orchard, garden, meadow and 20¼ acres of arable land, with a further two closes of arable land extending to 11 acres. Stephen Winkworth (*sic*) was stated to be the tenant from year to year.[43]

The outcome of neither of the above two sales is known, but by 1811, Anthony Guy was now proprietor of both of these properties, the land tax value was unchanged and by the time of the 1813 land tax assessments, Thomas Pinniger was the occupier up to and including the year 1824.[44] Therefore, the land which he was farming at Little Bedwyn extended to 738 acres and if the woodlands were included, a total of 800 acres.

Although the estate was offered for sale in 1819, it was unsold at that date and five years later, was again offered for sale by auction on

---

40 TNA: IR24/15, *Land Tax Redemption Office: Registers of Redemption Certificates, Registered numbers 6000-6999, 1799.*

41 Public Ledger & Daily Advertiser, 18 June 1810.

42 Standen – probably Chute Standen on the Wiltshire / Hampshire border. TNA: IR24/43, *Land Tax Redemption Office: Registers of Redemption Certificates. Registered numbers: Nos 31000-31999, 1799.*

43 Salisbury & Wiltshire Journal, 18 January 1808. However, the land tax assessments indicate that Stephen Wentworth was the occupier of both of these properties from at least 1784 to 1808, WSA: A1/345/27, *Land Tax Assessments: Little Bedwyn, 1780-1831, 1869-1881.*

44 WSA: A1/345/27, *Land Tax Assessments: Little Bedwyn, 1780-1831, 1869-1881.*

22 September 1824.[45] The land tax records list John Pain as proprietor and occupier from 1825 until 1828: in his diary, Thomas Pinniger makes reference to Mr Pain from December 1824 onwards and Pinniger left the farm on 30 June 1825 (see the section *Leaving Little Bedwyn* below).

The weather

In 1813, much as it is today, farming was influenced by the weather and this is reflected in the attention given to it in these diaries, particularly in the later years. That Pinniger was the owner of a barometer is made clear in his entry for 24 November 1813 when he states *Glass continue rising*. The type of barometer he owned was probably not dissimilar to that shown in Figure 3. Whether the display on his barometer included inches of mercury (inHg) is not certain, in that he would include entries such as those below when extremes of barometric pressure were recorded:

> 3 October 1820. GLASS unusually high to the R in fair
> 17 October 1820. Glass never lower to the V in very dry

In the early years of his diaries, he tended to just record the extremes of weather, such as the high winds experienced in early February 1822:

> 2 February 1822. WIND boisterous all day, and in the night & Sunday morn an HURRICANE, tops of two Wheat Ricks blown off, and six others, some half, some partly striped, - the Barns also much injured, and very heavy storm of rain with it

and the heavy rain experienced in December that year:

> 5 December 1822. TEMPESTUOUS=DARK=NIGHT – VERY=HEAVY=RAIN

This resulted in a flood recorded in the diary on the following day and on 19 December, it was noted that ploughing had re-commenced, the first time since 8 November, the land being too wet: however, ploughing stopped again the following day and did not re-start until 4 January, after which it was intermittently held up by freezing conditions.

The following October, continuous rain with winds and snow

45 Salisbury & Winchester Journal, 6 September 1819; Morning Chronicle, 7 October 1819; Salisbury & Winchester Journal, 13 September 1824.

caused *the great[es]t flood, ever known in any Persons memory*: the full details of this weather event and its effects can be read in the transcribed diary for 1823.

Land, buildings and equipment

Many of the field names listed in the 1809 sales particulars occur throughout the diaries up to 1825. The following entry names five of the listed fields, all of which were within the Little Bedwyn estate:

> 17 & 18 March 1813. VETCHES Spring sown in Gate Close TURNIPS sheep finish'd East field & began South field EWES & LAMBS 11 score & 9 EWES & LAMBS 15 left in Rick Yd FOLD dry sheep removed fm Marls to Jockey Gd

Similarly, there are references to the land which Guy acquired from Stephen Pearce, exemplified by the following entry from 19 May 1820: *Ploughing for, & sowing Bell TURNIPS, Pearces Piece.*

The impression given from the 1809 sale particulars was that there had been substantial expenditure on improving both the farmhouses and the farm buildings, suggesting that Pinniger had moved to a farm which was well served with buildings for the activities of the farm.[46] However, some small-scale building work did take place: indeed on 21 December 1814, he noted that the new stable was finished and that the horses were put into the stable for the first time. In 1818, an old carthouse was taken down (14 April) and seems to have been replaced by a new skilling (9 December): this might suggest that he was contemplating an increase in the number of cattle which he kept, but there are no obvious clues from his diaries that this happened.

However, with regard to the threshing equipment, it would appear that this was insufficient to meet the needs of the farm. In the summer of 1813, T Edwards began puddling the mill pound and by 8 July, he had also finished the machine hatches and pound, indicating that repairs had been necessary.[47] On 27 January 1814, Pinniger recorded that the water wheel

46 The following quotation comes from the sales particulars: The several buildings are substantial and in good state, a large sum of money having been recently judiciously expended in repairs and improvements, WSA: 315/28, *Printed sale particulars relating to properties, including the manor of Little Bedwyn, 1809.*

47 Puddling is the use of puddled clay to make a waterproof lining for a pond

was set going and he began to thresh the wheat.

As well as this water powered threshing equipment, there was also a portable horse-driven thrashing machine which would have been used inside one of the barns on the farm, powered by a horse-engine located just outside the barn and driven by one or more horses. Its movement between various locations confirms that there were at least four barns in use on the farm, as the entries below demonstrate:[48]

18 October 1813: PORTABLE=MACHINE, put to work, home Barn New wheat
6 December 1813: Thrashing Machine put down at Jockey
30 December 1813: Machine brot from Jockey to Stockwells
23 February 1814: Machine took to Cowleaze Barn

It would appear that these barns were based on the traditional threshing barn arrangement with one or more threshing floors and mows on either side of the floors. In 1823, wet weather required him to divert his men from hoeing to the carting of chalk and the flooring of the mows, as the following extracts from the diaries show:[49]

24 July 1823. Turnip land too wet to hoe, 10 Men no work for: - put them to Chalk Carting & new flooring 4 Mows
29 July 1823. Turnips want hoeing & 10 Men want employ, - but for Cart'g Chalk, floor'd 5 Mows, fill'g Mill tail with Chalk, to pave it
21 August 1823. Men making Chalk floors in the Barn mows Jockey, too wet to cut the Oats now ready

---

or mill pound: puddled clay is clay which that has had all the air pockets squeezed out of it to make a solid, immovable, watertight layer.

48 That there were barns at Jockey and Stockwells is confirmed by the following diary entries: 3 September 1813: OATS Madams ground carr'd to Jockey Barn; 11 September 1813: Oats carr'g from E F the Rick made by Stockwells Barn. There may have been as many as six barns, as described in the 1809 sales particulars: one at the Mansion (Lot 1), two at the Home Farm with a further barn at the cottage called Greenways (Lot 2), and two at Jockey Farm (Lot 3), WSA: 315/28. *Printed sale particulars relating to properties, including the manor of Little Bedwyn, 1809.*

49 Mow - the storage bay to either side of the threshing floor in a traditional barn: the purpose of the mow was to provide storage for the unthreshed sheaves and likewise for the threshed straw. Carole Ryan, *Farm and rural building conversions: a guide to conservation, sustainability and economy* (Ramsbury: The Crowood Press, 2013), pp. 65-66.

Further, in May 1824, his men lowered the earth in one of the barns for a new hollow floor. This was probably for the insertion of a suspended timber floor at ground level, which was considered to produce *a better and brighter sample of grain,* to replace the existing beaten earth floor (usually clay, if available).[50]

That there were two thrashing machines on the farm is confirmed in the entry for 19 April 1814 when the diary entry indicates that the horse machine was being used to produce clover seed, whilst barley was being threshed with the water machine. Whether it was due to the inefficiency of these machines, or the need for doubling the threshing capacity, new machines were installed in 1814 and 1815:

26 August 1814: MACHINE the new Water one complated.
6 September 1815: Machine, the Newhorse set going.

On 29 November 1815, the water machine was being used to thresh wheat, whilst the horse machine was threshing barley, although either machine could be used for any of the grains to be threshed.

In 1819, a new mill was constructed, as described in the following diary entries:

1 March. Odd Men, making hedge round the land for a Mill
24 March. 1 Team to the Mill for the Mill tackle
25 March. Sand 2 Waggon load from Sudden for the Mill[51]
29 March. MILLWRIGHTS, came to work
2 April. MILL, began the foundation
4 May. Masons began the Mill
13 May. T E_ began sinking the Brook to the Mill tail[52]
8 June. Mill, ground the first sack of Barley
13 July. MILL, began grinding grist

Thus, in just over two months, a new water mill had been built and commissioned for the grinding of corn, rather than driving a thrashing machine. Indeed, three months later, the Little Bedwyn estate was

50 Arthur Young, *The farmer's guide to hiring and stocking farms*, Vol. 1 (London: n.p., 1770), p. 50.

51 This is probably Sudden Farm within the tithing of Wolfhall in the parish of Grafton.

52 T E is probably T Edwards who had done other work for Pinniger relating to the mill pound and the water meadows.

advertised for sale by auction and the description included *a newly-erected water grist mill.*[53] However, the mill was not always available for work, as recorded on 25 October 1820 when wheat and barley were sent to Denford Mill, in that there was insufficient water to drive the mill at Little Bedwyn.

Earlier diary entries had indicated that bricks were being transported *from the kiln* but the location of the kiln was never specified, although Table 1 refers to a field known as Brick Kiln Dell: however, an entry in 1820 states: *Bricks & Lime from Down Kiln,* suggesting that there was a kiln within the farmland itself. On this occasion, the bricks were for a cottage and on 2 September that year, Pinniger recorded that the masons had finished building *Lewes's House.* Following the wet harvest season experienced at Little Bedwyn in 1821, in late December, bricks were carted from the kiln to make a new kiln for drying corn. The building of this new kiln started on New Year's Eve 1821 and by 15 February 1822, he had started to use it to dry corn. There is a reference to the sowing of kiln-dried oats on 13 April that year.

In addition to buildings, outside storage for crops was provided by way of ricks kept off the ground on timber structures supported on staddle stones. In 1817, these are referred to in the diary by number (there being at least 12 rick staddles available) and the following entry provides some detail about their construction:

> 7 October 1817. Barley carrying from the further side of N field, put on No.11 part of long staddle and the parting above. – the Rick 126 feet long 24 feet wide [...] feet high above the staddle on 45 stones

It is unfortunate that Pinniger did not return to the space he left for the height of the rick, but from the reference to 45 stones, we can deduce that these were arranged either as 5 rows of 9, or 3 rows of 15. For the former, then this would imply a spacing of 15ft 9ins between the staddles and 6ft between the rows, whilst the latter would have had a spacing of 9ft between staddles and 12ft between the rows.

A further entry describes the size of the rick, only on this occasion, the length of the rick is omitted:

> 8 September 1823. Wheat Dean heath finish'd carrg 24 Acres, - all put upon the high Staddle 1260 Tythings – is 52 pr A The Rick 20 feet from the ground to eves, - and 40 feet over the Rick on the eves,

53 Salisbury & Winchester Journal, 6 September 1819.

Looking after the land

The soils at Little Bedwyn were a mixture of clayey loam to sandy loam; chalky, silt loam; and clay to sandy loam.[54] The clay components would have produced a slow-draining soil which compacted easily, making it difficult to cultivate and requiring the addition of organic material to break it down. To improve the condition of the soil, a number of products were used. Each year, peat ashes were brought in by barge on the Kennet & Avon Canal, as the following entries describe:

> 3 June 1813. PEAT Ashes in 50 Tons Boat full with 13 hd Bushels (say holds 2000) the first Ashes made this season, Woolhamp[to]n[55]
> 25 February 1818. ASHES. Peat 800 Bushls land'd measured 880 @ 3d Bu & 1¼d pr Bu: [Transport?] from Green of Newbury

Peat ashes were sown [spread] on turnips, sainfoin, clover and wheat. Not only were ashes bought-in but in April 1817, Pinniger recorded the following: *PEAT diging in the Moors, for making Ashes* and in July 1820, the diary has an entry referring to more home-produced ashes (ashes from the down): here, spreading rates were given as between 60 and 100 bushels per acre.[56]

Other products used to increase the fertility of the land included soot brought from Newbury: on 20 March 1815, 67 bushels were spread on three acres of wheat. Chalk was also spread on crops along with lime flour and occasionally, pigeon dung was spread on the land, as recorded in October 1818.[57]

As well as the cultivation and harvesting activities on the land, there were other jobs necessary to maintain the land: these included hedging, ditching and draining which were mainly carried out during the winter months. For example:

---

54 Natural Environmental Research Council, *mySoil app – Little Bedwyn, Wiltshire* [accessed 17 April 2020].

55 Woolhampton is a village eight miles to the east of Newbury, close to the River Kennet.

56 The typical nutrient content of peat ashes is very low: 1.4%P (phosphorous) and 0.3%K (potassium). Mikko Moilanen, Jorma Issakainen and Klaus Silfverberg, *Peat ash as a fertiliser on drained mires – effects on the growth and nutritional status of Scots Pine, Working Paper of the Finnish Forest Research Institute, 231* (Vantaa, Finland: Finnish Forest Research Institute, 2012).

57 Lime flour is probably powdered calcium carbonate.

22 March 1815. Carting Stones off the Breach for drain in the 18 Acres

There were also a number of water meadows on the farm. These meadows were irrigated artificially, in order to increase the quantity and improve the quality of the grass produced, in particular providing feed for sheep or cattle in the early spring when fodder supplies were running low.[58] The water meadows required maintenance during the winter period as the following entry shows:

24 January 1818. T EDWARDS & 5 Men finish'd the Watermead, except the Carriage

Grass crops

As well as pasture land, Pinniger grew sainfoin and clover to provide fodder for all classes of livestock. At this time, all of this fodder was made as hay, which was mown by hand, the swathes turned and when the hay was dry, it was loaded on to carts and formed into ricks, either in the rickyard, or in some cases in the field itself, as seen in this diary entry:

21 June 1820. HAY, the first carried, made a Rick Nfield Up furlong

Having made the hay into ricks, these were then thatched to protect them from the weather.

There seems to have been a particular technique for making hay on the water meadows, as shown in the following entry:

12 May 1819. Hay, geting from the Watermeads, half made, to the old field Nfield to make it.

---

58 Hadrian Cook and Tom Williamson (eds), *Water Meadows: History, Ecology and Conservation* (Macclesfield: Windgather Press, 2007), p.1. The following quotation comes from the opening paragraph on the subject written by Eric Kerridge in 1953: *From the latter part of the seventeenth century to the latter part of the nineteenth the floated water-meadow was one of the greatest achievements of English agriculture and an integral part of the sheep-and-corn husbandry of the downlands of Wiltshire, Dorset, Berkshire and Hampshire. In Chalk Wiltshire in this period the overwhelming majority of the several and common meadows by the streamside were floated,* Eric Kerridge, 'The Floating of the Wiltshire Watermeadows', *Wiltshire Archaeological and Natural History Magazine*, 55 (1953), 105.

Hay will not make [dry] on wet land: the sun draws the moisture from the land through the hay laid on the ground with the outcome being that it never becomes properly dry such that it can be stacked without heating.[59] Given the explicit design of water meadows, then except in particularly dry seasons, it would appear as though the hay was moved in a part-dried state from the water meadows to a dry field, where the drying process could be completed.

Arable crops

Many of the entries in his diary describe the field operations required to prepare the ground for sowing a variety of arable crops. These including ploughing (or plowing as Pinniger sometimes spelt it), raftering, harrowing, thwarting, dragging and rolling.[60]

Ploughing used a lot of resources on the farm, both human and animal. Diary entries from 1821 record the following:

> 18 April 1821. Ploughing the dung N field finish'd 18 Acres in 3 day 6 Ploughs
> 23 April 1821. Ploughing Nfield for Barley with 8 Teams – 15 Horses – 12 Oxen

The first entry confirms a work-rate of one acre per day per plough-team, whilst the second shows that there were five teams of three horses each

59 When stacked, the heating of hay is a natural part of the curing (preservation) process. Heating occurs as a result of: (1) biochemical reactions from the grass itself as the hay cures, usually taking the temperature to 110°C; (2) activity of microorganisms in the hay. The latter thrive with extra moisture, increasing their activity and raising the temperature of the hay stack beyond 110°C: if the temperature rises above 170°C, then chemical reactions commence quickly raising the temperature above 400°C causing the wet hay to set on fire, Drovers, *Understanding wet hay* https://www.drovers.com/article/understanding-wet-hay [accessed 18 May 2020].

60 Raftering - Ploughing so as to leave a narrow strip of ground undisturbed, turning up a furrow on to it on each side, thus producing a succession of narrow ridges", G. E. Dartnell and Rev. E. H. Goddard, *Wiltshire Words: a glossary of words used in the county of Wiltshire* (Avebury: Wiltshire Life Society, 1991), p.129. Athwart – across from side-to-side, transversely, Lexico, *athwart* https://www.lexico.com/definition/athwart [accessed 13 May 2020].Thwarting - Probably therting, to plough land a second time, at right angles to the first ploughing, so as to clean it more thoroughly, Dartnell & Goddard, *Wiltshire Words,* p.166.

(which would have required a ploughman and boy each) and three teams of four oxen (again needing an ox-driver and boy).

The technique of double ploughing was described in November 1820, as follows:

> Double Ploughing the Pasture Pearses Piece,- the turf ploughd off and buryed in the furrow with the underfurrow, make good work for hoeing in of wheat

This approach was also used for ploughing the old sainfoin crop in the same year.

On occasions, larger teams of horses and/or oxen were required, such as in May and June 1818, when there was no rain for a whole month – indeed there was no significant rain until late September that year:

> 13 June 1818. Ploughing N field as yesterday & finish'd the Piece 40 Acres with say,- 47 days 1 Team, Plough very hard – 4 Horses & 6 Oxen to a Plough
> 19 June 1818. Ploughing as yesterday & finishd the 6 & 18 A (about 25A) say 26 Teams 1 day.

Both examples show a work-rate of slightly less than one acre per team per day, not unexpected given the difficult conditions: however, the usual horse/oxen power applied for ploughing was three horses or four oxen to each plough – here the power had been increased to overcome the hard ground.

Another technique used to cultivate the land was seven-shearing, as seen in the following diary entry:

> 12 June 1815. 7 shearing the remainer of lower furlong parlour field.[61]

The term seven-shearing comes from the cultivator with seven working tines or shares used to carry out this operation. In Mitchell's *Sketches of Agriculture* dated 1827 a 'Cultivater' is defined as: *Mister, Lester's seven-share plow or scuffler; it works upon three wheels*: Figure 4 shows one such implement.[62]

---

61 Furlong - Measure of length equal to 220 yds or 660 ft. Also used to describe a strip of newly-ploughed land lying between two main furrows, Dartnell and Goddard, *Wiltshire Words*, p.62.

62 James Mitchell, *Sketches of Agriculture, or Farmer's Remembrancer* (London;

The intermediate processes between ploughing and sowing included rolling, harrowing and pressing, the latter being a separate operation to rolling, as described below:

> 27 June 1823. SWEDES, began sowing, Sweardown & Lane Piece, sow'd 20A Dung Plough'd in, roll'd, and Press'd – ready for sowing too dry before.
> 28 June 1823. Swedes & Green rounds, sow'd 20 Acres, - 40 Acres in two days make good work, with frequent warm showers, harrowed fine after the 10 wheel Iron PRESSER, proveing great advantage[63]

First mentioned in this year, the ten-wheel iron presser appears to be a new piece of equipment being used by Pinniger.[64] A further entry on 20 September 1824 confirms that he was using both the presser and sheep to tread the ground: *Press rolling & tread'g Kite=hill finish'd. (wheat).* In sharp technological contrast, another field operation recorded is as follows:

> 21 March 1823. Bushing Dung on Sainfoin Jockey ground

This is referring to the use of a heavy gate with thorn cuttings or other brushwood interwoven between the bars of the gate, to act as a light harrow to spread the dung on the field: sometimes a large thorn bush may have been pulled behind a horse to achieve the same effect, hence the term.[65]

Baldwin and Cradock, 1828), p.54. W. Lester, *A history of British implements and machinery applicable to Agriculture, with observations on their improvement* (London: Longman, Hurst, Rees, Orme and Browne, 1811), Plate 6 between pages 180 and 181.

63 Green rounds - a variety of turnips, also known as Green Norfolk or Common Green Top White. These turnips have a regular round shape, flattened but not much hollowed on the upper and under surface – the upper surface is green coloured and the lower surface is white. This variety is hardier than the Norfolk White or Norfolk Red, The Farmer's Magazine, Vol. V, No.2, August 1836, p.116.

64 The presser is referred to in an 1835 agricultural report for Berkshire, as follows: *The land is now become too wet either to tread with sheep or to use the presser or wheel-roller, and it is now impossible to get it so firm at bottom as it ought to be, and we are fearful that on this account the late sown will not stand so good a chance for a productive crop as that which is worked down firm,* The Farmer's Magazine, Vol. III, No.6, December 1835, p.500. This is probably the forerunner of the modern-day furrow press which follows (or is used after) the plough.

65 Bush - Noun. A heavy hurdle or gate, with its bars interlaced with

The diaries record the sowing rates for the various crops and these are summarised in Table 4.

Table 4: Arable crops – sowing rates

Crop	*Sowing rate (per acre)*	*Date of entry*
Wheat	2½ Bu	20 September 1813
Spring Wheat	2¼ Bu	22 March 1813
Barley	Near 6 Bu	30 April 1822
Oats	5 Bu	26 March 1813
Rye	2 Bu	30 April 1822
Beans	3 Bu	12 March 1813
Clover	17 li	30 April 1822
Vetches	7 Bu 2½ Bu 3 Bu	16 March 1813 16 September 1813 1 October 1813
Peas	2½ Bu	19 March 1813
Swedes (Sweeds)	2½ li	8 June 1813

Wheat varieties planted included Talavera, Taunton, America and Golden Dun, whilst Black Siberian oats and Ringwood rye are referred to.[66] Varieties of turnips included Norfolk Whites, Aberdeens, Red Rounds, Tankards, Green Rounds, and Bells. Varieties of beans mentioned are Heligolands and Kidwells, whilst small gray [or grey] peas are the only such seed sown.[67] Interestingly, the first and only reference to the sowing of mangel wurzels does not appear in the diaries until July 1824.

Work rates for sowing are sometimes recorded as on 23 May 1814 when the sowing of barley in the 27 Acres was completed in 4 days, although the numbers of men, horses or oxen are not given. However, an

brushwood and thorns, which is drawn over pastures in spring, and acts like a light harrow; Verb. To bush-harrow a pasture, Dartnell & Goddard, *Wiltshire Words*, p.20.

66 Taunton - A variety of wheat. "With respect to different species of wheat, the grand divisions are into white, red, and bearded; the sub-divisions, or varieties, are numerous and very unimportant, in any point of view. The best in England, according to my experience are, the Essex and Kentish white and red, and the Taunton Wheat (Somerset) which appears a mixture of both", John Lawrence, *The New Farmer's Calendar; Or, Monthly Remembrancer, for All Kinds of Country Business: Comprehending All the Material Improvements in the New Husbandry ... By a Farmer and Breeder* (London: Symonds, Vernon, Hood & Wright, 1800), pp.116-117.

67 Gray peas or grey peas – a late-ripening variety of field peas, Wade Martins and Williamson, *The farming journal of Randall Burroughes*, p.128.

entry from 1818 is more informative:

> 28 & 29 May 1818. BARLEY, sowing finish'd Plough'd in, in the two Days, 22 Acres Whitley furlong, and harrow'd two tine. 6 – two Horse, 3 – three Ox } [Total] 9 Ploughs, - and two Horses and two Oxen harrowing[68]

Conditions must have been good at this time, or the soil was lighter, in that fewer horses and oxen were used for ploughing.

The traditional method of sowing seeds was by broadcasting: this involved the distribution of the seed by hand across the ploughed and harrowed land. Whilst many agricultural labourers would be skilled at producing an even distribution of seed across the field, this cannot be achieved if there is the slightest wind: given the acreages to be sown then they would not be able to wait for a still day. Around this time, seed drills were available and the first indirect reference found in the diaries to the use of a drill comes in the following entries:

> 1 May 1821. Oats – sow'd Jockey grounds 5 Bu pr Acre 6 Bu: rough Sainfoin pr A @ 32s pr Qr 60 li Mill'd Sainfoin pr A @ 40s € 12 li Mill'd Hop pr A @ 28s € = 4 holes Machine
> 4 May 1821. Sow'd Crab tree Piece to Oats 5 Bu Oats 1 Bu Chert Bents, 7 li Old Broad, 4 li Mill'd Hop } pr Acre hole in Machine bare half.

The phrases *4 holes Machine* and *hole in Machine bare half* are referring to the settings on the seed drill used to deliver the correct quantity of seed. At the time, this would have been achieved through a rudimentary mechanical device, as described here for the cup-feed type of drill: *the forward section [of the drill] contains the seed which drops through apertures, the size of which can be regulated by slides, to the bottom section*[69] These slides would have been fixed with steel pegs through the correct position in a series of holes in the slide, to regulate the size of the aperture opening. In the same year, it appears that Pinniger had acquired a new seed drill:

---

68 Two-tine - '2 tine the Harrows' means harrowed twice, Davis, *Agriculture of Wiltshire*, p.51.

69 Feed mechanisms for seed drills fall into two categories: cup-feed and force-feed, Claude Culpin, *Farm Machinery, 8th edition* (London: Crosby Lockwood, 1969), pp.198-203. Seed drill in Encyclopaedia Britannica Company, *Encyclopaedia Britannica, 11th edition* (New York: Encyclopaedia Britannica, 1911).

2 June 1821. Plough'd & sow'd about 4A Near side of Up: fur: Nfield to Tankard Turnips | SWEDES, began sowing Church fur: Nfield, 1 hole in new Machine[70]

This continuation of changes in technology, through the replacement of broadcast techniques with drilling, is recorded again in January 1823 when wheat was being sown (that is, broadcast), whereas by October of the same year, wheat was being drilled, apparently by a contractor:

13 October 1823. DRILL'G, Wheat, by W. Rummin

and in the following year, barley was being drilled as well:

8 April 1824. BARLEY, - began Drilling Nfield above the road

although six days later, Pinniger notes that barley sowing was finished in north field. Whilst a drill had been used to plant the barley seed, it may still have been commonplace to refer to the process as 'sowing'.

Beans were planted in 1815 by dibbling at a rate of three bushels per acre and in 1821, more detail can be derived from the following entry, which includes the row spacing and seed spacing within rows:[71]

26 February 1821. BEAN, planting, began in the Breach, Kidwells 9 In x 18 In 2 or 3 Beans in a hole – 1 Sack pr Acre, cost 13s 6d pr Sack

However, by 1823, as with wheat and barley, beans were being drilled, at a rate of 4 bushels per acre, rather than dibbled (planted using a dibble) or sown.

70 Tankard - White, Green and Red Tankards are all varieties of turnip. They are unsuitable for winter feeding, in that more than half the root is above the ground surface, and thereby exposed to frost: however, they generally mature earlier than other turnips. These turnips have oblong, or tankard-shaped roots which are often bent or crooked. The White Tankard is the largest of the tankards, but is also softer in texture than either the Red or Green Tankards. The colour of these latter varieties describes their colour, although they are white under the ground, Cuthbert William Johnson, *The Farmer's Encyclopædia, and Dictionary of Rural Affairs: Embracing All the Most Recent Discoveries in Agricultural Chemistry, Volume 1* (Philadelphia: Carey and Hart, 1844), p.1067.

71 Dibble – a pointed hand tool for making holes in the ground for seeds or young plants, Lexico, *Dibble* https://www.lexico.com/definition/dibble [accessed 18 May 2020]. Dibbling – to plant using a dibble.

It is of note that oats were often not sown on their own, as the following two entries demonstrate:

> 3 May 1814. Oatsowing Bushels ground finish'd 10 Acres 1½ Bu: Peaces Ray 2 li Honey suckle 20 li Broad, Tail } to A
> 8 April 1815. Oats sowing Potshard 5 B: Mr Potters 21 li tail Broad 10 li Old mill'd hop } pr A 7 Bu: Old Ray to the 10A

That is, hop clover, ray-grass (perennial ryegrass) and honeysuckle were all sown with the oats: further, sainfoin (both rough and milled) were recorded in 1821.[72]

Once the crops were sown, there were some operations required to keep the crop in a healthy state such as hoeing and weeding. In particular, there was a concern to stop turnips flowering and going to seed, in that this appears to have resulted in the sheep 'blowing'.[73] Somewhat primitive methods were used, both for the cause and the effect, as these two entries show:

> 31 March 1818. TURNIPS, in the Breach run'g to seed,- drawing a Gate over to break of[f] the green.
> 3 April 1819. Mowing Turnips the Lains yesterday to try to prevent blowing the Sheep, and took a tub of water to the Turnips to pour on the Sheeps backs when blown, for the purpose of cureing them

72 The oats would have acted as a 'nurse crop' for the other seeds, which would then act as a 'cover crop' over the winter. A nurse crop is an annual crop (such as oats) used to assist the establishment of leguminous plants such as hop clover (*Trifolium campestre*) and perennial grasses such as ray-grass (*Lolium perenne*), King's Agriseeds Inc., *Nurse Crops* https://kingsagriseeds.com/wp-content/uploads/2014/12/Nurse-Crops.pdf [accessed 7 July 2020]. A cover crop is a non-cash crop grown primarily for the purposes of protecting or improving the land between periods of regular crop production, Agriculture and Horticulture Development Board (hereafter AHDB), *Opportunities for cover crops in conventional arable rotations: Information Sheet 41* (Stoneleigh: AHDB, 2015) https://ahdb.org.uk/cover-crops [accessed 7 July 2020].

73 This condition is probably sheep bloat. Brassicas (including turnips and swedes) are most toxic when in flower and can lead to potential health issues. Further, brassicas generally are rapidly broken down in the rumen (stomach) producing excess gases which cannot be relieved quickly enough, resulting in bloating: fibre (hay or straw) fed alongside these crops helps to reduce this problem, Limagrain UK Limited, *The essential guide to forage crops* https://www.lgseeds.co.uk/uploads/Forage-Brochure_Singles.pdf , p.5 [accessed 23 March 2020].

As for the harvest, there are many entries covering yields, timing and the effects of the weather. A particularly late harvest (for both hay and arable crops) occurred in 1816 as the following diary entries record:

> 27 August. REAPING – cut a road through Northfield, the Wheatfull but not hard
> 5 September. damp morngs & dull days – bad haymaking
> 14 September. HAYMAKING FINISHD began 29th June
> 27 September. WHEAT HARVEST = FINISH'D
> 2 October. VERY UNKIND TIME FOR HARVEST | Many large Farmers, carried no Wheat.
> 11 October. OATS began carrying Marls in the evening, by Moon light till 11 OClock
> 14 November. BARLEY, made a Rick from lowerfurlong Swear down & carried the Barley fm Bonnings Corner & part of lane piece to Rick in Yard – till 11 OClock night when the Snowfrozen on the Barley - left of with 3 loads out to FINISH [on 16th] | BEANS, made a Rick with the Heligolands in morning
> 22 November. HARVEST finish'd.

The harvest home was held on 30 November, some six to eight weeks later than usual.[74]

With regard to yields, Pinniger recorded his wheat yield in 1814, as shown in Table 5.[75] Thus, the relative quantities per acre varied: this would relate to planting dates, weather, soil quality, amount of fertiliser (either sheep dung from the process of sheepfold or the use of artificial fertilisers) and the variety planted.

---

74 The year 1816 was known as the 'year without summer', attributed to the eruption of Mount Tambora on Indonesia in April 1815, Lucy Veale and Georgina H. Endfield, 'Situating 1816, the "year without summer" in the UK', *The Geographical Journal*, 182.4 (2016), 318-330.

75 Acreage - 1 acre = 4 roods, 1 rood = 40 perches. Tithing (spelt tything in the diaries) or tething: a shock of ten sheaves, for convenience in tithe-taking, Dartnell & Goddard, *Wiltshire Words*, p.168. Shock (or cock) - of wheat (or other grains), consisting of several sheaves set up together for carrying. The number of sheaves was formerly ten [also known as a tithing or tething], for the tithing man's convenience, but now varies considerably, according to the crop. From the definition of *hyle* or *hile*, Dartnell & Goddard, *Wiltshire Words,* p.84.

Table 5: Wheat yields at Little Bedwyn, 1814

Acreage			*No. of tithings*	*Tithings per acre*
Acres	*Roods*	*Perches*		
16	1	21	773	47.2
11	1	30	569	49.7
14	2	16	572	39.2
3	1	30	152	44.2
2	0	24	109	50.7
20	0	30	773	38.3

Other crops

As well as the traditional arable and grass crops, each year a crop of bark was produced: this is no surprise given that there were more than 60 acres of woodlands on the farm. The following entries provide information about the manpower required, the quantities of bark produced and the value of the bark:

> 6 May 1819. BARKING, finish'd 51 Trees @ 1s 6d ea | 453 Yards 3 rind thick, of 30 In lengths & the cover piece @ 4d pr Yd – 4 Waggon loads
> 24 May 1824. Barking, finish'd, 4 Men – 13 days each Trees, - small & large 217 Yards of Bark 3 Tons 4½ cwt @ 20s/Ton for throwing and Barking.[76]

As to whom the customers for the bark were, this is not recorded: in all likelihood, it would have been sold to leather tanners and in 1822, one load of bark was delivered to Beckhampton, whilst three loads went to Chippenham. Around 1815, the value of stripped bark in south-west England was between £9 and £13 per ton, so a payment of 20s per ton for throwing and barking the trees was not excessive.[77]

Livestock

The predominant livestock on the farm were sheep. On 22 March 1813, Pinniger records his stock of sheep, as follows: *Ewes & Lambs 12*

76 Throwing – felling (of timber), Dartnell & Goddard, *Wiltshire Words,* p.166. Barking – refers to de-barking or bark stripping of felled timber.

77 L. A. Clarkson, 'The English Bark Trade, 1660-1830', *Agricultural History Review*, 22.2 (1974), 142.

Sco 10 | Ewes not lambd 2 Sco | Dry sheep 4 Sco | [Total] 18 Sco 10.[78]. What is not clear is where these 370 sheep came from. This is not a large number of sheep for the farm: when he recorded his stock of sheep in 1823, the total number was 1,369. It has not been possible to determine where he lived prior to coming to Little Bedwyn some six weeks before this entry, but there are no details of sheep being bought-in during this period.[79] It is possible that the sheep were already on the farm and belonged to Anthony Guy, in that the Mansion part of the estate, which was occupied by Anthony Guy in 1811 and 1812, had been sold with 134 acres of land. Similarly, it is unclear whether the sheep continued to be owned by Guy or whether he sold them to Pinniger.

However, there may be some indication of ownership from a report in the Salisbury and Winchester Journal of November 1818. John Stone was committed to 12 months' imprisonment in the Devizes house of correction for stealing a quantity of wheat in the straw, from an open field in Little Bedwyn, the property of Thomas Pinniger and A. Guy Esq.[80] Assuming that this is referring to the wheat, rather than the field, being the property of both men, then this suggests some form of sharefarming arrangement between the two of them.[81] Possible arrangements may have included:

- Sharing the output: Sowing to halves which involved a division of the crop (such as wheat) instead of a fixed money rent for the land;
- Sharing the profits: A common arrangement was for the landlord and tenant each to take half the increase (or value added) – for example, this could be half of the profit on sheep let out on a grazing agreement to the tenant;
- Partnership, sharing crops and resources: Guy and Pinniger could have been working in partnership with half shares in equipment, stock and grain;
- Providing working capital through a stock and land lease: As landlord,

78 Sco – score.

79 Thomas Pinniger, as a single man, may well have been living at his parent's farm at Tytherton before moving to Little Bedwyn: with the exception of his brother Broome, his siblings had all married by 1813.

80 Salisbury and Winchester Journal, 2 November 1818. The diaries record that Pinniger attended the sessions for the trial of J Stone on 21 to 24 and 26 of October.

81 Elizabeth Griffiths and Mark Overton, *Farming to halves: The hidden history of sharefarming in England from medieval to modern times* (Basingstoke: Palgrave Macmillan, 2009), pp.14–20.

> Guy may have leased stock including cattle, sheep, horses, oxen and pigs, with the land to enable Pinniger as tenant to start farming.

Indeed the final option of a stock and land lease may explain how Pinniger was able to move to Little Bedwyn and almost immediately have some 400 sheep. With the end of the Napoleonic Wars in 1815 and the resumption of cheap imports of grain, landowners were looking for ways to attract tenants and to make a profit: although there is no known archival evidence of such arrangements in the nineteenth century, this absence of evidence does not mean that sharefarming arrangements were not practised at this time.[82]

On 25 April 1813, there was an influx of sheep on to the farm, as recorded in the diary: *TEGS – 58 came fm Chip[penha]m.*[83] Pinniger gives no more indication than this and given that Anthony Guy had land at Sambourne Farm and elsewhere in Chippenham, it is possible that these sheep were "transferred" to Little Bedwyn, rather than being purchased by Pinniger. However, on 5 October, there is the following diary entry: *56 6 tooth Ewes fm Mr Rendall.* This is possibly William Rendall of Grafton Farm (Manor Farm, East Grafton): he was the occupier of this farm in the adjoining parish, of just over 500 acres from 1812 to around 1841.[84] A further example of sheep transfer comes from an entry in the diary in 1822:

> 16 August 1822. 146 two tooth Wethers 54 Wether Lambs } sent Mr Guy for Nettleton Farm[85]

The particular wording of *sent Mr Guy for Nettleton Farm* does not suggest any form of sale, or movement for over-wintering (Pinniger had always stressed this in previous diary entries and it is too early in the year to be doing this): rather, it suggests a transfer of 200 sheep from one of Mr Guy's farms to another. Likewise, the record of sheep stock as at 1 September 1823 includes 30 wethers and 70 wether lambs and there is a pencil note of

82 Griffiths and Overton, *Farming to halves*, p. 133. This follows the well-known archaeological adage that the absence of evidence is not evidence of absence.

83 Teg - A sheep in its second year, Lexico, *Teg* https://www.lexico.com/definition/teg [accessed 1 May 2020].

84 WSA: A1/345/190, *Land Tax assessments, Grafton, East and West, 1780-1881*; WSA: 9/1/109, *Savernake Estate, Survey and valuation of farms near Savernake including Grafton, 1811-1812.*

85 Wether - In sheep, those in their first year from around Christmas to the first shearing, Davis, *Agriculture of Wiltshire*, p.260.

Mr Guy in the margin opposite these two items: see Appendix B.

It is likely that a similar arrangement occurred with cattle, as exemplified by the following entry:

> 7 March 1823. 2 Cows & Calves from Chip[penha]m | 1 Cow Vinny - retd to Chipp[enha]m

again suggestive of exchange or transfer, rather than sale. Likewise, movements of oxen were recorded as follows:

> 27 May 1824. 3 – 3 Yr & 3 – 4 Yr old Oxen Welch from Chipp[enha]m
> 5 June 1824. 8 Old Oxen retd to Chip[penha]m

A major aspect of the management of sheep was that of mating the ewes. Each year, a number of ewes were withdrawn from the main flock for sale at fairs later in the year: these were known as the *sale ewes*, with the remainder known as the *stock ewes*. From the diaries, it appears that these were put to the ram in the third week of August and were sold as in-lamb ewes. Some of the stock ewes were then put to the ram at the end of the first week of September to produce a few early lambs and the remainder from the middle of the month. In most years, Pinniger records the *time out* – the date of the first ewe due to lamb using a basis of 21 weeks (147 days) to determine this date: this is consistent with the modern-day gestation period of between 142 and 152 days. He seemed to bring rams into the farm each year to introduce fresh blood to the flock: for example, in 1813, he brought three two-tooth rams from Chippenham – once again, *brought* rather than *bought*.[86]

Each year all of the sheep were washed, usually in the middle of June: they were then shorn (or as Pinniger described it, sheared) around a week later. It appears from the diary that as well as their regular wages, he provided refreshment to his workers (this being referred to as *usage*), the sheep wash not being within the farm's curtilage, as follows:

> 10 June 1823. Washing Sheep, - Usage 5 Qts Best Ale 5 Qts Strong Beer
> 2 large loaves of Bread Bacon & Cheese, in proportion

In 1814, there were 550 sheep to shear and this was undertaken in

86 Two-tooth - In sheep, when the first central pair of permanent teeth present at 12-18 months old. Also, from shearing time after one year old, Davis, *Agriculture of Wiltshire*, p.260.

one day by 11 men, suggesting a work rate of 50 sheep per man per day: in the middle of July, around 360 lambs were also shorn. Similarly in 1815, 10 men took two days to shear 720 sheep and 320 lambs, a total of 1040, representing a work rate of 52 sheep shorn per man per day. As well as the shearers, other manpower was required to carry out this process, as described in the following entry from the diary:

> 14 June 1823. SHEEP SHEAR – 10 Men @ 2s pr day 2 Women @ 1s pr day winding 1 Man lifting Sheep out 1 Man Carry'g Sheep 1 Man Leading Sheep out to be markd 2 Women for the Fleeces 2 Children for the loose wool.

Given the large number of sheep kept on the farm, it is likely that the wool crop may have contributed significantly to the farm income. In 1815, the wool was weighed on 28 August: Table 6 shows the quantities produced. The number of fleeces needed to make up one tod was recorded: some of these numbers were written with one or more bars above or below the number, indicating that the particular tod was one or more pounds heavy or light: this has been represented in Table 6 by the plus sign (+) for heavy and the minus sign (-) for light.[87]

Table 6: Wool weights, 1815

EWES				*DRY SHEEP*			
12+	11-	10	10-	11	12+	10	11
12	13++	11-	11+	10-	13	10-	11
10	12-	11	11-	12++	10-	11	11+++++
12	11-	13	11-	10-	11+	11-	
12	12	13+	12+	10	10	10+	
13	13	12	12+	11	9	11	
12+	12+	12	11	12	11-	11+	
11- -	11- -	11-		11	11+	11	
12+	14	12+		10-	10	12	
398 fleeces				324 fleeces			
34 tods want 1 li				30 tod 5 li over			

Thus a total of 722 fleeces were weighed with a total weight of 1796 lbs (just over 16 cwt or 0.8 of a ton) with an average fleece weight

87 Tod - A former English weight for wool, about 28 pounds, Websters New World College Dictionary, *Tod* http://www.yourdictionary.com/Tod [accessed 1 May 2020].

of almost 2½ lbs per sheep. In addition, there was 114lbs of wool from the 320 lambs, being 5.7 ounces per lamb. When the wool was weighed in 1817, 654 fleeces weighed 1700 lbs, that being 2.6 lbs per fleece.

For whatever reason, Pinniger did not sell his 1818 wool crop until the following year when both crops were weighed together. A total of 1275 fleeces weighed 117 tod and 23 li (3299 lb, average fleece weighing 2.6 lb) which he sold for £276 17s 6d: this equates to just over 1s 8d per lb. The total received for the wool equates to £20,630 at 2018 prices: that makes the price of one pound of wool £6.25.

Similarly, in July 1822, he sold the wool crop for 1820, 1821 and 1822 to Mr Fielder at Newbury: on this occasion, 2,550 fleeces were sold weighing 220 tod and 6 li (6166 lb, average of 2.4lb per fleece) for £337 4s 8d.[88] However, he may have wished that he had kept his wool a little longer, in that on 5 August he recorded that wool was dearer, having advanced from 31s to 37s 4d per tod (1s 1d to 1s 4d per lb) – from £4.90 to £6.03 per lb at 2018 prices. By comparison, the average price for clean wool at the British Wool auction held on 10 March 2020 was 106p/kg (48p per lb). Therefore, two hundred years ago, the wool crop made a more significant contribution to the farm's income than it does today.[89]

Despite the extent of the farm, it was necessary to over-winter some of the sheep elsewhere, as the following entries show:

> 20 September 1815. SHEEP 100 Sale Ewes 70 Wether Lambs 160 Chilver Lambs 4 Ram Lambs Sent to Harnish Park 70 Acres took @ 24s Acre till Lady day. | SHEEP AT HOME. 545 Ewes at Fold 23 Chilver Lambs 20-Wether Lambs 8-2 tooth Lambs 2-Barren Ewes, 1-Ram Stag 15-Rams.[90]
>
> 25 April 1816. 254 Sheep came from WINTERING

It can only be assumed that the difference of 80 was a result of sales, transfers or death.

As for sales of sheep, the following diary entries give some idea of the number sold, the prices raised and where they went:

88 £100 was worth £7452 and £8951 in 1819 and 1822 respectively, at 2018 prices. Measuring Worth, https://www.measuringworth.com/calculators/ukcompare/relativevalue.php [accessed 23 March 2020].

89 Auction price from British Wool, *British Fleece Wool Price Indicator*, https://www.britishwool.org.uk/price-indicator [accessed 23 March 2020].

90 Chilver - A ewe lamb, Dartnell & Goddard, *Wiltshire Words,* p.27. Harnish – Hardenhuish, near Chippenham.

24 November 1817. Marlbro FAIR. 100 good mouth Ewes @ 34s Batton | 100 head wether Lambs @ 24s 6d | 50 Cull wether Lambs @ 20s Mr Brown, Sutton nr Battle Kent. 16 Coops @ 1s 6d ea for all

4 November 1824. 150 Ewes good mouth Appleshaw fair Mr Finch Chelmsford @ 28s 6d [91]

With a further sale of sheep in 1817 at Lambourn Fair, Pinniger then records the stock of sheep on the farm, as follows:

4 December 1817. LAMBOURN Fair, 6 Coops @ 1s 6d | 40 Cull wether lambs @ 18s [£]36 | 10 Cull Ewes @ 28s [£]14 | 20 Wether Sheep @ 36s [£]36 | 10 Cull @ 26s [£]13 | [Total] £99 | STOCK of Sheep to Winter – 472 Ewes wth the Ram 10 Ewes 2 tooth to pass the Ram 2 Ewes Wethering 4 Ewes Sickly [Total] 488 | 181 Ewe Lambs 9 Wether Lambs 4 Sickly Lambs 20 Rams 14 Ram stags 3 Wethers [Total] 719

Similarly, the following year, he had 723 sheep to over-winter, having sold 180 ewes and 150 lambs.

Diary entries for cattle are few and far between, suggesting that perhaps only one or two cows were kept to provide milk for the household. On 15 January 1814, the Norman cow calved: this is possibly of the Normande breed which originated in the Normandy region of France in the early nineteenth century and is a dual-purpose dairy breed. They produce milk with a high-fat content which is particularly suitable for making butter and cheese.[92] On the other hand, it has been reported that Alderney (Channel Island) cattle were usually known as "Normans" in Hampshire.[93] Occasionally, cattle appear to be brought to the farm to consume excess produce, as in December 1817, when 15 Scotch heifers were brought to straw.

Counting the number of pigs that farrowed in the six month period

---

91 Appleshaw is a village in Hampshire, between Ludgershall and Andover, just to the north of Weyhill. Batton may be one of Benjamin Batten, Benjamin Batten senior and Matthew Batten, sheep dealers listed in a notice for the Hungerford Sheep and Lamb Fairs, Salisbury & Winchester Journal 18 June 1821.

92 This breed of cattle is kept in England today: for example, until 2017, a herd of 100 Normande milk cows was kept at a Worcestershire farm, Worcestershire Farmsteads Project, *Building record for The Lingens Farm, Sledgemoor, Broadwas-on-Teme, Worcester. WR6 5NR* (Badsey: Worcestershire Farmsteads Project, 2020), p.98.

93 Idstone, 'The Alderney Cow' in *The Field,* 18 May 1872, 454.

between April and October 1813, there were at least 17 sows on the farm at that time.[94] The sizes of litters produced varied between four and eight with an average of six. Pigs were folded on vetches and clover. In October 1817, the stock of pigs is recorded in the diary, as follows: 11 stores, 1 boar, 4 sows with 35 young, 2 sows with 7 young (first litter), giving a total of 60 pigs. From this, it might appear as though having managed the farm for a few years, Pinniger had decided to reduce the number of sows kept: however, it seems that sows in pig were classed as stores, in that in the entry for 12 November 1818, he again recorded 17 sows in farrow.

Between January and March 1823, he made a series of extensive comments about the pigs which were killed on the farm, the costs of their keep and the value of the produce: he made further notes on the subject in 1824. Rather than leaving these entries to interrupt the flow of the transcribed diaries, they have been collected together in Appendix A.

As the primary form of motive power, horses were important at Little Bedwyn: with more than 700 acres of land to maintain, a large number of horses were needed as the following entry shows:

> 2 June 1814. 12 Horse's harrowing & 4 Horse's draging N field

As well as horses, Pinniger kept working oxen, as seen earlier (see *Arable Crops* above). Both horses and oxen were worked hard, as the following entries show:

> 16 December 1814. Rain incessantly, Horses in Stable all day The Oxen not at work The first rest day for them since last Winter.
> 4 October 1824. <u>7 Horses, - resting</u>, and turned out to Grass, - being nearly worn out, - no rest <u>the last 12 Years</u>

Of course, neither horses nor oxen would have worked on Sundays.

Livestock feeding

From time to time in the diaries, glimpses are provided into the management of livestock in terms of feeding: many entries relate to

94 The gestation period for pigs is 114 days (approximately four months) and common practice (until recent times) was to wean the piglets after eight weeks, at which point the sow comes into season some five to seven days later and is put to the boar, thereby starting the process of reproduction roughly every six months.

shortages of feed due to either wet weather in winter or dry weather in the summer, but occasionally there is quantifiable data to examine.

With regard to the feeding of hay to sheep during the winter, the following entry gives some idea of quantities:

> 23 February 1817. HAY, the Northfield Rick finish'd began 3rd Decr being 82 day with 610 Sheep 44 Tons

With 2,240 lbs to the ton, then this equates to approximately two pounds of hay per sheep per day. The other way of looking at this is that there was approximately half a ton of loose hay to be moved each day: cut from the rick and loaded on to a cart, moved from the rickyard to wherever the sheep were grazing or housed, off-loaded from the cart to the field or hayracks, all undertaken manually.

As well as hay, the sheep were folded on swedes and turnips. The following entry gives an idea of the acreage required to keep the sheep over a defined period:

> 6 May 1818. SWEEDS, 36 Acres finished feeding, began the Piece with 700 Sheep the 20th Decr 20 Weeks on every day.

This equates to 16 sq. ft. of swedes per sheep per day.

As for the horses, the following diary entries give some indication of the amount of feed required for them:

> 16 August 1817. HORSE'S, finish'd the Sainfoin began it the 30th June, 17 Horses 6 Acres
> 12 July 1819. Horses finish'd the Sainfoin Eastfield began it 25th May, 5 Acres – 16 Horses

That is, between 280 and 330 square feet of sainfoin was needed per horse per day over the summer period.

The total consumption of hay in the year 1817 to 1818 was recorded on 1 March 1818: 58¼ tons for the sheep, and 17¾ tons for the oxen and horses, leaving a stock of 120 tons. Therefore, if all of the hay in stock was consumed then this represents a total of almost 200 tons: in this year, 472 ewes had been put to the ram and there were around 10 oxen and 20 horses on the farm.

Livestock husbandry

All forms of livestock are prone to disease and in the early nineteenth century, the cures and remedies seem very primitive compared to what the modern-day livestock farmer has available to him or her. Further, some of the explanations of disease or death demonstrate the lack of understanding of the causes at that time: as ever, sheep in particular seem to be more susceptible than other animals.

On the whole, the sheep spent their time either grazing the pasture land or folded on the arable land. The main exception to this was when they were lambing, when a 'standing pen' or 'standing fold' was created to enclose the sheep and their lambs during this period. This would often be constructed in the field close to their current grazing, for example:

> 9 February 1818. made a STANDING FOLD in the middle of Sweeds the Ewes time being out the 14th

On some occasions, the pen was constructed in a more sheltered location, as the following entry shows:

> 6 February 1819. SHEEP, removed from the Fold N field to standing Pen, Stockwells Rick Yd having 40 Ewes cast their Lambs, change their Hay now from Ray grass to the New Sainfoin.[95]

One problem with the standing pen was that the manure produced by the sheep mixed with the straw started to rot and produce heat, necessitating the construction of another such pen.

In 1815, a number of ewes were aborting their lambs: the outcome is described in the following diary entries.

> 5 February 1815. 24 Ewes cast their Lambs.
> 9 February 1815. Ewes continue to Cast their Lambs, nearly at full time, some fall alive & die immediately
> 28 February 1815. 1 Waggon for Fellmongers refuse. 2 Tons 12£
> 1 & 2 March 1815. 2 Waggons Ea day for Fellmongers Refuse.

95 Cast – A sheep that is on its back and is unable to get up again unassisted; An older breeding ewe (also known as a draft ewe) which is sold off; Of ewes, to abort their lambs, National Sheep Association, *Dictionary* http://www.nationalsheep.org.uk/know-your-sheep/dictionary-corner/ [accessed 15 January 2018]. In this case, the third definition applies.

For those lambs which did survive that year, the danger was not over:

> 22 April 1815. 3 Lambs lost by their Blood becoming foul. & 2 Lambs saved by bleeding and giving dassys & Salt and Water.

Dassys probably refers to daisies – used traditionally as a spring tonic to 'cleanse the blood'.[96] It was not that 1815 was a unique year: in 1818, Pinniger recorded that the inclement weather at the end of March was resulting in significant losses of ewes and lambs, noting that one fellmonger had already bought 25,000 lamb skins.

However, not all lambing seasons were filled with despair. In 1820, the diary entries were very heartening, as follows:

> 19 February 1820. 150 Lambs this evening – 1 Ewe only cast her lamb & No Ewe died, and 10 doubles *
> 2 March 1820. 300 Lambs this evening. (and none has died)

The final entry for the lambing season came on 5 March when 330 lambs were recorded. However, it is of note that the proportion of twin lambs was very low: 10 doubles out of 150 lambs means that 130 ewes had given birth to a single lamb, giving a lambing percentage at this point of only 107%.[97]

One of the processes of sheep husbandry was the shortening of the long tails of the lambs, as on 20 April 1816 when 417 lambs tails were cut: this would have been carried out using a sharp knife. Lambing that year started on 19 January and by 12 March the majority (397) had been born: thus they were all at least one month old when their tails were removed, in contrast to modern day practices of applying a tight rubber ring within one or two days of birth. The practice was not without its problems, as the following entry shows:

> 4 May 1821. LAMBS Tails cut only a week a[re] very much swoln & raw

96 Anne McIntyre, *The Complete Floral Healer*, Gaia Books, 1996 in Positive Health Online, *A new look at Daisy (Bellis perennis)* http://www.positivehealth.com/article/herbal-medicine/a-new-look-at-daisy-bellis-perennis [accessed 19 March 2020].

97 This compares with the current national benchmark for lowland flocks of a lambing percentage of 183%, Eblex Ltd. *Target ewe fertility for better returns* (Huntingdon: Eblex Ltd, 2008), p.3.

= should be cut early'r or seered

A further process was the castration of the ram lambs, those not being kept as rams to serve the sheep. Once again, it appears that this was carried out using a knife a further month later, as on 24 May 1816 when 191 lambs were cut. This activity was also not risk-free, as seen in the diary entry recorded a week later:

> 31 May 1816. Lambs die 4 others making 16 - some People suppose <u>bad VERDIGREASE</u> the cause, or not useing enough good, 1 oz to 30 Lambs a proper quantity the best @ 1s/oz – the bad @ 4d – only a small quantity used now for 196.

Verdigrease (or Verdigris) is copper acetate and was used as an antimicrobial: it would appear that this was applied to the resulting wounds on the scrotum to prevent infection when the lambs laid down on the ground. In this case, it appears as though either poor quality or insufficient quantity (or both) was being blamed for the deaths of 16 lambs.

Injuries also happened during the shearing of sheep, as the entry made three days after shearing had taken place shows:

> 27 June 1815. Lamb died broken belly Young sheep died, prick in the Thigh wth Shears. | Lamb Kill'd, injured Shear day in the Back

When the sheep were shorn in 1820, Pinniger recorded a couple of recipes, one for treating damage done to the sheep with the shears and the other for the preparation of ruddle or marking 'paint':[98]

> 23 June 1820. Recipe for the cuts wth Shears. Equal quantity of Tar & grease Brimstone flour'd & salt mix'd to Keep off the fly. | Sheep mark. Equal quantity of Tar & grease Pitch & ruddle, continually simmering.

Similarly, horses and oxen were also not exempt from disease and death. The following entry describes the effect of heat on working animals:

> 31 March 1815. VERY WARM Ox died Rolling over come with heat Its killing – the Air fill'd its body – died soon after the others hitch'd off.

98 Ruddle - Red ochre paste applied to chest of ram which marks rump of ewe when the ram mates with her.

With horses being one of the two forms of motive power, their health was very important. However, they suffered a range of ailments for which the treatment was limited. The following extract from the diaries provides an example.

> 11 May 1817. HORSES thriving, their blood foul & itching, took from each horse – 3 Quarts of BLOOD after gave each horse, 2 oz of Salt petre dissolved in their water, to purify their blood

In this instance, thriving describes a horse rubbing the hair away from its body: this was one of many ailments for which bleeding was considered to be useful. Whilst three quarts may seem to be an excessive quantity, reference books of that time suggested that 'two and a half or three quarts may be sufficient'[99]

Another disease that is recorded in the diaries is that of murrain: however, this is not considered to be referring to the cattle plague or rinderpest of the second half of the nineteenth century, but rather to any unidentifiable disease.[100] The first reference in the diaries occurs at the end of 1818: *Ewe Kill'd, the Murrain this the 5th.*

Labour

The extent of the labour force on the farm at Little Bedwyn is best judged from the entry on 10 April 1818, when rain stopped all work: *RAIN, without ceasing all day. Horses in the Stable, 10 Men, 7 Boys, Carters*

99 Robert Pearson, *Every man his own horse, cattle & sheep doctor* (Leicester: J Browne, 1811). An adult horse weighing between 450 and 600 kg has a total blood volume of between 33.75l and 45l (30 to 40 quarts): the modern-day safe volume for a single bleed is considered to be between 3 and 4 quarts [adapted from Sarah Wolfensohn & Maggie Lloyd, *Handbook of Laboratory Animal Management and Welfare, 3rd edition* (Oxford: Blackwell Publishing, 2003)].

100 The definition of the term *Murrain* varies depending on the historical period. It is frequently found in medieval farming accounts: for example, see Mark Page (ed.), *The Pipe Roll of the Bishopric of Winchester 1301-2 (Hampshire Record Series Volume XIV)* (Winchester: Hampshire County Council, 1996), p. 133, the glossary to which defines *Murrain* as an unspecified disease or accidental death. In the period of these diaries, it was still referring to unspecified and unexplained infectious diseases in cattle and sheep, whereas by the late 1860s, it was primarily used to describe cattle plague or rinderpest, anthrax, foot-and-mouth disease and swine fever.

&c. in consequence no employ. Assuming that each carter looked after two horses and Pinniger had around 20, then he employed around 30 people, including the shepherd and his boy. As for the names of his workforce, he rarely ascribed men's names to individual tasks, but rather referred to groups of workers as "mowers", "reapers", "hoers" for instance. Individuals are named only when they leave their employment, join the workforce or commit some misdemeanour.

As well as his own labour force, Pinniger used a range of other workers to provide labour at peak times. For example, when the oats were ready for mowing in August 1813, this was carried out by *the Chippenham Men*. With the end of the Napoleonic Wars with France in 1815, large numbers of soldiers were demobilised and were looking for work. These two entries show that they were employed on the farm for specific tasks.

> 1 August 1816. Green rounds in N F: fit to hoe – 4 SOLDIERS put on them
> 6 January 1817. 8 Billet men repairing Roads by Picked mead, others throwing Trees.

The reaping of wheat was clearly carried out by a temporary workforce, given the numbers employed, as on 28 July 1818 when 42 reapers were employed. Leasing or leazing is the local word for gleaning: whilst farm workers were often allowed to do this to collect grain for their own use during the winter, the following entry suggests that Pinniger was paying women and children to do this and then selling the grain collected at a healthy profit.

> 23 September 1818. LEAZING Barley when fine the Corn now growing into the ground. 8 Women @ 7d pr day 4s 8d, 6 Girls @ 4d pr day 2s, 11 Childn @ 2d pr day 1s 10d [Total] 8s 6d, have leazed a fine day 6 Bu Best Barley, now selling at 72s[101]

Despite the size of his workforce, both permanent and temporary, there were times when the readiness of the crops to harvest and delays to harvesting due to bad weather resulted in too much work needing to be done all at once. This occurred in 1819:

> 11 August 1819. ALL THE CORN, ripening together the Pease, the Barley Pedlers & Oats Cowleaze } dead ripe, and 35 Acres Turnips want

101 Leaze (or lease) – to glean, Dartnell and Goddard, *Wiltshire Words*, p.92.

> hoeing the first time, no hands for either. The 40 Reapers get on very slow the Wheat being badlodged[102]

The heavy storms had lodged the wheat, making harvest a slow process. Barley, oats and peas were all ripe and ready for harvest, whilst the weeds were threatening to overwhelm the emerging turnip crop.

Labour relations appeared good in Pinniger's early years at Little Bedwyn, although on 3 May 1814, he recorded the following: *Barking finish'd Dean heath, Men disagree in prcie leave off.*[103] An earlier entry throws light on the differential between men and women workers:

> 2 June 1813. BEAN=HOEING Men began @ 10s Acre | PEASE=HOEING Women began @ 5s Acre

Of course, it may have been that pea-hoeing was easier work than bean-hoeing and it was possible to hoe twice the acreage in a given period of time – or perhaps not.

However, after 1820, it would appear that the workforce at Little Bedwyn was not so stable: in April 1821, Pinniger recorded that W. Law left Bedwyn and the following day, H. Edwards arrived. Later in the same year, three carters and their boys left their employment at Little Bedwyn. The position which Law and Edwards held on the farm becomes clear with the following entry for 2 January 1822: *Harry Edwards, Bailiff, discharged* and on 21 January, Robert Hacker "entered".[104] On 2 April in the same year, there is a diary entry which simply states *Burgess Notice*: who Burgess was, what role he performed and whether he was given or gave his notice is unclear, but this reflects a continuing trend of unrest between employer and employed.

Events recorded beyond the farm

Throughout the diary, reference is made to the various fairs held in the area: St Anns Hill (or Tanhill), Marlborough, Wilton, Appleshaw and Weyhill being amongst them. In 1817, he records:

102 Lodged - Of wheat (and other cereal crops) laid or beaten down by wind or rain, making it difficult to get dry and to reap or mow, Dartnell & Goddard, *Wiltshire Words*, p.95. Such conditions occur today and result in slow harvesting rates, even with modern machinery. Badlodged – a bad case of lodged.

103 prcie – price.

104 The term entered is used to describe the start of employment or service.

> 13 October 1817. WEYHILL,- hop fair 1st day the 12th, Sheep fair the 10th

This would have been an opportunity for farmers to purchase hops for brewing their own beer and ale, which fuelled the harvest in the following year: in 1826, Pinniger recorded 13,000 pockets of hops being grounded at Weyhill Fair with Farnham hops selling for between £6 10s 0d and £8 8s 0d, and country hops at around £1 cheaper.[105] However, he makes no note of the fact that cattle, horses, cheese and leather were sold at the fair in his time, as well as there being a hiring fair, a trade fair where allegedly anything could be purchased, and a pleasure fair.[106]

In 1817, we see the first entry relating to an external event, when Pinniger records on 19 November a church service for the Princess: this would have been for the death of Princess Charlotte of Wales.[107] On 16 February 1820, the burial of King George the third was noted in the diary.

Garden, family and church

The farm clearly kept Pinniger well occupied at this time: however, there are a number of references to him planting up his garden in the

105 Pocket – large strong jute sack, measuring some 6 feet long by 3 feet wide, holding about 1½ cwt of hops when full, Bromyard & District Local History Society, *A Pocketful of Hops: Hop growing in the Bromyard area, a new edition* (Bromyard: Bromyard & District Local History Society, 2007), p.91. Grounded (or pitched) – offered for sale. Farnham hops were grown specifically in the parish of Farnham in Surrey whereas country hops came from Alton, the Bentleys, Froyle, Binstead, Crondall and Odiham in Hampshire and other districts as far away as Kent. The considered opinion was that Farnham hops were the best for brewing and thus commanded a higher price, Anthony C. Raper, *The Ancient and Famous Weyhill Fair* (Andover: The Weydon Press, 2017), pp. 67-77.

106 Hiring (or mop) fairs were held in Wiltshire at Wootton Bassett (both Lady Day and Michaelmas fairs), Devizes & Wiltshire Gazette 6 April 1837, Devizes & Wiltshire Gazette 21 September 1843; at Cricklade, Devizes & Wiltshire Gazette 9 March 1837; and at Marlborough, Salisbury & Winchester Journal 7 October 1848.

107 Princess Charlotte (1796-1817) was the only child of George, Prince of Wales (later George IV). After a fifty-hour labour, which resulted in a still-born son, she died five hours later on 6 November 1817. These two deaths brought to an end the direct line of succession of George III and resulted in an outbreak of national mourning. National Galleries Scotland, *The Eye of Princess Charlotte of Wales* https://www.nationalgalleries.org/art-and-artists/63697/eye-princess-charlotte-wales-1796-1817 [accessed 21 March 2020].

springtime and late summer. Crops grown included peas, broad beans, kidney beans, celery, broccoli, cabbage and potatoes. He is also involved in the brewing of both beer and ale, hence his interest in the price and availability of hops.

In the first five years of the diary, there are no references to the church (apart from the above church service for Princess Charlotte) and the only reference to any form of relaxation comes in the following succinct entry in the diary on 11 July 1816: *LEISURE TIME*. Of course, at this stage in his life, he was a single man.

On 7 March 1820, the diary entry has a small heart symbol in the margin and another one at the end of the entry: see Figure 5. These heart symbols continue to appear intermittently throughout the entries for 1820: indeed, on 19 March (Sunday), a cluster of three hearts end the entry. In December of that year, the diary entries are annotated with place names or abbreviations written in red: the place names include Bath, Bremhill, Chalfield and Chippenham. Whether these were related to preparations for his forthcoming marriage to Mary Large, daughter of Abbot and Mary Large of Tockenham Court Farm, is unclear but they were married on 28 December 1820 in Lyneham church, for which date, the following diary entry appears:[108]

> Carting Road earth to meadow | VERY FINE CLEAR SUN SHINE DAY | Very cold freez'g winds

Although he was married that day, the importance of recording the weather was not far from Pinniger's mind. Thereafter, the heart symbol in the diaries ceased.

He recorded the birth of their first-born child, Ellen, on 5 April 1822 in his usual compact manner: *GOOD FRIDAY, 2 OClock A M Ellen born*. On 5 and 6 of June, the diary entry records *Lymington & Isle of Wight:* given that this was only two months after the birth of their daughter, it is considered unlikely that either Mrs Pinniger or the baby accompanied him on his visit to the New Forest and the island.

The first "religious" entries appear in October 1822 with the death of his wife's father:

> 2 October 1822. Mr Abbot Large, my Wifes Father OBIT 14 Chap Rev: 13 V. And I heard a voice from Heaven, saying unto me, Write, blessed

108 *Wiltshire, England, Church of England, Marriages and Banns, 1754-1916 for Thomas Pinniger* [accessed via Ancestry, 7 July 2020].

are the dead &c
7 October 1822. FUNERAL of Mr Abbot Large
13 October 1822. Sunday, Lyneham, Fun: Ser: 34 Ch Deutr 5 V.[109]

These three entries demonstrate all three of the major styles of emphasis used by Pinniger in the diary for such events (see EDITORIAL NOTE). In particular, he used the format of a thick black border across a double page spread in the diary, when he recorded the death of their daughter Ellen on 18 July 1824, aged 2¼.[110]

After a difficult harvest in 1824 due to excessive and prolonged periods of rain, and whilst farming operations continued at Little Bedwyn, Pinniger recorded a series of visits he made to Winchester, the Isle of Wight, Portsmouth and Southampton in early November that year: again, given that their daughter Mary was only three months old at this time and with the recent death of Ellen, it is likely that he again made this journey on his own.

Leaving Little Bedwyn

The Little Bedwyn estate was advertised for sale by auction on 22 September 1824, as it had been previously in 1819.[111] The particulars confirmed that there were all the necessary and convenient farm buildings along with a grist mill and thrashing machine driven by water and around 820 acres of land. It was stated that the purchaser could have early possession, potentially confirming that Thomas Pinniger was not subject to any period of notice arising from a tenancy agreement. However, no reports of the sale have been found.

---

109 Then I heard a voice from heaven saying to me, "Write: 'Blessed are the dead who die in the Lord from now on.' " "Yes," says the Spirit, "that they may rest from their labours, and their works follow them.", King James Version, 14 Revelation, v.13.
So Moses the servant of the Lord died there in the land of Moab, according to the word of the Lord, King James Version, 34 Deuteronomy, v.5.

110 In the period 1825 to 1837, the mortality rates for two agricultural parishes (Ash in Kent, and Morchard Bishop in Devon) were as follows: Infant Mortality Rate (IMR – death before the age of one): Ash, 106 per 1000 live births; Morchard Bishop, 82; Early Childhood Mortality Rate (ECMR – death between the ages of one and four): Ash, 78 per 1000 live births; Morchard Bishop, 48. See Romola J. Davenport, 'Urbanization and mortality in Britain, c.1800-50, *Economic History Review*, 73.2 (2020), 465.

111 Salisbury and Winchester Journal, 13 September 1824. Salisbury and Winchester Journal, 6 September 1819.

That something was indeed happening with regard to Thomas Pinniger and Little Bedwyn is confirmed in the entry for 2 December that year: *Plough'g as yesterday Mr Pain* which is repeated the following day. On 11 December, there is a diary entry which states *Mr Pains, first Goods, sent* which Pinniger underlined for emphasis. Three days later, he records that:

> 14 December 1824. Messrs Wooldridge & Rogers began valueing the Timber on Little Bedwin Estate for Messrs Guy & Pain

This is followed by a list of sheep marked to be valued (see Appendix B), whilst four days later:

> 18 December 1824. Plough'g Fold drift Whitley fur & finish'd Horses, work no more till they are Mr Pains

A valuation of the live and dead stock took place on 21 / 22 December as described in the diary:

> 21 December 1824. Valuation, by Mr Wm R: Brown of Broad Hinton Manor Farm and Mr [...] Lewes [...] of the Live & Dead Stock on Lt Bedwin Farm, from Mr Guy to Mr Pain
> 22 December 1824. Valuation finish'd by W R Brown & Anthony Lewes of Chilton Candover[112]

Thus, it is only towards the end of the diary covering his time at Little Bedwyn that it is confirmed that the livestock and the equipment belonged to Anthony Guy and that he was arranging for these to be sold to Mr Pain. It is however still unclear what the commercial relationship between Guy and Pinniger was, but a conventional tenancy arrangement appears unlikely. In particular, Pinniger and his family left the farm at what might be regarded as an unconventional date, being neither Lady Day nor Michaelmas.

On 15 January 1825, Pinniger recorded that the sale of the Little Bedwyn estate was completed and had been paid for at Marlborough: the estate was sold for £32,800 with a further £4,132 for the timber, £2,766 for the live and dead stock, £69 for the fixtures and £3,000 for

112 Chilton Candover is a village in Hampshire, 10 miles north-east of Winchester.

the makeout.[113] The latter item covered the valuation of growing crops and acts of husbandry completed: whether or not he received any portion of this figure is not stated or known.

Pinniger was clearly looking for another farm by this stage: on 31 January he looked over Rabson Farm, extending to 592 acres, for which the annual outgoings of rent, tithes and poor rates were £1000.[114] During this period, he continued to record the work on-going on the farm, but perhaps in less detail than previously. From 25 March (Lady Day), nearly all of the daily entries recorded the weather with little about the farm: the last fat pig was killed on 22 April and the last wheat rick was put into the barn for threshing on 9 May.

On 14 May, the first load of goods was sent to Chippenham and his mare, Kitty, along with two colts were sent to Mr Rawlings at Bremhill Grove, presumably on temporary livery until Pinniger had moved elsewhere. Two further loads followed and when the fourth and final load was despatched on 29 June 1825, Thomas Pinniger left Little Bedwyn and John Pain Esq became both owner and occupier.[115] As to why the farm was sold at this time is undetermined: however, given that it was previously offered for sale in 1819, and the subsequent events surrounding Anthony Guy, it can only be surmised that he needed the money.

FROM BEDWYN TO BECKHAMPTON VIA CHIPPENHAM (1825 TO 1829)

The farm

In that Thomas Pinniger and his family had to leave Little Bedwyn, following the sale of the estate, it would appear that Guy felt obliged to provide somewhere for them to live, as suggested in the entries in the diary for 29 and 30 June 1825:

> 29 June 1825. Wednesday, Fine day – 4th & last load Goods to Chippenham and left Bedwin | Fine day

---

113 Makeout - see glossary.

114 The likely parish location of Rabson Farm is Winterbourne Bassett. WSHC, *Wiltshire and Swindon Farmsteads and Landscapes Project: database extract from Wiltshire Historic Environment Record of Wiltshire farms as recorded on the Second Edition Ordnance Survey maps* (Chippenham: WSHC, 2014).

115 WSA: A1/345/27, *Land Tax Assessments: Little Bedwyn, 1780-1831, 1869-1881.*

30 June 1825. Thursday, - Storms | Entered Mr Guys Farm House Chippenham

This does, of course, beg the question, where was Mr Guy's farmhouse? The land tax assessment for the parish of Chippenham for 1826 records Thomas Pinniger occupying a 'Farm House &c.' for which the associated land tax was 7s 1d: the proprietor was Anthony Guy Esq. In 1827, 1828 and 1829, the property was described as a dwelling house and land with the same charge of 7s 1d for the land tax.[116]

Examining the details of the auction held on 8 April 1830 at the direction of the assignees of the estate of Mr Anthony Guy, a Bankrupt, the sale comprised 22 lots mainly cottages, gardens and small pieces of land (with regard to Guy's bankruptcy see *Events recorded beyond the farm* under THE BECKHAMPTON YEARS below).[117] Lot 1 was described as: *an elegant freehold burgage messuage pleasantly situate at the entrance of the town of Chippenham near the bridge*. This lot also included: *A close of rich pasture land containing about one acre, called The Isle of Rhe, immediately adjoining, surrounded by the River Avon.*[118] It is probably this piece of land which is referred to in the diary entry for 23 May 1826: *Fine day, little rain, eve'g | Horse, put on Mr Guys Island.*

Lot 2 was for Sambourne Farm, described as a compact farm consisting of a messuage, buildings and gardens along with land, as shown in Table 7.[119]

Table 7: Lot 2 – Sambourne Farm

Description	*Usage*	*a*	*r*	*p*
A messuage, barn, stable, skillings, carthouse, yards and three gardens		1	1	30
The Home Close	Pasture	4	1	30
The Ten Acres	Pasture	9	2	26
The Meadow	Pasture	3	2	16
Gaston's Mead and Brook	Pasture	2	1	32
		21	2	14

116 WSA: A1/345/99C, *Land Tax Assessments: Chippenham. 1816-1831.*

117 WSA: 415/53/1, *Sale particulars relating to property in Chippenham, of Anthony Guy, bankrupt, with valuation, 1830.*

118 The Isle of Rhe, also known as the Isle of Rea, is located in the River Avon, just to the west of the Town Bridge leading to the north end of the High Street.

119 With regard to the description, see Figure 6: The Meadow is probably the field marked on the plan as The Three Acres. Skilling – cowhouse.

Unfortunately, the sales particulars do not state the amount of land tax due for this lot: however, an annotated copy of the sales particulars shows that Lot 2 was purchased by Mr Rich for £2,740 and the Land Tax assessment for 1830 shows Richard Rich paying 7s 1d for 'a farmhouse &c, late Guys'.[120]

Figure 6 is taken from Powell's map of Chippenham dated 1784. This shows the buildings of Sambourne Farm below Home Ground (although somewhat rubbed at this point, it is possible to make out the letters _ _ NBOUR _ _ _ A R M) with three gardens to the left. Comparing the field names on the map with those in Table 7, Gaston's Mead and The Ten Acres are the same, Home Ground is probably Home Close, and The Three Acres is probably The Meadow (3a 2r 16p). Further, the Isle of Rea is not far (to the left in Figure 6) from the farm.

In all probability therefore, Thomas Pinniger was residing in the farmhouse at Sambourne Farm in Chippenham during the period 1825 to 1829.

The weather

The entries in the diaries are much reduced during this period, although daily records of the weather continue: indeed there are very few entries in the second half of 1825 which are not weather-related. Details are given of extreme periods of weather such as in July 1825, when there was an extended period of very hot days, as the following entries show:

> 17, 18 & 19 July 1825. The heat over powering, - many People & Horses died
>
> 19 July 1825. Warmer than the East Horses, many Kill'd in the Coaches, - and Men discontinued working

From 10 to 20 April 1827, Thomas went to London and Windsor, as well as visiting a farm in Kent: whilst he kept records of the weather there, his wife had made observations on the weather at Chippenham to enable him to compare the two.

Fog and thunderstorms were recorded along with flooding, in particular, the floods of August 1828 as recorded in the following diary entries:

---

120 WSA: 2811/3, *Sale particulars of freehold burgage house called Bridge House near the Bridge, Sambourne Farm, and other properties all in Chippenham, 1830.*

11 August 1828. Glass rising, dry morning, heavy storm after 12 O Clock, and frequent Storms till 4 O Clock, from this time Rain pouring all night for 12 hours
12 August 1828. FLOOD this morning to cover Mr Guy's Island, frequent Storms till 6 O Clock even'g, when fine, a fog rising. Stars very thick, and milk'y way seen, portend'g fine weather, WHEAT not very much sprouted, if dry from this time
13 August 1828. Fine fogg'y morn'g, dry, Glass steady Rain, all afternoon
14 August 1828. Rain in night, and thin steady rain all day | FLOOD, again on the Island | Wheat advanced to 40s pr Sack.

His experience as a farmer enabled him to observe the potential effects on the wheat if the wet conditions were to continue, yet he noted that the flooding at this crucial point in time just before harvest had caused the price of wheat to increase at the markets.

Pinniger the observer

So what was Thomas doing whilst he was living in Chippenham? From a number of his entries, it would appear that he was observing what other farmers were doing, along with the general progress of land, crops and livestock through the seasons. Examples of this characteristic include:

20 June 1826. Hay making, general the last week
9 April 1827. Mr Smith, Bremhill turned his Cows out to good Grass and, others, for want of Hay

He also followed local and national events. In 1826, the closure of the cloth manufactories in the town had led to widespread unemployment amongst the weavers and the effect on the poor rates was significant.[121] Elections for parliament were held in June that year and the diary has the following entry:

1 June 1826. WOOL: 8 Tons, - arrived at Chip[penha]m from Mr Gye of London, Candidate for the Boro: to employ the Poor

It would appear that Mr Gye was hoping that this action would assist his campaign as a candidate against Mr Grossett, the current member

121 Devizes and Wiltshire Gazette, 15 June 1826.

of parliament and Mr Maitland. The names of the successful candidates were recorded in the diary on 12 June 1826: *Chippenham Election Gye & Maitland*. Pinniger also recorded newspaper reports on the fires in the mountains of Scotland (4 July 1826) and the death of the Duke of York (6 January 1827 – recorded in his usual format for such events within a thick black border).[122]

However, Pinniger seems to have had plenty of spare time, in particular, during the winter of 1826-1827. Between the entries for 9 and 10 of October 1826, he included a list of names and monetary values entitled *Royal Allowances (Annual Acc[oun]t for 1825)* taking up six-and-a-half diary pages. A list of years, monarchs and monetary values entitled *Revenue of England since the Conquest* was also included, occupying a further page of the diary. On 14 December that year, he described the eclipse of the moon on 14 November 1826 on one diary page and the eclipse of the sun on 29 November 1826 on the next diary page, using diagrams as well as text.

Between the diary entries for 1826 and 1827, an Account of the Prices of Wheat from 1646 to 1826 is presented, taken from the Register kept in the Audit Books of Eton College: these occupy three double-page spreads in the diary. A further three double pages in the diary after the entry for 28 February 1827 relate to propositions made for the Corn Laws – these are reproduced in Appendix I. All of these entries demonstrate his continued interest in matters both agricultural and non-agricultural.

Crops and livestock

Pinniger's agricultural activities at Chippenham seem to have been confined to the garden at the farmhouse. On 1 May 1826, he recorded that potatoes, peas, cabbages and gooseberries had all been damaged by the late frost, whilst on the following day he was planting kidney beans. He was also growing cucumbers and indeed using a heated frame to do so, as noted in the following entry: *26th April 1827. Sharp, frost, Icecles, - 6 In long, & large as a finger, - from steam of Cucumber frame, dry day and warmer.*

As for livestock, the only references in the diaries for this period relate to horses: he seems to have kept a mare (Kitty) and two colts. However, the horses do not appear to have been kept at Sambourne Farm, as the following examples indicate that Pinniger was paying for their livery:

122 Prince Frederick, Duke of York and Albany KG GCB GCH (16 August 1763 – 5 January 1827) was the second son of George III.

26 December 1825. Colts removed from Cocklebury to Newlease @ 2s pr Week
20 April 1826. The Mare Kitty, taken from Cleveancey to Cockleberry at Grass and Hay - @ 3s 6d pr W[123]

He also records on 24 April in the same year that he was buying in hay from Cockleberry and there are no entries for haymaking during the 1826, 1827 or 1828 seasons. Thus, it would appear that he was occupying the farmhouse, but not the farm land, at Sambourne during this period: this is consistent with his diary entry for 30 June 1825 quoted at the beginning of this section – *Farm House* rather than *Farm*. As ever with livestock, he recorded incidents of illness as on 29 June 1826 when the two colts were unwell, suffering from the disease known as strangles.[124]

Family matters

As in earlier years, Pinniger continued to record births, marriages and deaths of various members of his family, as well as those of his friends and noted individuals in the area. The closest such events were the birth of his daughter Ann on 18 June 1826 and his son Thomas on 31 July 1828.

However, you can sense his loss through the diary entries when his own father died aged 79 on 20 August 1827. He records the details of the funeral service as well as the Sunday funeral sermon given by the Revd Mr Clifton at Bremhill. His father's will was proved in London on 7 March 1828.[125] As well as being appointed a trustee and executor, Thomas received a bequest of £5,000: this was to be raised by selling or mortgaging part of Cowitch Farm, which was occupied by his brother John.[126] He also received his father's secretary-bookcase.[127]

---

123 W - week. Pinniger uses *Cockleberry* and *Cocklebury* interchangeably: he was probably referring to Cocklebury Farm, Chippenham.

124 Strangles - One of the most common equine diseases in horses in the UK. It is a highly contagious infection of the upper respiratory tract caused by the bacteria *Streptococcus equi subspecies equi*, Horse & Hound, *Strangles* http://www.horseandhound.co.uk/tag/strangles#fzzIiHCL23gUIEKE.99 [accessed 1 May 2020].

125 TNA: PROB11/1738/65, *Will of Christopher Pinniger, Gentleman of Bremhill, Wiltshire, 7 March 1828.*

126 Cowage or Cowitch is Cowage Farm, Hilmarton.

127 Secretary-bookcase – a writing desk with drawers and a bookcase above and to the rear.

In 1825, his niece Catherine Pinniger, the eldest child of his brother John who lived at Cowitch (Cowage), was married and another niece, also called Catherine was born to William and Susannah Spackman at Chalfield: Susannah was his sister.

On a personal note, Pinniger made the following entry on 26 June 1828: *Wilson & Son, Dentist, 13 Paragon Buildings.* It does not give the town or whether he visited the dentist, but it is clear that it was important to him to have a record of this in his diary.[128] However, it probably explains the inclusion of Bath in his diary entries of 8 and 10 July that year.

Finding another farm

There are a number of entries in the diary recording details of farms in terms of their acreage, rent expected, tithes and poor rates, as well as crops being grown and the livestock which the farm was supporting at the time. The named farms were Rabson Farm, Beaversbrook Farm, Coomb Farm at Enford, Fosbury Upper Farm and Monckton Farm (see Appendix G).[129]

In the summer of 1826, Pinniger looked over Broadstock Farm three times, but on the third occasion, he abandoned the idea of purchasing this farm.[130] The following summer, he surveyed Grange Farm and Claverton Farm and between the entries in the diary for 21 and 22 July he recorded the sale of three farms located within a mile or two of Malmesbury (see

128 Gye's Bath Directory of 1819 lists Mr Willson as dentist to his R.H. the Duke of Clarence at 13 Paragon Buildings, Bath, H. Gye, *Gye's Bath Directory, corrected to 1819* (Bath: H. Gye, 1819), p.7. By 1846, the practice had expanded and moved premises in that Silverthorne's Bath Directory recorded Willson (Isaac) & Son (John) as dentists at 12 Paragon Buildings, Bath, H. Silverthorne, *Silverthorne's Bath Directory* (Bath: H. Silverthorne, 1846), p.184.

129 The possible parish locations of these farms are as follows: Rabson Farm – Winterbourne Bassett; Beaversbrook – Hilmarton; Coomb [now Combe] Farm – Enford; Fosbury [Upper] Farm – Tidcombe and Fosbury, WSHC, Wiltshire and Swindon Farmsteads and Landscapes Project: database extract. Mon[c]kton farm – Winterbourne Monkton: fields belonging to the farm include Hackpin Down and Windmill Hill Down (see Appendix G), with Hackpen Hill and Windmill Hill being recorded in the parish of Winterbourne Monkton, J. E. B. Gover, Allen Mawer and F. M. Stenton, *The Place-Names of Wiltshire* (English Place-Name Society, Volume XVI), (Cambridge: Cambridge University Press, 1939), pp 310-311.

130 Bradenstoke is marked as Broadstock on a 1773 map, Gover and others, *Place-Names of Wiltshire,* p.270.

Appendix K).[131] In his diary entry for 30 January 1828, it was noted that Littlecott Farm at Enford was for sale and he wrote some notes on the value of the farm at the end of Volume 4 of the diaries – see Appendix L. It is possible that he visited the farm on 9 February and had met Mr Moore on 19 February, with a view to purchasing the farm: however, he declined to do so. Four days later, he looked over Beckhampton Farm with Mr Guy, again with a view to making a purchase: the farm clearly met his requirements in that on 27 February, Thomas Pinniger took a mortgage and bought the farm.[132]

From Chippenham to Beckhampton – a protracted journey

At the time of purchase, the farm was occupied by William Philpot: he (and possibly his father also called William before him) had occupied the farm from 1782 or earlier according to the Land Tax records.[133] Pinniger returned to the farm on 28 March 1828 to look over it again for the first time since the purchase had been made and on 3 April he bought seeds for the farm.

On 5 April, Pinniger met with William Philpot and Anthony Guy to determine the terms for Philpot (as the current occupier) to quit Beckhampton Farm and an agreement was signed four days later: however, it is not known what period of notice Philpot was subject to. From the diary entries, it appears that Philpot was to carry out certain acts of husbandry in order to prepare the farm for Pinniger entering. For example, the diary entry for 10 April states: *Barley, Wm Philpot, began Drilling bottom of the 19 Acre Piece, and my Seeds, sowing after, heal'd with one tine of the Harrows.*

Given that his wife is rarely mentioned in the diaries, it could be concluded that the following entry records her first visit on 13 May 1828, to what was to become her new home: *Warm fine day look'd over Beckhampton, wth Mrs Pin*. The entries in the diary now start to become

131 These farms are probably Manor Farm and Cleverton Farm, in the parish of Lea & Cleverton, WSHC, *Wiltshire and Swindon Farmsteads and Landscapes Project: database extract*. The village of Lea is approximately two miles from Malmesbury.

132 It has not been possible to determine from whom Pinniger got his mortgage, or for how much. Although the farm may have met his requirements at this time, there is a note in the diary on 8 October 1829 which states: Sent a Letter of complaint to Mr G. Assuming that this was to Anthony Guy, there is no indication as to what the complaint was about.

133 WSA: A1/345/22A, *Land Tax Assessments: Beckhampton. 1782-1841.*

more extensive each day, detailing not just the weather, but the activities that were underway at the new farm, mainly undertaken by William Philpot: an example of this is recorded on 27 June 1828 (see diary for 1828). However, it was not until late November that the diary records:

> 25 November 1828. All Wm Philpots, Sheep now sold and off the Farm, Ewe Tegs at Keep

Philpot was still working for Pinniger in January 1829 and there is a suggestion that Philpot was dragging his heels in finding a new farm to move to. On 13 March, Pinniger went with Philpot to look at a farm at Nettleton: the following day, the diary entry records that: *Willm Philpot took that Farm* underlined by Pinniger for emphasis. Whilst the sale at Beckhampton on 25 March is noted in his diary, the Devizes and Wiltshire Gazette gave more details of the property of Mr Phillpott (*sic*) who was quitting his farm.[134] The following items were advertised for sale:

> Three capital well timbered Rick Staddles, portable Thrashing Machine, Horse Trough, Hurdles, Flakes, Carts, wheels and iron arms; Cheese Press and Tub, Brewing Tackle and Casks, two Brewing Coppers and Grates, iron Boiler; also various Household Furniture, &c.: comprising four post Bedseads, mahogany bureau ditto, and stumps; feather and flock Beds, Tables and Chairs, Cucumber & Melon Lights, with numerous other effects.[135]

On the final day of March, Messrs W. R. Brown and J. Stratton valued the ploughing, hay etc for Pinniger and Philpot.

Philpot quit the house on 3 of April and the removal of goods and furniture from Chippenham to Beckhampton took place over the period from 4 to 8 of April. The following entry demonstrates that moving house was no less stressful then, than it is today.

> 9 April 1829. Thursday, All Family removed from Chippenham to Beckhampton | Dry morning, Rain before the 2nd & last load Waggon

134 Devizes & Wiltshire Gazette, 19 March 1829.

135 Flakes – frames, barred with ash or willow spars, somewhat resembling a light gate, used as a hurdle where extra strength is needed. Flake hurdles are used to divide a field, or for cattle, the ordinary sheep hurdles being too weak for the purpose, Dartnell and Goddard, *Wiltshire Words*, p.56. Bedseads – should read bedsteads – a typographical error in the newspaper.

could be load'd, intervals of Rain, and dry till 6 O Clock, Waggon arrived at Beckhampton, half pass 12 O Clo at Night

At last, the Pinniger family had arrived at Beckhampton Farm.

THE BECKHAMPTON YEARS (1829 TO 1847)

The farm

According to the land tax assessments, William Philpot had occupied land owned by James Sutton with a land tax of £8 12s 4d since at least 1782.[136] Philpot also occupied a parcel of his own land during this period (Land tax £2 17s 8d), as well as land owned by the Duke of Marlborough in Beckhampton (Land tax £8 3s 6d). In 1788, a terrier and valuation of Beckhampton Farm belonging to James Sutton Esq was prepared by Richard Richardson of Devizes.[137] Table 8 summarises the content of this terrier.

Table 8: Summary of Terrier of Beckhampton Farm, 1788

Premises	*State*	*Quantity*		
		a	*r*	*p*
Homestead & Home Close	Pasture	1	3	2
Half of Bell Close	Meadow		3	7
Half of Long Close	Meadow		1	0
Half of New Inclosure	Meadow		3	7
Phelp's Close & Cottage	Meadow	1	1	0
4 Acres in Silbury Mead commonable (tithe free)	Meadow	3	1	38
1 Under Northmead Hedge (tithe free)	Meadow		2	36
¼th of six premises (named)	Meadow	5	1	16
	Sub-total	14	1	26
In Tenantry Field, 47 pieces		98	1	36
In Furze Down, now plow'd		7	0	0
Share of 315a 2r 25p of Down for 336 sheep in a flock of 1192	Pasture	88	3	36
	TOTAL	208	3	18

Assuming that the land in Tenantry Field was arable, along with the seven acres ploughed in Furze Down, then the proportions of arable and

136 WSA: A1/345/22A, *Land Tax Assessments, Beckhampton, 1782-1841.*
137 WSA: 248/195, *Terrier of Beckhampton Farm in Avebury, 1788.*

pasture/meadow are approximately equal. The tenant was to pay all taxes, except the land tax. It was noted in the terrier that the estate had *in some degree been benefitted by exchanges but would be very much improved, if the same were let out in severalty. The rent should be 120£.*

In the land tax assessment for 1789, James Sutton was listed as the proprietor of premises, occupied by William Philpot, with a land tax payable of £11 10s 0d.[138] From this, it would appear that Sutton had bought the parcel of land which William Philpot owned and that the above valuation of Beckhampton Farm had been undertaken to determine the rent which Philpot should pay for the whole farm. Quite why Philpot had sold his land to Sutton and then rented it back is unclear, but it is possible to surmise that Philpot needed the cash. At the time of the 1794 enclosure of Avebury parish, of which Beckhampton was a tithing, Sutton and Philpot were still the owner and occupier respectively of the land shown in Table 9.[139]

Table 9: Land at Beckhampton Farm, 1794

No	*Description*	*Quantity*		
		a	*r*	*p*
41	Field to the north of the farm on opposite side of the Turnpike Road	4	0	37
43	Land to the south and west of the farm on the Downs	187	0	26
44	Water meadow by Silbury Hill	8	2	24
	TOTAL	200	0	7

Table 10: Land at Beckhampton Farm, 1794, as recorded by Pinniger

	a	*r*	*p*
Whitelands	4	0	37
Pasture Waste	0	2	37
Arable & Down	187	0	26
Silbury Mead	8	2	24
	200	3	4
Old meadow & Buildings	6	1	24
TOTAL	207	0	28

138 This value equates to the sum of £8 12s 4d and £2 17s 8d, as previously noted.

139 WSA: EA95, *Avebury Enclosure Map, 1794.*

In addition, James Sutton's name was written across the following pieces of land on the map: Bell Close, Hoskins Close, Long Close, Home Close (and possibly Tookers Close). On opening Volume 5 of his diaries, Thomas Pinniger recorded details of the enclosure award for Beckhampton: see Table 10. Comparing Table 10 with Table 9, Pinniger's entries for Whitelands, Arable & Down and Silbury Mead correspond to numbers 41, 43 and 44: the other two entries in Table 10 provide the acreages of the five named closes taken from the enclosure map. As a result of whatever exchanges had taken place previously, along with any division of the downland through this enclosure process, the farm had remained at around 208 acres in extent.

It is of note that the water meadows around Silbury Hill were owned by three parties: the Duke of Marlborough (6a 1r 38p), Peter Holford Esq (13a 0r 1p) and James Sutton (8a 2r 24p). Again in the opening to Volume 5, Pinniger records the rights of watering Silbury Mead, as awarded by the Enclosure Commissioners: Peter Holford Esq was to have the water exclusively for three days, commencing 1 November each year, with the Duke of Marlborough having the next three days and finally James Sutton Esq having the next four days. As for the expenses of hatches and other control mechanisms to effect the watering of the meadows, he recorded that these were to be split as follows: Peter Holford Esq, 9s 6d in the pound; Duke of Marlborough, 4s 6d in the pound; and James Sutton Esq, 6s 0d in the pound. The effects of this arrangement will be observed later (see *Disputes and disagreements* below).

When James Sutton redeemed the land tax on the farm in 1799, the redemption certificate did not indicate the acreage.[140] He continued as the proprietor of Beckhampton Farm according to the land tax assessments until 1802, when it changed to Estcourt Esq. who continued to be recorded in this way until 1815, the year that Anthony Guy of Chippenham bought the farm. William Philpot was the tenant throughout this period.[141] The purchase of the farm from Guy by Pinniger continued the long-term business connection between these two individuals. From 1829 onwards, until the land tax records cease to provide useful information in 1840, Thomas Pinniger was both proprietor and occupier of Beckhampton Farm.[142] The other farmers in the village were John Wentworth and

140 TNA: IR24/16, *Land Tax Redemption Office: Registers of Redemption Certificates. Registered numbers: Nos 7000 – 7999, 1799*. Certificate number 7794 states "containing in the whole [...] acres (more or less)".

141 WSA: A1/345/22A, *Land Tax Assessments, Beckhampton, 1782-1841*.

142 The series of assessments in WSA: A1/345/22A, *Land Tax Assessments,*

Robert Philpot, both occupying Robert Holford's land.[143]

In the 1846 tithe award for the parish of Avebury, Beckhampton Farm comprised the land described in Table 11.[144]

Table 11: Land at Beckhampton Farm, 1846

No	*Description*	*State of cultivation*	*Quantity*		
			a	*r*	*p*
34	Down Furlong	Arable	82	1	32
31	Large Down	Pasture	56	1	20
32	Small Down	Pasture	13	2	25
33	Upper Down	Pasture	12	3	16
35	Left hand side of St Anne's Hill Road	Arable	23	0	24
37	Farther Meadow	Pasture	1	2	18
38	Farther Meadow	Pasture	1	1	0
39	Home Meadow	Pasture	1	2	8
40	House, farm, buildings, yards and gardens		1	0	3
41	Paddock	Pasture			29
42	Garden				27
43	Garden				36
44	Whitelands	Arable	4	2	0
52	Silsbury Mead	Pasture	8	3	13
	TOTAL		207	3	11

Beckhampton, 1782-1841 ends in 1840: in 1839 and 1840, Thomas Pinniger is listed as owner and occupier of land in Beckhampton but no value is given for the Land Tax and there is no assessment for 1841. The assessments for the years 1842 until Pinniger's death in 1847 do not include land where the land tax had been exonerated, as Beckhampton Farm had been in 1799, WSA: A1/345/22B, *Land Tax Assessments, Beckhampton, 1842-1884.*

143 John Wentworth had occupied a farm in Beckhampton tything (in the parish of Avebury) with Land Tax £28 6s 8d since 1818, Robert Holford Esq having taken over as proprietor from Peter Holford Esq in 1804, WSA: A1/345/22A, *Land Tax Assessments, Beckhampton, 1782-1841*: a lease dated 1818 from Holford to Wentworth named the farm as Great Beckhampton Farm, extending to 438 acres, WSA: 2027/2/1/855, *Avebury Deeds, 1741-1907.* Similarly, Robert Philpot had occupied a farm in Beckhampton tything with Land Tax £8 3s 6d from 1822, Robert Holford Esq having taken over as proprietor from the Duke of Marlborough in 1824, WSA: A1/345/22A, *Land Tax Assessments, Beckhampton, 1782-1841*: a lease dated 1824 from Holford to Philpot was for an unnamed farm (named in 1843 as Little Beckhampton Farm, in a notice extinguishing tithes on both farms) extending to 178 acres, WSA: 2027/2/1/855, *Avebury Deeds, 1741-1907.*

144 WSA: TA/Avebury, *Tithe Award and map for the parish of Avebury, 1846.*

The land on the farm at this time was split 58% arable, 41% pasture and 1% other: in comparison with the parish as a whole, the proportion of arable was 51%, with meadow, pasture and down making up the remaining 49%.[145] At this time, the rest of titheable water meadows, known as North Mead (Nos. 53 & 54 – a total of 15a or 24p) was owned and occupied by William Tanner.

The inn

According to his diary, Pinniger purchased the Inn from Anthony Guy on 18 June 1828: there is no record of the price paid. Evidence for the extent of the Beckhampton Inn comes from the Tithe Award: owned by Thomas Pinniger and occupied by William Treen at this time, Table 12 summarises the land associated with the property.

Table 12: Beckhampton Inn, 1846

No	*Description*	*State of cultivation*	*Quantity*		
			a	*r*	*p*
27	Beckhampton Inn, buildings, yards & gardens		1	0	6
28	Cuckoo Pen	Pasture		1	7
29	Paddock	Pasture	2	1	2
	TOTAL		3	2	15

In 1794, the Inn had been marked on the enclosure map as the Catherine Wheel Inn at the junction of three turnpike roads: Devizes to Marlborough, Beckhampton to Calne, and Beckhampton to Avebury.[146] As well as no doubt providing a rental income, the Inn created a demand

145 TNA: IR18/10895, *Tithe file for Avebury (parish), Wiltshire, 1836-1870.*

146 WSA: EA95, *Avebury Enclosure Map, 1794.* The Inn has been known by various names over time. In 1785, the Inn was known as Beckhampton House, Bath Chronicle & Weekly Gazette, 27 January 1785; in 1786 it was referred to as the Beckhampton Inn at Beckhampton, Salisbury & Winchester Journal, 25 December 1786; in 1796, it was described as Beckhampton House Inn in a set of sales particulars, WSA: 415/45, *Sales particulars for property in Bromham, Beckhampton, Avebury and Chippenham including The Beckhampton House Inn, Avebury Great Farm and Lower Sheldon Farm, 1796*; in 1799, it was known as Beckhampton Inn, when William Edmunds from Bath entered the Inn, Bath Chronicle & Weekly Gazette, 23 May 1799; and in 1809, the house of Mr Edmunds was referred to as the Catherine Wheel Inn, Salisbury & Winchester Journal, 9 October 1809.

for straw for the coach horses stabled there, as follows:

> 9 October 1832. COACH=HORSES, came to the Inn, to serve with Straw 15-Viz The White Hart, York House & Exeter Mail all Mr Halcombs[147]

and, in turn, a source of fertility for the farm land:

> 19 October 1829. Thwarting the two Couchey Lands & pick'g it off, - and the first Waggon Load of Dung from the Inn put on it, and the Fold follow

As landlord of the buildings (rather than the licensed premises), Pinniger was clearly responsible for their maintenance:

> 28 September 1829. Painting began Windows of the Inn
> 25 September 1830. 1000 Bricks to Inn & 16 Bu: Lime to Inn | Cart g Marter Earth to Inn[148]

Indeed, the later entry appears to be materials for an upgrade to the Inn as on 5 October, William and John Robbins were tiling the privy there.

On 19 January 1835, a Mr Treen inspected the Inn: it would appear that he liked what he saw, in that five days later, Pinniger served notice on Mr Watts to quit the Inn at Michaelmas 1835. That William Treen intended to extend the inn to provide racing stables, as well as the stabling for the coaching trade at the inn, is confirmed in the diary, as follows:

---

147 The White Hart coach to London left Bath at four every afternoon travelling through Devizes and Marlborough; the York House coach to London left Bath at a quarter to three every afternoon travelling through Chippenham, Calne and Marlborough, Pierce Egan, *Walks through Bath: Describing every thing worthy of interest* (Bath: Meyler & Son, 1819), p.311. The [Bath and] Exeter Mail left Bath at half past seven travelling through Devizes, John Cary, *Cary's New Itinerary* (London: G & J Cary, 1828).

148 Marter earth - Earth for the mortar used in the construction of buildings. From prehistory until the 19th century, earth was used in a variety of ways to construct buildings of every type across Britain. Typically, earth mortars are made from local natural subsoil, where clay minerals act as the binder to sand and silt particles. In such mortars, quicklime was added to the earth: analysis of historic earth mortars indicates the addition of around five per cent quicklime, which will have doubled in volume with slaking, Tom Morton, Nigel Copsey and Rebecca Little, *Earth Mortars* http://www.buildingconservation.com/articles/earth-mortars/earth-mortars.htm [accessed 30 April 2020].

> 23 February 1835. The 1st two loads of Stones fm Calne for building the Raceing Stables at the Inn for Mr Treen

In all, six loads of stone were brought from Calne during the month. Early March saw the start of work at the Inn with the waggon house being enclosed and stables being built: 32 bushels of lime and a further load of stone were brought from Calne and 2,000 bricks were carted from Devizes with five horses. The waggon house was enclosed by 28 March.

On 7 April, Pinniger wrote in his diary that Mr Watts had declined to pay his rent, following which he noted eight days later that Watts and Treen did not agree.[149] Watts appears to have been unhappy about the building work having started before the end of his notice period: not only was Watts still refusing to pay his rent on 14 May, but he was not allowing Pinniger to build for Mr Treen, which is not surprising in that he had already lost use of the waggon house for that purpose.

John Watts held his first sale on 18 September at the Inn and a range of items were sold including furniture, casks, mash tub, malt mill, hop strainer, wort tubs, as well as 36 bushels of malt and half a pocket of hops.[150] From this, it is clear that Watts was brewing, as well as selling, beer at the premises. A second sale was held on Thursday 1 October by his creditors, the sale closing on the following day: however, Watts persisted in keeping possession, insisting that he had paid his rent (after a threat of distress) up to Michaelmas. William Treen entered Beckhampton Inn the following Saturday as a result of John Watts giving him possession at 8 o'Clock in the evening, having extorted £5 from Treen to do so. Two days later, Pinniger sent a waggon and three of his horses for Mr Treen's goods.

With Watts gone, work on building the stables started in earnest, as on 7 October 1835, Pinniger recorded the following: *Began preparing for the foundation of the Racehorse Stables at the Inn. for Mr Treen.* Good progress was made with the new stables, the horses being sent to Honey Street for the new roof at the end of October: on 7 November, waggons went to Box and Pickwick for *Pitchen* stone probably for the floors, and again on each of two occasions later in the month, 1,000 slates were brought from

149 Presumably this was the rent due as at Lady Day (25 March).

150 Salisbury and Winchester Journal, 14 September 1835. Wort - The infusion of ground malt or other grain before fermentation, used to produce beer and distilled malt liquors, Lexico, *Wort* https://www.lexico.com/definition/wort [accessed 18 May 2020].

Devizes. Slating was finished on 3 December with window glazing and plastering of the ceiling and walls following on. Sarsen stones were used for pitching the floor of the eight stalls in the new stable.[151] A new road and a new well were also provided: the well was sunk 22 feet down before water was found.

On Christmas Eve 1835, Pinniger wrote that the new stables at the Inn were finished, apart from the boarding of four stalls and that they were occupied before they were finished by six racehorses brought from Marlborough. Whilst the carpenters finished the stalls on 4 January, they left Mr Treen to finish two of his boxes, in that they were now occupied by horses: these were finished five days later.

The Inn continued as a coaching inn and various references to the coaches passing through and the delivery of straw to the inn for the coach horses have been recorded in the diaries. However, on 26 January 1836, Messrs Halcombs (operators of some of the coaches with horses stabled at Beckhampton Inn) informed Pinniger that they had received notice from Mr Treen to quit the Inn: from this, it would appear that Treen was focussing more on race horses, rather than on coach horses. Later that year, work was completed on the construction of a shed over the well at the Inn.

In 1839, Mr Treen was recorded in the diary as building a stable by the Devizes door of the yard: this is in contrast to such work being done by Pinniger himself. However, by 1840, more changes were afoot at the Inn: on 4 January, the horses for the Star coach left the Inn and in September that year, those of the White Hart day coach were moved to Avebury *to meet the Railroad beyond Swindon*: the horses for the night coach remained at the Inn until 16 December, when along with the York House coach, they were routed through Avebury to the railroad at Hay Lane.[152]

Pinniger was still carrying out repairs at the Inn in 1845, when the thatching of the stables was completed on 7 January, except for those by the Calne Road: the work had started in the previous October and he records the costs as £5 12s 6d for labour, 12s for tar thread with an

151 Pitchin[g] – whilst paving is carried out using large flat stones, 'pitching' uses small uneven stones set on edge, Dartnell & Goddard, *Wiltshire Words*, p.119.

152 Pinniger uses the term *railroad* rather than *railway* in his diary entries. Hay Lane was a temporary terminus created when the Great Western Railway line was extended from Faringdon. Known officially as Wootton Bassett Road Station, it was four miles by road to the east of [Royal] Wootton Bassett. Edward Terence MacDermot, *History of the Great Western Railway, volume 1* (London: Great Western Railway Company, 1927). Hay Lane is located just to the south-east of Junction 16 of the M4 motorway.

estimate of £20 for the straw used.

The Inn was closed on 10 December 1846 as a result of the failure of Toms and Matthews who had been declared bankrupt.[153] A sale of household furniture and other effects including brewing equipment belonging to Toms and Matthews were sold at the Inn on 22 January following: however, Pinniger senior recorded that the inn was re-opened by Richard Hazle for William Treen on the following day with Pinniger junior noting that Mr Hazel (*sic*) was to remain until October 1847 to sell for Treen in order to save the latter the expense of acquiring a licence. From a local newspaper report, it would appear that the Inn was still a going concern in October 1849 when it was to be let or sold.[154]

The weather

In common with his time at Little Bedwyn, Pinniger continued to record details of the weather in his diaries: indeed, this intensified such that hardly a day was missed where a record was not made. In addition, in his first year at Beckhampton, he recorded the number of days rain in each month, as shown in Table 13.

Table 13: Days of rain at Beckhampton Farm by month, 1829

Month	Apr	May	Jun	Jul	Aug	Sep	Oct	Nov	Dec
Days of rain	27	6	16	25	22	24	18	9	12

The year 1830 started with a prolonged period of snow and ice with very sharp frosts recorded in the diary. However, conditions worsened on the twentieth of January with the snow drifting all night and all day. The only work carried out by the men was carting hay to the sheep, but when the snow reached the top of the hurdles in the fold, the sheep were brought home to the yard to avoid them being buried.

On the roads, the snow reached between two and three feet deep: as a result, the coach traffic was disrupted. The mail coach was delayed

153 William Toms and John Matthews, Brewers of Hungerford, Berkshire were listed as bankrupts in the London Gazette of 15 December 1846, Salisbury & Winchester Journal, 19 December 1846.

154 Applicants to contact Mr Pinniger, Devizes and Wiltshire Gazette, 4 October 1849. This is contrary to the history of Beckhampton Stables prepared by David Acock, in which he states that Billy Treen closed the Inn in the late 1820s. Beckhampton Stables, *The History of Beckhampton* https://rogercharlton.com/the-history-of-beckhampton/ [accessed 17 April 2020].

arriving at Beckhampton by five hours and indeed it overturned between Beckhampton and the White House on the Down, whilst the horses were being led, the passengers having already disembarked to walk. The coach known as the Monarch stopped at the same place and both coaches had to be retrieved by Mr Wentworth's cart horses. Coaches from Calne made it as far as Cherhill only to return, and those from Devizes reached Beckhampton but had to remain there.

Overnight into 21 January, only the mail through Devizes made it to Beckhampton arriving 3½ hours late: it tried to continue to Marlborough but only reached Silbury Hill before returning to the Inn. In the morning, two hundred men were digging out the snow between Cherhill and Marlborough with snow having drifted up to 5ft deep. The road was passable by three in the afternoon and the following evening and night numerous delayed coaches passed through Beckhampton. In his diary, he noted that he had made a similar record in 1814:

> 20 January 1814. SNOW drifed, the roads impassable – SHEEP DUGOUT of the Snow, Ewes fold in Gate Close Horses in the Stable

The cold weather continued in 1830: on 2 February, he recorded that the lid of the washing copper was frozen on, even with a good fire lit under it: the thaw did not come until six days later. However, the snow was not over for the year:

> 2 April 1830. SNOW, to cover the ground again before morning, and continued, large Blossoms wth hail & rain at intervals all Day, and keen blows the wind and piercing is the cold - more Snow fell this Day, than any one day this Winter & drifted 3 feet Deep

Coach traffic was similarly disrupted at the beginning of February 1831 and again in 1836, when the London coaches were almost buried and no coaches or mails had been through Beckhampton on the evening of 2 February. Both inns were full of travellers and two ladies had to remain in the York House coach overnight, in that they could not be rescued. On the following day, the coach from Calne to London went through the fields in order to continue its journey. Later in 1836, a distinguished person made his way by carriage through Beckhampton and again had to be diverted through the fields:

> Tuesday 27th Decr. The first Carriage since Sunday night (Xtmas day)

> pass'g our road was the DUKE OF WELLINGTON'S, on his way to Badmington, who could not pass the Turnpike Gate, the Road being x of SNOW (x full) turn'd into the Fields by Silbury hill and through Mr Wentworths Rick Yard and the back of our Stable Yard at 4 oClock this afternoon.

Severe frosts were not unknown at Beckhampton, even in May, such as the one recorded in the diary:

> 7 May 1831. This morning, the most severe Frost, none equal to it remembered by any Person living at the period of year, - The Potatoes under cover, cut to the ground, the Beans, Clover, cut & black, as well as the Ivy creeping the wall & the evergreen Shrubs thick ICE and Windows cased with frost like the mid of Winter, great loss sustained by the Nurserymen. Cold dry East Winds, the Glass riseing.

In contrast to the wet weather of 1829, Pinniger noted in 1834 that with few exceptions, it had been dry from mid-February to the end of May, as a result of which the wheat only came up to his knees, much of the spring corn had not germinated and the field grass was thin.

Other extremes of weather recorded including the thunder, lightning and hailstorms experienced in early June 1835:

> 7, 8 & 9 June. See heavy Thunder, lightening, hail & Rain Cattle & Sheep Kill'd, & great injury sustained by breaking Glass, 2'000 Panes broke in Mr Heales Green house Calne

It is interesting to note the size of a glasshouse in Calne with at least 2000 panes of glass.[155]

---

155 W. Heale & Son were nurserymen of Calne and Devizes, selling trees, shrubs, fruit trees and other plants, including a stock of 6 million quicks or thorns of two, three and four years growth: the main nursery was at [Potterne] Wick near Devizes, Devizes and Wiltshire Gazette, 27 October 1836. According to the 1841 Census, William Heale was described as a nurseryman living in Hermitage, Calne (the property is listed between Springfield House and Quarr Barton along Curzon Street), TNA: HO107/1168/4, *Census returns. 1841 census. Wiltshire. Hundred: Calne including Parish: Calne, Township: Calne, ED4* [accessed via Ancestry, 21 May 2020]. Hermitage Cottages are now Grade II listed buildings at 56-62 Curzon Street, Calne. Historic England, *National Heritage List for England: Hermitage Cottages, 56-62 Curzon Street, Calne* https://historicengland.org.uk/listing/the-list/list-entry/1247123

In 1838, there had been a long period of hard frosts starting early in the New Year: by mid-February, a slow thaw started such that by the twenty-fourth, the ground was still frozen apart from the top two inches. Heavy rain on the previous evening resulted in the top layer of soil being washed down the slopes taking the growing wheat crop with it. The lambing pen at Beckhampton Farm was covered with water and the sheep became very dirty. However, Pinniger did not suffer the fate of some farmers: he noted in his diary the report in the local paper regarding Watcombe Farm near Lambourne where a similar deluge of water and soil flowed over frozen ground into the farm yard, drowning 198 ewes out of the 200 in the yard, along with 240 lambs as well as damaging barley and oats stored in the barn. In a later entry, he notes that 10 acres out of his 18 acre piece of wheat had no growth, whilst the remaining eight acres was very thin: further, every swede and turnip had been destroyed or had rotted as a result of the frosts.

The wet haymaking, harvest and autumn sowing seasons of 1839 prompted him to record the number of days with and without rain in the final six months of the year: see Table 14.

Table 14: Number of days with and without rain at Beckhampton Farm, 1839

	Jul	Aug	Sep	Oct	Nov	Dec	Total
Rain	20	14	26	21	23	20	124
No rain	11	17	4	10	7	11	60

Buildings

When Thomas Pinniger bought 10 pieces of Baltic Dantzic timber on 15 January 1829, he neither stated where he had made his purchase nor for what purpose.[156] However, subsequent diary entries show that he was mounting an extensive programme of building and re-building. Whether the buildings were in poor repair when William Philpot vacated the farm, or whether they were not fit for the purposes which Pinniger intended is not clear.

[accessed 21 May 2020].

156 Dantzic timber is a variety of northern pine grown chiefly in Prussia, taking its name from the port through which it was shipped. It is strong, easily worked and durable if well seasoned: baulks are of 18 to 45 feet long and generally 14 to 16 inches square, Henry Fidler, *Notes on Building Construction, Part III, Materials* (London, Oxford & Cambridge: Rivingtons, 1879).

His purchase of timber was followed by his brothers William from Kennet and John of Cowitch (Cowage), carrying stone for him from Calne to Beckhampton in substantial quantities, as follows:

> 3 February 1829. STONES, - Brother W P_ Kennett began Carrying for me from Calne to Beckhampton, for Building
> 9 February 1829. STONES, Brother J P_ Cowitch, began carry'g for me from Calne
> 14 February 1829. Brother J P_ carry'g Stones for me all the week – 6 Loads, (18 Yards)

Chalk was dug for building and timber was landed and taken to the yard. At the beginning of March, slates were collected from Bristol: Pinniger's team of horses left at ten o'clock on Sunday night to collect a new waggon which was then to be loaded with slates from Bristol. However, all did not go well in that they returned empty at four o'clock on the Tuesday morning having damaged the waggon in Bristol: we see one possible consequence of this in the *Labour* section below.

The first building to be constructed was a new stable, using part of an existing skilling: although this was a conversion rather than a new-build, the work seems to have been completed in short order, in that by 9 March the stable was finished and the horses now remained at the farm to commence work there. The quality of the buildings was not being compromised in that 150 feet of ashlar stone was brought from Box in mid-March. Bricks were brought back from Devizes as return loads, having taken William Philpot's threshed wheat to market there (the diary records 1,600 bricks on 21 May and 1,800 on 3 August).

Work commenced on a range of buildings including the brewhouse: building work started on 28 April and by 25 June, it was completed and on the following day, Pinniger baked in the brewhouse and prepared for brewing. The brewhouse was fully utilised in March the following year, when he used 10 bushels of malt to brew one hogshead of strong beer, one of ale beer and two of table beer.[157]

Pinniger had already started building a new waggon house in mid-June, with Maslens the masons undertaking this work, yet no sooner was the brewhouse range finished, than work started on taking down the existing well-house and other buildings to make way for a cellar, dairy and other buildings.

---

157 One hogshead is 54 gallons or 432 pints, Colin Chapman, *How heavy, how much and how long?* (Dursley: Lochin Publishing, 1995), p.40.

Significant quantities of chalk were being brought in for another new stable: some sixty-nine loads were carted in the second half of July. The diaries tell us that the foundation of the cart stable was started on 18 August and that the eight masons and carpenters began sleeping over the nag stable from 25 August. We know that the cart stable had a hipped roof, from the following entry:

> 19 September 1829. 2 Sides & 1 end of Cart Stable Tiled, the other end waiting for Tiles from the end of Dweling House

from which it would appear that the house was to be modified as well. This was confirmed five days later when Messrs Strong and Ruttiford were noted as planning the front of the house.[158] However, progress with the house was such that the tiles were not re-used on the stables: on the twenty-sixth, a team of horses and waggon was sent to Pickwick for tiles to finish the stable and the horses occupied the stable on the same day.

With the focus turned to the house, work on the farm buildings was suspended until the middle of 1831 when work commenced on a new skilling next to the rickyard, replacing an existing wood-house. In 1832, another skilling was built in the Further Meadow: however, this is not marked in the Farther Meadow (Nos. 37 & 38) on the later tithe map. There is an interesting reference to the thatching of a *hurdle house*: it is presumed that this is where the hurdles were stored out of the weather when they were not needed. It is possible that it was to the same design as the one that exists at Priddy in Somerset, where the hurdles are stacked under a thatched roof: see Figure 7.

No further building work was recorded at the farm until 1838, when Pinniger notes chalk being carted for the barn and by 30 April, 70 loads had been brought: it is only on 5 May that it becomes clear that this was for a new barn in that he states that there will be no more threshing work undertaken until there is a new barn. Five days later the barn was being uncovered for rebuilding, from which it appears that the new barn was a replacement not an addition: the masons faced and sorted the stone from the existing building and 1,000 slates were brought from Devizes.[159] Once

---

158 Ruttiford is possibly Thomas Rutherford, stone mason of Quarr, Calne, J. Pigot, 'Commercial Directory 1830' in K. H. Rogers (ed.), *Early Trade Directories of Wiltshire (Wiltshire Record Society, Vol.47, indexed by J. H. Chandler)* (Trowbridge: Wiltshire Record Society, 1992) [hereafter WRS47], p.66.

159 The use of the term *uncovered* suggests that the original barn was thatched.

more, timber for the barn came from Honey Street and there is an entry referring to windows being set in the barn: this may have represented an improvement on the old barn. A further improvement comes from an entry on 30 July when a waggon went to Honey Street for *Roof of Machine house*: this would have been a small extension on the side of the barn to house the horse engine which was used to drive the thrashing machine and other feed preparation equipment housed within. One such machine was the chaff cutting machine in that it was recorded in March 1841 that the horse power to the machine had been fixed. The diary entry for 4 August 1838 records: *The building part of the Machine House finish'd, which with one side of the Barn to be Slated, and the Walls painted to make a finish* thereby confirming that the machine house formed an outshut located on one side of the barn and was covered with a catslide continuation of the main roof of the barn.[160]

A further 600 slates enabled the barn to be finished, as recorded on 18 August 1838, and subsequently:

> BARN, FINISH'D SLATING 18th July began Slating 18th May began Building 22nd March began Faceing Chalk 6th March began Diging Chalk 4 Carpenters 12th Septr finish'd the Barn[161]

At the same time, a potato steaming house was constructed: this was obviously a small building in that construction started on 1 September 1838 and was finished 11 days later. Such a building would have enabled potatoes to be boiled before use as animal feed: a diary entry from December 1838 records the fattening pigs being fed with barley meal and potatoes.

In July 1839, an engine pump was put into the well by Mr Hammond

---

It appears that Pinniger was replacing the thatched roofs on a number of his buildings: in 1835, the thatch was taken off the granary in order that it could be slated.

160 Outshut – subsidiary compartment at the side or end of a house or barn, under a roof: this is unlike an aisle which is open to the body of the building. Catslide roof - roof which covers one side of the main slope and an outshut in one continuous slope, N. W. Alcock, M. W. Barley, P. W. Dixon and R. A. Meeson, *Recording timber-framed buildings: an illustrated glossary* (York: Council for British Archaeology, 1996), pp. G12 & G15. The drive shaft from the horse engine would have passed through the side wall of the barn to drive the machinery.

161 The remainder of this page of the diary was left blank, thereby enabling the later entry of 12 September to be added.

of Marlborough.[162] This was clearly a labour saving device in that when Pinniger was brewing the following February, he recorded that the new pump had filled the copper with 102 gallons of water in five minutes. Later that year, high winds stripped the thatch from the skilling, as a result of which a team and waggon was despatched to Honey Street for roof timbers and the skilling was re-roofed with slates brought from Devizes.

As for building repairs, chalk was transported to the cowhouse, pig sty and skilling: this would either have been used to repair the walls if they were made from chalk, or else to improve the floors of these buildings. In addition, building maintenance was carried out, as on 7 April 1832, when Pinniger recorded the gas tarring of the granary, iron fencing etc.

As early as 1825, the preservative properties of coal tar, or gas tar as it was sometimes known, were being promoted by the gas works in Southampton: advertised as the best and cheapest preservative for timber buildings as well as implements of husbandry, particularly iron work, it was available from the works at 4d per gallon, or 3d per gallon if bought by the barrel.[163] Under the Devizes Paving, Lighting and Improvement Act, the Commissioners were raising funds for the construction of a gas works later in 1825.[164] By 1827, it would appear to have been operational, in that the accounts of the above act recorded some £2,800 paid towards the construction of the works and a further sum of £87 8s 0d for the lighting of 76 public lamps between 23 January and 1 May 1827.[165] It is likely that Pinniger would have sourced his gas tar from the works at Devizes.

As well as the construction of new buildings, Pinniger also invested in new fences: in 1843, he bought 105 iron hurdles weighing 3 tons and 13 cwt at a cost of £33. These were erected around the hedge in the meadow and were subsequently gas-tarred as well: in addition, 650 fir trees were planted round the meadow.

One issue that was experienced at Beckhampton was a shortage of water: in November of both 1832 and 1833, he noted that the wells had run dry and records the digging of a new well in late November 1833.[166] Having done this, water returned to the two existing wells on Christmas

162 Mr Hammond of Marlborough - probably Jeremiah Hammond, plumber, glazier and whiting maker of High Street, Marlborough, J. Pigot, 'Commercial Directory 1842' in WRS47, p.125.

163 Salisbury & Winchester Journal, 18 April 1825.

164 Devizes and Wiltshire Gazette, 24 November 1825.

165 Devizes and Wiltshire Gazette, 20 September 1827.

166 See diary entry for 27 November 1833.

Day 1833, having been dry since 7 November. By 19 January 1834, the springs were rapidly rising, the water meadows were flooded more than at any time before and the wells were over full: towards the end of January, it was recorded that it had rained almost continually since 18 December.

Building the new house

The foundations of the house were started on 25 September 1829 and timber for joists and other components was transported from Honey Street wharf, a distance of some seven miles by road:

> 29 September 1829. 2 Waggons to Honey St Wharf for 2 Logs of Memel Timber for the Joist of New House[167]
> 22 October 1829. 6 Horses to Honey Street for 3 Logs of Timber on Clarke & Robbins's Carriage

A further three logs were collected on the following day. It is of interest that the timber came from the sawmills at Honey Street. The partnership between J. Clarke and S. Robbins of Honey Street, Devizes, Pewsey and Burbage wharfs was dissolved in 1834.[168] However, it would appear that Samuel Robbins continued the business at Honey Street Wharf: reports of two fires in 1854 and 1858 name him as the owner of the premises.[169] By 1859, Robbins ran a chemical manure works, as well as a steam saw mills and in a newspaper notice, he thanked his customers for his support and announced that he was taking two partners.[170] The first was Ebenezer Lane, who had experience in chemical analysis and the manufacture of artificial manures: the second was Thomas Pinniger, who had been connected with all branches of Robbins' business for 14 years: the firm was to be known as Robbins, Lane & Pinniger.[171]

That the work on the house was indeed an extension to the existing house, rather than a new house *per se*, is confirmed in the following two

167 Memel - Memel timber is very similar to that from Dantzic, but is not as strong and the baulks are generally 13 to 14 inches square. It is also named after the port through which it was shipped, Fidler, *Notes on Building Construction, Part III, Materials*.

168 The Globe, 8 January 1834.

169 Salisbury and Winchester Journal, 8 April 1854; Devizes & Wiltshire Gazette, 16 December 1858.

170 Devizes & Wiltshire Gazette, 30 June 1859.

171 Thomas Pinniger the farmer and Thomas Pinniger of Robbins, Lane & Pinniger were related: see Appendix X for a full explanation.

entries:

> 9 November 1829. Up'r String course put on the new part of House
> 12 November 1829. Building began on the Walls of the Old House

By 28 November, the masons had finished and the house was ready for the remainder of the roof, which was completely slated by 8 December with the hips, ridge etc having been leaded: once again, the building had a hipped roof, matching that of the cart stable. At the end of the year, a team of horses and waggon was sent to Box for coping stones and the portico: this was clearly to be a house with an imposing frontage. However, it would be another year before it was completed.

At the end of March 1830, 1,000 bricks were brought from Devizes and by this time, the wall for the staircase was being built and the attic floor had been laid ready for lath and plastering. Still more timber was required and six horses were sent to Honey Street for a piece of Memel timber 34 feet long, along with 30 deals each 21 feet long: it would appear that this was for the flooring of both the first and ground floors of the house. Progress continued with completion of the house: stone for paving the hall floor was transported from Downend stone quarry, some 33 miles from Beckhampton. Pinniger recorded the route taken by his waggon, the journey there and back taking the best part of two days. From the diary entry, it appears that different routes were taken for the outward and return journeys:

> 16 July 1830. To Downend, Stone Quarry, from Beckhampton 33 Miles Chippenham to Ford 6 Miles Ford to Marshfield 4 Miles Marshfield to Bridge yeate 7 Miles Bridge yeate to Mangotsfield 3 Miles Mangotsfield to Downend Quarry 1 Mile | Bridge Yeate to Hall lane Colliery ½ Mile Hall lane to Wellsbridge Turnpike (HILL) and Brokam hill 1 Mile return to Bitton 1 Mile, thence through Upton to Lands down 4 Miles[172]

By August, work on the house was nearing completion:

---

172 Bridge yeate – Bridgeyate; Wellsbridge – Willsbridge; Upton is probably Upton Cheyney. For the outward journey, Pinniger has not recorded the distance from Beckhampton to Chippenham of 13 miles, making a total distance of 34 miles; the return journey seems to be a more circuitous route (probably to avoid Tog Hill with the loaded waggon) – Lansdown to Beckhampton being 28 miles, makes a total journey of 38½ miles.

> 2 August 1830. CARPENTERS, 2 remain'd at home, out of the 4 – work geting nearer a close | 3 Masons & 2 Boys, - Paving Court & Plaster g
> 19 August 1830. 2 Masons finish'<g>d, dress'g off the Portico having been 5 days at that & put'g the top on, Carpenters, lay'g best Parlour Floor
> 24 August 1830. 2 Men fixing the Spouting round the House
> 6 September 1830. Cart & 2 Horses to the Low, for Sand Stucco'g the Rooms

Pinniger gave the masons a finishing supper on 9 October, in that the house was nearly finished but it was not until 4 December that he could record: *James Hillier, the Carpenter having quite finish'd, left us, no Tradesmen from this day at work.* However, on Sunday 20 February 1831, he recorded in the margin of the diary: *New House first occupied, Parlour,* with a further marginal note on 21 March *Bedroom first occupied New House.*

Looking after the land

As one might expect when moving to a new farm, Pinniger would have found it useful to have observed what his neighbours were doing, in order to understand the seasonality of the various agricultural activities in his environs. The following entry from his first year at Beckhampton demonstrates this:

> 6 May 1829. Rt Philpots SWEDE'S finish d, and his Flock on the Down. | Mr Wentworths Flock finish'd SILBURY MEAD, - and began Field Grass. his Tegs on the Down.

It would appear that as part of the agreement between Guy, Pinniger and Philpot, Philpot was to undertake certain acts of husbandry: one of these was that of chalking the land to improve its fertility. The following diary entries present the costs and quantities involved in this operation:

> 10 January 1829. CHALKING, Wm Philpot began for me at Beckhampton, the 2nd Piece by furze Down 9 Acres @ 52s pr Acre with 3 Carts, 6 Horses, 6 Men & 2 Boys If done by Men from Wells with Barrows at 42s pr Acre
> 19 January 1829. Chalking wth Carts, cover 3 of 4 Pole squares which is 48 Pole @ 52s pr A, - pr day = 10 Carts Loads of 15 Buckets ea: 150 Buckets

to 16 Poles or a 4 Pole square, (9 Buckets pr Pole is 144 Buckets to 16 Pole.

It might seem unnecessary to apply chalk to this land: after all, Beckhampton lies in an area of chalk downlands and indeed the bedrock geology of the area is Holywell Nodular Chalk Formation.[173] Whilst the soil here is predominately chalky, silty loam, there is some sandy loam along the sides of the turnpike roads to Devizes and Marlborough.[174] Assuming that Furze Down was part of one of the pasture downs listed in the tithe award (see Table 11), then it is located on one side or the other of the Beckhampton to Devizes Road, up to ½ mile from the village of Beckhampton. Here, there is a band of soil on either side of the road which is sandy loam, the soil parent material being river terrace sand and gravel. These are alluvial deposits made by a former river which ran along the valley (the course followed by the A361 road at this point) between the higher downs on either side, where the soil is chalky clay to chalky loam, as one might expect. The application of chalk, as a source of lime, is required on grassland (pasture or meadow) to maintain the soil at an optimal level of acidity: soil acidity increases as calcium is leached from the soil into drainage water.[175]

From the former entry, it would appear that Pinniger thought that Philpot could have been more cost effective in this work had he used men with barrows, rather than horse and cart. On the other hand, his calculations of the chalk used seems somewhat excessive. At a rate of nine buckets per pole, then this equates to nearly 140 tonnes per acre: however, it is not known whether the chalk was dried before application.[176]

173 British Geological Survey, *iGeology app – Beckhampton, Wiltshire* [accessed 9 April 2020].

174 Natural Environmental Research Council, *mySoil app – Beckhampton, Wiltshire* [accessed 9 April 2020].

175 Lime is a calcium-containing inorganic material. AHDB, *Grass+ Factsheet 8: Grassland liming* (Stoneleigh: AHDB, 2012) https://dairy.ahdb.org.uk/resources-library/technical-information/grass-management/grassplus-factsheet-8-grassland-liming/ [accessed 14 April 2020].

176 One bucket of chalk equals 1½ bushels and one bushel is 1.28 cu. ft.: thus, one bucket has a volume of 1.92 cu. ft., Chapman, *How heavy?*, pp 38-39. The average density of dry chalk is 1.79 tonnes per m^3 or 111 lb per cu. ft., J. P. Bloomfield, L. J. Brewerton and D. J. Allen, 'Regional trends in matrix porosity and dry density of the chalk of England', *Quarterly Journal of Engineering Geology*, 28 (1995), S131. Thus, one bucket holds 214 lbs of chalk: at 15 buckets to the cart load, this would give a load of nearly 1½ tons. There are 160 square poles in an acre, Chapman, *How heavy?*, p.28. Thus, at 9 buckets per pole, this equates to 137.57 tons of chalk per acre.

The benefits of chalking were expressed in the diary in January 1835 in terms of reduced horsepower for ploughing, couch-free land, and in turn a better crop grown:

> 3 January 1835. FINISHED fallowing the further Down Piece all Chalk'd for Turnips. NB. The half of this Piece was Chalk'd in January 1830 with Barrows is now clean from Couch, Plough freely the other upper half chalk'd last month is full of Couch and Ploughs a Horse heavier on this half of the Piece was a bad Piece of Wheat & the lower half a very good Piece last year. such encouragement for Chalking *[177]

Digging the chalk was not without its dangers, as the following entry, written within a partial thick black border, records:

> 29 January 1830. Carting Road Earth to Rick yard till 10-O-Clock, when all hands call'd off, to assist in geting Wm Paradise out of the Chalk Pit, being buryed in the Earth fallen in upon him when filling the Bucket in chalking the land. Jno Chivers of Avebury & others got him out in very, great danger of their own lives at about half pass 12 O-Clock.

Whilst there is no record in the diary as to what happened to William Paradise, it could be concluded from the use of the black border, that the outcome was not good.[178] Indeed, in 1834, confirmation of his death appears in the diary:

> 6 December 1834. CHALKING began Wheat stubbles further Down Piece, the remaining of the Piece from where the Man was kill'd W Paradise see 29th Jany 1830

Whilst the use of peat ashes brought in by barge on the Kennet & Avon Canal was observed at Little Bedwyn, it would appear that coal ashes were used at Beckhampton. Early on, Pinniger observed that Mr Brown was sowing coal ashes on clover and the first record of his own use of coal ashes is made on 13 March 1830. It would appear that with

---

177 Fallowing - A frequent ploughing and pulverising of land to make it lighter, and clean from weeds when it has become foul by repeated crops, Davis, *Agriculture of Wiltshire*, p.55.

178 William Paradise was buried at Avebury on 4 February 1830, aged 23. The burial register states: Accidentally killed. *Wiltshire, Church of England, Deaths and Burials, 1813-1916* [accessed via Ancestry, 30 June 2020].

no canal passing close to the farm and the relative proximity to the town of Devizes, the switch from peat ashes to coal ashes was made: indeed at the end of January 1831, Pinniger recorded sending two waggons to Devizes for 120 bushels of coal ashes: these cost him 3d per bushel and were *turnpike free* – that is, there was no charge for using the turnpike road to transport this product. In that year, he records a further 300 bushels of coal ashes being brought in and on 18 February, work started to spread them at a rate of 30 bushels per acre, being sufficient for 14 acres that year, around 7% of the farm's acreage.[179] By 1833, the rate of application had increased to 40 bushels per acre, which in that year was spread across 13 acres.

One particular treatment used for poor pasture land was the application of salt: in 1831, Pinniger applied nine bushels of salt mixed with earth to an acre of *sour down.*[180] The application of salt to grassland provides sodium essential to livestock diets and also improves the palatability of the grass.[181] Another treatment was the application in 1832 of a mixture of materials - soot, ashes & hen dung - for poor growth in wheat.[182]

It would appear that William Philpot had not left the farm in a tidy state: in April 1829 Pinniger's men were grubbing the banks, cutting and laying the hedges around the meadow and clearing away hay left in the

---

179 The typical nutrient content of coal (anthracite) ashes is lower than peat ashes: 0.1-0.15%P (phosphorous) and 0.1-0.15%K (potassium). Nigel Davenport, *NPK value of everything organic!* (Rochdale: The Nutrient Company, 2019) https://thenutrientcompany.com/blogs/horticulture/npk-value-of-everything-organic-database [accessed 15 April 2020]. The current understanding of coal ash is that whilst it contains a number of nutrients beneficial to plant growth, it releases very little to the soil: further, it contains a number of potential toxic trace elements, including lead, cadmium, chromium and nickel. If it has a beneficial use, then it is as a liming agent reducing the acidity of the soil, Sabry M. Shaheen, Peter S. Hooda and Christos D. Tsadilas, 'Opportunities and challenges in the use of coal fly ash for soil improvements – A review', *Journal of Environmental Management,* 145 (2014), 249-267. However, it is likely that Pinniger used coal ashes, as a fertiliser in the same way as he had used peat ashes at Little Bedwyn, given their ready availability from either households or the gas works in Devizes.

180 See diary entry for 30 March 1831.

181 The current recommendation is for an application of 100kg/ha (88 lbs per acre) of agricultural salt, Scotland's Rural College (SRUC), *Technical Note 652: Fertiliser recommendations for grassland* (Edinburgh: SRUC, 2012), p.6. 1 bushel of salt weighs 56 pounds, Chapman, *How heavy?*, p.62. Therefore, Pinniger's application of nine bushels to one acre seems somewhat excessive by modern day standards.

182 See diary entry for 18 May 1832.

field from the previous season:

> 27 April 1829. Willm Philpots, last Years HAY-COCKS got out of Silbury, mead, the Grass kill'd

A week later, the foul headlands and parts of the barley land were being ploughed, harrowed and rolled.

On a number of occasions, the diary records co-operation between Pinniger and one of his neighbours, Mr Wentworth: this worked in both directions, as follows:

> 24 July 1829. 1 Team all Day – for Mr Wentworth | 1 Team after 12-O-Clock – for Mr Wentworth
> 21 October 1829. 3 Teams of Mr Wentworths (10 Horses) Plough'g Press'g & harrow'g for Wheat

This spirit of co-operation was once again observed in 1842 when on 11 May John Wentworth provided six teams to assist Pinniger with the ploughing of Tan Hill Piece for Barley sowing: over the following week, Pinniger returned two teams on two days followed by one team on two days, thereby balancing the favour. However, there are no records of a similar arrangement with his other neighbour, Robert Philpot.

As in the fields at Little Bedwyn, a number of weed infestations took over growing crops from time to time: in particular at Beckhampton, there seemed to be a lack of labour to undertake the necessary weeding in a timely fashion allowing docks and charlock to run rampant. This is well illustrated in a diary entry from 1 June 1831: *Hoeing the Pease 2nd time full of Charlock & no time yet to weed the Oats.* Weeding the peas and beans was not finished until ten days later when weeding of the oats commenced: by this time, the charlock was such that the women were cutting it from within the oats using reaping hooks. Another form of infestation was that of mice, such as observed on 31 May 1831: *Wheat Rick put into Barn full of MICE* to which Pinniger made the observation *N B Dress the Ricks after Harvest at any COST.*

On 18 July 1835, Pinniger recorded that the top piece of his vetches had been destroyed by rabbits, *from the Preserve* and that he had sent a letter to Mr Estcourt to complain. The use of the word *Preserve* suggests that there was a rabbit warren being managed in the area. A week later, Mr Estcourt inspected the damage done by the rabbits to the vetches and also turnips, and he promised to have the rabbits shot. Despite this, they continued to be a problem: in March 1836, 240 hurdles were erected to enclose the covert from where they were emanating, yet by April the wheat on Further Down Piece was eaten off to the ground. Mr Estcourt inspected the damage on 21 May but the outcome was not recorded, other than that Pinniger estimated the damage at £50 when it came to reaping time in early September.[183]

The rabbits were still a problem throughout the year 1837, as the following entry describes:

> 7 December 1837. RABBIT'S employ'd my Shepherd to destroy doing me a great injury, by no care taken by the Warren man for the last Ten Months in stoping the holes in the Hurdles to Keep them back from my land till this day, find'g I am distroying them.

Once again, Mr Estcourt promised to destroy the rabbits but it was only by November 1838 that the rabbits appear to have been brought under control, when the hurdles erected in March 1836 to keep the rabbits off the wheat were removed from Horton Down.

Evidence of improved husbandry is recorded in the diaries at the end of June 1842 when Daniel's Patent Manure was applied to the swedes: an equal quantity of road earth was mixed with the manure to prevent it killing the seed. The manure was supplied in sacks weighing 1½ cwt containing 4 bushels each and was applied at the rate of 30 bushels per acre: however, Pinniger left a blank space for the price.[184] Further evidence of the use of improved techniques comes towards the end of the same year:

> 1 December 1842. NO WHEAT on the OLD land having change from 4 to 5 field system to provide better Sheep Keep.[185]

183 Worth £4,623 at 2018 prices. Measuring Worth, https://www.measuringworth.com/calculators/ukcompare/relativevalue.php [accessed 14 April 2020].

184 A report on the use of Daniels Patent Manure on carrots shows the application rate as 32 bushels per acre at a cost per acre of 34s: thus it would appear that the manure cost around 1s per bushel, J. M. Aynesley, 'Experiments on different manures for carrots', *Journal of the Royal Agricultural Society of England,* (1843), 270. Mr Daniels Patent Manure was described by the inventor as consisting of ammonia, pulverised wood and bituminous matter. The Farmers Magazine, Series 2, Vol. 5, No. 6, June 1842, p.422.

185 As seen above, the soil on Beckhampton Farm is predominately chalky, silty loam with some sandy loam along the sides of the turnpike roads to Devizes and Marlborough. The five-field system observed by Davis for the sandy soils was as follows: 1. Wheat; 2. Turnips; 3. Barley and seeds; 4. Clover; 5.

Similarly, in June 1844, he recorded that *Ground Bones* or *Bone Manure* was being applied at a rate of two quarters per acre. It appears that Pinniger was buying the bone manure from Devizes, Bristol, and the Poor Law Unions of Chippenham and Calne at a cost of between 18s and 20s per quarter: the latter two sources would have been supplied by the respective workhouses where the grinding of bones provided employment for the inmates.[186] It appears that he was still uncertain of the benefits of using bone manure in that he described the drilling of swedes in 1845 with bones as a *trial.*

In 1846, a new waggon house was being constructed in the Mead and the diary entry for 1 April records the carting of chalk to this new building *to make a Floor to mix Ash's & Guano.*[187] Indeed, on 13 April, 50 bushels of ashes and 6½ cwt of guano were brought from Devizes. It would appear that Pinniger was applying guano in a controlled experimental manner: the diary entry for 24 June 1846 recorded that half of the turnips were drilled with ashes & guano, whilst the other half had no manure applied, and similarly on 1 July, the drilling of turnips was carried out, with one part having ashes & guano applied, a second part with ashes only and the third part having dung only.

In order to carry out the tillage and other operations on the farm, Pinniger bought machinery and equipment soon after taking on the farm: on 19 March 1829, he bought *2 Ploughs @ £4 1 Sowing Machine £3 3s*

Fallow, ploughed early, and folded as close as possible, preparatory to wheat. The turnips in the rotation provide green feed to winter the lambs at home, which was Pinniger's intention. Davis, *Agriculture of Wiltshire*, p.63.

186 The strenuous task of crushing old bones to turn them into fertilizer was at that time one of the activities undertaken by workhouse inmates, alongside stone-breaking, corn-grinding, gypsum-crushing and wood-chopping. However, in 1845, men in the Andover workhouse bone-yard were so hungry that they were eating the tiny pieces of marrow and gristle attached to the bones they were crushing. Fighting had broken out when a particularly succulent bone was found. This led to the Poor Law Commissioners forbidding bone-crushing as a workhouse activity from 8 November 1845: the incident became known as the Andover Scandal. The Workhouse, *Andover, Hampshire* http://www.workhouses.org.uk/Andover/ [accessed 28 April 2020].

187 Guano is the excrement of seabirds and bats, used as a fertiliser. The use of guano as a fertiliser was reported in John Joseph Mechi, *Letters on Agricultural Improvement* (Longman, Brown, Green and Longmans: London, 1845) see for example, pp.54–57.

Od} Tasker.[188] Robert Tasker, the founder of the company, was at this time producing an original design of plough suited to the conditions in the chalk fields of the south of England: whilst no details of this plough have survived, the coulter and share would have been made of iron, whilst the beam and mouldboard were made of timber.[189] It is of note that throughout the nineteenth century, Taskers transported loads of imported timber from Messrs Robbins, Lane & Pinniger of Honey Street, to their factory at Abbotts Ann, near Andover.[190]

In April 1829, Pinniger bought three rick staddles at a sale at Compton and later that month, three of his brother William's horses were rolling the clover with a new iron roller. Hurdles for folding the sheep were bought from Ramsbury Park: there are records of two lots of 12 dozen hurdles each.

Other developments in the use of machinery includes a reference in June 1841 to *Nine sharing, the Down Piece for Turnips*: this would have been a development of the seven-share cultivator as used at Little Bedwyn and may well have enabled a higher work rate by the addition of two further shares to the cultivator. The advantage of drilling, rather than sowing, turnips was also described that year in that it was possible to hoe between the drilled rows of plants. He also bought a chaff waggon from Honey Street at the end of 1841 which was placed in the field to provide chaff for the sheep.

Unlike the farm at Little Bedwyn, it does not appear that there was a thrashing machine on the farm initially, either water-powered or horse-powered: the farm at Beckhampton was only a quarter of the size and possibly did not warrant the ownership of one. Indeed, it appears from the diary entries before 1832 that the crops were threshed by hand, but in that year, the following entries were recorded:

> 6/7 March 1832. Thrashing Machine, fix d by 1,=O-Clock, & thrash by 1-O-Clock – the 7th by [...] Sainsbury of Marlbro
> 18 June 1832. 2 Horses to Burbage for Thrashing Machine
> 20 June 1832. 4 Horses finish'd Thrashing the Wheat at half pass'4

188 An 1813 billhead for the firm of Taskers of Andover offers Tasker's improved ploughs, drills and broadcast machines, amongst a range of other items. R. T. C. Rolt, *Waterloo Iron Works: a history of Taskers of Andover, 1809-1968* (Newton Abbot: David & Charles, 1969), p. 38.

189 Rolt, *Waterloo Iron Works*, pp. 30-31.

190 Rolt, *Waterloo Iron Works,* p. 102. The company name is given as Messrs Robbins, Lane & Pinnegar of Honeystreet Wharf.

O-Clock, began 7 ea morn 2 days, 39 Sacks
23 July 1832. The Thrashing Machine, brot from Marlbro
2 August 1832. Thrash'd, the BEAN RICK 28¼ Sacks with the Machine

Thus, it would appear that Pinniger was dependent either on an agricultural contractor who would arrive in his own time or the opportunity to borrow a machine when no-one else wanted it. The June entries confirm that the machine was horse-powered and the fact that it was put to work immediately also suggests that it was on loan / hire. It is possible that this lack of a thrashing machine kept on the farm may have contributed to Beckhampton Farm not being targeted directly during the Swing Riots of 1830 (see section on *Law and Order* below).

However, on 2 April 1836, three horses and a waggon were sent to Bath for a thrashing machine, thereby enabling Pinniger to thresh his corn when he was ready to do so, as demonstrated by this diary entry:

4 April 1836. Wheat Rick, the remaining part put into Barn see 15th Feby began it, have taken all the rain since, a Waggon load of sheaves spoild for want of Barn room, have a Machine in consequence

The thrashing machine was put to work shortly thereafter: eight days later he recorded the first day of threshing with the machine when around 20 sacks of wheat were threshed with four horses driving the machine. He continued to improve the equipment on the farm: in December 1844, a new chaff engine was installed and connected to the horse-powered drive and in February 1845, a new thrashing machine was being used to thresh the wheat. It would also appear that by 1846 a new seed drill had been purchased, in that he records in his diary the cog wheel that needed to be fitted to the drill for a given seeding rate, as shown in Table 15.[191]

Table 15: Settings for seed drill, 1846

No. of cogs on wheel	*Seeding rate (Bushels per acre)*
29	3¼
27	3½
25	4
23	4 full

191 The seeding rate in cup-feed drills is determined by the size of the cups and the speed at which the seed barrel rotates: the latter is varied by fitting a drive cog wheel with differing numbers of teeth, as in Table 15. Culpin, *Farm Machinery*, p. 198.

Pinniger was also trying out a mechanical method of weeding the beans in 1846, when Thomas Pinniger junior had been using a horse hoe in the beans, to which his father made the following entry:

> 9 May 1846. BEAN'S, hand hoeing, after Horse hoe very, good Implement

Grass crops

Hay was made from the crops of sainfoin and ray, as well as meadow grass.[192] However, in his first year at the farm, the weather was unfavourable for haymaking, as the following entry shows:

> 6 July 1829. Hay, or rather Dung, moved the Cocks Silbury Mead, not moved since 26 June

The danger of hay catching fire was as likely in the 1830s as it is today: see earlier notes on the heating of hay in the *Grass crops* section of THE BEDWYN YEARS above. On 22 September 1838, Pinniger recorded:

> Dry night, and as beautiful fine drying day for hay making as any this Summer. | HAY, carried 9 loads, extraordinary weather for 22nd of September

Perhaps it was due to the good weather, the fact that he had not finished carting all of the hay, or that he did not have the necessary manpower, but the hay rick had not been thatched, before it rained two days later. As a result, the hot hay rick was opened and a hole cut into it to provide ventilation: more rain required two further holes to be cut.[193]

In 1834, Pinniger wrote the following (undated, but between two entries for 8 August) entry in his diary:

> Trefolum=incarnatum a New French Grass to be Draged into Wheat & Barley Stubble 10 li pr Acre @ 1s pr li ready to feed 14 days before Clover, in the Spring [194]

---

192 Ray or ray-grass is taken to be ryegrass.

193 It is also possible that this good weather so late in the season encouraged Pinniger to collect the hay and build the rick before the hay was properly dry: an example of the Yorkshire farming saying that "there is more bad hay made in a good season, than good hay in a bad season".

194 *Trifolium incarnatum* - known as Crimson Clover, Carnation Clover, Italian Clover or Scarlet Clover, IUCN Red List of Threatened Species, 2020-1, *Crimson Clover, Trifolium incarnatum* https://www.iucnredlist.org/

Whether or not he planted this new "improved" crop is not clear from the diaries: however, he does record Mr Brown dragging the seed into some barley stubble at a rate of 15 lbs per acre that year.

In 1839, there appears to be a move by Pinniger to increase the proportion of arable land on the farm, when his men started on 21 January to grub out the furze on a piece of downland, extending to between 12 and 14 acres. The furze was made into faggots for burning and the turf itself was burnt to provide ashes for incorporation into the soil. This was followed by the operations of breast ploughing, harrowing, raftering, pressing, ploughing and rolling until it was ready for sowing on 2 September.[195] At the end of the 1840 harvest, he recorded that the 13 acres in New Down piece had produced *the heavyest Wheat ever remember, the Wheat particulary heavy & good in general*: the yield was 96 tithings per acre compared with between 70 and 78 tithings per acre across the other 37 acres of wheat that year, demonstrating the increased yield achievable from ploughing up old pasture.[196] The New Down was then ploughed for oats the following year. Later in 1844, he records that his neighbours, Mark Sloper and Thomas Brown were similarly breaking up their Down land, starting by breast ploughing it and again in 1847, when his nephew Jacob was breaking up 50 acres of down (out of a farm acreage of 900) using the same technique.

Arable crops

The only wheat variety mentioned in the diary is that of Talavera, until October 1837 when Pinniger refers to sowing Golden Drop thick set wheat which he had bought from Mr Waters of Upavon, along with thick set Worcester wheat: Red Essex wheat was sown in 1839 for the 1840

species/176390/7231548 [accessed 12 May 2020]. Clover is a fixer of nitrogen in the soil.

195 This process of reclaiming land for arable cultivation is known as burn-baking.

196 The stored fertility beneath permanent grassland was also put to good effect in the early years of the plough-up campaign of the Second World War, Brian Short, Charles Watkins, William Foot and Phil Kinsman, *The National Farm Survey 1941-1943: state surveillance and the countryside in England and Wales in the second World War* (Wallingford: CABI Publishing, 2000), pp.33-35. However, after two to three years, these reserves were exhausted, Brian Short, *The Battle of the Fields: Rural community and authority in Britain during the second World War* (Woodbridge: Boydell Press, 2014), p.229.

harvest.[197] The use of seed saved from previous harvests is evidenced in the diaries: for example, on 10 September 1830, wheat sowing began using 17 bushels of his own wheat on 6 acres of the middle down piece, which had received applications of lime and urine. Good progress was made with wheat sowing that year, in that:

> 8 November 1830. Wheat Sowing FINISHED, the Turnip land 4¾ Sacks Old Wheat 17 Sacks New Mrs Edwards of Wilton ½ Sacks Mr Crook, Corton [Total] 22¼ on 27 Acres

The reference to Mrs Edwards would have been Mrs Ann Edwards, widow of John Edwards who had farmed Wilton Manor Farm since 1788: the land tax assessment for Wilton tithing in Great Bedwyn parish shows Ann Edwards as the occupier of the farm in 1830.[198] In all probability, Thomas Pinniger would have known them from his time at Little Bedwyn.

As for barley, the variety Shewclear was mentioned in 1835 and Shaveleer in 1836: these are probably the same and in all likelihood, he was referring to Chevalier barley.[199] By 1844, Moldevia barley was recorded in the diaries. Oats were often described as either black oats or white oats, with occasional reference to black Tartery oats. However, for the first time in the diaries, Pinniger recorded the drilling of Georgian Oats in April 1831: whilst he tells us that they were drilled after swedes and that

---

197 Talavera - A variety of spring wheat: by 1842, it was not considered to be as prolific as other varieties, in that it comes late to harvest and is liable to be damaged from wet during harvest, The Farmer's Magazine, Series 2, Vol. 5, No. 5, May 1842, pp. 341-342.

198 John Edwards of Great Bedwyn married Ann Rushly of Great Bedwyn on 19 December 1801, *Wiltshire, England, Church of England, Marriages and Banns, 1754-1916 for John Edwards* [accessed via Ancestry, 11 August 2019]. John Edwards of Wilton was buried 13 December 1826, *Wiltshire, England, Church of England, Deaths and Burials, 1813-1916* [accessed via Ancestry, 11 August 2019]. WSA: A1/345/25, *Land Tax Assessments, Wilton, 1780-1881.*

199 Chevalier - a variety of barley. In the United Kingdom, up until the early 1800s, barley consisted of so-called land races, which were mixtures of varieties grown from saved seed, and known by exotic names such as "Old Wiltshire Archer." By gradual selection and re-selection, a number of varieties of malting barley had been developed, including Chevalier. This variety dominated the English crop for around 60 years from c.1820, and only really started to fall out of favour when William Gladstone repealed the Malt Tax in 1880. Craft Beer & Brewing, *The Oxford Companion to Beer definition of Chevalier (barley)* https://beerandbrewing.com/dictionary/ATZOtv7Wj1/ [accessed 30 April 2020].

W. Rumming did the work using three horses and harrows, he does not record the acreage sown with ten sacks each weighing 161 lb.[200] Later that year, 12 sacks and one bushel were sown on 8¾ acres: this equates to 225 lbs, or just over two hundredweights, per acre.[201] Rye is first mentioned at Beckhampton in 1833: however, in this year Pinniger was unable to graze it early enough and on 24 July, reaping of the rye began, although it is possible that the crop was grown both for grazing and as a grain crop.

Again, the importance of thatching the ricks to keep out the rain is emphasised in the diary entry for 2 September 1830, when the thatcher started thatching a rick of Talavera wheat, and having been interrupted by an eclipse of the full moon, worked on by candlelight and moonlight to complete the job at some time after midnight.

It is possible to derive some crop rotations from the diaries by identifying the crops planted in each year in each named field: however, in 1835, Pinniger lists the crops planted for eight years of cropping the three parts of the down: this is summarised in Table 16. He notes that turnips, vetches, grass and hay were grown in between the arable rotation: it appears that over a six year period, the rotation was wheat, other crop, wheat, other crop, oats or barley, other crop. This is consistent with conditions found in some farm lease agreements, that two 'white' crops (wheat, barley, oats) should not be grown in consecutive years.

Table 16: Crop rotation on the down at Beckhampton Farm, 1829 to 1836

Year	*Home Piece*				*Midd: Piece*				*Furr Piece*			
	Crop	*a*	*r*	*p*	*Crop*	*a*	*r*	*p*	*Crop*	*a*	*r*	*p*
1829									Wheat	9	0	32
1830	Wheat	8	3	6								
1831	Oats	8	3	6	Wheat	8	2	39	Oats	8	3	0
1832					Rayseed							
1833	Wheat	8	3	6								
1834					Wheat	8	3	19	Wheat	8	2	21
1835	Wheat	8	3	7								
1836					Oats*				Wheat			

* - pencil note *Barley*. Diary entry for 13 September confirms that Barley was harvested from Middle Down Piece that year.

200 Georgian oats – a report for Perthshire in 1835 noted that the variety of Georgian oats had *long since disappeared*, The Farmer's Magazine, Vol. III, No.6, December 1835, p.501.

201 One bushel of oats weighs 39 pounds, Chapman, *How heavy?*, p.52.

A new variety of swedes, Arial, appears in the diaries in 1830 and in 1831 Pinniger records drilling Kidwell beans and marrowfat peas: in 1837, he was drilling dwarf marrowfat peas. Once again, as at Little Bedwyn, Tankard turnips were sown:

> 26 June 1829. TURNIPS, began sow'g Tankards 2 holes every other place & small Cog Wheel

or rather, being drilled, given the instructions in the diary for the setting of a drill. From the entry for 29 May 1830, we see that a setting of *2 holes & blank* achieves a sowing rate of 2 pounds of seed per acre: this may not seem to be much, but there are approximately 170,000 turnip seeds in a pound, and at this sowing rate equates to 70 seeds per square yard. By 1832, Norfolk White and Green Round turnips were also being grown: Globe turnips were planted in June 1845. However, germination of swedes and turnips was not always guaranteed, as the following diary entry shows:

> 7 August 1832. TURNIP green round seed, sow'd over the swede field where fail'd twice before in places, and rolled it in

Much has been written elsewhere about the effects of the fly on the turnip and Pinniger does not fail to do likewise:

> 2 June 1824. Turnips, - sown 17th May shewing rough leaf attack'd by the Fly, very doubtful, their standing

However, in 1835, another pest of the turnip makes an appearance:

> 22 August 1835. The young Turnips hoe'd out Down Piece was the 16th very gross, are now almost quite destroy'd except the rib of the leaf, by a black Maggot an Inch long, call'd the Negro Maggot, bre'd by a Yellow Fly depositing Egg's, like the yellow Dung Fly, many methods of destroying them suggest'd, but the most likely to succeed is that of removing them from the leaf by two Men drawing a Waggon line over a land at a time 50 Maggots being on one Turnip (the leaves) Some People say 20 others 40 Years, since we were visited by this destructive blight Insect

Within a week of this entry, more turnip seed was being harrowed in to the land where the black maggot had destroyed the turnips. However, by 12 September, the maggot had returned to destroy the turnips which had been hoed. There is no record of the black maggot in 1836, but the following year it had returned to once again destroy the turnip crop.

It also appears that Pinniger was writing down husbandry rules which he felt were successful and that he would want to follow in succeeding years, for example:

> 22 September 1829. Barley moved the Ears off the ground not moved before since Mowed the 4th Inst and under Ears of the Swath as bright as when standing, - this plan generally approved, - the Barley often moved by some People is become black [The following note is written in the margin] Note this is a good Rule

This characteristic also applies to the making of hay. Turning hay on dry days moves the wet grass on to the top of the swathe, in the hope of making the crop into fodder. When followed by wet days, this results in the hay becoming black and full of dust from the soil being beaten up into the hay by downpours of rain. A further husbandry rule was as follows:

> 4 April 1831. N B Observe in future, in the begining of Feby draw the Swedes a little, by breaking the tap root, after which will keep good and not run to seed

However, by 1842, he had started the practice of drawing the Swedes fully, placing them in a heap and covering them with earth: in that year, he had found them to keep well and these were then cut them up to feed to the lambs.

Pinniger was growing a larger crop of potatoes than at Little Bedwyn, as well as a wider range of varieties, including Prolifficks, White Apple, Red Apple, Black, Purple and Magpie.[202] In 1842, Pinniger planted dwarf rape amongst his vetches, applying 10lb by machine and he did so again in 1843, describing it as an experiment.

Finally, there are no records of barking at Beckhampton, in that there was no extensive area of woodlands as there was with the 60 acres at Little Bedwyn. On occasions, trees were felled as in 1831, when six elm trees were taken down in the rick yard: these were moved to the saw pit and therefore appear to be for use on the farm.

202 See diary entries for 20 October 1830 and 21 May 1832.

Livestock

The predominant livestock on the farm were sheep, but it would appear that Pinniger did not "inherit" any with the farm, as he did at Little Bedwyn. On 3 June 1829, he went to Ilsley market (the fourth that year) and bought 100 ewe tegs at 28s each from John Hewett of Mapledurham and a further 130 at 29s from George Castle of Laten Down.[203] It appears that the sheep were rested at Winterbourne on their journey from Ilsley, in that 229 ewe tegs arrived at the farm on 6 June and were folded on the second down piece for Swedes. They were washed at Cherhill two days before they were sheared, as shown in this entry in the diary:

> 10 June 1829. 5 Men Shear'd 11½ Score Tegs @ 1s pr S 1 Winder 2s Allowed over 1s | From 8 O Clock in morn'g to 8 in eve'g | Gammon for Breakfast, Fillet Veal & Bacon & Plumb Pudding for Dinner

Recording what victuals he had provided them, it would appear that he was keen that they should return in future years.

He also bought three rams (two 6-tooth and one 8-tooth) from his neighbour Mr Wentworth at 50s each along with 37 two-tooth ewes at 16s each from Mr Coward of Roundway. This enabled him on 18 November 1829 to turn out two of Mr Wentworth's rams to 265 stock ewes.[204] It would appear that the ewes Pinniger had bought from Mr Coward were already in lamb, as we have seen was the tradition for sale ewes, in that the first lamb was born to one of them on 24 January.

By 1831, there were 366 ewes in the flock and when the sheep were counted on 9 June 1834, there were 379 sheep and 189 lambs. As for shearing the sheep, the following entry again gives an indication of the manpower required:

> 27 June 1831. <u>SHEEP=SHEAR</u>, 18 Score & 6 – at 1s pr Sco 6 Men & a Boy, - charged 1s for Boy as Colt began 8 in morn'g – finish d – half pass

203 Ilsley – East Ilsley and West Ilsley are villages and civil parishes in Berkshire on the downs 10 miles north of Newbury. Mapledurham – a small village, civil parish and country estate beside the river Thames in southern Oxfordshire, four miles north-west of Reading. Laten Down – possibly Latton, a village and civil parish in Wiltshire, 1½ miles north of Cricklade.

204 This represents a somewhat tall order for Mr Wentworth's rams, even though they were mature and experienced. The average ram to ewe ratio for lowland flocks is one ram to 40 ewes, Eblex Ltd. *Target ewe fertility for better returns*, p.9.

6 – eve'g

As for their work rate, allowing 1½ hours for meal breaks, then six men sheared 366 sheep in nine hours, giving an average of 6.8 per man-hour: however, it appears that at one point they were clipping them much faster at a rate of 11.4 per man-hour, but this was not to Pinniger's satisfaction: *Shear'd 8 each in 42 minutes, at finishing, too hurried, to make good work.* Indeed, in 1834, he employed nine men to shear 379 sheep in a day lasting 13 hours: again allowing 1½ hours for breaks, then the work rate is 3.6 per man-hour. He observed that the shearing would be better by employing nine men at 2s 6d per day (cost: 22s 6d) than 6 or 7 men at 1s per score (cost: 19s 0d). From this, it appears again that he would prefer a proper job to be made of shearing his sheep – quality rather than speed.

Livestock improvements were noted in August 1841, when Pinniger bought two rams from Thomas Mills of Fittleton: these were recorded as *half Sussex,* suggesting that he was trying a different breed. However, with the poor lambing seasons experienced in 1841 and 1842, and ewes dying during these years, Pinniger notes that he had not sold the sale ewes in 1842 or 1843 and that some of his stock ewes were now older than were normally kept for breeding.

In February 1830, he bought an Alderney cow and calf from Mr Line: this is likely to have been one of the three breeds of Channel Island cattle, known for their small size, low feed consumption and creamy milk.[205] Having only the one cow, there would have been a shortage of milk for the house when she was near calving, as the entry for 4 February 1831 demonstrates: *The Cow, leting dry, having calved 12 Months since, and Calve very soon.* The cow calved on 24 March, having been bulled on 14 June previous: the calf was sold three months later for 22 shillings.[206]

205 There were three distinct breeds of Channel Island cattle, Jersey, Guernsey and Alderney, Anon, 'The Alderney Cow' in *The Saturday Magazine*, August 1 1835, 47-48. However, Felicity Crump states that there was an archaic practice of naming all Channel Island cows as "Alderney": further, the real Alderney cow was present on the island only between the 1860s and the German invasion of 1944, BBC News, *Alderney cow: the breed Jerseys and Guernseys overshadowed* https://www.bbc.co.uk/news/world-europe-guernsey-20994463 [accessed 10 April 2020].

206 This date represents a gestation period of 283 days which is consistent with the average period of 280 days expected today, AHDB, *Pd+ Factsheet 1 – Fertility terms and definitions* (Stoneleigh; AHDB, 2012) https://dairy.ahdb.org.uk/resources-library/technical-information/fertility/pdplus-factsheet-1-fertility-terms-and-definitions/#.XpRGRi-ZPus [accessed 13 April 2020]. This date also represents a dry period of 49 days: current recommendations

From the diary entries the cow was served by the bull at *Hail Farm*: this is probably Hayle Farm near Cherhill. In 1838, this cow was retired: she had cost £12 at the age of four and was replaced by a three-year old Alderney at a cost of £12 10s 0d. On one occasion, there was no dry period for the cow: bought on 16 March 1843, the cow calved on 22 April 1844 having been milked every day. This was admitted in the diary as a *mistake in her time*, that is, there was an error in the calculation of the calving date.

With a nod towards the improvements generally attributed to "high farming", in November 1841 there is a diary entry which records the feeding of oil cake to the cow that was being fattened. Pinniger had bought 60 oil cakes and was feeding half an oil cake to the cow (about 1.7 lb in weight) along with barley meal and chaff twice a day, as well as swedes: the cow clearly enjoyed the oil cakes, in that they were all eaten by 6 January 1842. They were priced at 10s 6d per cwt with 1s carriage which brought the cost to £1 2s 0d. When the cow was killed on 24 February, it weighed 1,012lb (9 cwt) and sold for £27 16s 6d: from the diary entry, it turns out that this was the Alderney cow purchased in 1838.

As with other livestock, there were no pigs on the farm until he bought two little pigs from Kennet (probably from his brother William) in August 1829. However, at the end of November that year, he records a sow being brimmed by Mr Hitchcock's boar, so he was intending to breed his own pigs, which is confirmed in his diary entry for 12 April 1830 when the sow farrowed seven piglets.[207] Once again, Pinniger records the profit or loss made on the fattening of pigs: see Appendix A.

A new addition to his livestock not mentioned in the Little Bedwyn years was that of bees: on 25 March 1829, an entry in the diary shows that bees were removed to Beckhampton with a yoke: it is not clear where these came from, but if carried any distance, the carrier would want a balanced load and it is reasonable to assume that two skeps were brought to the farm. However, on 9 June the bees swarmed: there are no further references to bees in the diary thereafter.

As well as the mare and two colts which Pinniger had kept during his time in Chippenham, he bought a further two colts in mid 1830,

---

are for a period of at least 42 days. The dry period is the most important phase of a dairy cow's lactation cycle, allowing the cow and her udder to prepare for the next lactation, Tom O'Dwyer (ed.) *Teagasc Advisory Newsletter, Dairy* (Carlow, Ireland: Teagasc, November 2018) https://www.teagasc.ie/media/website/publications/2018/DairyNews_November2018.pdf [accessed 13 April 2020]. However, in 1834, Pinniger only discontinued milking the cow 14 days before she was due to calve.

207 Brimmed – this refers to when the boar has mated the sow.

thereby increasing the available horse power on the farm and reducing his dependency on Mr Wentworth and his brother William to provide him with horses at times of peak cultivation. It appears that this enabled co-operative working to take place in both directions as on 28 September 1830, when three horses and two waggons were helping Mr Low with his barley harvest and four horses were assisting work at his brother William's farm at West Kennet: on the following day, Mr Low's team were helping to plough the right-hand piece of Tan Hill. Notwithstanding this degree of co-operation, an account was kept as shown in the entry from 18 October 1830: *2 Teams Plough'g for Mr Wentworth now out of Debt & 1 day over.*

Thereafter, Pinniger continued to refresh his stock of horses, as the following entries from 1835 show:

> 24 April 1835. The 5 Year Old Black Gelding exchanged with Mr Hillier Reaves for a Grey Gelding rising 3 Years payd 6£, (the Colt 26£) the Horse £23 10s at 3 Years old
> 25 May 1835. Swindon Fair, with Mr Hillier Reeves Wroughton { Sold may Gray Mare for [...] Bot my Bay Horse at [Pencil note] 31£ | Sold the 6 Yr Rone Cart Horse [...] Bot A 3 yr Grey Cart Gelding [Pencil note] 28 Gui[208]

As well as the cart horses used for agricultural purposes, Pinniger also kept nags for riding and driving purposes. In 1845, he bought a horse once again from Hillier Reeves:[209]

> 2 July 1845. NAG, Bay gelding, bot of Mr Hiller Reeves, got by Leronzo, who challenged England to Trot for 100 Sovereigns, rise'g 6 year 28£

However, the nags were not always reserved for non-farm work, as the entry for 19 December 1845 shows: *The 2 NAG'S, cuting CHAFF, the 1st time the Plough'g being behind.* That is, when there was a need to catch up with ploughing that had been delayed by the wet weather and late harvest that year, the nags were put to work driving the horse machine which

208 may - my. Gui – Guinea, a unit of currency equal to one pound and one shilling (£1 1s 0d).

209 In the 1841 Census, Hillyer Reeves is described as a horse dealer, TNA: HO107/1176/13, *Census returns. 1841 census. Wiltshire. Hundred: Elstub and Everley including Parish: Wroughton, Tithing: Wroughton ED2* [accessed via Ancestry, 16 April 2020].

powered the chaff cutter.

There are no oxen mentioned in the diaries whilst Pinniger was at Beckhampton: the soil here is predominately chalky, silty loam with some sandy loam along the sides of the turnpike roads to Devizes and Marlborough. The only clay-type soil is the clay to sandy loam found around the north and west sides of Silbury Hill: this is where the water meadows were located and these were not ploughed. On the other hand, the soils at Little Bedwyn were a mixture of clayey loam to sandy loam; chalky, silt loam; and clay to sandy loam: this greater proportion of clay-type soils may have made the use of oxen more appropriate.[210] However, there is one entry on 18 July 1846 in which he describes how *6 Horses & a Bull & 4 Waggons took 15 loads dung pr day.*

Livestock feeding

In the early part of 1829, there was more grass keep than his own sheep could keep up with: to stop the grass being wasted, Pinniger took in sheep to graze from other farmers:

> 19 September 1829. Silbury Meadow, the last tie gave the Sheep, began 19th Augst 228 Tegs – our own 115 Tegs – Messrs A & R Larges [Total] 343[211]

However, on 22 March 1830, he was concerned that there may be insufficient feed for the sheep: *The Swede greens, growing fast, and the Lambs want more food than Milk yet fear Swedes will be gone before have Meadow Grass.* He was fortunate to be located close to his brother at West Kennet:

> 5 April 1830. 2 Horses to Kennett wth a load of Hurdles and a load of Hay for my Ewes & Lambs to feed of Swedes for Brother Wm

210 Natural Environmental Research Council, *mySoil app – Beckhampton, Wiltshire* [accessed 9 April 2020]. Natural Environmental Research Council, *mySoil app – Little Bedwyn, Wiltshire* [accessed 17 April 2020].

211 The term 'tie' refers to part of a field, which had been temporarily divided up in order to manage the effective grazing of the crop (examples of tie are found in the diaries referring to grass, clover and turnips): the modern equivalent is strip grazing. For example, on 14 June 1832, the diary entry states: *The grass in the Old field, is become very heavy | the Sheep tread it under foot in the double fold | began tieing it out*, whilst on 10 August 1833, the diary entry states: *The 98 Wether Lambs put in Silbury mead for Marlbro Fair, the grass hurdle'd in ties.*

Of course, his brother benefitted from this arrangement in that the sheep dung fertilised his land ready for the planting of the next crop in the rotation.

Again in 1832, he was concerned as to whether there would be sufficient winter feed:

> 12 January 1832. SWEDES, The EWES, began feeding and reduced their HAY nights from 7 to 4 Truss's, the Hay probably not enough to keep the Sheep all the Spring, but the Swedes probably enough The Ewes do well with Oat Straw & Water mornings, Swedes after & Hay in Fold | The Tegs (104) Swedes all day & 2 Truss's of Meadow Hay in Fold

This balance of excess and dearth of feeding stuff for his livestock was a constant concern for Pinniger, but is no different to the situation that farmers find themselves in today, given that farming is an industry subject to the vagaries of the weather.

Livestock husbandry

With sheep again being the main livestock on the farm, their performance was important to the farm's income stream. In his first year at Beckhampton, Pinniger achieved a lambing percentage of 85%, being 226 lambs from 265 ewes. In the following year, he made the following observation:

> 15 March 1831. 150 Lambs, in 10 Days 4 Ewes gone by, not keep a Lamb 4 Ewes want Lambs No Ewe died, no Lambs bought and only 4 doubles

His emphasis on *only 4 doubles* within this positive comment on lambing suggests that doubles were seen as a problem and that there being only four doubles was a good thing. In 1832, whilst 330 sheep were round-tailed (of which 220 were breeding ewes), only 188 lambs had their tails seared off, again giving a lambing percentage of 85%.[212]

It is of note that 226 lambs were tailed and cut in the middle of May and by early July, 225 lambs were weaned: it would appear however that whoever did the cutting undertook it in a more hygienic manner than the

212 In 1840, Pinniger noted that it had been a *very successful lambing season*: 212 lambs were produced from 228 ewes, a lambing percentage of 93%. Clearly, a percentage of even 100% was not expected at this time.

operative at Little Bedwyn.[213] Indeed, in 1831, Pinniger names Abednego Blackman of Charlton who in that year cut 93 lambs at 1s per score: Blackman was employed again in 1834, suggesting that he did a good job. However, in 1835, an incident regarding the penning of sheep for round-tailing, tail-cutting and ear-punching, prompted Pinniger to write himself a reminder not to pen the sheep too close together: a ewe and two lambs were suffocated with several others coming close to death.[214] Whilst Blackman carried out the castration of lambs with success, the outcome was less certain for older rams: when five of them were castrated in 1839, two of them died within two days of the operation, but the application of *Farmers Oils* both internally and externally seems to have been successful in that no further deaths are recorded.

The majority of one whole page of the diary at the start of the entry for 8 May 1830 is headed *WOOL, weigh'd, Mr Geoe Bailey* with a table of weights and a series of calculations. This would have been the wool from the *11½ Sc Tegs* sheared on 10 June 1829 and therefore was the first wool clip for Pinniger at Beckhampton: the yield from 223 fleeces was 30 tod 17 li – that is, 857 lb or 3.8lb per fleece. This is some 50% heavier than the fleeces recorded at Little Bedwyn, suggesting that a different breed of sheep was now being kept. Although demand for wool may have affected market prices, the price received was 21s per tod, or 9d per lb, might suggest that the quality of this wool was not as good. There are further records on wool in the diaries during this period and these are presented in Appendix E.

As ever with sheep, there were losses throughout the year: when Pinniger recorded his stock of sheep as at 18 November 1830, he noted that 12 of his original flock of 265 ewes had died, one of which had died recently as a result of *a bite of the Sheep DOG, poisen'd & mortifyed in the Shoulder and another dieing.* There is no record of what became of the dog. In 1838, a mad dog attacked Pinniger's dog in the farm yard and both were shot.

When sheep died on the farm, the cause was often uncertain and the remedies were still archaic by modern standards:

> 31 January 1831. Yesterday, a fourth Teg Sheep died with the Murrain, and this morn'g a Ewe died of the same know no remedy, but puting <u>Tar</u> about the Fold to smell [...]

---

213 In this context, the term 'cut' refers to the process of castration of ram lambs.
214 See diary entry for 27 April 1835.

There was clearly a belief in the preventative properties of tar, as the following entries suggest:

> 8 August 1831. SHEEP, gave a Table Spoon full of TAR to each in hope, to prevent their dieing by the MURRAIN, & tar'd sticks in the Fold = 20 died since Shear day
> 25 August 1836. 400 Sheep, gave a Table spoon full of TAR to each to prevent their die'ing by the MURRAIN consume'd and fix'd Tar'd Stakes every night in the Fold

Indeed, such a belief would have been consolidated by the observation made on 11 February 1832: *Two tooth Ewe DIED, murrain, the first since the 14 Augst.*

Similarly, the destruction of sheep ticks was carried out using chemicals that would not be used today:

> 17 October 1832. LAMB'S, dipd in a solution of Arsnick & soft Soap, to kill the Ticks, and prevent ther tucking, Viz 1 li of Arsnick & 2 li of Soft Soap to 30 Lambs[215]

When the sheep were dipped in 1845, the diary entry for 24 June noted that the sheep were unusually full of ticks, causing the sheep not to thrive, resulting in Pinniger noting that one should never save the expense of dipping sheep.

Whilst animal diseases took their toll, shortages of keep, human illness and the weather all had an effect on livestock at Beckhampton. In particular, the weather at the beginning of January 1837 was very frosty and was followed later in the month by a spell of heavy rain which meant that it was not possible to transfer ricks of wheat and barley to the barn for threshing: on top of this, the men were ill and unable to work for three weeks into February meaning that no threshing was possible and there was no straw for the sheep. The turnips were too dirty to feed and the sheep were kept on hay and water. The outcome was described in the following entries:

> 10 February 1837. Wind & rain in night & thin driving rain all day heavy rain evening, exceeding boisterous winds during the day, very

215 Tuck - "In a tuck and go" - in convulsions; "All of a tuck" - with nervous hasty movements, Dartnell & Goddard, *Wiltshire Words,* p.SA517. Tucking probably refers to the action of the lambs, when irritated by the ticks.

> DISHEARTENING, prospect for the Sheep, almost starved more by the weather than Keep, yet but little Keep & hay 6£ pr Ton
> 11 February 1837. Rain heavy night, DREADFUL WINDS, blowing a HURRICANE and very heavy rain before day to 12 o-Clock, impossible to serve the Sheep wth Straw from the Barn, had HAY, 3 times in Fold Tegs brot from Turnips, hay 3 times | CALAMITUS, time, my Ewes do so ill, expect many will die, can have no hope they will keep a Lamb, they appear hunger baned[216]
> 13 February 1837. Standing Pen made for Ewes, being unable to walk from bottom of Tan hill Piece to Straw in Yard, through weakness, their time up the 14th, the 1st Lamb this morn'g – and dead

The effects of the inclement weather and the lack of feed was observed by Pinniger in many flocks in the area. Ewes were lambing but had no milk with which to feed them: not only were the lambs dying, but the ewes as well, being *reduced to skin and bone only.* He also noted that there was a suspicion that the sheep had been infected with influenza from humans. The cold weather continued, with the water barrel at the standing pen needing to be defrosted using hot water from the tea kettle and the sheep continued in the standing pen until 19 April that year.

Whilst the weather improved, the lambs continued to die in early May. Pinniger was at a loss to understand the cause having fed them on differing diets: turnips & hay, swedes & hay, water meadow grass, and folded on fallows with hay, and yet the lambs continued to die. At this point, he turned to his well-tried remedy – tar – with the following consequences:

> 13 May 1837, TAR gave a Pap spoon full to eack the 1 <...> Lambs about 4li in ¾ hour John & Jas Davis the Boy & myself, made them exceeding SICK & ill much alarm'd though half of them would die

A few lambs died after this, but at a slower rate and just over a week later, he observed that the lambs were doing better, thereby reinforcing the belief in the power of tar. Overall though, it was not a good year for sheep: of the 230 ewes put to the ram in the autumn, there were only 102 lambs for tailing on 26 May, with 50 ram lambs to be castrated.

There are few references to the husbandry of pigs: on 23 July 1834, he records that 9 pigs were cut: again, this is taken to mean castrated and

---

216 Hunger-bane - To starve to death, Dartnell & Goddard, *Wiltshire Words*, p.196.

he observes that the proper time to do this is not before the pigs were eight weeks old.

Labour

Following the incident with the damaged waggon in early March 1829 (see *Buildings* section above), the diary records that John Freegard, carter, left employment at the farm: he was replaced by Thomas Millerd at 9s per week with a bonus of 1s per week *if behave well.* Pinniger's difficulties with his workforce continued through his first year at the farm. In late April, the Dobson brothers, John and William, charged for a day's work although they were at home and later went on strike on 1 May. On 13 May 1829, he employed Isaac Maslen and Aaron Andrews as chalk diggers, but they absconded two days later, having been busy robbing highway users on the turnpike road, instead of digging the chalk.

Pinniger clearly did not have any suitable permanent workforce at this time. When the field grass was ready for mowing, he borrowed George Brewer from Mr Wentworth *not having a Mower.* Similarly, when it came to reaping on 4 August, he had no reapers on the farm. The reaping did begin on the following day, but only through the employment of four temporary labourers, known as taskers.[217] When his second crop of clover needed cutting, he employed the two Whites from Lyneham, but five days later, they struck for *2s & Beer, instead of 1s 6d* – it appears that Pinniger had no option but to concede to their demands in that they began the mowing. It would appear that either William Philpot did not have a full and competent workforce when he was tenant at the farm, or perhaps that he had persuaded the best of them to move with him to his new farm at Nettleton.

This lack of competent labourers is a continuing theme. In August 1831, two labourers left their service in that neither could mow nor hoe turnips and Pinniger had to employ men from outside the parish, some of whom did not stick with the job:

> 18 August 1831. <u>OAT'S</u>, began Mowing by Silas Wiltshire & Jacob Minchin, having no Men in the Parish to do it. | Very drying some little rain | TURNIP'S, Old field, hoeing, spoild for want of hoeing before harvest, neither of our 4 Men could do it
> 19 August 1831. BEAN'S, 3 Taskers, began draw'g - & left it

217 Tasker - A tramping harvester or casual labourer who works by the piece, Dartnell & Goddard, *Wiltshire Words,* p.164.

> 2 September 1831. Willm Pierce, our only man & two Strangers from Cliffe, began Thrashing Wheat yesterday

By 1835, he had eight of his own staff who could reap the wheat, five men and three women: on 8 August he noted that no competent reapers called at the farm looking for work: however, *18 impudent ones, call'd* demanding between 13s and 16s per acre, but he did not need to take them on, in that his wheat was not quite ripe.

There were certain jobs such as the preparation of the water meadows where Pinniger brought in men to do the work: for example, on 28 November 1831, Isaac Gilbert began levelling the bank in Silbury mead but a flood on 10 December stopped the work such that three days later *the leveling Men return'd home, to Grafton* – they would have been men he knew from his time at Little Bedwyn.[218] Another task was that of mole catching for which he employed Simkins of Littlecote for two days in February 1832.

From time to time, he would provide meat for his workforce and labourers, of which the following present two examples:

> 9 July 1829. Sheep giddy Kill'd 36 li for Masons & Carpenters Dinner – for put'g Beams on Waggon House
> 5 June 1830. Giddy Sheep kill'd for Labourers

Quite what disorder the sheep were suffering from that made them *giddy* has not been discerned, but it is hoped that whatever the cause, it was not transmitted to the workers. It also appears that he provided his shepherds with suppers during lambing time:

> 16 March 1835. The 3 Shepherds discontinued their Suppers having it a Month in lamb'g

At this time, agricultural labourers were hired for a fixed term and were contracted to complete their period of service, usually one year. On 15 June 1830, the diary entry recorded that *Jos Bromham, left his place & engagement* and eight days later Pinniger obtained a warrant for him: three

---

218 This is probably Isaac Gilbert of East Grafton, who in the 1841 Census was aged 65, his occupation being stated as Agricultural Labourer, TNA: HO107/1180/1. *Census returns. 1841 census. Wiltshire Hundred: Kinwardstone, including Parish: Great Bedwin, Township: East and West Grafton, ED4* [accessed via Ancestry, 8 July 2020].

days after that Bromham was committed to Marlborough Bridewell for leaving Pinniger's service.

There are a number of references to labourers with the surname of Davis throughout this period at Beckhampton. Quite what was going on in 1837 is not explained further, but Pinniger recorded that he had found two hurdles and six stakes in the bedroom of W Davis, his carter.[219] In September that year, when there was reaping work to be done, both Jacob Davis and James Davis absented themselves from their work, much to Pinniger's inconvenience. However, in October 1838, two entries in the diary relate to the Davis family:

> 12 October 1838. WILLIAM DAVIS, - Carter & 2 Sons left
> 13 October 1838. JOHN DAVIS discharged himself, the last of the Family, Father Mother 6 Sons & Daughter happy release

It would appear that Pinniger was not unhappy to see the Davis family leave Beckhampton Farm, although this represented the loss of 12 labourers from his workforce. The labour situation was not helped in March 1839, when his labourer William Chandler and family emigrated to Australia. However, this was mitigated to some extent with the employment of David Richens as carter and his three sons, although two months later, Richens was implicated in an attempt to steal barley:

> 8 December 1838. Frost and fine day, Horses Thrashg Barley Saturday eve, before closeing the BARN I discovered SECRETED in a Chaff Basket by Geoe Timcomb and Willm Hitchens the two Thrashers my BARLEY, three parts full covered with some rudderings with intention to give the Horses, the CARTER, being accessory to it, admiting it his first offence[220]

In July 1839, Richens was in charge of a return load of 2¼ tons of stone from Atworth when he neglected to drug the waggon wheel at the top of Cannings Hill.[221] The waggon over-ran the horses, forcing them over the bank at the bottom of the hill into a wheat field, spilling the loaded stone. Nobody was killed but one horse was injured and the waggon was damaged: the horse returned to work five weeks later. Ten

219 See diary entry for 1 June 1837.

220 Rudder – a sieve, Dartnell & Goodall, *Wiltshire Words,* p.135. Rudderings – sievings.

221 Drug - To drug a wheel - to put on some kind of drag or chain, Dartnell & Goodall, *Wiltshire Words,* p.48.

days after the accident, Pinniger noted in his diary that Richens had been insolent. However, on 12 August, Richens was still in Pinniger's employ and made his first journey since the accident: he returned from Devizes in a drunken state, was very abusive and discharged himself, probably saving Pinniger having to do so himself.

Labour disputes continued into the following year, when Francis Richens, under-carter, absconded and a warrant was issued for leaving his service. Pinniger then records that Thomas Bailey and Thomas Bullock did similarly and it appears that they had all gone to work on the railway. Later in the year, Richens was committed to Marlborough prison, serving six weeks for breaking his service.[222] As at 11 May 1840, this left Pinniger with a workforce comprising the shepherd (David Fishlock), carter (George Titcombe) and three boys: eight days later George Titcombe was insolent and Pinniger gave him his notice to quit at Michaelmas. Titcombe seems to have continued at Beckhampton Farm beyond his notice date but continued to be impertinent and eventually on 7 November, he discharged himself from Pinniger's employ: John Marsh was employed as carter in his place. However, towards the end of November, Pinniger discovered his own coat, along with sheep gates under the bed in the house of Fishlock the shepherd: at his trial on 4 January, Fishlock was committed to four months imprisonment with hard labour.[223]

Pinniger's difficult relationships with his carters continued: three years later, it appears that John Marsh had left and on 11 October 1843, John Bird left as carter being fined 21s by the magistrates for disobedience etc., with Stephen Rodbourne replacing him in the role.

Disputes and disagreements

As observed above in the section on *The farm*, the water meadows around Silbury Hill were split into three parts originally with three different owners: in turn, there were in 1830 three different occupiers, John Wentworth, Robert Philpot and Thomas Pinniger himself. In order to "share" this resource, there was a need to divide the meadows, as the following entry demonstrates:

> 28 April 1830. Mr Philpots Swedes, finish'd, and began his Watermeads 8½ Doz HURDLES required for him & me each to divide the Meadow, and 30, for him, me & Mr Wentworth ea

222 Devizes & Wiltshire Gazette, 10 September 1840.
223 WSA: A1/125/67, *Calendars of prisoners, 1841.*

As confirmed in the tithe award, part of the water meadows at Silbury were occupied by William Tanner: however, his actions were often detrimental to Pinniger's use of his allotted share of the meadows, such as in May 1831 when Silbury Mead was too wet to cut, as a result of Tanner keeping the water on the mead.[224]

William Tanner's belligerence seems to have continued: on 20 October 1832, the Grafton men started cutting out his water course.[225] However, two days later, Pinniger wrote: *WILLM TANNER of KENNETT, forfeit'd his word, and careless of his Honour forbid the Grafton Men, proceeding with the Work in his Meadow as agreed with me on the 18th Instant, to clean out his Water carriage for a sum, not exceeding 3£, and in all probability would not cost 40s.* The Grafton men finished clearing the carriages in Pinniger's water meadows and then returned home.[226] However, there is further evidence that Tanner was not working his water meadows in the agreed manner, such that water was provided to all occupiers in turn and fairly:

> 24 October 1832. Mason, making new Hatch work in Silbury mead, to keep back the flooding caused by WILLM TANNER'S unlawfully useing Hatches, & fixed 18 Inches to high to let the water draw off Silbury mead[227]

Tanner's management of his part of the water meadows continued to interfere with Pinniger's: in April 1833, William Tanner had flooded part of Pinniger's land causing the grass to lodge, whilst the other side was in good order.[228] The overall outcome was summarised in the diary on 14 May when it was noted that whilst Messrs Wentworth and Philpot had finished feeding the water meadows that day, Pinniger was just starting to feed his mead at Silbury, leaving rye and meadow grass to spoil for want of grazing. Similarly, following a period of continual rain during December 1833 into January 1834, the flooding eventually subsided but not on the

224 See diary entry for 21 May 1831.

225 Reference to men from Grafton working on the water meadows has already been noted in the *Labour* section above.

226 Carriage – the main carriage provides the feed water from the river to the water meadow system, Michael Cowan, *Wiltshire Water Meadows* (East Knoyle: The Hobnob Press, 2005), pp.31-36.

227 Hatch – in this context, a mechanism for controlling the flow of water in a water-meadow, Cowan, *Wiltshire Water Meadows*, pp.109-116.

228 See diary entry for 12 April 1833.

water meadows as a result of Tanner's hatches being too high.[229]

By the end of 1834, Pinniger was clearly fed up with Tanner's actions in that he commissioned his brother, Broome Pinniger, lawyer of Chippenham, to proceed against Tanner for *injury sustained in Silbury md.* On 12 January following, Broome Pinniger went with Thomas Pinniger, John Wentworth and George Maslen (the latter two from Beckhampton) to survey the hatches and other arrangements in Tanner's meadow. The following entry provides details from their visit in terms of the layout and dimensions of the water meadows:

> 28 January 1835. The level taken from the Sill of the Hatch in Mr Wentworths Silbury Mead to the Sill of the Hatches in Mr Tanners meadow which is 600 feet in length, and the fall only 2¾ Inch's | The fall from the Sill in Mr Wentworths Meadow to Mr Tanners Ditch is 12 Inch's, from the Ditch Mr Tanners Carriage rise 7½ Inch's in the first 10 Feet | It is 116 feet from the Ditch between Mr W and Mr T_ and Mr T_ Sill and the rise is 9¼ Inches from the Carriage in Mr Wentworths Mead to the Sill in Mr Tanners Mead | 484 feet from the Sill in Mr Ws mead to Ditch 116 feet from Ditch to Mr Tanners Sill. [Total] 600 | It is nearly a dead level round Silbury Hill to Hatch in Mr Wentworths Mead

At around the same time, a dispute was developing with Robert Philpot who had enclosed the waste at the bottom of Pinniger's Tan Hill Piece. On 3 May 1834, a meeting was held at the office of Mr Merriman in Marlborough with Baverstock Merriman, W. R. Brown, George Brown, Robert Philpot and Thomas Pinniger in attendance to decide the right to the piece of land at the bottom of Tan Hill Piece: W. R. Brown was acting as referee for Pinniger, whilst Messrs Merriman were acting for Robert Holford Esq, proprietor of the land which Robert Philpot occupied.[230] The matter was resolved in Pinniger's favour, in that: *Mr Bavk Merryman for & in behalf of Mr Halford, relinquished all claim to the right of a Road, through my Tan hill Field to Mr Wentworth's Meadow, which was 35 Years since possessed by Mr Jefferies to a Barn in the said Meadow. & not since that period.*

On 17 May, W. R. Brown wrote indicating that he would direct Mr Halford's (*sic*) tenant (Robert Philpot) to give him possession of the said

229 See diary entry for 14 February 1834.

230 Baverstock Merriman – Thomas Baverstock Merriman was one of the partners of Messrs Merriman, attorneys in Marlborough, J. Pigot, 'Commercial Directory 1842' in WRS47, p.124. WSA: A1/345/22A, *Land Tax Assessments, Beckhampton, 1782-1841.*

corner of land: however, it was not until 10 June that Philpot surrendered the land, despite being directed by Mr Merriman (described in the diary entry as Steward – presumably land steward to Robert Holford Esq) to do so and Pinniger having complained to both Merriman and W. R. Brown about Philpot's obstinacy.

Later, a boundary dispute arose in 1843 between Pinniger and Messrs Sotherton and Whorell, landowners of adjoining property. This was settled with Pinniger losing a strip of his land for a ditch, 3ft wide: however, it appears that there had been a ditch there previously, but in 1820 William Philpot had ploughed it in. The boundary was marked with hillocks and large stones and indeed later that year, Pinniger himself erected four large boundary stones, one of which was dug up and thrown over. Pinniger re-erected the stone on March 14 1844, only for it to be thrown down and put up again over the following two days. All seemed to be settled until in November that year, Thomas Brown made a bank and ditch which Pinniger regarded as an encroachment on his land.

The outcome of this boundary dispute seems to have encouraged a similar quarrel when in March that year, Samuel Wentworth claimed a ditch in Whitelands (glebe land - one of Pinniger's fields): however, Samuel subsequently admitted to his brother John (Pinniger's neighbour) that he was not justified in his claim. Later that year, George Brown unjustly accused Pinniger of filling in a ditch by the Glebe. Pinniger might reasonably have felt the victim of a series of contentious claims.

Disputes were not limited to the farm land and its boundaries. As part of the farm, there were two cottages and on 16 September 1843, Pinniger gave both of his tenants notice to quit. Michael Wilshire went willingly four days later, but Ann Edmunds had to be ejected on 8 November by a police officer: Edmunds remained in the road with her belongings for two days. In early December, a writ was issued by William Copeland for Ann Edmunds for her ejection and she served a declaration on Pinniger towards the end of the month. The next diary entry relating to this incident was recorded on 21 May 1844, when: *PROCESS served me a SUIT, to eject me from my COTTAGE, late occupied by Edmonds.*

However, it was not until 22 July 1846 that the case was brought before the assizes as a *nisi prius* case.[231] Pinniger pasted in a newspaper

231 *Nisi prius* – translates as "if not before" or "unless before". Prior to 1873, all civil cases at common law were begun in one of three London courts. Unless the case was heard at assize in the claimant's county before a given date, it would be heard in London – hence the hearing of the case between Edmunds and Pinniger at the county assizes was described as a *nisi prius*

cutting into Volume 7 of his diary describing the case. Mrs Edmunds was still claiming that the cottage was hers: however, Mr Crowder (acting for Pinniger as defendant) explained that the strip of land on which the cottage was built, along with Beckhampton Inn, were owned by Mr Edmunds, husband of the plaintiff.[232] In 1802, Edmunds mortgaged this property to Anthony Guy of Chippenham for £100 and between then and 1810, he built the cottage on the strip of land: in 1810, he conveyed all of the property including the cottage to Mr Guy for £495.[233] After the property was conveyed subsequently to Pinniger, he allowed the old man Edmunds, and his wife after his death, to remain in the cottage for 15d per week. Pinniger maintained the cottage including putting on a new roof after which he wished to raise the rent to 2s per week. Mrs Edmunds refused stating that she had a right to the cottage and that she only paid 15d per week for the cellar. The judge stopped the case prematurely before further evidence was brought, finding in favour of Pinniger.

There was also a disagreement regarding the use of the so-called *back gate*: this appears to have been an alternative to using the turnpike gate. In May 1844, Samuel Wentworth (it has been seen that Pinniger had previously had a good relationship with John Wentworth) had crossed the bottom of one of Pinniger's fields with his roller and scarifier when it was ready for planting: the field was then planted with potatoes but Wentworth once again crossed the land with his roller. Later in the year, Wentworth

---

proceeding.

232 It is likely that Mr Crowder was Mr Crowdy. Wm Morse Crowdy & Alfred Southby were attorneys in Swindon: they were also clerks to the board of guardians of the Highworth & Swindon Poor Law Union and to the turnpike trust. J. Pigot, 'Commercial Directory 1842' in WRS47, p.138. Further, Crowdy & Brace were attorneys of Highworth and Henry Crowdy Crowdy of Highworth was clerk to several turnpike trusts. J. Pigot, 'Commercial Directory 1842' in WRS47, p.117. Given Pinniger's involvement as a Poor Law guardian for the parish of Avebury and as a commissioner for the Beckhampton Turnpike Trust, then he may well have come into contact with one or both of these firms.

233 In October 1809, the Beckhampton turnpike tolls were to be let by auction at the house of Mr Edmunds, being the Catherine Wheel Inn, Salisbury & Winchester Journal, 9 October 1809. In April 1810, Beckhampton Inn was to be let with immediate occupation, by applying to the Chippenham offices of Guy & Michell, Salisbury & Winchester Journal, 9 April 1810. NOTE: Whilst Pinniger uses the name Edmunds and Edmonds interchangeably, and the newspaper cutting in the diary uses Edmonds, the name has been standardised to Edmunds in this volume.

continued to lock the back gate to prevent waggons passing, causing them to go via the turnpike gate instead. The disagreements with Samuel Wentworth continued into 1845 when Wentworth objected to Pinniger constructing a bank on his own land by Tan Hill Road. Exchanges of angry words resulted and on 31 May, Wentworth brought George Brown along as an "independent" person to view and condemn the bank: the outcome was that Brown approved the bank and advised Wentworth to enclose the road on his side in a similar manner. However, two diary entries from February 1846 record the Rev. Merrick becoming involved as mediator between Wentworth and Thomas Pinniger, which the latter viewed as having little likelihood of success: notwithstanding this, there are no further entries relating to disputes between the two parties.

Of course, it should be remembered that, with the exception of those cases reported in the press, the diaries present only one side of the story. However, it is of note that on 26 April 1846, Pinniger recorded the death of Mr William Kemm of Avebury whom he described as *much respected, being a quiet and peaceable neighbour.*

Events recorded beyond the farm

On 24 November 1829, Anthony Guy was declared a bankrupt, as a result of the action taken by a Mr Bush of Trowbridge. The seriousness of this matter prompted Pinniger to write the following entry within a thick black border, normally reserved in his diary for recording a death:

> Mr GUY of Chippenham declared, absented from his Crediters & the Docket struck by Mr Bush of Trowbridge. Myself & Brothers met at Mr Guys Office to consult with Messrs Goldneys &c. on the best means to be taken, till 10-O-Clock at Night[234]

Further, on 17 December, John Pinniger of Cowitch (Cowage) similarly struck the docket against Guy. Six days later, he attended a meeting of

234 The first step in bankruptcy proceedings consists of the creditor making an affidavit of the amount of the debt, stating that he believes the debtor has become a bankrupt and paying a bond to follow up the proceedings: when these are delivered at the bankrupt office, an entry is made in what is called the "docket-book", upon which the petitioning creditor is said to have struck the docket. Robert Henley Eden, *A practical treatise on the Bankrupt Law, 2nd edition* (London: Joseph Butterworth & Son, 1826), pp. 51-52.

commissioners at Bath, along with Mr. Godwin of Beaversbrook, Mr Ruming of Draycot and Mr Salter of Langley, to prove and declare Anthony Guy of Chippenham, a Bankrupt. On 30 December, the local paper reported that Guy, described as a money scrivener, broker, dealer and chapman had been declared bankrupt.[235]

Anthony Guy appeared before the Commissioners in January 1830: indeed, Pinniger records that he could not attend the second day of the proceedings due to the heavy snow. The bankruptcy proceedings rumbled on with Pinniger receiving a summons from the commissioners in Guy's bankruptcy on 27 August, but it was not until 21 December 1830 that: *A=GUY'S, Estate this day Paid his Crediters 1s in the Pound which require Twelve Thousand Pound* [236]

On a smaller scale, one of Pinniger's two neighbours, with whom it has been seen he had some disputes, was finding it difficult to make his farm pay its way. An entry dated 11 June 1833 states that Philpot had been distrained for his rent of £185 tithe-free, suggesting that Philpot was struggling to pay his rent.[237] From the diaries, it appears that Robert Philpot continued to struggle making ends meet at his farm: in January 1840, 260 ewes were distrained for unpaid rent and sold by auction, making an average of 27s each. On this occasion, the outcome was that from Lady Day 1840, Pinniger's other neighbour, John Wentworth, took over the tenancy of Philpot's farm – both were owned by Robert Holford Esq.[238]

In 1832, it appears that there was a move to standardise the measure for wheat, as shown in the following entry:

> 12 April 1832, CONTENTION in Devizes MARKET The Mealman, persist in buying Wheat by the Imperial Bushel only & the Farmers

235 Devizes and Wiltshire Gazette, 30 December 1829. Money-scrivener – a person who raises money for others (Webster's English Dictionary).

236 It is not clear from this entry whether it was (a) the payment (at 1s in the pound) which required £12,000 to be found at that point or (b) the creditors who were owed £12,000. Assuming the former, then the full value owed to the creditors was £240,000, or £21.11m at 2018 values (real wealth value), Measuring Worth, https://www.measuringworth.com/calculators/ukcompare/relativevalue.php [Accessed 11 April 2020]

237 Distrain - to seize (someone's property) in order to obtain payment of rent or other money owed.

238 This is confirmed in a notice extinguishing tithes dated 1843 for both Great Beckhampton Farm (438 acres) and Little Beckhampton Farm (180 acres), WSA: 2027/2/1/855, *Avebury Deeds, 1741-1907*.

> in selling by the Winchester Bushel only but little business done in consequence
> 19 April 1832. DEVIZES, Market, attend'd by Dealers from Newbury & Reading, in consequ of the contention

This change was clearly important to Pinniger in that he opened his diary entries for 1835 by describing the impact that the introduction of imperial measure was having on his farming business: see Appendix U.

On 7 January 1836, Pinniger noted that the poor were now being relieved under the new system and that the relieving officer was Mr Budd. The Poor Law Amendment Act had been signed into law by King William IV on 14 August 1834 and the Marlborough Poor Law Union was formed on 24 November 1835. Its operation was overseen by an elected Board of Guardians, 16 in number, representing its 14 constituent parishes including Avebury. Thomas Pinniger recorded in his diary on 25 November 1835 that he had been elected Guardian for the parish.[239] It would appear that he did what he could to employ the poor, thereby reducing the burden on the parish who contributed to the costs of those in the workhouse and those receiving outdoor relief. However, such benevolence was not always as productive as it might have been:

> 20 January 1836. BARLEY, Winnow'd 14 Qr & 2 Bu, thrash'd by a Man aged 67 and a under Carter aged 19 at 5s pr Week ea: to keep them off the Parish Books 15 days each, prices given pr Qr vary 1s 1s 2d to 1s 4d N B. Detected the one dictating to the other writing a Letter instead of working

His diary entries for the early part of 1839 were concerned with the market prices for both wheat and barley, as a consequence of the Parliamentary debate on the Corn Laws.[240] He recorded that a petition

239 Anthony Brundage, *The English Poor Laws, 1700-1930* (Basingstoke: Palgrave, 2002), p.69; The Workhouse, *Marlborough, Wiltshire* http://www.workhouses.org.uk/Marlborough/ [accessed 20 April 2020].

240 Government intervention in the sale of grain crops for food was driven by the need to balance three different concerns: (a) cheap bread for the consumer; (b) farmers desiring high prices for their produce; (c) the government wishing to achieve a balance between civil unrest (if bread were unaffordable) and encouraging agriculture to produce more home-grown food. The Corn Laws were concerned with both domestic grain trade and foreign grain trade, both imports and exports, depending on the success of each harvest. Restrictions on low-priced imports were achieved through the

to both Houses of Parliament was signed at the Castle Inn, Devizes on 7 February against any alteration to the existing Corn Laws. He also noted that on 1 March the repealers lost a vote in Parliament by 172 votes to 361: further, there were 921 petitions against any alterations whereas there were only 120 such petitions in favour of a free trade in corn. By mid-March the markets declined as a result of large importations of wheat and the on-going debate in Parliament. The campaign against the Corn Laws rumbled on: in February 1844, there is a diary entry which records a demonstration to oppose the Corn Leaguers held in Devizes. In January 1846, Pinniger sent a petition to both Houses of Parliament, signed by himself, his son (Thomas Large) and five labourers for the continuation of protection of agriculture, that is, of the Corn Laws: Pinniger later recorded Sir Robert Peel's proposed measures on the Corn Laws up to the year 1849, as well as the passing of the Free Trade Corn Bill in July 1846.[241]

The year 1841 saw the commutation of tithes in the parish of Avebury: Pinniger notes two meetings held on 24 May and 14 June that year and that the tithes were commuted by Aneurin Owen on 9 August, at the Castle Inn, Marlborough.[242] In August 1843, Pinniger noted that Mr Crook's estate in Avebury had been sold at the Beckhampton Inn for £3,750: the freehold farm extended to 124 acres, there was no land tax to pay, the tithes had been commuted but was described as being let at a very inadequate rent of £150 per annum, currently occupied by Robert Philpot.[243]

Whilst the tithes may have been commuted, there was still the poor rate to be paid: indeed, in April 1844, John Stratton of Milton and Mr Bramble of Devizes looked over Beckhampton Farm to make a new Poor Rate assessment. Pinniger met them on his new downland, suggesting

use of a sliding scale of duty, determined by the price of home-grown wheat, thereby protecting Britain's farmers. Mingay, *Agrarian History of England and Wales, Volume VI, 1750-1850*, pp. 209-213.

241 Sir Robert Peel (1788-1850) was prime minister at the time, up until 29 June that year.

242 Aneurin Owen was one of the Assistant Tithe Commissioners. His report on the parish of Avebury to the Tithe Commissioners starts: *I attended, pursuant to your instructions August 9 1841 and by adjournment May 9 1842 to award rent charges in commutation of the tithes in this parish.* He then attended further meetings on 6 April 1843 and 26 August 1845 to hear objections to the Award: Pinniger notes the 1843 meeting in his diary stating that the matter was settled. The sealed copy of the apportionment of the rent charge was not received by the Commissioners Office until March 1846, TNA: IR18/10895 *Tithe File: Avebury (parish), Wiltshire, 1836-1870.*

243 Wiltshire Independent, 13 July 1843.

that this change in the use of the down from pasture to arable land would result in a change in the Poor Rate to be paid.[244]

In May 1845, he recorded progress with the construction of what is now known as the Lansdowne monument:

> 23 May 1845. TOWER, build'g on Calston hill by the Marquis of Lansdown, it is 20 feet square the first part, and is now 30 feet high, the Wall'g near 6 feet thick

As well as his farming activities, Pinniger was also an investor in stocks: on 19 August 1841, he recorded the sale of 1,000 3½ stocks receiving £980 2s 6d and commission: he also noted that he had sold similar stock in July 1840, receiving around £430.

Turnpikes and railways

Within the parish of Avebury, there were a number of toll gates which were let by the trustees of the Beckhampton Turnpike Roads.[245] In 1834, the Beckhampton tolls were advertised as including the toll gates known as Beckhampton Gate, Kennet Gate, Avebury Gate and Avebury Bar as well as the weighing engines at the first two gates.[246]

244 Whilst Pinniger made no observations of the outcome of this assessment, it is possible that the Poor Rate due may have decreased, in that arable land provided more labour opportunities than pasture land.

245 Devizes and Wiltshire Gazette, 15 May 1834.

246 According to Robert Haynes and Ivor Slocombe, there are four known toll houses in the parish of Avebury: (1) Avebury: On the A[4]361 in the centre of the village of Avebury near the Red Lion Inn (National Grid Reference (NGR) SU 103 700); (2) Beckhampton, Waggon & Horses: On the A4 Calne to Marlborough road opposite the Waggon and Horses pub (NGR: SU 090 689); (3) Beckhampton, Cross roads: On the A4 Calne to Marlborough road at the Beckhampton roundabout (NGR: SU 088 690): this was built in 1857 to replace the one opposite the Waggon and Horses pub; (4) Kennet: On the south-east side of the A[4]361 Beckhampton to Swindon road to the south of Avebury on the corner where the main road meets the B4003 leading south-east to the A4 towards Marlborough (NGR: SU 104 698): this was built in 1797, Robert Haynes and Ivor Slocombe, *Wiltshire Toll Houses* (East Knoyle, The Hobnob Press, 2004), pp.6-8.
It should be noted that: (a) whilst they state that, in 1804 a new weighing machine was erected at Beckhampton Gate and the existing one moved after repair to Kennet Gate (p.7), they record that the weighing machine was moved to Avebury Gate (p.6) rather than Kennet Gate (p.8); (b) these four toll houses do not align exactly with the four named toll houses from

The auction was to take place at the house of Mr John Watts, called the Beckhampton Inn: the notice stated that they were last let for £1,362. On 22 May 1834, Pinniger recorded that: *BECKHAMPTON Tolls let to Ezel Evans for 1387£ Let last year at 1362£ to Isaac Selman Low.* This was a considerable amount to pay (worth between £127,000 and £131,000 at 2018 prices): the tenant would then have had to charge tolls in order to recover the rent and potentially to make a profit.[247]

On 14 November 1836, he was appointed one of the commissioners of the Beckhampton Turnpike Trust. Having been sworn in at his first meeting, he notes that the tolls had been let in 1836 to Evans for £1395 and were let to James Coller of Bath at £1,400 from 1 January 1837.

An entry from 26 December 1836 gives an indication of the toll charges made: *Not a Passenger through the Turnpike Gate from 1-o Clock in Night to 3 o-Clock this afternoon, when a Horse paid the first Penny.* At that rate, it would have taken a lot of traffic to recoup the amount of £1,400 paid. However, in 1840, Pinniger recorded his personal experience of the fines exacted, over and above the toll, when he sent a team of four horses to Atworth for paving for his gig house. The load was overweight at 5 tons and 19 cwt: the weight allowed over the bridge at Melksham in summer being 4 tons and 15 cwt and Table 17 shows the calculation of the fine.

Table 17: Calculation of overweight fine at Melksham Bridge toll gate, 1840

For the first 2 cwts	3d per cwt	6d
For the next 3 cwts	6d per cwt	1s 6d
For the next 5 cwts	2s 6d per cwt	12s 6d
For the next 14 cwts	5s per cwt	£3 10s 0d
	Total fine[1]	£4 4s 6d

He noted with regret that Cox, the renter of the turnpike gate at Melksham Bridge, had exacted the full fine, given that the cost of the load of stone (£3 12s 0d) was less than the fine. Further, he suggested that his waggon had been overloaded either thoughtlessly or wantonly.

In 1840, the tolls at Beckhampton had been let in the previous year for £1,452 with the weighbridges, but they were then let for only £1,205,

1834: (1) may have been either Avebury Gate or Avebury Bar; (2) and (4) were named in 1804 and had weighing machines at that date – therefore, it is likely that these were Beckhampton Gate and Kennet Gate respectively; (3) was not in existence in 1834.

247 Measuring Worth, https://www.measuringworth.com/calculators/ukcompare/relativevalue.php [accessed 18 April 2020].

this time without the weighbridges: further, he recorded that they were let for £651 on September 1 1841. This reduction in letting charges is probably a reflection of the re-routing of coaches through Avebury to link with the railway at Swindon, as noted above in the section on *The Inn*: further, on 11 December, he noted that the carrier's waggons had now started to go through Avebury to the railway station at Hay Lane, instead of along the usual Marlborough road.[248]

The railway clearly interested Pinniger in that on 31 May 1841, he walked through Box Tunnel before it was completed and then went to Chippenham and saw the London train arrive and depart: at this time, he noted that the four-horse coaches were discontinued through Beckhampton with them going instead to meet the railway at Hay Lane.[249] Further, he noted that the railway opened through to Bristol at the end of June: in particular, on 16 July, the last such coach (Miss Fromont's) was discontinued due to a lack of passengers and the last mail coach finished on 3 August. The re-routing of the coaches away from Beckhampton would have meant that replacement coach horses were no longer housed at the Inn, depriving Pinniger of the loss of income from providing them with straw and the loss of dung from the Inn for his land. Further, this would have produced a reduction in trade for the publican and may well have been a factor leading to the eventual closure of the Inn.

Law and order

The year 1830 saw the culmination of unrest amongst agricultural labourers, resulting in the so-called Swing Riots. Named after the fictitious Captain Swing whose name was used to sign off threatening letters to farmers, the riots started with the destruction of thrashing machines in Kent and spread across counties in the south of England including Wiltshire. Whilst attacks were focussed on the thrashing machine, regarded as a labour-displacing device, their actions included incendiarism, demands for increased wages and better poor relief, along with objections to the tithe system.[250] In his diary, Pinniger wrote an article entitled *Alarming Times* and this is reproduced in full in Appendix

248 Pinniger used the term *railroad* rather than *railway*.

249 However, the first recorded travel on the railway in the diaries came on 30 May 1842, when Pinniger travelled by rail from Bath to Bristol, having gone from Bradford[-on-Avon] to Bath by canal: the diary does not record how he made the return journey from Bristol to Beckhampton.

250 Carl Griffin, 'The Violent Captain Swing?', *Past and Present*, 209 (2010), 149.

P. He recorded the impact on his own life in his diary entries for 25 to 27 November: on the first day, he noted that the market at Devizes was poorly attended as a result of the destruction of thrashing machines in the previous week and the fact that some farmers were riding with the cavalry in order to scour the country for the ringleaders of the rioters.

Indeed, there had been several arson attacks in the Marlborough district between 17 and 22 November, in particular, at Collingbourne, Ludgershall, Oare, Amesbury, Everly, Stanton St Bernard and Winterslow, where barns and the machinery therein, along with ricked crops on the farms, had been the target. This was followed by the destruction of thrashing machines in the county, with sixty-one destroyed over a four day period.[251]

On 26 November, Pinniger personally rode out with two hundred horsemen and the cavalry, who succeeded in taking four rioters: they rode *from Rockley to Winterbourne, Cliffe, Corton, Hillmarton, Lyneham & Wootton=basset.* On the following day, they rode in the direction of Oare, Pewsey, Milton and Easton taking a further seven prisoners.

Whilst Hobsbawm and Rudé only record six incidents in Wiltshire between the start of December 1830 and the end of December 1831, Pinniger continued to note incidents in his diary.[252] On 12 March 1831, he wrote the word *Threat* in pencil in the margin of his diary with no further explanation and similarly, the word *Fires* on 16 April. On the first Sunday in June, he recorded the following:

> 5 June 1831. Incendiary, Mr Neat of Monkton last night (Saturday) a large Wheat Rick, Bean & Oat Rick, with Three Barns, Sow & Pigs, & 40 Sacks Wheat destroy'd by Fire, between 10 & 11-O-Clock 3 Waggons, Gig &c & 3 Ricks of Straw[253]

251 Eric Hobsbawm and George Rudé, *Captain Swing* (London and New York: Verso, 2014), pp.322-336. Two thrashing machines were destroyed on 21 November, followed by 25, 23 and 11 machines on the following three days.

252 Hobsbawm & Rudé list the following incidents between 1 December 1830 and 31 December 1831: 29 December 1830, Kilmington, Riot (incitement); 10 January 1831, Upton Lovell, Arson targeted at the parson, value of damage £400; 11 January 1831, Potterne, Arson targeted at the parson, value of damage £400, again; 11 January 1831, West Lavington, Arson targeted at a publican; 22 January 1831, Amesbury, Arson targeted at a farmer; 28 August 1831, Broad Chalke, Arson targeted against a farmer in respect of thrashing machines, Hobsbawm & Rudé, *Captain Swing,* pp. 344-358.

253 This incident is not listed in Hobsbawm & Rudé, *Captain Swing,* p.356.

According to a local newspaper report, the fire had started in a rick of straw close to a barn which housed a regularly-used thrashing machine. However, the paper was unable to speculate as to why the farm had been set on fire: there were insufficient agricultural labourers in the parish to do the farm work, even if all of the wheat grown there had been threshed by machine. Further, Mr Neate had employed 20 men from other parishes throughout the previous winter. This may suggest that those involved were not locals.[254]

The diary includes further records of incendiary actions in 1833 and 1834:

> 11 November 1833. FIRES, frequent, one at Urchfont last night & one at Wroughton in the night Monday the 4th Incenderies
> 21 January 1834. Tuesday night Incendiary FIRE, at Lower Bupton, Barn & Buildings, distroyd

By comparison, the latest dated incident recorded in Wiltshire was of arson at Broad Chalke on 28 August 1831.[255] From the diaries, it would appear that all was not as settled as the background to the national form of prayer in November 1832 perhaps suggested – see section below on *Church and charity*.

No sooner had the impact of the Swing Riots passed, than concerns about the New Poor Law led to further unrest. Incendiary attacks were recorded at Christian Malford and Calne in late 1834 / early 1835 and entries in Pinniger's diary recorded a fire at Bushton and one at Chute around this time, which may relate to this issue. [256] Indeed, on 9 February 1836, the firing of a bean rick at the farm of John Brown of Chiseldon was recorded in the diary, with Pinniger noting that John Brown was a Guardian of the Poor.[257]

From the following diary entry, it would appear that a member of Pinniger's extended family acted unlawfully, in that the entry is written within a thick black border: *29th October 1832. John Pinniger alias Willm*

254 Bath Chronicle & Weekly Gazette, 9 June 1831.

255 Hobsbawm & Rudé, *Captain Swing*, p. 357.

256 Adrian Randall and Edwina Newman, 'Protest, Proletarians and paternalists: Social conflict in rural Wiltshire, 1830-1850' *Rural History*, 6.2 (1995), 216. See diary entries for 24 October and 7 November 1834.

257 Chiseldon – part of the Highworth and Swindon Poor Law Union, The Workhouse, *Highworth and Swindon, Wiltshire* http://www.workhouses.org.uk/Highworth/ [accessed 20 April 2020].

Golding of London, Transported for life for Horse & Gig stealing.[258] Indeed, he was committed before the Middlesex justices in October 1832 for horse stealing.[259] He was transported for life to Van Diemen's Land on the ship *Lotus* with 215 other convicts, leaving on 20 December 1832 and arriving nearly five months later.[260]

Pinniger was the victim of minor thefts from the farm whilst at Beckhampton, including the theft of 10 sacks of potatoes in April 1837 after a long, cold and wet winter that year: such a crime was probably carried out in desperation to stave off hunger. More potatoes were stolen in November that year, this time from the home skilling. Lawlessness continued into 1838 with Pinniger recording a number of incidents: fence stakes were taken from his skilling in the meadow, a fire destroyed a skilling at Bupton Farm and the shooting of a farmer on his return from market.[261] Brian Rumbold was a farmer at Lyneham: having been shot he was confined to bed for five weeks and at the time of the trial in August he could still not use his left arm.[262] George Maskelyne (aged 35) was indicted for shooting at Mr Rumbold with the intent to murder him. The report of the trial records that the jury consulted for only a short time before reaching a verdict of guilty and the judge sentenced him to death: Maskelyne was executed on 6 September.[263] On the following Sunday, Pinniger attended church in Lyneham where the Revd Mr Young's sermon was based on the ninth verse of Psalm 119

258 Whilst this individual could have been an unrelated person with the same surname, which caught Thomas Pinniger's attention, the use of the thick black border suggests that he was a relative.

259 House of Commons, *Accounts and papers of the House of Commons, Volume 29* (London: House of Commons, 1833), p.29.

260 State Library of Queensland, *British convict transportation register – John Pinniger* https://convictrecords.com.au/convicts/pinniger/john/118790 [accessed 15 April 2020].

261 See diary entries for 16 and 17 January 1838.

262 Although the Devizes & Wiltshire Gazette, Salisbury & Winchester Journal and Wiltshire Independent all reported the injured party as Brian Rumbold, he was actually Bryan Rumboll. In 1841, Bryan Rumboll was a farmer, aged 60 at Lyneham Court, TNA: HO107/1179/7, *Census returns. 1841 census. Wiltshire. Hundred: Kingsbridge including Parish: Lyneham or Lineham, ED 11.* In 1851, Bryan Rumboll was the farmer of 880 acres employing 11 men and 5 boys at Lyneham Court, TNA: HO107/1834. *Census returns. 1851 census. Wiltshire. Registration District: 251, Cricklade. Registration Sub-district: 1 Wootton-Bassett, including Parish: Lyneham or Lineham, ED 2c.*

263 Wiltshire Independent, 16 August 1838.

relating to Maskelyne's death.[264]

Garden, family and health

In March 1830, having concluded his programme of building, it appears that the materials for making a garden were being ordered, the foundation for the garden wall having been completed by 15 October the previous year: coping stones from Box, 181 yards of bordering stone from Pickwick, as well as flints picked off the fields and earth for the garden paths. Mid-April saw his men wheeling earth into the garden and levelling it and the masons were putting the coping stones on the garden wall. Finally, on 21 April, Pinniger was able to sow *all small Seeds & Pease*: however, he must have already planted his fruit trees or they were already there, in that they were in full leaf at the end of that month.

Garden crops that he planted included potatoes, peas, onions, kidney beans, summer and spring cabbage, and broccoli: in 1830, he set 200 spring cabbage plants and in 1831, the same number of cabbage plants for the summer and winter. As well as vegetables, there is reference in his diary to *Flour (sic) seeds, planted in Garden*. Towards the end of 1831, he lists the trees and climbers planted in the yard, in 1833 he listed the flowering trees and pot flowers in the garden and in 1834 the trees planted by the front gates. Details of all these are given in Appendix O.

There is an interesting reference relating to the garden in February 1835: *BOG=EARTH, Waggon & 4 Horse from the Sicily Clump Bedwin Common for American Plants.*[265] The diary records that the *American plants* were planted in the bog earth at the end of March: Pinniger left the rest of the diary page empty – perhaps he intended to return to this and fill in details about these plants, but unfortunately we have no further details of them.[266]

There are occasional references to matters domestic, as the following examples show:

> 25 August 1832. the Men diging holes in Garden for Clothes Posts
> 19 November 1833. 6 li Ginger @ 1s 4d, & 19 li Treacle @ 3d with Flour enough to make 28 li of Gingerbread @ 8d – is 18s 8d, put into 4

---

264 Psalm 119, v.9: How can a young person stay on the path of purity? By living according to your Word.

265 See diary entry for 17 February 1835.

266 See diary entry for 28 March 1835.

Hogsheads Cider & 2 Hogsheads of Cider=Kind[267]

With regard to family, there are numerous references to his brother William who, according to the diaries, lived at Kennet. When William's son Jacob married in 1842, William was described as being of *West Kennet*, less than two miles along the turnpike road to Marlborough from Beckhampton.[268] In 1843, Jacob took over Kennet Farm, signing the agreement for taking the farm on 3 March: this suggests that Kennet Farm was tenanted, as opposed to Beckhampton Farm which Thomas owned. By 1845, William had moved to Prospect House, Devizes, dying there some twelve months later, one year before his brother Thomas.

In 1843, horses from Kennet were once more assisting at Beckhampton Farm, either due to a lack of horse-power or manpower. The 1851 census shows Jacob as a farmer of 970 acres, employing 24 men, 10 boys and 12 women (the only other farm in the village extended to 975 acres and was farmed by Thomas Kemm).[269] It is therefore not surprising that from time to time William was able to lend men and horses to his brother in his early days at Beckhampton, and likewise Jacob was able to assist his uncle in later years.

The provision of advice and guidance within the Pinniger family is seen in 1840 when, according to the diary, Thomas went with his cousin Samuel Smith to look over Oakhanger Farm near Hungerford.[270] They then looked at Broad Hinton Farm and on 31 March an agreement was signed transferring the lease of the farm to Samuel Hale Smith, whose occupancy of the farm commenced on 7 April. On 19 May, he records that Samuel married Miss Emma Crook.[271] Similarly, he looked over Pen

267 Ciderkin - the washing after the best cider is made, Dartnell & Goddard, *Wiltshire Words,* p.28.

268 Oxford Journal, 7 May 1842.

269 TNA: HO107/1838. *Census returns. 1851 census. Wiltshire. Registration District: 255, Marlborough. Registration Sub-district: 1 Marlborough, including Hamlet: West Kennet, ED1* [accessed via Ancestry, 12 May 2020].

270 Samuel Hale Smith was actually his nephew being the sixth child of Richard Sadler Smith and Elizabeth Smith neé Pinniger (Thomas's elder sister), born 19 February 1807, WSA: 4381/2/1, *Pinniger genealogical memoranda book, c.1820-1943*, p.14. On the 1899 Ordnance Survey map, Oakhanger House lies one mile west of Shefford Woodlands, which is four miles north of Hungerford. The adjacent farm is named Wickfield Farm: in 1840, this may have been known as Oakhanger Farm.

271 Emma Crook was the fifth child (twin to Ann) of Henry Crook and Ann Crook neé Pinniger (one of Thomas's younger sisters), born 24 August 1814, WSA: 4381/2/1, *Pinniger genealogical memoranda book, c.1820-1943*,

Farm for Henry Crook.[272]

There are a few entries in the diaries which provide an insight into the physical appearance of Pinniger and some of his family members. On the inside cover of Volume 5, he recorded their heights at various dates: see Table 18.[273]

Table 18: Heights of Pinniger family members

Date	*Name*	*Height*	
		ft	*in*
1 March 1829	Mary	3	2½
	Ann	2	10
11 May 1828	Samuel Smith	6	0
	Thomas Pinniger	5	11¼
2 February 1828	Thomas Crook	5	9⅜
20 March 1829		5	10½
3 March 1830		5	10⅜

Likewise, he recorded their weights on 14 December 1837: see Table 19.[274]

Table 19: Weights of Pinniger family members as at 14 December 1837

Name	*Weight*	
	Sco	*li*
Thomas Pinniger	8	5
Mrs Pinniger	7	16
Miss Broxholm	6	4
Mary Pinniger	4	4
Ann Pinniger	4	4
Thomas Pinniger	3	4
Martha Pinniger	2	4
Abbot Pinniger	No weight recorded	

p.22. Samuel and Emma Smith had seven children, WSA: 4381/2/1, *Pinniger genealogical memoranda book, c.1820-1943*, pp. 259-260, 279-280.

272 Henry Crook was the seventh child of Henry Crook and Ann Crook neé Pinniger (one of Thomas's younger sisters), born 2 October 1816, WSA: 4381/2/1, *Pinniger genealogical memoranda book, c.1820-1943*, p.22. Pen Farm may be either Lower Penn Farm in the parish of Hilmarton or High Penn Farm in the parish of Calne without. WSHC, *Wiltshire and Swindon Farmsteads and Landscapes Project: database extract.*

273 WSA: 4381/1/5, *Thomas Pinniger's farming diary, Beckhampton farm, Avebury, 1828-1832.*

274 WSA: 4381/1/6, *Thomas Pinniger's farming diary, Beckhampton farm, Avebury, 1832-1838.*

It is of note that the weights are given in scores and pounds, rather than stones and pounds: this might suggest that everyone was weighed on the same scales as the sheep and pigs.

Pinniger's daughter Martha was born on 9 December 1832, the first of their children to be born at Beckhampton: she was followed by their final child, Abbot Large, born on 8 July 1835. Three months later, on 28 October, there is an intriguing entry in the diary: *VAXINATED – the 2 time for safety Mary, Ann & Tom from their Brother Abbots Arm*. It is likely that this vaccination was for smallpox and it would appear that Abbot, at this stage aged only 3½ months, had the disease: their daughter Martha was not vaccinated, so it is possible that she had already contracted smallpox.[275] Abbot Large was christened at Lyneham church on 3 June 1836 but died from measles on 20 March 1839, aged 3¾ and was buried at Lyneham on 26 March.[276] In the entry for 10 October 1840, Pinniger records that Dr Bowen attended their daughter Mary, but left no hope: Mary died on 16 November, aged 16 years, being the third of their six children to die before reaching adulthood. In 1844, Thomas junior was described as very ill with Scarlet Fever, from which he clearly recovered.

In January 1837, a pencil note in the diary says that the influenza was universal: his labourers were ill such that there was nobody to work the thrashing machine resulting in no straw for the ewes. Later in the month, the following obituaries were recorded in the diary:

> 25th Obit Mr James Young Hayward of Tufton aged 32 Yrs, of the Influenza, now prevailing through the Kingdom, 3 Men died in this Parish the last 3 days, my Men ill have not Men to do my work (in great difficultys)
>
> Feby 4th Died of the Influenza, Calne Mr Jas Fry, Aged [...]

---

275 Developed by Edward Jenner of Berkeley, Gloucestershire in 1798, his vaccination method was to transfer the blister fluid from a person with the disease to a person who had not yet had smallpox: hence *from their Brother Abbots Arm*. See case VI in Edward Jenner, *An inquiry into the causes and effects of the variolae vaccinae* (London: Edward Jenner, 1798) where Jenner rubbed the variolous matter into two slight incisions made in her arm. There is no record in the diary as to when the children were vaccinated the first time, but it is of note that the accepted practice was for a second vaccination to be made *for safety*.

276 *Wiltshire, England, Church of England Births and Baptisms, 1813-1916 for Abbot Large Pinniger* [accessed via Ancestry, 3 July 2020]; *Wiltshire, England, Church of England Deaths and Burials, 1813-1916 for Abbot Large Pinniger* [accessed via Ancestry, 3 July 2020].

From this it is clear that the influenza was rife, and in some cases, resulted in death.

As for Thomas Pinniger himself, between the entries for 30 September and 1 October 1836, he made three entries between double lines in his diary: the first two were of deaths, whilst in the third he records: *TP_ Aged 56, mercifully spared a little time*, from which it is possible to conclude that he had been seriously unwell. Further, in the margin of the diary around 27 June 1837, there is the following note: *TPs accidt*, but there are no details of the accident or how it affected him. Accidents on farms were not uncommon in the early nineteenth century: the death of William Paradise was noted above (see *Looking after the land*) and in January 1840, Pinniger recorded an accident on a farm at Urchfont, where the farmer died having shattered his arm, when starting a new chaff cutting machine. Two accidents occurred at Beckhampton Farm when the meadow gate fell on John Rodbourne the carter's boy in 1844 and broke his thigh and the following year, young Rodbourne dislocated his shoulder.

As for his children's education, a note in the margin of the diary at the end of March 1837, states *30 Mary to school*: it does not say where she went to school, but she was aged 12. Similarly, on 15 May 1838, an entry in the diary notes that their son Thomas attended his first day at school with Mr Cornwell at Avebury: Thomas was nearly ten years old at this time.[277] There was clearly no urgency for the children to attend school at this time but by 1841, Thomas was attending Rev. Mr. Jacob's school.[278] Likewise, Martha's first day at Devizes School was on 28 March 1844, when she was aged 11.

Finally, an unusual entry was made in the diary on New Years Day 1838, as follows: *Mr P_ Cook WYETH, commenced Painting our Portraits*: the five portraits were completed by 14 January.[279]

277 William Cornwall, dissenting minister & school master, TNA: HO107/1185/2. *Census returns. 1841 census. Wiltshire Hundred: Selkley, including Parish: Avebury, ED1* [accessed via Ancestry, 6 July 2020].

278 Revd. William Bowman Jacob ran a gent's boarding school in Silver Street, Calne, J. Pigot, 'Commercial Directory 1842' in WRS47, p.106. In the 1851 Census, he is recorded as William Boreman Jacob, a clergyman without cure and a proprietor of a school in Silver Street, Calne, TNA: HO107/1837. *Census returns. 1851 census. Wiltshire. Registration District: 254, Calne. Registration Sub-district: 1 Calne, including Parish: Calne, ED2c* [accessed via Ancestry, 6 July 2020].

279 This is probably Peyton Cooke Wyeth (1811-1862) who was an American

Involvement in public life

It has already been noted that Pinniger was elected to the Marlborough Poor Law Union as guardian for the parish of Avebury in 1835, and appointed commissioner of the Beckhampton Turnpike Trust in 1836. Prior to this, he had entered the office of High Constable, presumably of the hundred of Selkley in which Beckhampton was located, on 22 April 1831. He recorded his attendance at the Marlborough sessions on 18 October 1831, but on 17 April 1832, he resigned his Office to James Looker of Winterbourne Monkton.

Finally, on 4 November 1839, Thomas Pinniger and John Wentworth measured the roads in the parish of Avebury: the recorded outcome is shown in Table 20.[280]

Table 20: Road measurement, Avebury, 1839

	Miles	*Furlongs*	*Chains*	*Links*
Hard Roads	3¾	0	1	95
Green Roads	2	0	5	51
	5¾	0	7	46
Church Road, hard Foot Path	0	3	7	97
	6	1	5	43

Church and charity

From 1830 onwards, Pinniger starts to make more reference to activities at his parish church of Avebury, Beckhampton lying within that parish. Examples include Good Friday that year and the general fast held two years later:

> 9 April 1830. Good Friday, All, Labourers wth Masons, Carpenters & Sawyers attend d at Church, Avebury
> 21 March 1832. General FAST, Mild fine day | Ploughing as yesterday,

portrait painter, known to be in Liverpool, England in 1835. *England, United Grand Lodge of England Freemason Membership Registers, 1751-1921 for Peyton Cooke Wyeth* [accessed via Ancestry, 3 July 2020].

280 There were 100 links to one chain; 10 chains to one furlong; 8 furlongs to one mile. One mile = 1760 yards; one furlong = 220 yards; one chain = 22 yards; one link = 7.92 inches, Colin Chapman, *How heavy?*, pp. 26-27.

> before & after Church service

Later in 1832, a diary entry provides evidence of the delivery of a national prayer:

> 18 November 1832. Sunday, dry day, FORM PRAYER, read in Church's for abundant Harvest.

This form prayer was one of a number of national prayers prescribed by the Church. The decision to give national thanks for an abundant harvest followed the particularly poor harvest in 1831. However, the text of the prayer made a connection between the good harvest, the decline in cholera and the end of social disturbances and political discontent (as had been seen in the recent Swing Riots), as being signs of divine intervention. There were no such national prayers for the good harvests of 1833 to 1836. The prayer was to be said in all churches, after the general thanksgiving at morning and evening services, on each of the three Sundays after 6 November 1832.[281]

As might be expected for a farmer of some 200 acres, there would have been a pew for Pinniger and his family in the church: this is confirmed when he records in October 1830 that the pew was finished, having taken one man eight days work. He clearly felt that there was a need for a new pew in the church and there is no reference in the diaries to any payment of pew rent.

From 1834 onwards, he starts to include more detail about church services in some of his entries. For example, on the death and burial at Lyneham of his mother-in-law, he recorded the details of the funeral sermon.[282] He also started to record the readings at church on Sundays, or the basis for the sermon, as for example:

> 2 August 1835. Sunday at Chippenham 4th Chapr Hebrews 9th Verse by the Revd Mr Morgan

From time to time, Pinniger noted the recovery from potential

281 Philip Williamson, Alasdair Raffe, Stephen Taylor and Natalie Mears, *National Prayers: Special Worship since the Reformation, Volume 2: General fasts, thanksgivings and special prayers in the British Isles 1689-1870 (Church of England Record Society 22)* (Woodbridge: Boydell Press, 2017), pp.798-800.

282 *England, Select Deaths and Burials, 1538-1991 for Mary Large* [accessed via Ancestry, 6 July 2020]. See diary entry for 13 March 1835.

farming disasters arising from the weather, attributing the satisfactory outcome to God, as exemplified below:

> 24 June 1837. MIDSUMMER=DAY, Sultry warm, the WHEAT in general just bursting the EAR, having made rapid progress since the RAIN the 11th -12 & 13th Inst, the universal opinion till lately was, that harvest would be a month later than usual, yet now forwarder than last year see 24th June 1836, the Crop of Field grass doubled since the rain with the heat = let us acknowledge and admire & bless the wonderful Power and goodness of the Almighty towards us, in giving us plenty when we expected a scarceity. this extraordinary effort in the Vegitable nature, is the Theme of every one, see the Newspapers May we ever again put our Trust in God

Likewise, he put his trust in God on 23 April 1838 after the severe damage caused by a long period of frosty weather:

> W P_ Ploughing at East Kennett where the Wheat Fail'd, being Kill'd by the Frost for sowing Oats, many other People have Plough'd their Wheat Field and sown Spring Corn, half my Wheat Field is nearly distroyed, will wait till the 6th of May KNOWING GOD ALMIGHTY can cause a great change and give me a good Crop in whom I trust, as he did (See the 2nd Kings 19th Ch 29th V) to the good King Hezekiah by giving him a second self sown Crop, spring up from the first.

From his diary entries, it would appear that he did not plough up his wheat, but instead, his wheat harvest was not finished until 2 October with the report of *a very bad Crop & bad Corn.* Similarly in 1840, after a severely cold and wet winter and early spring, the wheat harvest was finished on 4 September for which he gave thanks *to a bountifull Providence for an abundant crop.*

Pinniger also performed charitable acts such as on 30 December 1830 when he took barley to Devizes market and returned with 40 bushels of coal for the poor. Again, in 1836, he spent £50 on coals for the poor, sending four horses to Devizes to collect the coal and in 1839, 2 tons, 2¾ cwt of coal was brought on 10 January. Further, as a result of clearing the down to plant corn, he was able to distribute 300 furze faggots amongst the 20 poor families in the village and an additional 250 in March that year.[283]

283 Faggot - a bundle of sticks bound together as fuel, Lexico, *Faggot* https://www.lexico.com/definition/faggot [Accessed 30 April 2020].

At the end of November 1844, Thomas Pinniger devoted a whole page of his diary to an entry in which he paraphrased and personalised the twelfth chapter of the Book of Ecclesiastes, in which he reflected on his current state of health and his own mortality:[284]

> Oh remember now my Creator in my evil day now drawing nigh, my eyes growing dim with age (three score and five years) and my grinders no more (23th Novr 44) my slow feeding making me unfit for other mens Tables, and the white blossoms cover'g my head, the weight of the Grasshopper a burden unto me. | Oh remember the Silver cord will soon be broken, the Wheel also broken at the Cistern when no more water can be drawn, and soon shall I see all is vanity, and my dust will return to the earth as it was. | O that I fear'd God, and kept his Commandments

From 1846, his diary entries include a number of references to his son, Thomas Large Pinniger, who was starting to become more involved in the work of the farm, such as bean-hoeing by machine, and helping with the sheep where a younger man would be better able to handle them for jobs such as tailing, castrating, washing and shearing.

THOMAS LARGE PINNIGER

With his father recognising his own mortality, Thomas Large Pinniger (hereafter Thomas junior) started to write his own diary

284 1. Remember now thy Creator in the days of thy youth, while the evil days come not, nor the years draw nigh, when thou shalt say, I have no pleasure in them; 2. While the sun, or the light, or the moon, or the stars, be not darkened, nor the clouds return after the rain: 3. In the day when the keepers of the house shall tremble, and the strong men shall bow themselves, and the grinders cease because they are few, and those that look out of the windows be darkened, 4. And the doors shall be shut in the streets, when the sound of the grinding is low, and he shall rise up at the voice of the bird, and all the daughters of musick shall be brought low; 5.Also when they shall be afraid of that which is high, and fears shall be in the way, and the almond tree shall flourish, and the grasshopper shall be a burden, and desire shall fail: because man goeth to his long home, and the mourners go about the streets: 6. Or ever the silver cord be loosed, or the golden bowl be broken, or the pitcher be broken at the fountain, or the wheel broken at the cistern. 7. Then shall the dust return to the earth as it was: and the spirit shall return unto God who gave it. 8.Vanity of vanities, saith the preacher; all is vanity. 13. Let us hear the conclusion of the whole matter: Fear God, and keep his commandments: for this is the whole duty of man. King James Version, 12 Ecclesiastes, vv. 1–8, 13.

(Volume 9).[285] Starting in 1844, the entries in this volume are almost identical copies of the entries in his father's diary. Although Thomas junior had attended school since 1838, these early entries could be regarded as handwriting practice, or perhaps his father felt that it was important that his son should acquire the skills of diary writing for himself. That the entries in Volume 9 were indeed written by Thomas junior is confirmed on 17 January 1845 when the road widening at the foot of Tan Hill Piece was measured. Whilst Thomas senior wrote the following in Volume 7:

> Measured and witness'd by T P_ Junr and James Collier the banker

Thomas junior wrote in Volume 9:

> Measured by Father and witness'd by myself and James Collier the banker

As time went on, Thomas junior provided additional detail in his diary. As already seen (see *The Inn* above), the Inn was re-opened on 22 January 1847 by Mr Treen, with Mr Hazel remaining until October to sell for Treen: on 23 January, Thomas junior added in his diary that this was to save Mr Treen the expense of a licence.

Whilst the late-life entries in Thomas senior's diary set the initial tone for Thomas junior's entries in his own diary, he started to write more expansively later in the volume. However, there is no record of his father's death in his diary. Following his entry on 5 April 1847, the death is marked only by a blank page, with the entries thereafter starting at the top of the following page. Further, whilst Thomas junior records the life events of close family and those in the parish, he does not adopt the "black-border" style favoured by his father nor does he record life events beyond his immediate environment.

1847 ONWARDS

Thomas Pinniger had made his will on 5 April 1834, some thirteen years before his death. He appointed his brother-in-law Robert Large of Lyneham and his nephew John Crook of Corton (son of his sister Ann) as trustees and executors.[286] The bequests in his will were as follows:

285 WSA: 4381/1/9, *Thomas Pinniger's farming diary, Beckhampton farm, Avebury, 1844-1857.*

286 TNA: PROB11/2063/389, *Will of Thomas Pinniger, Gentleman of Avebury,*

£500 to his son, Thomas Large Pinniger, to be paid on him attaining the age of twenty-one;
His household furniture, plate, linen, china, books and all other household effects to his wife Mary until her decease: thereafter, to be split between the surviving children;
His messuages, cottages, inn, farm, lands, tenements and hereditaments in Beckhampton and Avebury along with his farming stock, agricultural implements, hay and corn, money and all other personal estate and effects to his trustees.

The trustees were to manage the farm and lands at Beckhampton with the assistance of his wife, until Thomas Large attained the age of 21. The will extended to seven sheets, the majority of which were concerned with the management of the trusts comprising monies, stocks, funds, and securities.

Whilst the last entry in Thomas Pinniger's diary was dated 20 March 1847, this was followed by the entries on the loose sheet of paper enclosed in Volume 8 (described in INTRODUCTION above).[287] The final entry on 1 April suggests that he was moving quantities of money: *500 fm Calne Bank | 700 - Check C P.*[288] Given the amounts in his will, it is reasonable to assume that these figures are pounds. It is possible that the sudden death of William Tanner on 18 March or his own declining health caused Pinniger to transfer these sums of money. With no indication of any illness or injury up to his final entry, he died just four days later on 5 April 1847.[289] The local newspapers reported his death at Beckhampton, aged 67, and noted that he was sincerely lamented by his friends and family.[290] He was buried on 12 April in Lyneham churchyard.[291]

His will was proved in London on 5 October 1847, by which time John Crook was the surviving executor. One week after probate was granted, his widow purchased a two acre close of land for £210 from the executors of William Tanner of Blackland House, Calne: the land was

---

Wiltshire, 5 October 1847.

287 WSA: 4381/1/8, *Thomas Pinniger's farming diary, Beckhampton farm, Avebury, 1847 Jan-Mar.*

288 C P – this is probably Christopher Pinniger.

289 WSA: 4381/2/1, *Pinniger genealogical memoranda book, c.1820-1943*, p.7.

290 Devizes & Wiltshire Gazette 8 April 1847; Salisbury & Winchester Journal 10 April 1847.

291 *Wiltshire, England, Church of England Deaths and Burials, 1813-1916 for Thomas Pinniger* [accessed via Ancestry, 9 July 2020].

occupied at the time by William Treen as tenant.[292] In the 1851 Census, Mary Pinniger was the head of household and described as 'Farmer of 213 acres employing 9 men, 3 boys & 7 women': Thomas [Large] (aged 22) was described as farmer's son, whilst Ann (24) and Martha (18) were described as farmer's daughters.[293]

Thomas Large Pinniger's diary ends on 31 December 1857. His mother Mary died on 18 August 1859, at her residence in Silver Street, Calne.[294] By the time of the 1861 census, he had married his first wife, his cousin Mary Large of Clevancy, and he was described as a farmer of 824 acres employing 23 men and 18 boys.[295] He had clearly expanded his farming enterprise beyond the 200 or so acres at Beckhampton – but that is another story, albeit not as well documented as that of his father.

CONCLUSION

What has become clear is that these diaries record three very different periods in Thomas Pinniger's life. Whilst residing at Little Bedwyn, he was farming some 800 acres but he did not own the farm. Further, he probably did not own the livestock or machinery: rather there appears to have been some sort of sharefarming arrangement in place between himself and Anthony Guy. Whilst he was living in *Mr Guy's farmhouse* in Chippenham, the diary entries become shorter and almost entirely focussed on the weather and the goings-on of others in the agricultural world. More detail appears in the diaries as he started to look for, and eventually purchased, another farm.

The final stage of his career and life was as the owner of a 200 acre farm in Beckhampton, along with one of the village inns. Here he was master of his own destiny: he owned the livestock and machinery and was selling his own produce. The diary entries become much more extensive, detailing building work, labour relations, his involvement in public life

292 WSA: 1171/144, *Conveyance of a close of land at Beckhampton to Mary Pinniger, widow. 1847.*

293 TNA: HO107/1838, *Census returns. 1851 census. Wiltshire. Registration District: 255, Marlborough. Registration Sub-district: 1 Marlborough, including Parish: Avebury, Hamlet: Beckhampton, ED1* [accessed via Ancestry, 12 May 2020].

294 Devizes & Wiltshire Gazette 25 August 1859.

295 TNA: RG9/1288, *Census returns. 1861 census. Wiltshire. Registration District: 255, Marlborough. Registration Sub-district: 1 Marlborough, including Parish: Avebury, ED2* [accessed via Ancestry, 12 May 2020]. WSA: 4381/2/1, *Pinniger genealogical memoranda book, c.1820-1943*, p.60.

as well as his direct involvement with the day-to-day management of the farm.

The farming diaries of Thomas Pinniger are an important source for the study of agricultural practices in the first half of the nineteenth century, in particular for two Wiltshire farms, much of which is not available elsewhere. However, they also contain much of interest for social historians and those researching meteorology, vernacular architecture and gardens. The range of content demonstrates the value of farming diaries to inform agricultural history in a local, detailed manner, which can then be set against other county or national accounts.

BIBLIOGRAPHY

Primary Sources (Manuscript)

The National Archives (TNA), Kew.

TNA: IR18/10895, *Tithe file for Avebury (parish), Wiltshire, 1836-1870.*

TNA: IR24/15, *Land Tax Redemption Office: Registers of Redemption Certificates. Registered numbers: Nos 6000 – 6999, 1799.*

TNA: IR24/16, *Land Tax Redemption Office: Registers of Redemption Certificates. Registered numbers: Nos 7000 – 7999, 1799.*

TNA: IR24/43, *Land Tax Redemption Office: Registers of Redemption Certificates. Registered numbers: Nos 31000-31999, 1799.*

TNA: PROB11/1140/178, *Will of William Kent, Gentleman of Little Bedwin, Wiltshire, 21 March 1786.*

TNA: PROB11/1738/65, *Will of Christopher Pinniger, Gentleman of Bremhill, Wiltshire, 7 March 1828.*

TNA: PROB 11/1835/77, *Will of Henry Crook, Gentleman of Hillmarton, Wiltshire, 22 August 1834.*

TNA: PROB11/2063/389, *Will of Thomas Pinniger, Gentlemen of Avebury, Wiltshire, 5 October 1847.*

TNA: PROB11/2065/124, *Will of George Neate, Farmer of Winterbourne Monkton, Wiltshire, 12 November 1847.*

Wiltshire and Swindon Archives (WSA), Chippenham.

WSA: 9/1/109, *Savernake Estate, Survey and valuation of farms near Savernake including Grafton, 1811-1812.*

WSA: 79b/9-12, *Title deeds, Kent family, 1575-1810.*

WSA: 79b/26, *Estate accounts, Kent family, 1792-1808.*

WSA: 79b/40-41, *Papers relating to the Kent family estate, 1790-1810.*

WSA: 248/195, *Terrier of Beckhampton Farm in Avebury, 1788.*

WSA: 315/27/5, *Copy will of William Kent of Little Bedwyn, 1785.*

WSA: 315/28, *Printed sale particulars relating to properties, including the manor of Little Bedwyn, 1809.*

WSA: 415/45, *Sales particulars for property in Bromham, Beckhampton, Avebury and Chippenham including The Beckhampton House Inn, Avebury Great Farm and Lower Sheldon Farm, 1796.*

WSA: 415/53/1, *Sale particulars relating to property in Chippenham, of Anthony Guy, bankrupt, with valuation, 1830.*

WSA: 939/1-9, *Farm account books: Wyndham family estates in Dinton, Teffont Magna and Sutton Mandeville, 1821-1850.*

WSA: 1033/89, *Deed of the manor house in Little Bedwyn and Jockey Farm and Little Bedwyn Farm, 1792.*

WSA: 1126/65, *Avebury tithe award and map (parish copy), 1845.*

WSA: 1171/144, *Conveyance of a close of land at Beckhampton to Mary Pinniger, widow. 1847.*
WSA: 1409/14/13, *Copy will of George Neate, farmer of Winterbourne Monkton (proved PCC), 1847.*
WSA: 1955/25, *Parish records, Little Bedwyn, 1730-1813.*
WSA: 2027/2/1/855, *Avebury Deeds, 1741-1907.*
WSA: 2811/3, *Sale particulars of freehold burgage house called Bridge House near the Bridge, Sambourne Farm, and other properties all in Chippenham, 1830.*
WSA: 4381/1/1, *Thomas Pinniger's farming diary, Little Bedwyn, 1813-1816.*
WSA: 4381/1/2, *Thomas Pinniger's farming diary, Little Bedwyn, 1817-1819.*
WSA: 4381/1/3, *Thomas Pinniger's farming diary, Little Bedwyn 1820-1823.*
WSA: 4381/1/4, *Thomas Pinniger's farming diary, Little Bedwyn and Chippenham, 1823-1828.*
WSA: 4381/1/5, *Thomas Pinniger's farming diary, Beckhampton farm, Avebury, 1828-1832.*
WSA: 4381/1/6, *Thomas Pinniger's farming diary, Beckhampton farm, Avebury, 1832-1838.*
WSA: 4381/1/7, *Thomas Pinniger's farming diary, Beckhampton farm, Avebury, 1839-1846.*
WSA: 4381/1/8, *Thomas Pinniger's farming diary, Beckhampton farm, Avebury, 1847 Jan-Mar.*
WSA: 4381/1/9, *Thomas Pinniger's farming diary, Beckhampton farm, Avebury, 1844-1857.*
WSA: 4381/2/1, *Pinniger genealogical memoranda book, c.1820-1943.*
WSA: A1/125/67, *Calendars of prisoners, 1841.*
WSA: A1/345/22A, *Land Tax Assessments, Beckhampton, 1782-1841.*
WSA: A1/345/22B, *Land Tax Assessments, Beckhampton, 1842-1884.*
WSA: A1/345/25, *Land Tax Assessments, Wilton, 1780-1881.*
WSA: A1/345/27, *Land Tax Assessments, Little Bedwyn, 1780-1831, 1869-1881.*
WSA: A1/345/99C, *Land Tax Assessments, Chippenham, 1816-1831.*
WSA: A1/345/190, *Land Tax Assessments, Grafton, East and West, 1780-1881.*
WSA: A1/345/392, *Land Tax Assessments, Tidcombe, Martin (in East Grafton) and Fosbury, 1780-1831, 1869-1881.*
WSA: D1/60/2/7, *Chippenham. Addition to churchyard, 1826.*
WSA: D1/62/4/1820/Pinniger, *Marriage Licence Bonds, 1820.*
WSA: EA95, *Avebury Enclosure Map, 1794.*
WSA: G19/1/53L, *Map of the parish of Chippenham by John Powell, 1784.*
WSA: P1/1823/70, *Will of Isaac Salter senior of Langley Fitzurze, Kington St Michael, clothier, 1823.*
WSA: TA/Avebury, *Tithe Award and map for the parish of Avebury, 1846.*

Museum of English Rural Life (MERL), accessed via the internet

MERL: FR WIL 11/2/4-6, *Farm account books of Aldbourne Farm, Baydon Farm and Warren Farm, Aldbourne, Co. Wilts, 1836-1856.*
MERL: FR WIL 11/2/18-20, *Account books of corn sold from Aldbourne Farm, Baydon Farm and Warren Farm, Aldbourne, Co. Wilts, 1831-1859.*
MERL: FR WIL 11/2/23, *Account book of sheep bought and sold for Aldbourne Farm*

and Warren Farm, Aldbourne, Co. Wilts, 1831-1879.

Accessed via Ancestry
TNA: HO107/1168/4, *Census returns. 1841 census. Wiltshire. Hundred: Calne including Parish: Calne, Township: Calne.*
TNA: HO107/1176/13, *Census returns. 1841 census. Wiltshire. Hundred: Elstub and Everley including Parish: Wroughton, Tything: Wroughton.*
TNA: HO107/1179/7, *Census returns. 1841 census. Wiltshire. Hundred: Kingsbridge including Parish: Lyneham or Lineham.*
TNA: HO107/1180/1. *Census returns. 1841 census. Wiltshire Hundred: Kinwardstone, including Parish: Great Bedwin, Township: East and West Grafton.*
TNA: HO107/1181/8. *Census returns. 1841 census. Wiltshire Hundred: Malmesbury, including Parish: Crudwell.*
TNA: HO107/1184/7, *Census returns. 1841 census. Wiltshire. Hundred: Potterne and Cannings including Parish: Potterne.*
TNA: HO107/1185/2. *Census returns. 1841 census. Wiltshire Hundred: Selkley, including Parish: Avebury.*
TNA: HO107/1185/15. *Census returns. 1841 census. Wiltshire Hundred: Selkley, including Parish: Monkton Winterbourne.*
TNA: HO107/1834. *Census returns. 1851 census. Wiltshire. Registration District: 251, Cricklade. Registration Sub-district: 1 Wootton-Bassett, including Parish: Lyneham or Lineham.*
TNA: HO107/1837. *Census returns. 1851 census. Wiltshire. Registration District: 254, Calne. Registration Sub-district: 1 Calne, including Parish: Calne.*
TNA: HO107/1838. *Census returns. 1851 census. Wiltshire. Registration District: 255, Marlborough. Registration Sub-district: 1 Marlborough, including Parish: Avebury, Hamlet: Beckhampton; Hamlet: West Kennet.*
TNA: RG9/1288, *Census returns. 1861 census. Wiltshire. Registration District: 255, Marlborough. Registration Sub-district: 1 Marlborough, including Parish: Avebury.*
TNA: RG9/1290, *Census returns. 1861 census. Wiltshire. Registration District: 256, Devizes. Registration Sub-district: 1 Bishops Cannings, including Parish: Stanton St Bernard.*
England, Select Births and Christenings, 1538-1975.
England, Select Deaths and Burials, 1538-1991.
England, Select Marriages, 1538-1973.
England, United Grand Lodge of England Freemason Membership Registers, 1751-1921.
Wiltshire, England, Births and Baptisms, 1813-1916.
Wiltshire, England, Church of England, Baptisms, Marriages and Burials, 1538-1812.
Wiltshire, England, Church of England Deaths and Burials, 1813-1916.
Wiltshire, England, Church of England Marriages and Banns, 1754-1916.
Wiltshire, England, Marriages, 1538-1837.

Secondary Sources

Alcock, N. W., M. W. Barley, P. W. Dixon and R. A. Meeson, *Recording timber-framed buildings: an illustrated glossary* (York: Council for British Archaeology,

1996).
Anon, 'The Alderney Cow' in *The Saturday Magazine*, August 1 1835, 47-48.
Aubrey, John, *The Natural History of Wiltshire: written between 1656 and 1691, with notes by John Britten* (London: Wiltshire Topographical Society, 1847).
Aynesley, J. M., 'Experiments on different manures for carrots', *Journal of the Royal Agricultural Society of England,* (1843), 270.
Banfield, Edwin, *Barometers: Wheel or Banjo* (North Curry: Baros Books, 1985).
Blaine, Delabere, *A Domestic Treatise on the Diseases of Horses and Dogs, 4th edition* (London: T Boosey, 1810).
Bloomfield, J. P., L. J. Brewerton and D. J. Allen, 'Regional trends in matrix porosity and dry density of the chalk of England', *Quarterly Journal of Engineering Geology*, 28 (1995), S131-S142.
Bowie, G. G. S., 'Northern Wolds and Wessex Downlands: Contrasts in Sheep Husbandry and Farming Practice, 1770-1850', *Agricultural History Review*, 38.2 (1990), 117-126.
British Geological Survey, *iGeology app – Beckhampton, Wiltshire* [accessed 9 April 2020].
Bromyard & District Local History Society, *A Pocketful of Hops: Hop growing in the Bromyard area, a new edition* (Bromyard: Bromyard & District Local History Society, 2007).
Brundage, Anthony, *The English Poor Laws, 1700-1930* (Basingstoke: Palgrave, 2002).
Cary, John, *Cary's New Itinerary* (London: G & J Cary, 1828).
Chapman, Colin, *How heavy, how much and how long?* (Dursley: Lochin Publishing, 1995).
Clarkson, L. A., 'The English Bark Trade, 1660-1830', *Agricultural History Review*, 22.2 (1974), 136-152.
Clater, Francis, *Every man his own Farrier* (Newark: J. Tomlinson, 1783).
Cobbett, William, *Rural Rides* (London: William Cobbett, 1830).
Cook, Hadrian and Tom Williamson (eds), *Water Meadows: History, Ecology and Conservation* (Macclesfield: Windgather Press, 2007).
Cowan, Michael, *Wiltshire Water Meadows* (East Knoyle: The Hobnob Press, 2005).
Culpin, Claude, *Farm Machinery, 8th edition* (London: Crosby Lockwood, 1969).
Dartnell, G. E. and Rev. E. H. Goddard, *Wiltshire Words: a glossary of words used in the county of Wiltshire* (Avebury: Wiltshire Life Society, 1991).
Davenport, Romola J., 'Urbanization and mortality in Britain, c.1800-50, *Economic History Review*, 73.2 (2020), 455-485.
Davies, Richard, Alan Petford and Janet Senior (eds), *The diaries of Cornelius Ashworth 1782-1816* (Hebden Bridge: Hebden Bridge Local History Society, 2011).
Davis, Thomas, *General View of the Agriculture of Wiltshire* (London: Board of Agriculture, 1811).
Eblex Ltd. *Target ewe fertility for better returns* (Huntingdon: Eblex Ltd, 2008).
Eden, Robert Henley, *A practical treatise on the Bankrupt Law, 2nd edition* (London: Joseph Butterworth & Son, 1826).
Egan, Pierce, *Walks through Bath: Describing every thing worthy of interest* (Bath:

Meyler & Son, 1819).
Encyclopaedia Britannica Company, *Encyclopaedia Britannica, 11th edition* (New York: Encyclopaedia Britannica, 1911).
Fidler, Henry, *Notes on Building Construction, Part III, Materials* (London, Oxford & Cambridge: Rivingtons, 1879).
Flood, Susan (ed.), *The diary of John Carrington, Farmer of Bramfield, 1798-1810: Volume 1, 1798-1804* (Rickmansworth: Hertfordshire Record Society, 2015).
Flood, Susan (ed.), *The diaries of John Carrington, Farmer of Bramfield, 1798-1810: Volume 2, 1805-1810 and John Carrington junior, 1810-1812* (Rickmansworth: Hertfordshire Record Society, forthcoming).
Gover, J. E. B., Allen Mawer and F. M. Stenton, *The Place-Names of Wiltshire (English Place-Name Society, Volume XVI),* (Cambridge: Cambridge University Press, 1939).
Griffin, Carl, 'The Violent Captain Swing?', *Past and Present*, 209 (2010), 149-180.
Griffiths, Elizabeth and Mark Overton, *Farming to halves: The hidden history of sharefarming in England from medieval to modern times* (Basingstoke: Palgrave Macmillan, 2009).
Gye, H., *Gye's Bath Directory, corrected to 1819* (Bath: H. Gye, 1819).
Haynes, Robert and Ivor Slocombe, *Wiltshire Toll Houses* (East Knoyle, The Hobnob Press, 2004).
Hobsbawm, Eric and George Rudé, *Captain Swing* (London and New York: Verso, 2014).
House of Commons, *Accounts and papers of the House of Commons, Volume 29* (London: House of Commons, 1833).
Idstone, 'The Alderney Cow' in *The Field,* 18 May 1872, 454.
Jenner, Edward, *An inquiry into the causes and effects of the variolae vaccinae* (London: Edward Jenner, 1798).
Johnson, Cuthbert William, *The Farmer's Encyclopædia, and Dictionary of Rural Affairs: Embracing All the Most Recent Discoveries in Agricultural Chemistry, Volume 1* (Philadelphia: Carey and Hart, 1844).
Kelly's Directories Ltd., *Kelly's Directory of Wiltshire* (London: Kelly's Directories Ltd., 1898).
Kerridge, Eric, 'The Floating of the Wiltshire Watermeadows', *Wiltshire Archaeological and Natural History Magazine*, 55 (1953), 105-118.
Lawrence, John, *The New Farmer's Calendar; Or, Monthly Remembrancer, for All Kinds of Country Business: Comprehending All the Material Improvements in the New Husbandry ... By a Farmer and Breeder* (London: Symonds, Vernon, Hood & Wright, 1800).
Lester, W., *A history of British implements and machinery applicable to Agriculture, with observations on their improvement* (London: Longman, Hurst, Rees, Orme and Browne, 1811).
Little, Edward, 'Farming of Wiltshire', *Journal of the Royal Agricultural Society of England*, V (1845), 161-180.
MacDermot, Edward Terence, *History of the Great Western Railway, volume 1* (London: Great Western Railway Company, 1927).
Marsh, A. E. W., *A history of the Borough and Town of Calne* (Calne and London: Robert S. Heath, 1903).

Mechi, John Joseph, *Letters on Agricultural Improvement* (Longman, Brown, Green and Longmans: London, 1845).
Mingay, G. E. (ed.), *The agrarian history of England and Wales, Volume VI, 1750-1850* (Cambridge: Cambridge University Press, 1989).
Mitchell, James, *Sketches of Agriculture, or Farmer's Remembrancer* (London; Baldwin and Cradock, 1828).
Moilanen, Mikko, Jorma Issakainen and Klaus Silfverberg, *Peat ash as a fertiliser on drained mires – effects on the growth and nutritional status of Scots Pine, Working Paper of the Finnish Forest Research Institute, 231* (Vantaa, Finland: Finnish Forest Research Institute, 2012).
Natural Environmental Research Council, *mySoil app – Beckhampton, Wiltshire* [accessed 9 April 2020].
Natural Environmental Research Council, *mySoil app – Little Bedwyn, Wiltshire* [accessed 17 April 2020].
Page, Mark (ed.), *The Pipe Roll of the Bishopric of Winchester 1301-2 (Hampshire Record Series Volume XIV)* (Winchester: Hampshire County Council, 1996).
Parslew, Patricia, *Beckhampton: Time present and time past* (Salisbury: The Hobnob Press, 2004).
Pearson, Robert, *Every man his own horse, cattle & sheep doctor* (Leicester: J Browne, 1811).
Randall, Adrian and Edwina Newman, 'Protest, Proletarians and paternalists: Social conflict in rural Wiltshire, 1830-1850' *Rural History*, 6.2 (1995), 205-227.
Raper, Anthony C., *The Ancient and Famous Weyhill Fair* (Andover: The Weydon Press, 2017).
Rogers, K. H. (ed.), *Early Trade Directories of Wiltshire (Wiltshire Record Society, Vol.47, indexed by J. H. Chandler)* (Trowbridge: Wiltshire Record Society, 1992).
Rolt, R. T. C., *Waterloo Iron Works: a history of Taskers of Andover, 1809-1968* (Newton Abbot: David & Charles, 1969).
Ryan, Carole, *Farm and rural building conversions: a guide to conservation, sustainability and economy* (Ramsbury: The Crowood Press, 2013).
Scotland's Rural College (SRUC), *Technical Note 652: Fertiliser recommendations for grassland* (Edinburgh: SRUC, 2012).
Shaheen, Sabry M, Peter S. Hooda and Christos D. Tsadilas, 'Opportunities and challenges in the use of coal fly ash for soil improvements – A review', *Journal of Environmental Management,* 145 (2014), 249-267.
Short, Brian, *The Battle of the Fields: Rural community and authority in Britain during the second World War* (Woodbridge: Boydell Press, 2014).
Short, Brian, Charles Watkins, William Foot and Phil Kinsman, *The National Farm Survey 1941-1943: state surveillance and the countryside in England and Wales in the second World War* (Wallingford: CABI Publishing, 2000).
Silverthorne, H., *Silverthorne's Bath Directory* (Bath: H. Silverthorne, 1846).
Sinclair, John, *The Code of Agriculture* (London: Sherwood, Neely and Jones, 1817).
The Asiatic Annual Register, Volume 7, 1807 (via Google Books).
The Asiatic Annual Register, Volume 8, 1809 (via Google Books).
The Farmer's Magazine, Vol. III, No.6, December 1835.
The Farmer's Magazine, Vol. V, No.2, August 1836.

The Farmer's Magazine, Series 2, Vol. 5, No. 5, May 1842.
The Farmer's Magazine, Series 2, Vol. 5, No. 6, June 1842.
The London Gazette, part 1, 1811 (via Google Books).
Thomas, Susan, *The Bristol Riots* (Bristol: Bristol Branch of the Historical Association of the University of Bristol, 1974).
Veale, Lucy and Georgina H. Endfield, 'Situating 1816, the "year without summer" in the UK', *The Geographical Journal*, 182.4 (2016), 318-330.
Wade Martins, Susanna and Tom Williamson (eds), *The farming journal of Randall Burroughes (1794-1799)* (Norwich: Norfolk Record Society, 1995).
Williamson, Philip, 'State Prayers, Fasts and Thanksgivings: Public Worship in Britain 1830-1897', *Past & Present,* 200 (2008), 121-174.
Williamson, Philip, Alasdair Raffe, Stephen Taylor and Natalie Mears, *National Prayers: Special Worship since the Reformation, Volume 2: General fasts, thanksgivings and special prayers in the British Isles 1689-1870 (Church of England Record Society 22)* (Woodbridge: Boydell Press, 2017).
Wiltshire and Swindon History Centre (WSHC), *Wiltshire and Swindon Farmsteads and Landscapes Project: database extract from Wiltshire Historic Environment Record of Wiltshire farms as recorded on the Second Edition Ordnance Survey maps* (Chippenham: WSHC, 2014).
Wolfensohn, Sarah & Maggie Lloyd, *Handbook of Laboratory Animal Management and Welfare, 3rd edition* (Oxford: Blackwell Publishing, 2003).
Worcestershire Farmsteads Project, *Building record for The Lingens Farm, Sledgemoor, Broadwas-on-Teme, Worcester. WR6 5NR* (Badsey: Worcestershire Farmsteads Project, 2020).
Yang, X. J., E. Albrecht, K. Ender, R. Q. Zhao and J. Wegner, 'Computer image analysis of intramuscular adipocytes and marbling in the longissimus muscle of cattle', *Journal of Animal Science*, 84.12 (2006), 3251.
Young, Arthur, *A six week tour through the southern counties of England and Wales, 2nd edition* (Dublin, Ireland: J. Milliken, 1771).
Young, Arthur, *The farmer's guide to hiring and stocking farms*, Vol. 1 (London: n.p., 1770).

Websites consulted

Agriculture and Horticulture Development Board (AHDB), *Grass+ Factsheet 8: Grassland liming* (Stoneleigh: AHDB, 2012) https://dairy.ahdb.org.uk/resources-library/technical-information/grass-management/grassplus-factsheet-8-grassland-liming/ [accessed 14 April 2020].
Agriculture and Horticulture Development Board (AHDB), *Opportunities for cover crops in conventional arable rotations: Information Sheet 41* (Stoneleigh: AHDB, 2015) https://ahdb.org.uk/cover-crops [accessed 7 July 2020].
Agriculture and Horticulture Development Board (AHDB), *Pd+ Factsheet 1 – Fertility terms and definitions* (Stoneleigh; AHDB, 2012) https://dairy.ahdb.org.uk/resources-library/technical-information/fertility/pdplus-factsheet-1-fertility-terms-and-definitions/#.XpRGRi-ZPus [accessed 13 April 2020].
BBC News, *Alderney cow: the breed Jerseys and Guernseys overshadowed* https://www.bbc.co.uk/news/world-europe-guernsey-20994463 [accessed 10 April

2020].

Beckhampton Stables, *The History of Beckhampton* https://rogercharlton.com/the-history-of-beckhampton/ [accessed 17 April 2020].

British Executions, *Henry Fauntleroy* http://www.britishexecutions.co.uk/execution-content.php?key=2364&termRef=Henry%20Fauntleroy [accessed 18 July 2020].

British Wool, *British Fleece Wool Price Indicator,* https://www.britishwool.org.uk/price-indicator [accessed 23 March 2020].

Craft Beer & Brewing, *The Oxford Companion to Beer definition of Chevalier (barley)* https://beerandbrewing.com/dictionary/ATZOtv7Wj1/ [accessed 30 April 2020].

Chettle, H. F., W. R. Powell, P. A. Spalding and P. M. Tillott, 'Parishes: Bishop's Cannings' in *A History of the County of Wiltshire*: Volume 7 (1953), 187-197 http://www.british-history.ac.uk/vch/wilts/vol7/pp187-197 [accessed 7 September 2018].

Clergy of the Church of England Database, *Buckerfield, Bartholomew* https://theclergydatabase.org.uk/jsp/search/index.jsp [accessed 22 July 2020].

Clergy of the Church of England Database, *Chippenham with Tytherton Lucas* https://theclergydatabase.org.uk/jsp/locations/index.jsp?locKey=1705 [accessed 20 July 2020].

Clergy of the Church of England Database, *Palmer, John (1799-1829)* https://theclergydatabase.org.uk/jsp/search/index.jsp [accessed 22 July 2020].

Davenport, Nigel, *NPK value of everything organic!* (Rochdale: The Nutrient Company, 2019) https://thenutrientcompany.com/blogs/horticulture/npk-value-of-everything-organic-database [accessed 15 April 2020].

Drovers, *Understanding wet hay* https://www.drovers.com/article/understanding-wet-hay [accessed 18 May 2020].

Historic England, *National Heritage List for England: Hermitage Cottages, 56-62 Curzon Street, Calne* https://historicengland.org.uk/listing/the-list/list-entry/1247123 [accessed 21 May 2020].

Horse & Hound, *Strangles* http://www.horseandhound.co.uk/tag/strangles#fzzIiHCL23gUIEKE.99 [accessed 1 May 2020].

IUCN Red List of Threatened Species, 2020-1, *Crimson Clover, Trifolium incarnatum* https://www.iucnredlist.org/species/176390/7231548 [accessed 12 May 2020].

King's Agriseeds Inc., *Nurse Crops* https://kingsagriseeds.com/wp-content/uploads/2014/12/Nurse-Crops.pdf [accessed 7 July 2020].

Lexico, https://www.lexico.com [accessed on various dates for definitions of various words]

Limagrain UK Limited, *The essential guide to forage crops* https://www.lgseeds.co.uk/uploads/Forage-Brochure_Singles.pdf , p.5 [accessed 23 March 2020].

Measuring Worth, https://www.measuringworth.com/calculators/ukcompare/relativevalue.php. [accessed on various dates].

McIntyre, Anne, *The Complete Floral Healer,* Gaia Books, 1996 in Positive Health Online, *A new look at Daisy (Bellis perennis)* http://www.positivehealth.com/article/herbal-medicine/a-new-look-at-daisy-bellis-perennis [accessed 19 March 2020].

Morton, Tom, Nigel Copsey and Rebecca Little, *Earth Mortars* http://www.buildingconservation.com/articles/earth-mortars/earth-mortars.htm [accessed 30 April 2020].

National Galleries Scotland, *The Eye of Princess Charlotte of Wales* https://www.nationalgalleries.org/art-and-artists/63697/eye-princess-charlotte-wales-1796-1817 [accessed 21 March 2020].

National Sheep Association, *Dictionary* http://www.nationalsheep.org.uk/know-your-sheep/dictionary-corner/ [accessed 15 January 2018].

Oxford Dictionary of National Biography, *Cobbett, William* https://doi.org/10.1093/ref:odnb/5734 [accessed 27 July 2020].

Oxford Dictionary of National Biography, *Hunt, Henry* https://doi.org/10.1093/ref:odnb/14193 [accessed 27 July 2020].

Oxford Dictionary of National Biography, *Somerset, Lord Granville Charles Henry* https://doi.org/10.1093/ref:odnb/26008 [accessed 27 July 2020].

O'Dwyer, Tom (ed.) *Teagasc Advisory Newsletter, Dairy* (Carlow, Ireland: Teagasc, November 2018) https://www.teagasc.ie/media/website/publications/2018/DairyNews_November2018.pdf [accessed 13 April 2020].

State Library of Queensland, *British convict transportation register – John Pinniger* https://convictrecords.com.au/convicts/pinniger/john/118790 [accessed 15 April 2020].

The Workhouse, *Andover, Hampshire* http://www.workhouses.org.uk/Andover/ [accessed 28 April 2020].

The Workhouse, *Highworth and Swindon, Wiltshire* http://www.workhouses.org.uk/Highworth/ [accessed 20 April 2020].

The Workhouse, *Marlborough, Wiltshire* http://www.workhouses.org.uk/Marlborough/ [accessed 20 April 2020].

UK Government, History, *Past Prime Ministers, George Canning* https://www.gov.uk/government/history/past-prime-ministers/george-canning [accessed 12 May 2020].

UK Parliament, *About Parliament, Living Heritage, The building & its collections, The Palace of Westminster, Architecture of the Palace, The Palace's structure* https://www.parliament.uk/about/living-heritage/building/palace/architecture/palacestructure/ [accessed 27 July 2020].

Walford, Edward, 'Blackfriars Road: The Surrey Theatre and Surrey Chapel', in *Old and New London*: Volume 6 (London, 1878), pp. 368-383 *British History Online* http://www.british-history.ac.uk/old-new-london/vol6/pp368-383 [accessed 24 July 2020].

Websters New World College Dictionary, *Tod* http://www.yourdictionary.com/Tod [accessed 1 May 2020].

EDITORIAL NOTE

Pinniger's handwriting was even and legible and the entries are generously spaced on the page, albeit heavily abbreviated. In cases where he wished to be more expansive, such as for the death of family members or examples of extreme weather, the diary was turned through 90 degrees and the whole page was used. These are indicated in the text by [Diary turned landscape for this entry] or similar note. When recording deaths, funeral services, funeral sermons, and similar "religious" events, three major styles of emphasis have been variously used in the diaries for such events: within a thick black border on all four sides; within a partial thick black border on two or three sides, and between two thick black lines. These are indicated in the text by [This entry written within a thick black border] or similar note.

Within the diaries, Pinniger has inserted a number of tables (for example, wool weights recorded when he weighed his wool). The various tables have been put into appendices in order to maintain the flow of the writing.

On the whole, the spelling is much as one would expect today. However, double letters in the middle of words are often absent, such as draging, rather than dragging, and cuting rather than cutting. There are two particular spellings which persist in the diary: sweeds rather than swedes (usage ended between 1818 and 1819) and remainer rather than remainder. Except where quotations are taken from reference documents, or contained in archive catalogue references, modern-day spellings of place names and surnames have been adopted: for example, Bedwyn rather than Bedwin and Merriman rather than Merryman.

The following editing conventions have been used in transcribing these diaries:

- Editorial notes and comments on the text (for example, explanation regarding missing entries in the diaries) are printed in normal type and enclosed in square brackets;
- Where appropriate, footnotes have been provided to explain an entry in the diaries;
- To provide clarity with respect to the use of ditto contractions and ditto marks, these have been replaced with the actual repeated

words;

- Where brief extracts from the diaries have been cited within sentences in the Introduction, these are presented in italics.
- Words in a bolder hand have been capitalised;
- Given the narrow width of the pages in the diaries, the original layout has not been preserved. Entries for each day are headed with the cardinal day of the month and entries within a single day are separated with a vertical stroke;
- The spelling, capitalisation and punctuation of the manuscript have all been preserved: however, double underlining has been replaced with single underlining;
- An asterisk (*) placed at the end of an entry indicates that some or all of the entry is highlighted with single, double or shaded lines either in, or just outside, the margin;
- In a very few instances, punctuation (a space or comma) has been added silently to improve clarity;
- Any erasures or alterations in the diaries have been retained in that they may illuminate Pinniger's thought process: words erased or crossed through in the manuscript are indicated in the transcription by being enclosed in <>;
- Any words which have been overwritten have been treated as an erasure, transcribed within <> and the correction transcribed as normal text;
- Where whole words have been missed out in the manuscript, these are indicated by [...] with a space on either side and where parts of words are missing, these are indicated by [...] but with no space between. In some cases, it is obvious what the omission was intended to be – in others it is less so. No attempt has been made to conjecture such missing words or letters, leaving the reader to come to their own conclusions;
- Contraction of words have been rendered as in the diaries – in particular, words ending in "ing" often end in "'g" or "g" and words ending in "ed" often end in "'d" or "d" and these have been rendered as written: contracted words have only been extended to clarify meaning, where necessary;
- Words including superscriptions have been converted to abbreviations – for example, wth becomes wth;
- Abbreviations have not been expanded – for example wth is not expanded to with;
- Pinniger developed an idiosyncratic and inconsistent method of

abbreviation: thus a word might be shortened in three ways – for example, folded could appear as foldd, fold d and fold'd;

- There are a number of occasions where Pinniger has opened a bracket, but has failed to close it – these have not been corrected;
- In a very few instances, intended insertions (originally marked by * or X) have been made silently to provide clarity;
- Where currency values are given, a number of formats have been used in the diaries: for consistency in the transcription, the following format has been adopted - £5 5s 0d;
- Fractions are represented by the appropriate fraction character (for example, 3/4 is represented by ¾);
- The long s (f) has been converted to a single s - thus 'fs' in the manuscript is transcribed as 'ss'.

LIST OF ABBREVIATIONS

The following abbreviations are to be found in this volume.

A	Acre
B, Bu, Bush	Bushel
Bot	Bought, bottom
C, €, Cwt, €wt	Hundredweight
Cl, Clo	Clock
d	Pence (penny – singular)
Doz	Dozen (12)
ea	Each
Ef, EF, Efield	East Field
f	Field, furlong
fm	from
fu	Funeral
fu, fur, furl, furl'g	Furlong
Geoe	George
Gro'd, gd, grd	Ground
Gt	Great
Hd	Hundred, head
Jno	John
Kid	Kidney (as in beans)
Ld	Load
li	Pound
m	Thousand
md	Mead(ow)
Nf, NF, Nfield	North Field
Nt	Net
Qr, qr, Qur	Quarter
R	Rood
Robt	Robert
Rt	Robert, right
S	Sack
S, Sc, Sco	Score – either number (20) or weight (20 lbs)
s	Shilling
Sf, SF, Sfield	South Field
Sk	Stock
Su, SU	Summer
Tithg	Tithing
Upr, up	Upper
W	Week, Winter
Whe	Wheat

GLOSSARY

Acre	Measure of area (of land) comprising 4,840 square yards. One acre comprises 4 roods or 160 perches. Abbreviated as A.
Arial	Variety of swede.
Athwart	As in "to plough athwart" – that is, to plough land a second time, at right angles to the first ploughing, so as to clean it more thoroughly. See Thwarting.
Awn	A stiff bristle, especially one of those growing from the ear or flower of wheat, barley, rye, and many grasses.
Badlodged	See Lodged – a bad case of lodged.
Bait	To feed horses and cattle on green herbage, clover &c.
Baiting	Green food for cattle. Also Green Bait.
Barking	Barking refers to de-barking or bark stripping of felled timber.
Bavin	An untrimmed brushwood faggot.
Beer	Also spelt Bear, Bere or Bae. A kind of barley hardier than the ordinary kind but of inferior quality. Ordinary barley has two rows of grain on the head, bear four. It is also called Big or Bigg.
Berry	The grain of wheat.
Bigg	Or Big. See Beer.
Black Siberians	A variety of oats. 'The Siberian black oat is a very large, long grain, but liable to drop out in the field: it requires as good land as the white oat; but, from its colour, is not quite so valuable. From the large size of the grain, it is necessary to sow five or six bushels per acre. It is sown early, and is a very forward sort'.
Blighted	Wheat, of which the stalk is a little twisted and ricketty, the blade being of a bluish green and curled up, the grain also is green and tubercled (warty appearance).
Blighting	Of rain, which has the potential to result in the blighting of wheat.
Brimmed	This refers to when the boar has mated the sow.
Brimming	This refers to when a sow is ready for mating or when the boar is mating with the sow.
Brit	To drop out of the husk, as over-ripe grain.
Britted	Of crops, where the over-ripe grain has dropped out of the husk.
Bush	Noun. A heavy hurdle or gate, with its bars interlaced with brushwood and thorns, which is drawn over pastures in spring, and acts like a light harrow. Verb. To bush-harrow a pasture.
Bushel	Measure of volume (of grains). Abbreviated as B or Bu.
Cage	Lambs' cages are cribs for foddering sheep in the fold: usually

	made semi-cylindrical with cleft ash poles about six or seven feet long, and about one foot diameter.
Carriage	The main carriage provides the feed water from the river to the water meadow system.
Cast	A sheep that is on its back and is unable to get up again unassisted; An older breeding ewe (also known as a draft ewe) which is sold off; Of ewes, to abort their lambs.
Catslide	Roof which covers one side of the main slope and an outshut in one continuous slope.
Cavings	[Also cave, cavin or keavin] The chaff of wheat and oats; in threshing, the broken bits of straw.
Chevalier	A variety of barley.
Chilver	A ewe lamb.
Churly	Dry, stiff, hard, as applied to the soil.
Ciderkin	The washing after the best cider is made.
Clog-weed	Hog weed, also known as cow parsnip in North Wiltshire, *Heracleum sphondylium*.
Cock	See Shock.
Cooler	A large open tub.
Couch	Also cooch. Couch grass or twitch grass, *Elymus repens*.
Couching	The removal of couch from the ground.
Cul	Also cull. Sheep or lambs picked out of the flock, as inferior in one way or another (size, lack of teeth, barren etc.).
Cut	1. Removal of the tail of a lamb with a knife; 2. Castration of a ram lamb, also with a knife.
Dantzic	Dantzic timber is a variety of northern pine grown chiefly in Prussia, taking its name from the port through which it was shipped. It is strong, easily worked and durable if well seasoned: baulks are of 18 to 45 feet long and generally 14 to 16 inches square.
Dibble	A pointed hand tool for making holes in the ground for seeds or young plants.
Distrain	Seize (someone's property) in order to obtain payment of rent or other money owed.
Drag	Alternative name for a harrow.
Dragging	Also draging. Harrowing.
Drug	To drug timber - to draw it out of the woods under a pair of wheels; To drug a wheel - to put on some kind of drag or chain.
East Field	One of the common fields, located to the east of the settlement. Abbreviated as Ef, EF or Efield.
Elm	Small bundles of fresh straw, damped and laid out straight for the thatcher's use. Also Helm or Yelm.
Faggot	A bundle of sticks bound together as fuel.
Fallowing	A frequent ploughing and pulverising of land to make it lighter,

	and clean from weeds when it has become foul by repeated crops.
Fellmonger	A preparer of skins or hides of animals, especially sheepskins, prior to leather making.
Flakes	Frames, barred with ash or willow spars, somewhat resembling a light gate, used as a hurdle where extra strength is needed. Flake hurdles are used to divide a field, or for cattle, the ordinary sheep hurdles being too weak for the purpose.
Flisk	A slight shower.
Fodder	Feedingstuff for cattle or sheep.
Fother	Alternative form of fodder; fothered – foddered.
Four-tooth or 4-tooth	In sheep, when the second pair of permanent teeth present at 21-24 months old. Also, from shearing time after two years old.
Furlong	Measure of length equal to 220 yds or 660 ft. Also used to describe a strip of newly-ploughed land lying between two main furrows. Abbreviated as fur, furl or furl'g.
Full mouth	In sheep, a complete set of 8 permanent teeth present at 42-48 months old – also known as whole mouth. Also, from shearing time after four years old.
Georgians	A variety of oats.
Gray or grey peas	A late-ripening variety of field peas. See Small Grays.
Green Bait	See Baiting.
Green Rounds	A variety of turnips, also known as Green Norfolk or Common Green Top White. These turnips have a regular round shape, flattened but not much hollowed on the upper and under surface – the upper surface is green coloured and the lower surface is white. This variety is hardier than the Norfolk White or Norfolk Red.
Griddle	To put peas, beans or oats through a machine with iron rollers to crack them for animal feed.
Griddled	Cracked.
Grip (or Gripe)	Noun. A grip of wheat is the handful grasped in reaping: it is then laid down in gripe when laid ready in handfuls untied; Verb. To grip wheat is to divide it into bundles before making up the sheaves.
Grittle	See Griddle.
Grittled	See Griddled.
Grounded	Also Pitched. Offered for sale.
Guinea	Unit of currency equal to one pound and one shilling (£1 1s 0d).
Hack	To hack is to loosen the earth round potatoes, preparatory to earthing them up; to hoe.
Hacking	Hoeing.
Hatch	Noun. A wallow or line of raked-up hay. Also a mechanism for controlling the flow of water in a water-meadow. Verb. 'To hatch up' - to rake hay into hatches.
Helm	See Elm.

Hogshead	54 gallons or 432 pints.
Hop and ray	Hop clover and ray-grass, sown together, a very common and very good custom; hop clover is yellow-flowering trefoil, or nonsuch; ray-grass is taken to mean ryegrass.
Hundredweight	Measure of weight comprising 4 quarters (Qr) or 112 pounds (li). There are 20 cwt in a ton. Abbreviated as C or Cwt.
Hunger-bane	To starve to death.
Iron Presser	This is probably the forerunner of the modern-day furrow press which follows (or is used after) the plough: diary entry for 28 June 1832 refers to "the 10 wheel Iron Presser".
Jag	The awn and head of the oat. Oats are spoken of as 'well-jagged', 'having a good jag', 'coming out in jag'.
Keep	Feed for animals.
Kid	The cod or pod of peas, beans &c.
Kin	See Ciderkin.
Ladies Slates	These are slates which measure 16 inches by 10 inches.
Lease	To glean, Also Leaze.
Leaze	See Lease.
Lodged	Of wheat (and other cereal crops) laid or beaten down by wind or rain, making it difficult to get dry and to reap or mow.
Long dung	Fresh unrotted dung with straw therein.
Makeout	The value of growing crops, along with acts of husbandry in terms of the labour involved in land preparation (ploughing, harrowing, drilling), manures and fertilisers applied, seeds sown etc., thereby providing compensation to the outgoing farmer for the time and materials expended, which provide the incoming farmer with crops to harvest.
Marter earth	Earth for the mortar used in the construction of buildings. From prehistory until the 19th century, earth was used in a variety of ways to construct buildings of every type across Britain. Typically, earth mortars are made from local natural subsoil, where clay minerals act as the binder to sand and silt particles. In such mortars, quicklime was added to the earth: analysis of historic earth mortars indicates the addition of around five per cent quicklime, which will have doubled in volume with slaking.
Memel	Memel timber is very similar to that from Dantzic, but is not as strong and the baulks are generally 13 to 14 inches square. It is also named after the port through which it was shipped.
Mizzle	Noun. Dialect word for light rain used in various counties, including Wiltshire and Yorkshire. Verb. To produce light rain, as in *It is mizzling*. Adjective (mizzley). Description of light rain.
Mow	The storage bay to either side of the threshing floor in a traditional barn: the purpose of the mow was to provide storage for the unthreshed sheaves and likewise for the threshed straw.
Muckle	Noun. Long straw dung from the stable which was the

	preferable manure for turnips; Verb. To apply muckle on a field.
Murrain	Unspecified and unexplained infectious diseases of cattle or other livestock. Also Murrin.
Murrin	See Murrain.
North Field	North Field – one of the common fields, located to the north of the settlement. Abbreviated as Nf, NF or Nfield.
Oldlay	Permanent pasture.
Outshut	Subsidiary compartment at the side or end of a house or barn, under a roof: this is unlike an aisle which is open to the body of the building.
Pence	Used in currency and indicated with d. There are 12 pence in one shilling and 240 pence in one pound (currency). Singular – penny.
Perch	Linear measure: 16½ feet or 5½ yards - see also Pole and Rod. Square measure: Measure of area (of land) comprising 30.25 square yards (equivalent to an area 5.5 yards in both directions). There are 40 perches in one rood and 160 perches in one acre. Abbreviated as P.
Pip	The bud of a flower.
Pipe stave	Each of the staves which are hooped together to make a cask or barrel.
Pitched	See Grounded.
Pitchin[g]	Whilst paving is carried out using large flat stones; 'pitching' uses small uneven stones set on edge.
(Hop) Pocket	Large strong jute sack, measuring some 6 feet long by 3 feet wide, holding about 1½ cwt of hops when full.
Pole	Linear measure: 16½ feet or 5½ yards - see also Perch and Rod.
Pot dung	Noun. Rotten dung (which with the sheep-fold) is the preferable manure for wheat. Verb. To apply pot dung on the field.
Pound	Measure of weight comprising 16 ounces. Abbreviated as li; Used in currency, usually indicated with £, sometimes as li. One pound comprises 20 shillings or 240 pence.
Quarter	Measure of weight comprising 28 pounds (li) Measure of volume: one quarter equals 8 Bushels: in weight terms, one quarter of Barley weighs 4 cwt and one quarter of malt weighs 3 cwt. Abbreviated as Qr or qr.
Raftering	Ploughing so as to leave a narrow strip of ground undisturbed, turning up a furrow on to it on each side, thus producing a succession of narrow ridges.
Ram Stag	A gelded (castrated) ram.
Raves	Waggon rails, sometimes applied to the flat woodwork projecting over wheels from side of forward part of waggon.
Ray	See Hop and Ray.
Ridgling	A horse which has not been 'clean cut', that is, is only half gelded, owing to one of its stones never having come down.

Rig	See Ridgling
Roarer	A horse which produces the sound of turbulent air within its larynx when worked hard.
Rod	Linear measure: 16½ feet or 5½ yards - see also Perch and Pole.
Rood	Measure of area (of land) comprising 1,210 square yards. One rood comprises 40 perches: there are 4 roods in one acre. Abbreviated as R.
Rough	To make rough, applied to horses' shoes when they are made rough to prevent them slipping in frosty weather.
Rowet-grass	The long rough grass in hedge &c. which cattle refuse; rowan or coarse aftergrass.
Rudder	A sieve.
Rudderings	Sievings - see Rudder.
Ruddle	Red ochre paste applied to chest of ram which marks rump of ewe when the ram mates with her.
Rugget	Ridged.
Scoop	A shovel.
Score	Measure of weight comprising 20 pounds. Abbreviated as S, Sc or Sco.
Scour	(In livestock) Noun. Diarrhoea; Verb. To suffer from diarrhoea.
South Field	One of the common fields, located to the south of the settlement. Abbreviated as Sf, SF or Sfield.
Shilling	Used in currency and indicated with s. There are 20 shillings in one pound and 12 pence in one shilling.
Shock	Of wheat (or other grains), consisting of several sheaves set up together for carrying. The number of sheaves was formerly ten [also known as a tithing or tething], for the tithing man's convenience, but now varies considerably, according to the crop. Also cock.
Shore	A broken place or gap in a fence, e.g. Shepherds' Shore in Wansdyke.
Six-tooth or 6-tooth	In sheep, when the third pair of permanent teeth present at 30-36 months old. Also, from shearing time after three years old.
Skilling	Cowhouse.
Small Grays	A variety of pea. For grey, as opposed to white or green, peas " ... the best time of sowing is about the beginning of March when the weather is pretty dry; for if they are sown in a very wet season, they are apt to rot, especially if the ground be cold".
Spit Dung	That sort of manure which has undergone complete fermentation or putrefaction and is reduced into a somewhat earthy state, so as to be dug or taken up by the spade or shovel in a sort of spit manner.
Squinting	Ploughing land for a third time, working in a diagonal direction to the first and second ploughings.
Stean	To stean a well is to line its sides with stone.
Store	Generally animals sold or bought for fattening.

Strangles	One of the most common equine diseases in horses in the UK. It is a highly contagious infection of the upper respiratory tract caused by the bacteria *Streptococcus equi subspecies equi*.
Talavera	A variety of spring wheat: by 1842, it was not considered to be as prolific as other varieties, in that it comes late to harvest and is liable to be damaged from wet during harvest.
Tankard	White, Green and Red Tankards are all varieties of turnip. They are unsuitable for winter feeding, in that more than half the root is above the ground surface, and thereby exposed to frost: however, they generally mature earlier than other turnips. These turnips have oblong, or tankard-shaped roots which are often bent or crooked. The White Tankard is the largest of the tankards, but is also softer in texture than either the Red or Green Tankards. The colour of these latter varieties describes their colour, although they are white under the ground.
Tasker	A tramping harvester or casual labourer who works by the piece.
Taunton Wheat	A variety of wheat. "With respect to different species of wheat, the grand divisions are into white, red, and bearded; the sub-divisions, or varieties, are numerous and very unimportant, in any point of view. The best in England, according to my experience are, the Essex and Kentish white and red, and the Taunton Wheat (Somerset) which appears a mixture of both".
Teg	A sheep in its second year.
Thiller	The shaft-horse of a team.
Throwing	Felling (of timber).
Thwarting	Probably therting, to plough land a second time, at right angles to the first ploughing, so as to clean it more thoroughly. Also see athwart.
Tie	Part of a field, which has been temporarily divided up in order to manage the effective grazing of the crop.
Time out	[Of livestock] Also time up. The date of the end of a pregnancy, when an animal was due to give birth.
Tod	A former English weight for wool, about 28 pounds.
Truss	[of straw] 36 pounds; [of old hay] 56 pounds; [of new hay] 60 pounds.
Tuck	"In a tuck and go" - in convulsions; "All of a tuck" - with nervous hasty movements.
Tuffet	A tuft of grass: large tufts of grass which cannot be broken up by harrowing and prevent sowing of seed. Also Tuftet.
Tuffetey	Ploughed land with tuffets remaining after ploughing and harrowing, resulting in an uneven surface, unsuitable for sowing or drilling.
Tuftet	See Tuffet.
Two-tine	'2 tine the Harrows' means harrowed twice.
Two-tooth	In sheep, when the first central pair of permanent teeth present

or 2-tooth	at 12–18 months old. Also, from shearing time after one year old.
Tything	Also tithing or tething. A shock of ten sheaves, for convenience in tithe-taking.
Vetches	Vetches (Latin name *Vicia sativa*) were also known as tares. "If intended to be sown in autumn, to stand the winter, care should be had to obtain the proper seed; there being two sorts of tares in the market, one called the *winter tare*, which should only be sown in autumn, and the other the *spring tare*, which would be destroyed in winter were it sown in autumn". Modern seed suppliers offer winter vetch and summer vetch seed (both *Vicia sativa)*, the latter described as not winter hardy.
Wake	Noun. The raked-up line (broader than a hatch or wallow) of hay before it is made up into pooks; Verb. To rake hay into wakes.
Wallow	Noun. A thin line of hay; Verb. To rake hay into lines.
Waned	Weaned, as of lambs.
Wether	In sheep, those in their first year from around Christmas to the first shearing.
Whelm	To engulf, submerge or bury.
White-eared wheat	Wheat with long awns.
Whole mouth	See full mouth.
Wort	The infusion of ground malt or other grain before fermentation, used to produce beer and distilled malt liquors.
Yean	(of a sheep or goat) give birth to (a lamb or kid).
Yelm	See Elm.

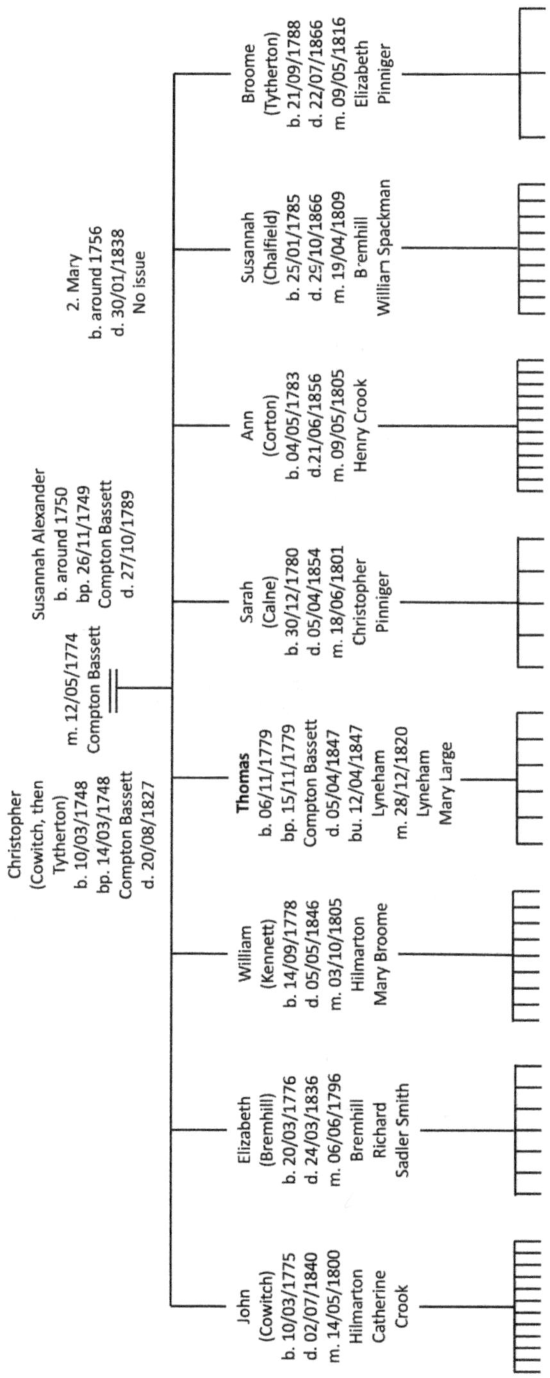

Key: b. – Born; bp. – Baptised; m. – Married; bu. – Buried
Sources: WSA 4381/2/1, otherwise accessed via Ancestry, as described in the Introduction.

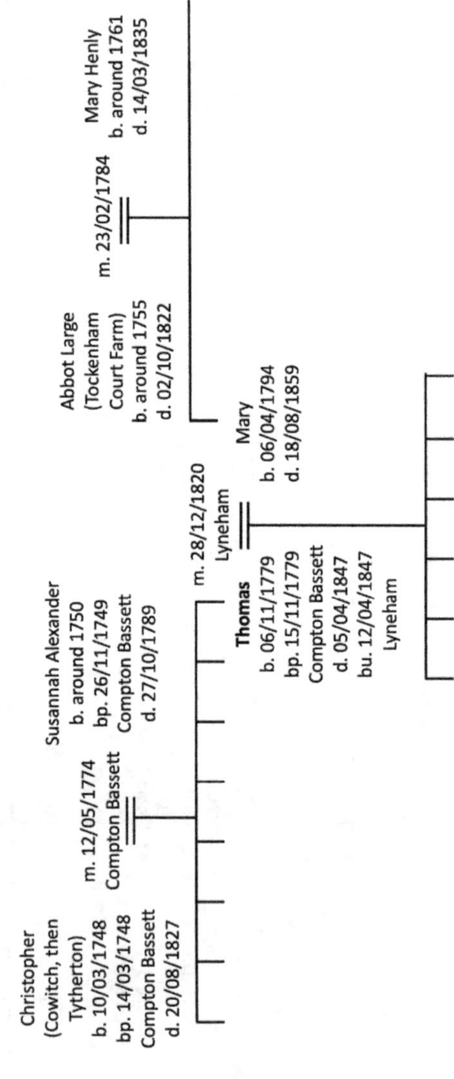

Key: b. – Born; bp. – Baptised; m. – Married; bu. – Buried
Sources: WSA 4381/2/1, otherwise accessed via Ancestry, as described in the Introduction.

Children of Thomas Pinniger and Mary Pinniger (neé Large)

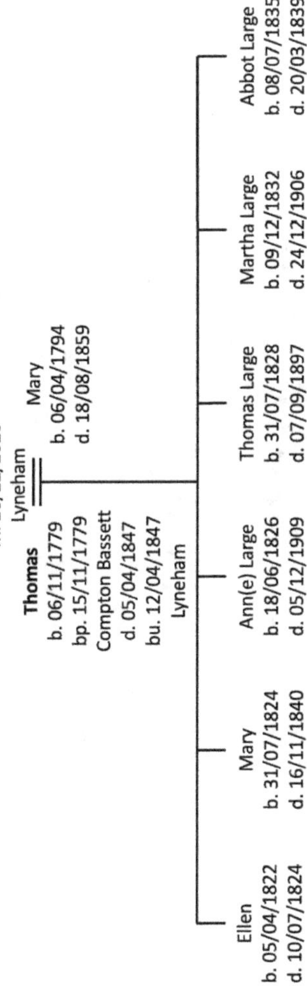

Key: b. – Born; bp. – Baptised; m. – Married; bu. – Buried
Sources: WSA 4381/2/1, otherwise accessed via Ancestry, as described in the Introduction.

ILLUSTRATIONS

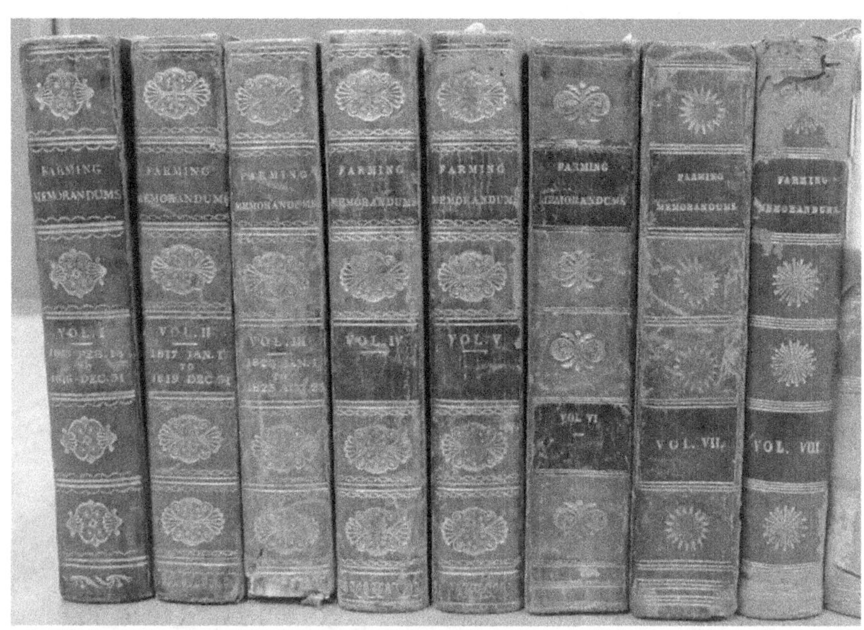

Figure 1: The farming diaries of Thomas Pinniger - eight volumes
Source: WSA: 4381/1/1-8.

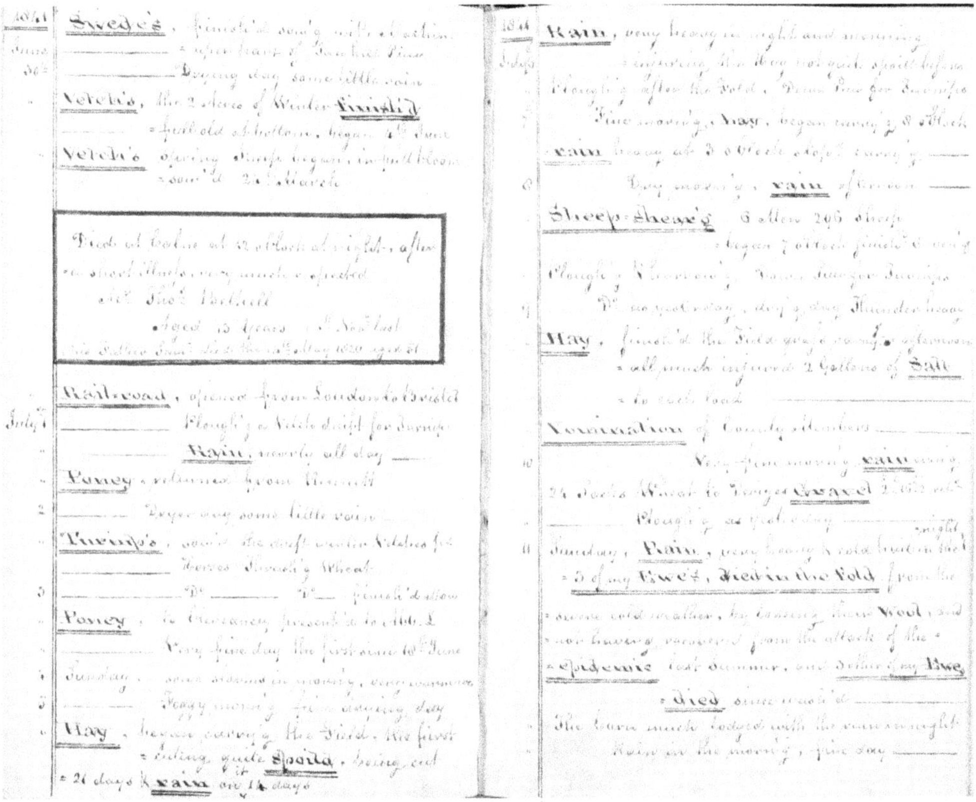

Figure 2: Extract from Volume 7 of the diaries
Source: WSA: 4381/1/7.

Figure 3: Dial from a barometer made c. 1815
Source: Edwin Banfield, *Barometers: Wheel or Banjo* (North Curry: Baros Books, 1985), p.71, by kind permission of Sue Ashton of Baros Books.

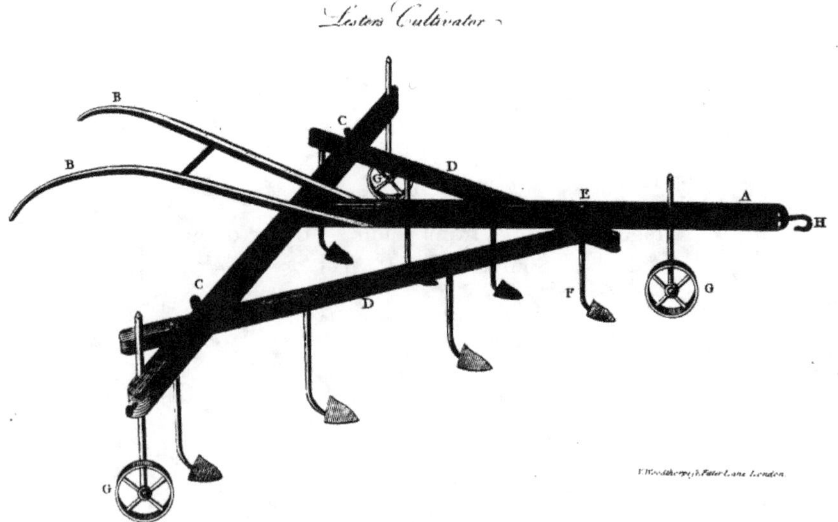

Figure 4: Lester's cultivator or seven-share plough
Source: W. Lester, *A history of British implements and machinery applicable to Agriculture, with observations on their improvement* (London: Longman, Hurst, Rees, Orme and Browne, 1811), pp. 180-181.

Figure 5: Extract from diary for 7 March 1820
Source: WSA: 4381/1/3.

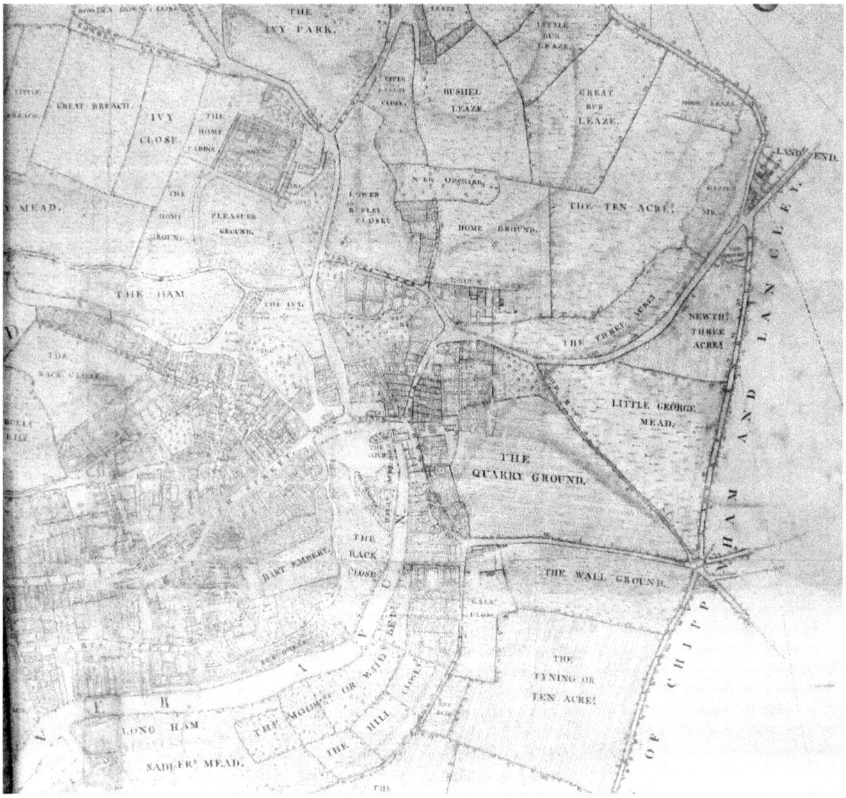

Figure 6: Extract from 1784 Powell's map of Chippenham
Source: WSA: G19/1/53L.

Figure 7: Hurdle stack with thatched roof at Priddy, Somerset
(Photo: Alan Wadsworth)

1823

January

1. Mild frost | Carting dung as yesterday
2. Carting dung as yesterday | THAW, rapid, with mizzle'y rain
3. THAW, very mild | Carting earth from road to Bonning's mead, - too hard for Ploughing - too soft for dung Cart'g
4. Ploughing Stubbles, Batts ground | Very mild, mizzle'g rain all day | Porker 62 li | BOAR, turned, at large to the Sows
5. Sunday, - dry, mild day
6. Dry day, - foggy | OLD XTmas WHEAT sowing, began since the frost, Turnip land Dean heath too wet before the frost, - now make good work the last sow'd - 8th NOVR, - both the old White ear'ed
7. Ploughing & Sowing as yesterday (7 Ploughs - 6 Harrows) | Foggy - fine day, very mild
8. Fine, mild clear day | Ploughing & sowing as yesterday and finish'd
9. Frosty morn'g & freezing all day | Ploughing Breach, rafter'g
10. Ploughing Breach fallowing Stubbles Shepherds Ground, - frost being too hard for Breach. | Very, sharp frost freezing all day
11. Very sharp frost | Carting dung behind Publick house
12. Sunday, - sharp freezing all day.
13. Carting Dung as 11th | Sharp frost
14. Sharp frost 2 LAMBS | Carting Dung as yesterday
15. Carting Dung & finish'd from home yards. | Snow, night & morn'g - and sharp frost | 14th - Oats brot the last from Cowlease Barn [Whole page of calculations on fattening a pig – see Appendix A]
16. Sharp = frost | Barley - part of Nfield Rick put into Barn
17. Barley - finish'd the Rick | Frost
18. Frost very sharp | Carting Dung from Cowlease Barn - to Pedlers Piece in heap
19. Sunday, - exceeding cold sharp frost
20. Thick fog & exceeding cold sharp frost | Carting Dung as the 18th | TURNIPS, discontinued, the young Sheep Parlour bottom, - the Hurdles frozen, to prevent removing.
21. Young sheep removed to Nfield fold'd on oldlay Hay & Water only | Carting dung finish'd from Cowlease Barn, and began Carting Dung from Jockey to the young Sainfoin | Very cold sharp frost
22. Very cold sharp frost | Carting Dung as yesterday
23. Carting Dung as yesterday | PIERCE'G COLD FREEZING WINDS all night, & all day [Notes on 2nd Pig killed – see Appendix A]
24. Carting Dung finishd the Jockey Yard, cover'd Jockey Gd | Cold freezing

Winds as yesterday

25. Frost & cold winds in the morning, - in the afternoon heavy SNOW till the evening 9 In deep, the level | Carting Wood for Labourers the morning
26. Sunday. - Cold freez'g winds
27. RAIN & thaw after the morning | Horses in the Stable
28. Horses in the Stable | Rapid THAW & FLOOD, the land covered with Water & the Wheat much strand'd | Mild & damp air
29. Mild & damp air, some heavy rain, the Horses geting Wood from Copse | 28th. - The Ewes removed from Nfield being too wet, to the stand Pen, Court Mead
30. Very MILD damp Air | Horses in the Stable
31. Plough'g - began first since the 10th. - Raftering the fold d land | Fiting up Burnt mill mead | Mild damp day

February

1. Mild damp day thick Rain great part of day, glass very low to V in very dry | Ploughing as yesterday & finishd | OXEN, one team began Plough'g
2. Sunday, - damp day, lower to R in dry
3. Thick rain nearly all day | Plough'g Breach, till the rain stop d them
4. Plough'g the OldSainfoin for Oats | Sharp frost & freezing all day
5. Very sharp frost Ploughs stop d on the Sainfoin | Plough'g - turning the furrow Copes grounds
6. Sharp frost to stop the Plough | Wheat Rick, the first new, <put> began puting into Barn, at mid day SNOW, commenced, & cover'd the Rick, half housed | Ewes removed from Ct mead to Stockwells for lambing, and began the Hay at Stockwells
7. Thaw, and mizzle g rain all day | Carting the Dung from the Standing Pen Ct mead to an heap Nfield fur: furlong
8. Carting Dung as yesterday & finished | Frost, morn'g - Thaw & storm Snow after
9. Sunday, frost morning rain all evening.
10. Rain very thick, all the day after the morning | Plough'g Sainfoin till the rain stopd the Ploughs, mid day | Hay Nfield 44 Acres 45½ Ton, finish'd began 14th Decr | Young Sheep removed from Nfield to the Turnips P=Bottom fold' wandyke, too wet on Turnips [1]
11. Thick mizzle'g rain nearly all day | Plough'g Sainfoin as yest'y
12. Plough'g Sainfoin as yest'y | Some storms, - dry'r day [More calculations on fattening pigs – see Appendix A]
13. Dry day - rain evening | Ploughing as yesterday
14. Plough'g as yesterday | Cold Storms
15. Dry day, some little Snow | Ploughing as yesterday & finished | (14) Ewes

---

1 P=Bottom – this is the field known as Parlour Bottom: see entry for 20 January above. Also referred to as Parlour Field or Parlour Piece. Wandyke – Wansdyke.

time up – 80 Lambs

16. Sunday, - some mizzle'g rain
17. Dry day, land geting dry | Ploughing Sainfoin Wansdyke for Oats | 140 Lambs
18. Very thick rain all day, and rough Winds | Horses in the Stable
19. Plough'g Sainfoin Wansdyke | Drying day | Fat Pig 9S 10 li | POTATOES, Planting in garden
20. Ploughing as yesterday | Wheat Rick, put remainer of one in Barn, began the 6th Inst | Very fine warm, sunshine clear day * | 200 Lambs
21. Very thick driving RAIN all day * | Plough'g as yesterday in morn stopd with rain after
22. Thick rain, wth storm of HAIL nearly all day. | Ploughing finishd Wansdyke Piece
23. Sunday, - thick driving and heavy rain all day
24. Fine drying day | Ploughing Stubbles, behind the Public House | 300 Lambs | Fat Pig 10 Sco 10 li
25. Fine fog'y morn'g, mizzle'g rain, nearly all day after and very heavy storm 3 OClock | The old Vetch Rick put into Stockwells Barn | Ploughing the Breach
26. Ploughing finished the Breach | Fine drying day
27. Ewes & Lambs 350, - turned to the Swedes Pearces Piece, the first time, in the yard, nights | Cold dry day, except a storm of Snow | Plough'g - fallow'g, behind the Public House.
28. Plough'g - fallow'g & finish'd fallowing | Fine day, except some little Snow

March

1. Sharp frost, morning, very fine day | Ploughing, further side of Nfield fold & rafter'd, for Vetches. | ASHES Peat, - the first Boat load 700 Bushels land'd, - in Meadow by Workhouse for Whitley furlong 2¾d pr Bushl 1d pr Bushl freight
2. Sunday, Dry cold winds
3. Mizzle'g rain, wth driveing winds | Ploughing Nfield as 1st [More notes on pigs – see Appendix A] Pease & Beans, Planting garden | Wheat sow'd 6th Jany just shew'g
4. Ploughing as yesterday | Very high WIND, in the night part of Graden Wall blew down and Ricks Strip'd [2] [More notes on pigs – see Appendix A]
5. VETCHES, began sowing sow'd 5¾ Sacks about 8 Acres fur: side of Nfield lower furlong Spring - from Forebridge Piece 1821 not quite sound | ASHES, - landed - 2nd load 700B | Dry windy day wth some Hail storms
6. Cold drying day little frost | VETCHES Winter sow'd - Nfield next the spring Vetches to succeed about 8 Acres - 7½ Sacks not good seed of 1821 | MARKETS - advancing Wheat 33s 21s about Xtmas Barley 38s 30s about

2 Graden – this is possibly the garden wall.

Xtmas Barley at other Markets 42s Qr Summer Vetches 27s Sack, at Harvest 10s

7. Very sharp frost, morning and heavy Snow all day all day after 9OClock | Ploughing Shepherds Ground in morning for Oats | (6) 2 Cows & Calves from Chip[penha]m | (7) 1 Cow Vinny - retd to Chip[penha]m

8. Cold drying day | Carting Faggots & Wood from Dell Copse, the Snow on the ground, stop'd all field work | SEED, Hop clover Rick, put into Jockey Barn.

9. Heavy & deep Snow, in the night dry day | Sunday

10. Fine frosty morning, rain & Snow nearly all day after | Carting Wood from Westboro Copse

11. Storm of rain hail & Snow | Carting Wood as yesterday | 400 - or nearly Lambs, 50 to lamb

12. Ash's - began sowing, Parlour field | Very fine drying day, Glass high

13. Very fine drying day, Glass high | Ashes, sowing Whitley fur | Barley Rick, Stockwells put into Barn

14. Ashes, finished sowing Whitley fur: [...] Acres | SOW, farrowed, - the first wth 8 | Mizzle'g, rain morn'g, fine mild, grow'g afternoon | OAT'S, began, sowing Kite hill on the back of oldSainfoin, black Siberians - 6 Bu pr A, - seed from the Breach

15. Oats finish'd sowing Kite hill 16 Acres wth 24 Sacks | Ashes, sowing Nfield lower Church furlong, own burnt | Damp morning, fine day

16. Sunday, - fine day

17. Very fine day | Oats, sow'd Wansdyke furlong 14 Acres 20½ Sacks Blk Siberians

18. BEANS, began Planting drilling 4 Bu pr Acre, Up: fur: Nfield, - where Vetches fail'd Seed, fm Mrs L -Tockenham [3] | Plough'g Shepherds ground | Mizzle'g rain at times all day | (15) Mare cover'd, by Mr. Vaizeys gray Horse, Golden dun, was got by the famous horse Golden dab, his Dam by old Hugh Capet.

19. Drilling Beans as yesterday | Ploughing as yesterday | Very drying, - cold day | Dry Sheep treading the Oats sow'd on the back of OldSainfoin begin to hollow under furrow, do much good * | Heavy storm of hail, and of Snow in the morn'g

20. Snow, heavy all the morning Rain, - after till 3 OClock | Horses in the Stable

21. Mizzle'g rain nearly all day | Ashes, - finish'd sowing Nfield 1 land, left unsown lower fur: opposide, new watermead hedge | Bushing Dung on Sainfoin Jockey ground

22. Heavy storms | Ploughing Shepherds ground finish'd for Oats

23. Sunday, fine day, some rain

24. Very fine drying day frost, morng | Ploughing Lains for seed Vetches |

---

3 Mrs L – this is probably Mrs Large, Pinniger's future mother-in-law.

Drilling Beans Nfield | TURNIPS, the young Sheep finishd Parlour bottom, began 18th Decr | SWEDES, the Young Sheep began Parlour bottom

25. Foggy morning, fine spring day | Ploughing as yesterday | Drilling Beans | Wheat Rick the last of 1821 damaged put into Barn
26. Drilling Beans, finish'd 18A 2R 16P with 18 Sacks | Ploughing finish'd Lains & Plough'g oldlay Nf: for Vet: | Foggy - morn'g fine day | Swede greens, just growing not fit to gather, all Turnips rotten. - See - 1st March 1822 Turnips in bloom [4]
27. Cold dry day | PEASE, Small Grays - sowing Nether side of Up furlong | Vetches of 1820 - sowing below them [Note in pencil] Winter| Ploughing in the Pease & Vetches
28. Good + friday | Pease finish'd sowing small grays to the hedge. | Vethches of 1820 sowing below them to the Road [Note in pencil] Summer| 6¼ Sacks Pease @ 14s 6d pr S | Foggy morn'g - very fine day
29. Foggy morn'g - warm Summers day | Vetches of 1821 - sowed 4 SU Sacks the lains next the hedge & 2¼ W Sacks to Copse to finish the Piece for Seed[5] | [Note in margin] Smith Pd for his land | Vetches of 1821 – Winter, Ploughing & Sowing lower furlong Nfield withing - 1 land of Mead gate for feeding & Wheat
30. Sunday. - dry & dull day till even'g, when little rain
31. Dull, drying day | Oats, - black Siberians Ploughing in Marls | Rolling began Clover | (30th) Swedes, the Ewes & Lambs finish'd the first Pearces Piece 6A 1R 34P began the 27th Feby 13 Acres - the next Piece of Swedes began the 30th

April

1. Oat Sowing finish'd Plough'g in Marls 10¼ Acres 6 Bu: pr Acre 1½ Bu own hop ½ Bu own oldray 8 li Broad – Wm Laws } hole full ¾ the Machine | The Dry sheep, left the Swedes P Bottom, to feed the Turips in 18 Acres. | Dry sheep Fold'd on Wheat Mushings, not dung'd when sown. | Hay, dry sheep none whilst at the 18 Acres | Hay, the good oldSainfoin, Ewes & L finish'd
2. Oats – black siberians sowing & Plough'g in 6 Acres of Shepherds ground | Cold storms Very warm yesterday & grow'g
3. Cold winds in morn'g, rain afternoon | Fallow'g Smiths Bonnings | Dry sheep retd from the 18A to Parlour bottom
4. Dry morn'g, - rain all day Teams drove home from Plough'g Cowlease | Dung heating in Yard under the Ewes & Lambs, - another standing Pen, made behind the Public house

---

4 Entry from diary for 1 March 1822 – "TURNIPS, East field running rapidly to seed...".

5 SU – Summer, W – Winter: referring to summer vetches and winter vetches.

5. Rain all the morning, the Horses in the Stable
6. Sunday, - cold damp day some rain
7. Rain at times, all day. Horses - in Stable, and geting Woods from Dells Copse
8. Dull day cold winds | Ploughing, Cowlease, athwart | Roots <&> from Marls, hedge & Ploughing Bank | Seeds Black, sent to be mill'd
9. Fine drying day, cold winds | Ploughing & Sowing began the 18A 1 Bushl small gray Pease under furrow 4 Bushls Blk Siberian Oats pr A for horse Corn.
10. Ploughing & Sowing the 18 Acres as yesterday | Very fine drying day
11. Very fine drying day & warm | Ploughing & Sowing the 18 Acres as yesterday, and finish'd, made good work – in 3 day – 25¾ Acres 7 Ploughs – 5 Harrows, - Oats 28 Sacks - Blk, Siberians Pease 8¼ Sacks small grays Hop, - best Tail, 1¾ Bu Ray <old> ¾ Bu Broad 8 li} pr Acre hole Nr full Sown in the 6 & 18 Acres | Barley Rick Stockwells put into the Barn | Dry Sheep began fold'g, swear down, next Snail hill, - feed'g Swedes Parlour bottom, eat but little Hay | Swedes, - hurdled off – for Seed green very short yet
12. White frosty mornings, warm clear days | Harrow'g Cowlease & Plough'g 2nd time for Oats | Barley Rick, thin put into Barn for Seed.
13. Sunday, - warm fine morning cold afternoon
14. Fine dry day, - frosty nights, and cold days, - the Wheat & grass going backwards, - looking not so well as 10 day since | Oats – 30 Acres first sowing, look'g green, and | Vetches Summer sow'd 5th March up fine, - and | Beans & Pease, break'g ground | Wheat harrow'g began Dean heth, two tine, - the ground dash'd and bound, hard [6] | 5 Ploughs, thwarted Batts ground 5 Acres, and began thwarting Marls | Harrow'g Cowlease, very couchy & thrash'g thin Barley for Seed with Horse Machine
15. Harrow'g Wheat Efield bottom | Plough'g Marls, athwart as yesterday | Swedes finish d Parlour bottom, - all Sheep in Pearces | Warm fine day
16. Mizzle y rain all morn'g | Plough'd & Sow'd Cowlease to Oats – 8¾ Acres – 8¾ Sacks COLT, sent to be broke (14th) Knightly Smith (foal'd 31st May 1821)
17. Drying winds Plough'd & Sow'd Batts ground to Barley 3 Bu: of Tail hop, & Ray
18. Some, rain, hail & Snow, in afternoon, - fine morning | Ploughing couchy part of Copes grounds athwart
19. Continued storms of HAIL Snow & rain | Ploughing & Sowing Copes grounds to Blk siberian Oats half under furrow, half at top 5 Bushls pr Acres 60 li Mill'd Sainfoin pr Acres @ 37s 6d pr € 20 li Mill'd Hop pr Acres @ 1¾d li | Our own hop mill'd by Thos Cowley Ramsbury Viz 16 Qr Black seed sent, and returned Screenings / Seeds &c. very bad 9 Bushls Not mill'd out, fit to sow 4 Bushls [Total] 13 Bushls | Best mill'd seed 18€ 0qr 6li Tail

6 Dean heth – this is the field known as Dean Heath.

mill'd seed 1€ 0qr 2li Hulks 2qr 23li Mill'd [Total] 19€ 3qr 3li @ 4s £4 The 16 Qr Seed @ 14s/Qr £11 4s [Total] £15 4s 0d (is 24s/- more than 1½d pr li | Round tailing Sheep, and Cuting Lambs, tails – 4 Men Viz 450 dry Flock 493 Ewes & 12 Rams [Total] 955 all 450 Lambs, just half Rams

20. Sunday – fine day
21. Very fine day | Ploughing & sow'g – as the 19th | Harrow'g Wheat | Thrash'g Barley wth Horses for seed, very thin & want'd
22. Ploughing &c as yesterday, and finish'd Copes grounds | Barley Rick – thin put into the Horse machine Barn for seed | Very fine day | Ewes & Lambs, - removed from Swedes Pearces Piece to WaterMeads, fold on oldlay Nfield for Vetches & Wheat after | Hay – discontinued, all the sheep this morning | Straw, - Barley, - given to the Ewes, to this morning, with Hay, do well, not a Ewe or Lamb, die, or do ill, - since the first few days of Lambing, & 50 Lambs, rear'd more than expected
23. Ploughing & Sowing, Barley Parlour bottom | Mizzle'g rain at times - all day.
24. Fine day | Ploughing & Sowing as yesterday | Swedes, - finish'd, by the dry Sheep
25. Watermeads, Burntmill, dry sheep began | Very fine | CUCKOO | Plough'g Parlour bottom and sow'g Barley
26. RAIN, fast nearly all day, at times very heavy | Horses, Carting wood, part of day, - and in the Stable | Sheep, - kept out of Watermead on oldlay, and Down.
27. Sunday, - - Cold dry day
28. Fine day, - Ploughing & Sowing Barley, Parlour bottom 5 Bu: thin Barley 1 Bu old Ray 7 li Broad Clover 10 li Mill'd hop} near half hole in Machine
29. Ploughing & Sowing as yesterday | Frosty morn'g's – fine days.
30. Frosty morn'g's – fine days. | Ploughing & Sow'g as yesterday

May

1. Very fine, mild day | Ploughing & Sow'g Barley, finish'd Parlour bottom, & began Pearces Piece | Harrow'g Wheat
2. Warm, mild, grow'g day | Plough'g & Sow'g Barley, - as yesty | Wheat Rick (long one) put into the Barn, - the 2nd -/- 8 now out
3. Ploughing & Sow'g Barley as yesterday | BEAN'S, - 7 Women began hoeing yesterday | Warm fine day
4. Harsh, cold, day, - Sunday.
5. Very warm fine day | Plough'g & Sow'g Pearces Piece to Barley | BARKING, Smith began | Harrow'g in 2½ Sacks Barley among the thin Wheat Snail hill, and rolling it
6. Very warm fine day Ploughing & Sow'g Nether Pearces Piece to Barley & FINISH'D | Harrow'g Wheat finish'd
7. Harrow'g BEANS, before hoers | Plough'g & Sow'g White Oats Parlour

field | Fold removed from Parlour field to Bonnings Common for Turnips | Warm, some mizzle'y rain

8. Warm, some mizzle'y rain, heavy storm even'g | OATS, Ploughing & sowing FINISH'D, Parlour field 20 Acres to Sainfoin, 60 li Mill'd pr Acre & hop 14 li Mill'd
9. Rough drying Winds | Plough'g the Breach athwart for forward Turnips | Carting Muckle, before fold Bonnings Corner for Turnips | Carting Wood <&> for drains.
10. Mizzle'g rain at times in morn'g – fine grow'g rain evening. | POTATOE'S, Plough'd 40 Pole in, Nf | Carting Dung for Potatoes & Vetches Nf | Ploughing the Breach as yester'y
11. Sunday OG=N Rain in night & morning, the land now, moisten'd, to fetch the Barley up together, dry before.[7]
12. Thwarting the Breach, finishd Showers of mizzle'g rain, & some hail | WATER=MEAD, - the Ewes & Lambs finish'd, & began Field grass forebridge Piece
13. WATER=MEAD, the Dry Flock, finishd and began the Down &c | VETCH'S, - Summer, Plough'g & Sowing about 6A Nf | Dry cold winds
14. Dry cold winds storm, evening | Ploughing & Sow'g Vetches as yesy [8] | Beans, finish'd hoeing | RAM=LAMB'S, cut 215 @ 5s Hd Kept for Rams [In pencil] 9 [Total] [...] [9]
15. Barking, finish'd for Mr Smith & began for Mr G [10] | Ploughing & sowing Vetches as Yesterday, - and Sow'd – summer Vetches, on 5A of winter Vetches appear to thin 1 Bu pr A and 2 tine the Harrows | Fine day
16. Blighting, mizzle'g rain morng fine day BA=H | Plough'g Shepherds Gd & sow'g Summer Vetches for Horses, the first
17. Mizzle'g rain morn'g fine day | Ploughing for Turnips Breach
15. BARLEY, the last Rick put into Barn
18. Sunday, - very fine day
19. Very fine day | Plough'g, rolling, harrow'g <&> the Breach
20. Plough'g, rolling, harrow'g <&> the Breach | Some mizzle'y rain morn'g fine after
21. RAIN in the night, to enable TURNIP. begin sowing, sow'd 6 Acres of Tankard in the Breach | Fine drying day
22. Heavy rain in morn'g, land too wet to Plough | Carting Oats & Hop seed from the Jockey
23. Finish'd carting Oats & Hop seed from the Jockey & Carting, Wood Hurdles & Timber home, fm Westboro Copse | Some showers, - the land full wet, for Plough'g.

---

7 OG=N - this possibly indicates that Pinniger was at Ogbourn[e] that day.

8 yesy - shortened form of yesterday.

9 Hd - in this case, per hundred, not per head.

10 Mr G - this is possibly Mr Anthony Guy.

24. Ploughing athwart, Sweardown lower furlong for Swedes | Bark carrying – for Mr Smith | Some mizzle'g grow'g rain.
25. Sunday, - heavy storm, morng warm, grow'g day, TOCK[enha]M
26. Warm grow'g day, partial storms. | Ewes & Lambs, - finish'd feed'g Forebridge Piece, & removed to Cowlease, - Fold removed fm Nfield to Pedlers Piece for Turnips | Ploughing as 24th
27. Ploughing as 24th | Fine warm grow'g day
28. Very fine warm day | Bark carry'g Mr Smiths | Ploughing & harrow'g for Swedes, Sweardown | Harrow'g Breach, for Turnps
29. Very warm day | Plough'g harrow'g &c as yest'y | GRASS, watermead, Horses began | Bark – remain'r – carried for Mr Smith – 13½ Tons @ £9 10s pr T Trees [...]
30. Bark – carried own 22 Cwt | Plough'g harrow'g & rolling as yesterday | Warm fine day
31. Very warm fine day Tyth[erto]n | Ploughing & Harrow'g Bonnings Corner. <u>HORSES=BLED</u> Chipp[enha]m

June

1. Sunday – very warm fine day
2. Ploughing & harrow'g the Breach for Turnips | Fine morn'g rain all evening
3. Fine day | Ploughing & Sow'g <u>Vetches</u> Nf
4. Ploughing & Sow'g & finish'd the low'r fur: for Sheep 44A | Fine morn'g, - & grow'g rain in the evening
5. MOWING, - ray grass Nfield Church fur: 7 Men began | Ploughing & Sow'g Vetches the remain'r of Shepherds Ground for Horses | Carting Dung for Swedes began Swear down lowr | Fine day
6. Fine day and warm | Carting Dung & Plough'g for Swedes as yesterday
7. Carting Dung & Plough'g for Swedes as yesterday | Fine day rain, in evening
8. Sunday, - some warm growing showers
9. Heavy, showers, warm, very grow'g | Carting dung, lane Piece &c. for swedes, - Plough'g, - harrow'g &c
7. <u>Field grass, the Horses</u>, began <u>2 Sacks of Oats less</u>, Now 3 Sacks of Oats 1 Sack Beans 14 Horses
10. Carting Dung &c as yesterday Heavy Thunder showers, and some Hail – very growing | Washing Sheep, - Usage 5 Qts Best Ale 5 Qts Strong Beer 2 large loaves of Bread Bacon & Cheese, in proportion *
11. Carting Dung to lane Piece, and Plough'g it in, Harrowing, roll'g &c | Hoeing Beans – 2nd time. | Warm fine day
12. Warm fine day | Carting Dung, Plough'g &c as yesty
13. Carting Dung, Plough'g &c as yesty | Warm fine day | SHEEP SHEAR – 9 Men @ 2s pr day
14. SHEEP SHEAR – 10 Men @ 2s pr day 2 Women @ 1s pr day winding 1 Man lifting Sheep out 1 Man Carry'g Sheep 1 Man Leading Sheep out to

be markd 2 Women for the Fleeces 2 Children for the loose wool. | Carting Dung Swear down and Plough'g it in, Roller follow, - the Dung – dry very fast. The Iron Press'r follow | Vetches – sow'd – two lands in Shepherds, last

15. Sunday, warm fine day | 2 Ewes lost, - died no cause.

16. 6 Lambs – died, with Scour. Fold'd in Pedlers Piece, which being full of Thou Thistles suppose the cause, fed in the Dells. Fold removed to Bonnings Corner, and fed in Bartletts.[11] | HAY, began carry'g Nfield, 60 Acres field grass cut Nf & Ef and began mowing Sainfoin

15. WHEAT=EARS, just bursting

17. Very fine day | Hay, finish'd carry'g Nf | Plough'g in Dung, Sweardown

18. Plough'g in Dung, Sweardown and Carting Dung | Fine day

19. Fine day | Carting Dung &c as yesterday | Mowing finish'd, all but Meads

20. Hay, began carrying Efield | Ploughing Dung in &c as yesterday | 8 Sows wth 7 Pigs each.

21. Carting Dung, - finish'd Lane Piece – 43 Acres Pot Dung'd South field, - Swear down, up'r & lower: fur: Lane Piece and Bonnins Corner. | Smiths Bonnins, - fold'd. | Ploughing Bonnings Corner finish'd | Ewes & Lambs Fold removed, fin Bonnings finish'd for Turnips to Hitchens for Wheat

22. Sunday, fine day Wheat Ears, - many quite out very uneven, and very slow about as forward as 29th May last year [12] | Beans, - blooming well, | Pease, not in Bloom | Vetches, first sow'g just Bloom'g

23. TURNIP. Tankard, Plough'g & Sow'g Breach, - where 1st sow'g fail'd, sow'd 21st May | Hay, finish'd finish'd Carr'g Whitley fur: Efield | Pressing finish'd Sweardown, with 8 Oxen for Turnips | Dull morn'g = fine day

24. VERY SHARP WHITE FROST, Potatoes turned Black | MIDSUMMER DAY, very unusual | Ploughing & Sowing Turnips as Yesterday. | Fine day

25. Fine day Storms of rain partially | Ploughing Pedlers Piece

24. Wheat Rick, put into Barn 7 Wheat Rick now in the Yard | LAMBS, waned, and began the Summer Vetches, very good sow'd the 5th March.

26. Fine day, RAIN in the evening | Plough'g Hitchens for Wheat | Sainfoin Hay, - carried Parlour field, - Jockey ground only to carry & finish | Market, - smallest remember'd at Devizes, advanced 2s pr S one lot 35s Sack

27. RAIN, heavy storms, frequent warm & growing | SWEDES, began sowing, Sweardown & Lane Piece, sow'd 20A Dung Plough'd in, roll'd, and Press'd – ready for sowing too dry before. | Ploughs, - stiring Larks lease for Wheat

28. Ploughs, - stiring Larks lease for Wheat | Swedes & Green rounds, sow'd 20 Acres, - 40 Acres in two days make good work, with frequent warm showers, harrowed fine after the 10 wheel Iron PRESSER, proveing great advantage

29. Sunday, - Thunder, and warm heavy thunder storms, very acceptable, for

---

11 Thou thistles - this is probably referring to Sow thistle (*Sonchus oleraceus*).

12 Entry for 29 May 1822 reads "WHEAT, Shepherds & Lane Piece many EARS, quite out".

the Turnip land, - the Barley and Oats want rain, both shewing the Ear, and very short on the poor Land * | 2 Lambs, - died with scour, on the wet Vetches, and many more in danger

30. VETCHES, mowing, for the Lambs and put in the leazes to wither and prevent their scouring | Ploughing and sowing Tankard Turnips – Breach, make very good work after the rain | Turnips sow'd 23rd just appearing | Very warm, growing day.

July

1. Drying after the morning Thunderstorm afternoon | Plough'g & sow'g Breach to Turnips finish'd | Rolling, Plough'g, Harrow'g Pedlers Piece | Dry Sheep finish'd folding Pedlers Piece, - and the WETHERS AND SALE EWES, drew and put on Vetches after the Lambs
2. LAMBS, continue to scour, on the Vetches, - Kill'd 3 more, in all 10 Kill'd & died | Hurdles removing to make a road by the Beans, for the Lambs to the aftergrass, - for a change of victuals | Ploughing & Sowing Bonnings corner to Greenround Turnips Mr Gaby's seed | Fine warm day
3. Fine warm day | Finish'd Ploughing & Sowing Bonings Corner to Greenrounds | Hay, - carried the remainer of Sainfoin, in good order & finish'd, - except the dry meadow, by Forebridge, - which began mowing to day
4. Mizzle'y rain at times | Ploughing Larks leaze | Carting Dung Pedlers Piece
5. Ploughing & Sowing Pedlers Piece to Greenrounds. | Wheat Rick, put into Barn | Very fine day
6. Very fine day some mizzle rain evening, - | Sunday
7. Ploughing & Sowing Pedlers piece finish'd, - | Warm Showers
8. Frequent Showers | Ploughing Nfield Vetches fed off and sowing Turnips, treading after with Sheep, to keep firm & moist
9. Very fine warm day | Treading Pedlers piece, sown to Turnips, with the Sheep to keep it moist and firm | Fallow'g Westboro
10. Fallow'g Westboro | Turnips – Ploughing & Sowing in Farm garden | Some mizzle'y rain
11. Some mizzle'y rain Blighting all the morning dry after | Ploughing finish'd Westboro | Carting Chalk to Rickyard
12. Carting Chalk to Rickyard | Ploughing Larks lease athwart | Storms of mizzle'y blight g rain, no good to Turnips & prevent Hay making
13. Sunday. Thin driving rain nearly all day, - very acceptable, the land being too dry for the Turnip seed to vegitate, in spots | Turnips & Swedes, but just shewing the rough leaf, other Farmers, in general hoe'd out.
14. Dull day, dry after morn'g | Ploughing Larks lease athwart
15. Ploughing Westboro athwart | Dry morning rain all the afternoon to stop the Ploughs
16. Heavy showers, and cold for season of Year | Oxen Ploughing Hitchens

after the Fold athwart | Horses, Plough'g & sowing New Tankard Turnip Seed after Vetches, fed off, bottom of Nfield, for Wheat after.

17. Ploughing &c as yesterday | Drying day
18. Frequent rain | Ploughing Hitchens | Ploughing the New field, forebridge Piece athwart for Wheat, to be dung'd & Plough'd again when firm for sowing. | Horses, - finish'd field grass ⋆ | Oxen finish'd Meadow grass and both began Watermeads ⋆ | Mow'd Burntmill mead for Hay
19. Plough'g, - Forebridge Piece. | Rain heavy, at times. | SEED Swede, - cut & put into the Barn.
20. Sunday, - dry day.
21. Storms. Ploughing Forebridge Piece.
22. Ploughing up Swedes, part of Lane Piece & part of Swear down lower fur: the land not moist enough to vegitate the Seed, sowed the 27th June, - and part of each Furlong the land loamy, a good plant of Swedes, and began Hoeing. ⋆ | Fine morn'g Glass high, Storm, mid:day. Very drying after. } not time to make Hay
23. Violent storm of RAIN. HAIL & Thundr Plough'g Forebridge Piece & Hitchens
24. Plough'g as yesterday | Turnip land too wet to hoe, 10 Men no work for: - put them to Chalk Carting & new flooring 4 Mows | Very fine day, - every appearance of a change to fine weather.
25. Frequent storms, - Hay thrown back again & nearly spoild | Plough'g N field & Sow'd Turnips | Plough'g Westboro
26. Heavy storms, Westboro too wet to Plough, Oxen resting | Ploughing Nfield for Turnips | Carting Chalk &c as yesterday | Hay removed from Forebridge meads spoiling the grass to Stockwells mead to finish making
27. Sunday, fine drying day
28. Frequent storms | Plough'g & Chalk'g as 26th
29. RAIN at times all day. The land, wet like Winter. The Sheep about the Pastures too wet to feed the Vetches. Turnips want hoeing & 10 Men want employ, - but for Cart'g Chalk, floor'd 5 Mows, fill'g Mill tail with Chalk, to pave it ⋆ | Plough'g Forebridge Piece
30. Plough'g Forebridge Piece | Hay brot from Burntmill mead, in a muck state, to make it Hay in Stockwells mead. | Frequent Rain, but dryer than some days pass'd | HOEING SWEDES again 23 hoers stop'd hoe'g by the rain since 22nd
31. Very fine day, except 2 or 3 time a damp air | HAY, carry'd the dry mead quite dry, - nearly a month mowed [13] | Plough'g as yesterday & Harrow'g | 2 Loads of green Sand from Wolfhall

---

13 The hay in the dry mead[ow] was mown on 3 July (see entry above) – hence, it had been mown for nearly a month.

August

1. HAY, - carry'd the Watermead in very fair condition, before 6 OCl RAIN immediately after. | Fine day, - except damp air at times | Ploughing Forebridge Piece finishd

2. Carting Dung from Jockey Yard to Westboro. | RAIN, in Night & all the morning to stop the Turnip hoes.

3. Sunday RAIN. almost, incessant, the whole day, a gloomey prospect of Harvest

4. Fine drying day after morn'g Turnips growing very fast 26 Hoe'rs on, - afternoon | Carting Dung from the Jockey to Westboro.

5. Mizzle'y storms of RAIN, frequent | Carting dung as yesterday | Harrow'g Hitchens <&> wth Oxen

6. Harrow'g Hitchins & Westboro | Carting dung as yesterday, and finish'd | RAIN, heavy storms.

7. RAIN, heavy storms. | Ploughing the unclean parts of Hitchens & Larks lease. | Vetches finish'd feeding Nfield to the Gate, - the Summer ones not quite ready, just blooming.

8. Clover, Whitley furlong, Sale Sheep and Lambs began feeding, and fold'd Forebridge Piece | Carting Chalk to Jockey Yard Westboro being too wet to Plough | Oxen resting | RAIN, frequent Storms

9. RAIN, a little in morn'g, warm & fine after | Carting Chalk as yesterday

10. Sunday, RAIN, frequent mizzley storms. TY[therto]N

11. Mizzle'y RAIN, morn'g & heavy even'g | Carting Chalk as the 9th TOC[kenha]M

12. Carting Chalk as the 9th & finish'd fill'g the Mows of the Barns. | Very fine warm day, few drops of RAIN in the evening | Hay, the dry mead & Watermead, put together & 2½ Bu: SALT, about 8 Tons

4. Jos Large Died at Purton Aged [...] [14] [This is the end of Volume 3 and the beginning of Volume 4]

13. Mizzle'y rain all day | Ploughing Westboro

14. Ploughing Westboro | Partial heavy storms, - fine at Devizes, - and the Market 1s per Sack lower.

15. The finest morning the last month, mizzle'g rain three hours mid=day, very fine evening | Wheat=rick – put into Barn white ear'd seed | Oxen Plough'g Westboro | Horses began Plough'g Whitley fur: after Sheep feed'g 2nd Clover

16. Mizzle'g morn, Pelting storms of RAIN & HAIL mid:day, fine after Market sinking, many began reaping. | Plough'g as yesterday, and load of Soot fm Hungerford.

17. Sunday, fine day

---

14 Joseph Large was buried at Purton on 8 August 1823, aged 73, *Wiltshire, England, Church of England Deaths and Burials, 1813-1916* [accessed via Ancestry, 16 July 2020].

18. Mizzle'y rain frequent | Turnips finish'd hoeing 1st time | Turnips began hoeing – 2nd time Stock Ewes, removed from Pasture to feed the Clover Efield, to fallow | Plough'g as yesterday (16th)
19. Turnips, hoe'g the young, where fail'd & sow'd again. | Soot fm Hungerford = some rain | Plough'g as yesterday
20. Plough'g as yesterday Foggy morning, appearance of a fine day, and change of weather hoe'g & harrow'g Turnips do well in the afternoon, then a succession of heavy storms, as wet and as dirty as the middle of winter
21. 2 Rams, - turn'd to Sale Ewes | Fine morning, - Storms in market time at Devizes, notwithstanding was a very dull sale highest wheat 32s pr Sack * | Men making Chalk floors in the Barn mows Jockey, too wet to cut the Oats now ready | Plough'g Clover lay after the Sheep fed off = the fallows too wet. | Heavy rain evening & night
22. Marlbro fair, heavy RAIN all the morning, dryer even'g | 2-four tooth Rams of Mr Hillier of Granham 9£. | Ploughing as yesterday. | Oats, - Wansdyke & Kite hill dead ripe, & Wheat fit to begin, stop d with wet. | Men throw'g up Dung in the Yds
23. Ploughing athwart the furrow Nfield after Vetches for Wheat | RAIN heavy in morn'g mizzle'y after | RAM, turned one of Mr Hillier's to 30 good Ewes, time out 10th Jany for the purpose of breeding some Rams. | Changed 5 Rams with Mr Rumboll
24. Sunday, rain mizzle y all the morn'g – dry afternoon | WHEAT=SPROUT'G upright, and lodged every prospect as bad Harvest as 1821 [15]
25. Rain early in morning, very warm and very fine day after, - the best prospect since ST SWITHENS day. | Turnip hoeing finish'd, - and watermead hay Rick top'd and thatch'd, - no oppertunity before. | Dung carting began for Wheat Forebridge Piece. | Clover, the Sheep, discontinued feeding, the Vetches being full large, - about 8 Acres Clover left for Hay or Seed. | Oxen Plough'g Larks lease.
26. Carting Dung, and removing Hurdles, fm Clover to Vetches Nf too wet to put Sheep on them. | Reapers – prepared to begin RAIN commenced at 9 OClock, continued all day, - to stop Dung carting | Labourers, riding Mill pond of Weed; and Masons, repairing the thorough.
27. Wind N, in morning, and the Glass, gradually rise'g, - the best appearance of fine weather | a few drops of rain, morn'g | OATS, reaping, - began even'g | Vetches, - Sheep began again Nfield | Plough'g, raftering Whitley fur:
28. Plough'g, raftering Whitley fur: & stiring green spots | Oats finish'd reaping Wansdyke Piece 12A, - mow'd 2A: | Fine foggy morn, quite harvest day.
29. REAPING, Wheat, began, Efield bottom | Wheat Rick, Shepherds

---

15 The harvest in 1821 was completed on 10 October with the diary entry "Barley finish'd carry'g E field and Harvest completed", whereas in 1822, the harvest was recorded by Pinniger as finished on 12 September. The 1823 harvest actually finished on 17 October: see later.

ground put into the Barn. | Rain, - a few drops in morning very fine day. | Harrow'g Turnips & fallows and stiring the Vetch fallows Nfield.

30. Plough'g & harrow'g as yester'y | RAIN, mizzle'g from 10 OClock to 3 OClock, - stop'd reapers began again at 4 OClock.
31. Sunday, - frosty morn, perfectly fine ripening day, - starlight evening, the milky way, appear'd cheering prospect of a fine harvest *

September

1. Very fine harvest day | Harrow'g Turnips, - Pedlers & Breach | Plough'g – thwarting Vetch's fed off Nfield | [Pencil note in margin] Men in House [Pinniger recorded his stock of sheep at this date – see Appendix B]
2. Fine day, - lowering evening | Oxen Ploughing as yesterday | Harrow'g fallows – Efield, and finish'd harrow'g – Turnips Turnips, - now grow'g rapidly | OATS, carried Wansdyke Piece
3. OATS, carried, - Kite hill | WHEAT=RICK, the first began | Finish'd – Plough'g as yesterday | Harrow'g &c. | Dean heath Wheat blighted reaping and laying in grip The White eard, sow'd 6th Jany a good Crop, - and good Berry if not blighted | Lowering, hollow wind, and some mizzle'g RAIN, mid=day.
4. Very drying fine day with a Cloudless Sky. | Carry'g Wheat all day, finish'd Efield bottom 25A | Harrow'g Nfield | Oxen Plough'g Westboro 5th time
5. Oxen Plough'g Westboro 5th time &c | Carri'g Wheat Clay furlong | Very fine harvest day, damp air in the evening
6. Warm fine day | All Horses bringing the Wheat from Dean heath | Oxen, Ploughing as yesterday | Vetches. Shepherds ground the Horses began, in Kid, sow'd 16th May * | Vetches – Nfield last sow'g – 4th June Sheep began, very good, full bloom *
7. Sunday warm fine day.
8. Sharp frost, foggy morn, fine day. | Wheat Dean heath finish'd carrg 24 Acres, - all put upon the high Staddle 1260 Tythings - is 52 pr A The Rick 20 feet from the ground to eves, - and 40 feet over the Rick on the eves, - 120 large sheaves the outside laying. | Plough'g Larks lease, 5th time
9. Plough'g Larks lease, 5th time | Carting Dung to Forebridge Piece | Very cold frosty & fogg'y morn & warm day.
10. Very cold frosty & fogg'y morn & warm day. | Carting Dung & Ploughing as Yesterday | Reaping – finish'd | Barley, - began mow'g | Oats, - mow'g Shepherds Gd | Hay, - mow'g – the 2nd Clover for the Horses, at seed time Clover, - the 2nd cut, began for the Horses, green, the Vetches Shepherds Grod too old.
11. Wheat, carried, the remainer of E field, made the 6th and last Rick, by the Ganiry. | Oxen Plough'g Hitchens, in spots green. | Cold, foggy, morning, and warm day | VETCHS. Seed, began cut'g Nfield
12. Warm fine day | Carting dung & Plough'g as yes: | Wheat. Burnt mill Gd

carr'd

13. Wheat, - harvest, finish'd Snail hill, - put into Barn for Seed. | Clover Hay, - carried 4A – without turning the Swath no time for it, - cut 10th, - goodhay & good crop. | Fine but dull day,- Glass sinkg threat'g for rain. | Removd Hurdles fm Nfield to Vetches Shepherds Gd Vetches Nfield, the Sheep finish'd | Plough'g Hitchens for Seed. | PEASE, began cuting
14. Sunday.- Glass, sinking very fast, dull day, in the even'g RAIN, heavy storms, some Thunder & lightening | Vetches Shepherds Gd the Sheep began.
15. The Glass, sunk, nearly half the Circle | Heavy storms in the night & in the morning. | Mowers, stop'd. winnow'g Seeds and thrash'g Oats. | Carting dung from Stockwells Barns doors | Ploughing as yesterday
16. Ploughing & Carting Dung as Yesty | Oats, - began carry'g Shepherds | Drying day | RAMS, turn'd, Mr Hilliers two to 450 Ewes, - being very thick at Ram, - the usual day, 21st Inst
17. RAIN, in the night, and mizzle'g rain till mid:day, dry'g Sun and wind till night. | Carting Dung & Ploughing as Yes | Oats black carry'g even'g. | BEANS, - began cuting Blk, Kid
18. Oats, - carried, Copes ground to Jockey Barn | Barley, - the first carried, part of Parlour bottom. | Frost very sharp, strong due & very fine day | Plough'g as yesterday
19. Plough'g as yesterday | Frost, dew, & fine day | Vetches, Nfield, carried, morn'g | Pease, Nfield, - carry'g all day and Night, to 3 OClock morng
20. Pease, finish'd, carry'g and about [...] loads | Barley, - in afternoon, carry'g Parlour bottom, and in night till 12 OClock, finish'd carr: | Oats, - Shepherds ground. | Frost, and very fine day. | Plough'g Forebridge Piece for for Seed. | Rams, - turned, 3 others to Stock Ewes, - viz – Kennet Osmond & Neale
21. Sunday, - rain after morning dry evening.
22. RAIN, - storms, all day | Carting dung, and Ploughing Forebridge Piece for Seed.
23. Carting dung, and Ploughing Forebridge Piece for Seed. | Fold removed fm Whitley fur: to finish Larks leaze, - and Sainfoin began feeding. | Dry day, - fine afternoon, the Barley, Pearces Piece began carryg 3 OClock, - finish'd 12 OClock night
24. Dry, but not bright day. | Barley, carrd a Rick from Pearces Piece in the afternoon straw dry but good not condition | Carting dung & thrashing & Ploughg as yesterday
25. Mizzle'g rain 6 OClock morn'g Finer day, - damp eve'g 10 OClock | Carting Dung, till mid day & Plough'g as yesterday. | Barley, - finish'd carrying Pearces Piece & began carry'g Parlour bottom.
26. Fine drying day | Barley carry'g began the day finish'd Pearces Piece, and carr'd Marls. Oats in good condition. Hither Pearces Piece [...] A in Barn Further Pearces Piece [...] A in 2 last Ricks Parlour bottom [...] A 3 Ricks near the Barn

27. Very fine drying wind & Sun the face of the earth in one hour in the afternoon, - covered with the Spiders Web a token of fine weather | Moon & Starlight morn'g, began carry'g Oats – Marls 4 OClock, it being too dark last night to finish the Piece. | Oats, carr'd, the Cowlease after breakfast, put into Barn, and in the afternoon, carried about 8A Oats from the 18 Acres, great Crop Finish'd mowing the 18 Acres Oats and began mowing the last piece of Oats, Parlour field. | Beans, - finish'd cuting, and Vetches, - began cuting the lains WATER=MEAD behind the mill began mowing for HAY. | Oxen Plough'g as yesterday.
28. Sunday, - very fine day.
29. Unusual sharp Frost, - and very fine drying day. | Oats, - carry'g the 18 Acres from 4 OClock morn'g till 9 OClock at night, when too dark to finish leaveing – 3 Loads to finish. | Oats, - finish'd mowing. | After=grass, Horses finish'd Efield | Ploughing as yesterday
30. Ploughing as yesterday | Fold – finish'd Larks lease, began Westboro | Dry night & morn'g till 7 OClock red morn'g & rainbow, and Glass sinking very low, when began raining & continued till mid day. | Oats – carr'd – the 3 loads left yesty and topd the Rick | Removeing Hurdles to Smith Turnips in afternoon.

October

1. RAIN, very heavy from 3 OClock in morning to mid:day, and storms after, - Glass fallen from G in change to E in very dry.[16] | Horses in Stable all day, and the Carters nearly all day 18A Beans 10A Vetches 12A Barley 8A Oats } not harvested
2. Fine drying day | 7 Ploughs for seed Forebridge Piece and all Men spreading dung.
3. Ploughing as yesterday. | Sharp frost, and fine day rain 7 OClock evening | Barley, carr'd about 8 loads fin Marls in evening, before the rain.
4. Fine drying day, - the Corn too wet from the rain fallen in the night, to carry. | Oxen Ploughing Whitley fur: for Seed. | Carting Dung to Whitley fur:
5. Sunday damp air
6. Rain in night, dry morn'g WHEATSOWING, - began, - Forebridge Piece, - Mr Budds Seed, - from below hill, 2½ Bu pr Acre. | Rain, - heavy, in morn'g for two hours, to prevent, to stop wheat sow'g, made good work afterwards. | Oxen Plough'g as the 4th
7. Oxen Plough'g as the 4th | Carting dung to Whitley fur: | Barley in afternoon, finish'd carrying fin Marls, - very dry, the best condition of any, -

16 Whilst Pinniger states "Glass fallen", assuming that his observation was correct, then the barometric pressure had actually increased. "Glass fallen" may simply be a reflection of the movement of the indicator around the dial, in that the word Change appears at the top of the instrument – see Figure 2.

(except under the Copse, removed to an open place to dry) | Rain in night, fine dry'g day

8. Fine drying day | Beans, - carrying all day | Oxen Plough'g as yesterday
9. Rain, - heavy in the night and storms of rain, hail, and Thunder in the day. | Carting dung, to Whitley fur: and Plough'g Whitley fur:
10. Plough'g Whitley fur: | Drying winds, and heavy storms all round us, - yet pass us. | Oats, - carried Parlourfield after 5 OClock, in fair order to Jockey Barn. | Barley, - carr'd remainer of Marls, removed before to dry
11. Rain, - very heavy in the night dry day, and the glass very low | All Ploughs in Whitley fur:
12. Sunday, - dry day
13. RAIN, very heavy, after the morning, - the Horses, retd from Sow'g Wheat Hitchens | DRILL'G, Wheat, by W. Rummin – uper part of Hitchens run together in the rain totally ruin'd to appearance the Seed Mr Budds, above hill
14. Fine day | Barley, - carried from Batts Gd to Stockwells, - no prospect of its dry'g in Batts, full of Clover: | Plough'g Nfield Vetch's fed off athwart for Wheat | Drilling Wheat Nfield low'r fur Mr Budds Seed, above hill. | Fine day, some mizzle'y rain. | TURNIPS, began feed'g – Ab: Smiths wth Sale Ewes, for the fold, - find'g Hurdles.
15. Fine day, - partial Storms | Plough'g as yesterday | Drill'g – as – yesterday – good work abt 9 Acres in 1¾ days 2¼ Bu pr Acre
16. Fine day, - partial storms | BEANS, finish'd carry'g | BARLEY finish'd carry'g Batts ground removed - see the 14th | Ploughing as yesterday.
17. Ploughing as yesterday. | Drilling Wheat Nfield one earth after Vetches, fed off | Frosty, morn'g, very fine day, but not without some mizzle'g rain very clear, mild Moon light night | VETCHES, carried the lains Nf in the afternoon, and finish d | HARVEST, 12 OClock at Night
18. Fine day, close rain, evening. Drilling finish'd, – both pieces. Ricks, - finish'd, thatch'g. Plough'g – Vetch gd Nf – athwart
19. Sunday, - fine day
20. Fold, - finish'd, Westboro, & removed to Whitley fur: | TURNIPS, N f – Sale Ewes began and drew Wether Lambs, put to Turnips Ewe Lambs, finish'g Smiths Turnips | Plough'g & Sow'g Larks lease Very fine time for it, very fine day.
21. Very fine time for it, very fine day. | Carr'd Smiths, Clover seed
22. Larks lease, finish'd sow'g Westboro, - began sow'g | HAY, Sale Ewes began. | Clover, Horses, finish'd, & Watermead grass. began | Finest Harvest weather now, Glass, - very high | Potatoes, began diging field
23. Ploughing & Sow'g as yesterday Very fine day, like middle June
24. Very fine day, like middle June | Plough'g & Sow'g finish'd Westboro
25. Dung, Cartg to Nf: Potatoe land and Cartg & Hay from Watermead to make it at home. | Plough'g Whitley fur: | Very fine day

26. Sunday, - Very fine day Glass very high to F in Fair
27. Very fine warm day Watermead Hay, - made two loads dry, see 25th | Plough'g as 25th, - and Carting Hurdles fm Smiths, Turnips finish'd | Earth, remove'g fm Cottage Yd
28. TURNIP'S, Breach, Sale Ewes began Fine morning, Rain in evening | Hard Wheat, 5 Sacks of Wm Laws sow'd - the Picked lands Whitley fur | The ground stale furrow to dry to make good work.
29. Plough'g & Sow'g next to yesterday, make good work after the rain. | Hay, the Sale Ewes began the Cowlease Rick | Fine day, - the Odd Men gardening, & removeing earth fm Cottage.
30. RAIN from 4 OClock morn g to the evening & continue'd | Plough'g Vetch land Nfield part of day.

[31.] The heavy rain in the Night, caused the water, to come rapidly down to the Lawn to overflow it, and fill the Cellar to the Ceiling in the morn'g of the 31st of Octr, - 20 Barrels, full as well as empty floating, some Hogsheads full, in the greatest confusion, - yet all uninjured and safe, except about 15 Gall: Raison Wine lost, by loose'g its bung. | Tremendous, boisterous night with heavy fall of Snow, 8 or 9 In deep on the level in the morning, - and continued very heavy in the morning, with heavy rain the remainer of the day, - to cause the great't flood, ever known in any Persons memory. | The heavy Snow lodging upon the Trees, with the boisterous Winds, blew down innumerable large Trees, and the weight of Snow upon others, splinter'd of the limbs, and injured vast numbers of others, 2 & 3 hundred Trees blown down in many single Parks. | Great loss in Sheep drowned | Our Sheep in distress for want of food, - no Hay or Cribs provided, and the rain too heavy for the Men to be in it. Turn'd the Sheep to the Hay Ricks to help themselves, and lay about the Pastures by night, - The Turnips cover'd wth Snow & too wet to put the Sheep upon them. [Diary turned landscape for this entry]

November

1. Oat Straw, - Stock Ewes began Cowlease Barn, - when dry to follow the Sale Sheep on Turnips the Breach. | Dry day | Horses, - geting Timber home and Splinter'd Limbs
2. Sunday, - frosty morn'g fine day
3. Dry day. | Plough'g finish'd Vetch land | Plough'g – furrows, Whitley fur | Wood, geting home.
4. Mizzle'y morn, – Rain drove Teams home | Ploughing, began Bean Stubbles
5. Rain, - all day, Horses in Stable | Vexatious, time, wth the Sheep.
6. Fine day | Ploughing, Bean Stubbles
7. Ploughing, Bean Stubbles | Fine day, - rain, in night | The wind fall Wood, continue geting home.

8. Some mizzle'y rain, fine day | Plough'g – Potatoe land Nfield for wheat, all the land too wet for sowing
9. Sunday, - very fine, the Glass rose unusually high to A in fair.
10. Very fine day | Wheat sowed the middle Piece of the Vetch land Nfield, the first sown since 29th Octr, much better, than if sown before the rain.
11. Very sharp frost & very fine day, glass rose to R in fair and the land cover'd with Spiders webb. | Plough'g & Sow'g Turnip land Nfield to wheat, a very fine time for sow'g. | Larks, lease & Westboro Wheat just appearing – the first sowings look'g well | All the Sheep, on Turnips, Breach
12. Plough'g & Sow'g as yesterday The land in the morning too hard to harrow, very sharp frost & fine day
13. Very sharp frost & fine day Finish'd Nfield sow'g wheat, and began Plough'g the Turnip land Breach for Wheat | 299 Young Sheep, sent to Littlecote to feed off - Mr R- Turnips, have'g too many, - find Bay & Hurdles no charge.[17]
14. Plough'g & Sow'g the Breach to Wheat, - in the morning the Frost almost too hard for the Plough, - Thaw in the evening & very mild.
15. Mild, fine day | Plough'g, the Clover stubble, after the Fold Whitley furlong, for Wheat, - very mild & kind
16. Fine warm Summers day Sunday.
17. Plough'g & Sow'd the Fold drift, Clover stubble Whitley fur: Fold'd after | Raw foggy morning, fine day Triming 200 Wether Lambs Triming 30 Wether [...] tooths for Mr Guy by 2 men & the Shepherd. | Mr Potters, Furniture, brot to the Cottage, Mr Potters Man slept in the House – 15th | Chalk removeing from Cottage Garden | Wheat Rick, old, - put into the Barn, - 3 now out.
18. Wheat sowing the Clover Stubble Whitley furlong. | Chalk Cart'g &c | Very mild, foggy day. | 200 Wether Lambs 30 two tooth Wethers sent to Chip[penha]m
19. Teams, at Work as yesterday. | Fine mild day.
20. Fine mild day. | Ploughing the last fold drift on Clover lay, for Wheat | Raftering Smiths, long ground &c | Carting Chalk as yesterday
21. Carting Chalk finish'd | Sow'd the ground Plough'd yesty Taunton Wheat | Raftering as yesterday | 5 Men Triming Sheep for Marlbro Fair 200 Ewes 100 Wethers [...] Culs | Beautiful, - fine clear mild day.
22. Dry dull day, Glass sink'g | Wheat sowing, - finish'd last lands, - Breach, wth Taunton Wheat. | Turnips, - finish'd Breach Turnips – began Sweardown. | Wood, - Carting, windfalls
23. Sunday, - very fine day.
24. Marlbro Fair, - very mild fine day. | Ploughing Bean Stubbles Nfield
25. Ploughing Bean Stubbles Nfield |Cart'g Wood, - and Bricks | Unusually,

17 Mr R – Bryan Rumboll, WSA: A1/345/210, *Land Tax Assessments, Hilmarton, 1780-1833.*

fine & mild Glass rise'g again to Fair | STRAW Pease, the Horses began haveing finish'd the Clover Hay | Oxen, - discontinued working except'g one Team.

26. Damp air, - very mild fine Weather. | Plough'g as yesterday. | 90 Bu: Soot @ 9d fm Hungerford large Waggon full, board'd raves.[18]
27. Ploughing as yesterday | Wood, Carting fm Dells Copse | Fine dry day, damp evening
28. Rain mizzle'y all day at times heavy'r | Ploughing &c as yesterday.
29. Ploughing &c as yesterday. | Mizzle'y rain all the night and all day | Mr Potters House finish'd.
30. Sunday, - dry morning, mizzle'y afternoon.

December

1. Plough'g Nfield Upr Furlong | Some mizzle'y Storms, Miss P[...]r slept at Cottage 1st Night
2. Some mizzle'y Storms, | Plough'g as yesterday
3. Plough'g as yesterday | Frequent Storms, and very boisterous Night.
4. Very heavy cold storms | Plough'g finish'd Nfield up fur | Plough'g Wansdyke Fur: began | Oats, - Jockey, - 2 Men, began Thrash'g | Ewes, - at the Straw, Jockey Yard | Ewes, - Hay, - began, Nights, the Straw morn'g's, Turnips Mid:day
5. Ploughing as yesterday | Carting, - strand Gravel to Mr Potters | Dry day.
6. Rain heavy in night damp day. | Plough'g as yesterday, and Cart'g Sand to Mr Potters. | Glass very low
7. Glass – rise'g very high to R in fair Frost and fine day, Sunday
8. Plough'g Wansdyke furlong & Cart'g dung from Colts Pen, & Carting Stones | Mizzle'y rain at times
9. Fine day | Ploughing & Carting Stones as yesty
10. Frost, sharp, in morn'g damp day | Ploughing as yesterday | Inquest, held on Wm Breadmore who slept in our Straw house Wednesday night 3rd Inst and died Friday morn'g 5th Witness's Robt Hacker Jno Barnard Jno French Rd Westall Wm Lea Jas Cannon Jno James Wm Chandler Our own Labourers And 5 others Verdict Died a natural death from bodily & mental debility. A false report propogated by Jno Bird, that myself knockd him down, & kickd him about.
11. Plough'g began Kite hill Fine day, mizzle rain in evening
12. Very fine day cold air | Plough'g as yesterday
13. Plough'g as yesterday | Frost morn'g, fine day
14. Sunday Frost morn'g, fine day Chip[penha]m
15. Fine day | Plough'g Kite hill | Bricks fm Kiln
16. Very sharp frost last even'g rain morn'g – fine day | Plough'g as yesterday |

---

18 Raves - Waggon rails, sometimes applied to the flat woodwork projecting over wheels from side of forward part of waggon.

2 Loads, Gravel, for Mr Potter

17. Plough'g as yesterday till mid:day when stop'd wth the Rain | Damp morn'g, rain all day after very boisterous evening.

18. 299, - Young Sheep, returned from Littlecote Turnips, 5W [19] | Frost, - & calm morning, a Snow storm - & fine day. | MARKET, improveing, sold old Wheat. 12 Sco: Nt @ 33s worth last Year this time – 20s only | Ploughing, finish'd Kite hill

19. Ploughing began Snail hill after the Fold, | Snow, - in the night to cover the land, remain all day. | Very fine clear freezing day | Wheat Rick, small old, put into Barn.

20. Unusual sharp FROST all last even'g till 12 OClock when rapid Thaw commenced with mizzle rain and continued steady till 3 OClock afternoon Horses in Stable all day. * | After 3 OClock, - put a cut of Barley into Stockwells Barn, - having no straw for Oxen or Horses. | Young Sheep & lambs HAY, twice a day, - old Ewes, - Straw Yard Jockey mornings, hay evening all Turnips mid=day. Bonnings.

21. Sunday, - slight frost, fine day Watermeads, - began dress'g the 18th Inst Oxen & Cows feed'g till then

22. Damp morn'g, - dry & mild after. | Plough'd Burntmill Gd and Plough'g Snail hill | Sow farrow'd 4

23. Plough'g began Dean heth | Mizzle'g rain morn'g, heavy'r at mid:day, Horses retd fm Plough'g

24. Thick rain all the morning to keep the Horses in Stable

25. Mild dry morn'g, mizzle'g rain mid day, - and heavy'r in even'g

26. Mild, dry morn'g mizzle y rain in evening | Plough'g Dean heath

27. Plough'g Dean heath | Rain in night, mild damp day, - thick rain in even'g

28. Sunday, - frequent storms of mizzle'y rain. Tock[enham]. Thunder & light'g in night.

29. Cold dry winds after the morn'g light'g in the evening. | Barley Rick put into Barn Plough'g – Dean heath

30. Plough'g – Dean heath | Storms of Mizzle'g rain continue frequent

31. Fine drying day, - fine evening Glass riseing | Plough'g Shepherds ground

1824

January

1. RAIN, heavy in the Night to cause a flood, - Storms in the day boisterous evening. | Ploughing Shepherds ground fallow'g for wheat. | Sheep all feed'g Turnips Bonings corner, - very wet time for it now, and all the last month, - yet the Ewes do well, the Straw yard, mornings help them much, - and the Lambs improved since returned from Littlecote. – The Ewes make good work with the Turnips, after the Straw yard, - follow'g the Lambs

2. Very boisterous drying Winds | Ploughing as yesterday.

---

19 5W - 5 weeks.

3. Ploughing as yesterday. Very fine day [Note in pencil] 4 Porkers 180 li 4 Porkers, say 40 – 60 - 40 & 40 li
4. Sunday, - very fine day, frost
5. Frost'y morn'g fine day. | Plough'g Shepherds Gd.
6. Plough'g Shepherds Gd. | Frost, morn'g, some mizzle'y rain after
7. Sharp frost morn'g Thaw even'g | Plough'g finish'd Shepherds, began the Cowlease.
8. Plough'g as yesterday. | Sharp frost morn'g, Thaw even'g
9. Damp mild day. Mr G_y, Tock[enham] | Plough'g as yesterday
10. Plough'g finish'd Cowlease, began plough'g Lawn Piece | Mizzle y rain all day.
11. Sunday, - fine mild day
12. Very sharp frost, - to nearly stop the Ploughs, freezing all day in shade. | Plough'g Lawn Piece.
13. The Frost, too hard to admit the Plough | Carting dunt to oldlay Nfield Church furlong for Turnips and Wheat.[20]
14. Carting dunt to oldlay Nfield Church furlong for Turnips and Wheat. | Sharp frost & very fine clear day. | Old Wheat Rick by Barn, took in
15. Frost, & Carting Dung as yesterday
16. Frost, & Carting Dung as yesterday | Sheep, finish'd fold'g Snail hill | Oxen taken to Jockey Straw Yard
17. Sheep removed, from Turnips Southfield, to Pedlers Piece, Turnips The Young fold'd after, - The Ewes pen'd Cowlease Barn Yard, Straw mornings, - Hay evenings. | Carting dung as yesterday | Mild frost, - some Snow
18. Very mild thaw and fine day | Sunday
19. Very fine mild day. | Carting dung up'r Church fur for Vetches & Wheat.
20. Carting dung up'r Church fur for Vetches & Wheat. | Slight frost morn'g, very mild fine day
17. [Note in pencil] Barley Rick in Barn
20. 4 Young store Pigs put to fatten 4 Large 3 yr old Pigs put to fatten
21. Plough'g the Dung in, the oldlay Nfield for Turnips & Wheat | Very fine mild day,
22. Very fine mild day, heavy rain even'g | Plough'g as yesterday
23. Plough'g as yesterday | Cold dry winds
24. Plough'g as yesterday and finishd | Very fine mild day, T Crook to Newbury
25. Sunday, - fine mild day
26. fine mild day | Plough'g finish'd Snail hill
27. Plough'g Lawn Piece, - Cartg Stones, and 20 Qr Black seed to be mill'd | Very fine mild May like evening
28. Cold winds, some rain, hail & Snow. | Ploughing as yesterday, and Carting Stones
29. Ploughing as yesterday, and 20 Qr Blk seed to Mill | Cold drying day

---

20 Pinniger wrote *dunt* but meant *dung* – see also entry for January 14.

30. Cold drying day very sharp frost in morning. | Fallowing gt bottom Efield for Swedes.
31. Fallowing gt bottom Efield for Swedes. | Cold fine day

February

1. Sunday Cold fine day
2. Very fine weather, - the land very healthy, - put all the Oxen to work, makeing 7 Ploughs fallow'g the great Bottom for Swedes *
3. Fallow'g the great Bottom for Swedes. | Very fine mild, - day * | The Sheep, - enjoy'g themselves on Turnips Pedlers Piece, the Ewes do very well in Straw Yard Hay & Turnip Shels, - the Young Sheep much improved *
4. Rain, in night, fine drying mild day | Ploughing Clay furlong for Vetches & Beans. | Fat Pig 7 Sco: 15 li the first
5. Fine day Plough'g as yesterday <u>Markets, advancing</u> Sold 19 S[acks] old thin Talavera 11 S[core] 5 li Nt 23 S[acks] old red Wheat 12 S[core] Nt 24 S[acks] old red Wheat 12 S[core] 5 li Nt [Total] 66 S[acks] @ 34s 6d The Talavera not worth 14s the year it grew 1822
6. Dry day | Ploughing as yesterday, and Carting Stones.
7. Ploughing as yesterday | Mizzle'g rain at times all day
8. Sunday, - fine mild day
9. Mizzle'y rain morn'g, fine & mild after | Plough'g , as yesterday. | 31 Lambs from the 30 Ewes the new Rams, put to to save Rams from | Pease, Beans & all small Seeds, - and Cabbage Plants put in the Garden. the first
10. MARE, foal'd, - in the Night from Mr Vaiseys Gray | Ploughing as yesterday | Rain in morn'g, fine afternoon
11. Very fine mild day | Ploughing finish'd Mushings | 60 Lambs
12. 80 Lambs | Ewes & Ewes wth Lambs brot from Pedlers Piece to Standing Pen Stockwells Young Sheep, remain on Turnips Pedlers Piece, - a few days to finish | Plough'g Nfield oldlay for Vetches | Timber geting home, and Hurdles removeing | Cold winds, and rain'y evening
13. Ploughing as yesterday, finish'd | Carting Timber & Wood [More notes on the economics of fattening pigs – see Appendix A] Dry day, - Glass sinking very low, - <u>SNOW</u> began 6 OClock eve'g
14. Snow, in the night, 8 & 9 In deep in the morn'g, on the level, and rain or snow all day | Horses & Oxen, in Stable &c. Men & Boys assist'g Shepherd.
15. Cold dry day. Sunday 120 Lambs
16. Warm, clear fine day, very sharp frost morn'g | Young Sheep at Turnips, absent two days | Ewes & Lambs, put to Turnips | Carting Wood from Dells Copse.
17. Carting Wood from Dells Copse. | The Bean, rick on the ground put into the Barn, and began thrashing it wth Horses | Very fine day.
18. Frost, & dry morn'g, heavy storm of rain & Snow, - afternoon. | Carting Stones from Copes Grounds to the Road, through the Copse.

19. 200, Lambs. 1 Ewe died after Lamb'g 8 Ewe Cast their Lambs &c. | Dry morn'g rain even'g | Plough'g Labourers Potatoe Land, Nfield
20. Seeds – 10 Horses brot from Ramsbury | Rain, - all night, and all day with some intervals of dry, to cause a flood.
21. Fine dry day | Plough'g the dung'd ground Nfield for Vetches.
22. Sunday, fine day
23. Glass rise'g, nearly at top, and wind <wind> North, with mizzle'y rain all day. | Horses, - moveing Hurdles from Pedlers Piece, others in Stable. | 240 Lambs – with only 2 doubles. 3 Ewes - died after lambing. | Turnips Pedlers Piece, - finished began feeding 17th Jany. The Ewes removed - 12th Febr – 8 Acres. | About 30 Acres, - of Swedes & Turnips remain to feed off, - The Turnips growing fast to green, - part nearly in bloom. The Swedes, - not much to green enough to gather for boiling *
24. Fine drying day | Barley rick, by Barn put into Barn | Horses in the Stable, the land being to wet for Plough'g or Carting of Stones
25. Carting Stones, from Pearces Piece Very cold drying day | Pease, Rick, put into Barn,
26. Vetch Rick, put into Barn, | Carting Stones from Marls. | Cold storms of hail & Snow
27. Snow, - all day, - light'ly Thrash'g - both Machines wth Carters & Boys
28. Fine warm Summers day after the morn. | Horses in the Stable, - the Men put'g Oats into Cowlease Barn
29. Sunday, - cold dull day. 350 Lambs.

March

1. Fine mild, calm Sunshine day | Carting Stones to Copse road finishd
2. Carting Chalk to Copse road finshd | Very dry'g cold freez'g wind,.
3. Very cold storms of Hail & Snow and high winds in evening. | Carting Chalk as yesterday
4. Carting Chalk as yesterday | Very sharp frost & fine day.
5. Storms of Rain | Carting Earth from Bagshot lane to Dean heath, to prevent its being Indicted [21]
6. Carting Earth as yesterday | Some storms, - but fine day,
7. Sunday, - Hurricanes of Wind & rain all morning, - calm mild evening.
8. Hurricanes of Wind, and driveing rain all the morning, - calm even'g | Horses in the Stable, - Men & Boys Thrash'g wth Machine
9. 2nd Boat load of Coal Ashes | Ashes, - began sow'g about 30 Bu: pr A @ 4d | Mild, - fine day, - some light storms.
10. Dry morn'g – began sow'g Ashes Rain & Snow, all day after 10 OClock, - Horses, returned to the Stable,

---

21 It is possible that the earth was causing an obstruction which may have required action by the Justices at the Quarter Sessions. Bagshot is a tithing to the north of Shalbourne, around five miles to the east of Little Bedwyn.

11. Sharp Frost, & dry morning. | Rain heavy all afternoon | Ashes – sow'g
12. Ashes – sow'g Fine day except, light storms of Snow
13. Very cold & heavy storms of Hail & Snow. | Ashes – finish'd sowing 17 Hd Bushls @ 4d Deliv fm Bath | Stone Cartg finishd
14. Sunday. – fine day
15. Fine day, - rain in evening | Plough'g – 6 Teams, thwarting Kite hill for Turnips, - all the Land too wet for Plough'g for Corn No Plough'g the last 3 Weeks. | Rolling, - began the Sainfoin
16. 7 Ploughs as yesterday | Heavy rain in night, dry day
17. Very mild, Summers day | Ewes & Lambs, removed from the standing Pen, & fold'd, - in Pearces Piece, - wth a standg Pen, when wanted adjoinging | Ploughing as yesterday
18. Ploughing Wansdyke furlong athwart, - very foul after Sainfoin for Turnips | VETCHES, - began Sowing Nfield the land too wet & cold before. 3 Sacks of new, 3 Bu pr A | Oat Rick from Marls put into Barn. | Barley Rick from Batts gd put into Barn. | Very fine mild day.
19. Very fine mild day. | Plough'g as yesterday | Vetches sow'd 3 Sacks Nfield tail 4 Bu pr A. | Sheep: Ewes & Lambs, discontinued Hay = have'g a load of Oat Straw Mornings, - better than the Hay, & Evenings – a load of Barley Straw in the standg Pen & fold'd with plenty of Turnips The Young Sheep, - Straw evenings Hay mornings, at Cowlease Barn & Turnips, Sweardown. ★ | About 15A: Swedes & 3A Turnips remaining.
20. Foggy morn'g, fine day. | Oat Rick the last part put into Cowlease Barn | Vetches, spring sow'd 8 Sacks of tail new - 1 Sack pr A – and 11 Bu of head old 3 Bu pr A, - in Mushings & Crab tree Piece. | 3 Ploughs in Wansdyke Fur:
21. Sunday, - thin rain all the morn'g – dry afternoon:
22. Dry Starlight night, dry morn'g till 7 OClock, - when commenced small rain till 8 OClock, - and succeed'd with heavy Snow, in the largest blossoms ever remembered, which continued all day with storms of Rain & Hail in afternoon. Sheep in the standing Pens all day Horses in the Stable. Men in the Barn. ★
24. Thick Snow in morn'g, - Sunshine for 3 or 4 hours after, - heavy hail storm & Rain afternoon. | Horses in Stable, some attend'g Sheep[22]
24. Carting Dung to Nfield for Vetches | Very cold damp day
25. Very cold drying winds | Ploughing Wansdyke Furlong & | Rolling Field grass
26. Plough'g as yesterday & finish'd & Rolling grass | Ploughing, began the Lains for Oats | BEANS, - began Drilling, Clay fur: New 3 shear | Wheat Rick, put the last Old into Barn, - 6 new in Yard | Fine day, - except, light hail Storm
27. Fine day, - except, light hail Storm. | Drilling and Ploughing as yesterday

22 This should probably be the entry for March 23.

28. Sunday, - some slight cold hail Storms. M: lent
29. Very sharp hard frost fine day. | Plough'g Nfield for Oats. | Beans, - finish'd drilling Clay fur: 10 Sacks [...] [Pencil note] Acre
30. Fine day, - with some slight storms of hail & rain | Barley, Rick Marls, put into Barn. | Drill'd Small gray Pease to to try the Machine and Vetches Clay furlong
31. The Turnips, finish'd feed'g nearly in bloom & good, - the Swedes not forwd | Pierce'g cold, freeze'g winds all day | 5 Ploughs, in the Lains for Oats | 4 Horses & Waggon to Newbury with 6 Fat Pigs

April

1. Sharp frost, and cold dry'g winds till 2 OClock, - when commenced Snow & Rain, continued all even'g | 6 Ploughs – stiring Nfield for Barley, - the Turnip land Bonning Corner, too wet to Plough for Oats
2. Rain in night and storms in day dry even'g The Land too wet to Plough. finish'd Cartg Wood from Dells Copse
3. Fine day | 7 Teams, Plough'g Nfield Up furl: athwart, - for Drill'g Barley. [More notes on fat pigs – see Appendix A.]
4. Sunday, very fine mild day Cow[...]h [23]
5. Fine warm, Sunshine day | Plough'g finish'd Nfield
6. OATS, - began Sowing Up: fur: Nfield 16A – about 6 Bu pr A Dutch seed from Cowlease. | Harrow'g Barley Gd for Drill'g | Barley Rick, put into Barn | Fine day, cold dry'g wind | Hay, the young Sheep discontid hav'g Oat & Pease Straw twice a day, and Swedes, - do well.
7. Mizzle'y rain all morn'g, dry eve'g. | Plough'g Bonings corner for Oats, - (Turnips fed off)
8. BARLEY, - began Drilling Nfield above the road | Ploughing as yesterday | Fine day, - cold air.
9. Fine drying day, - cold air | Drilling as yesterday | Plough'g Sweardown for Oats = Turnips fed off – too stiff to sow | Fat Pig, Kill'd 9 Sco.
10. Bitter cold storms of hail & Snow | Glass sink'g, - , - stop'd – drilling Barley | Plough'g Sweardown | Horse from White hart, Chip[penha]m to work for its keep.
11. Sunday, - Snow, to cover the ground in the morn'g, and heavy storms of Hail and Snow, afternoon, very cold backward Spring, great want of Hay & sheep Keep with many Farmers.
12. Fat Pig 10 Sco 2 li | 7 Ploughs in Sweardown, & Lane Piece after Turnips for Oats to dung to work yet. | Rain in night & morn'g drying day. | 14 li Mill'd hop ½ Bu good Ray } Sow'd in Lains with the Oats pr Acre, - the hole in Machine about a Qr
13. Barley, Drilling Nfield upfur 8 li own Mill'd Hop 5 li Mr Rumbolls Broad

23 Cow[...]h - probably Cowitch – Pinniger had relatives there – modern-day name is Cowage Farm in Compton Bassett parish.

½ Bu Mr Guys Ray } pr A ¼ hole Machine | Ploughing Lane Piece | Harrow'g Wheat Nfield | Thrashing Oats Cowlease Barn finish'd 25A: (The 6 & 18 A) – began the 1st Novr, - the Sheep only served with the Straw & do well with it, - being 5½ Months at Straw. * | Sharp frost, cold air Sunshine day

14. Sharp frost, mild'r air Sunshine day Barley sow'g finish'd Nfield [In pencil: 24] Acres wth 22 Sacks of Barley from Parlour bottom, seeds and above. | Plough'g as yesterday. Labourers Potatoe Land alloted. | Dry Sheep removed from Cowlease Barn & fold'd East field bottom for Swedes. & put to Straw Stockwells Yd Night & Morning
15. Fine pleasant morning, - cold harsh winds, evening. Market advance'd sold our last old Wheat 12 S 6 li Nt @ 37s, - such sold when harvest'd @ 20s & under pr Sack | Ploughing Sweardown lowerfur: after Turnips, for Barley. | 1 Team Carting Dung from Barn Dean heath to Cowlease Gd for Turnips
16. Teams at work as yesterday till 10 OClock, when stop'd wth Rain | Dry night & morn'g, - very cold driving Rain all day, backward Barren Spring, - no grass except Watermeads | Good Friday
17. Mizzle'g rain in the night & Morn'g to keep the Horses in the Stable till 10 OClock, - drying remainer of day. | Plough'g dung in oldlay Nfield for Vetches & Wheat.
18. Sunday, - fine mild, - glass rise'g Cold frosty morn'g.
19. Very fine drying day, | Rolling Barley & grass, | Plough'g – Swear down, low'r fur | 2 Teams, to Marlbro to change 40 Sacks seed Barley wth Mr Budd
20. Barley & Oats, Roll'd Nfield up: fur: 40 Acres 2 days 2 Horses | Ploughing Dean heath athwart for Barley. | Wheat Rick, the first New put into the Barn | Very, fine warm sun
21. Very, fine warm sun Cart'g dung – as the 15th | Plough'g &c Dean heath, & Drilling Barley Dean heath began Mr Budds, Seed 14 li Mill'd hop, 2 Pecks Mr Guys Ray
22. Rain in the night, dry & mild morn'g, - at 12 OClock commenced heavy storms of Hail & Rain at Marlbro, & below, with Thunder and Lighten'g, frequent, hail stones size of Damsons, - break'g Windows At Bedwin very little rain Plough'g Dean heth & Drilling Barley all day | Carting, dung as yesterday.
23. Very heavy storms of Rain all day Horses, - between the Storms, Cart'g Dung from Barn Dean heth.
24. Glass riseing, fine drying day | Plough'g Dean heth, and Cart'g Dung as yesterday | Cow Calved | Cow died 21st | 2 Sows farrow'd
25. Sunday, - fine mild grow'g day, - watermead grass much improved | Fold the Ewes & L: finish'd Nether Pearces Piece
26. CUCKOO, - first heard | Fold removed, to Lawn Piece for Swedes. | Plough'g, Cowlease, for Turnips & Lawn Piece for Vetches. | Thick –

driveing rain all day

27. Fine mild morn'g, cold rain in the even'g | Plough'g Dean heth for Barley | HAY, the Sheep discontinued having [...] Tons of Sainfoin, and [...] Tons of Meadow only left for the Horses & Oxen. | Vetches up fine first sowing. | Pease Beans, Barley & Oats appear'g

28. Lambs, Tails cut 404 | Round tail'g dry Sheep Round tail'g Ewes &c. by 4 Men | Thick mizzle'y rain all day | Plough'g – began thwart'g Nfield for forward Turnips, being too wet for Barley work, - Teams drove home mid:day

29. Nag mare, - put to Wornham, young (3 yr old) Gray Cart Horse | New Wheat, - the first have sold Devizes Market 12 Sco 4 li Nt @ 32s nearly the best in the Market, old sold for 37s 6d | Mizzle'y rain in morn'g, very drying remainer of day | Ploughing Nfield for Turnips

28. Standing Pen removed from Pearces Piece to E field bottom & Ewes & Lambs fold'd Lawn Piece for Swedes.

30. Glass low, - rough drying winds with partial storms | Barley the last Rick put into Barn. | Swedes, - the last tie to E & L | 320 Dry Sheep to Littlecote watermeads [...] pr A | Grass watermead began mow'g for Oxen & Horses behind the Mill. | Vetches, - drill'd about [...] Acres Nfield with [...] Bushls 3 Yrs Old | Barley, drilling, - Dean heath stop'd sow'g a week, with the wet weather. | Fat Pig Kill'd 12 Sc 3 li

May

1. Some mizzle'y ran, - fine grow'g day. | WATERMEADS, - Ewes & Lambs began | Ploughing as yesterday Lane Piece after Swedes, very stiff. | Drilling Barley as yesterday. | The Watermead grass grown the last few days, - rapidly, - the beds before quite bare, - now very thick, & lodged, - the Wheat also, very much improved *

2. Sunday, - Beautiful, - mild warm, grow'g morn'g, rain in evening.

3. Very cold winds all day, with mizzle'g rain. | Drilling Barley finish Dean heath. – and sow'd 12 Sacks Oats – Swear down & 4 Sacks Bonings Corner, Mr Washbournes Dutch 37½ li pr Bu @ 27s 6d Qr very short & good 21 Sacks Mr Budds Barley 14 li Mill'd hop, own seed 3 Pecks Mr Guys Ray } pr A ⅓ of hole in Machine

4. Fine mild day 4 Sacks the remainer of Mr Washbournes Oats, sow'd in Bonings Corner to mid: of Pit | 12 Sacks our own Oats continued sow'g Lane Piece | Seeds, - sow'd Mr Guys Ray Swear down to withing 1½ Lands of the Corner of Copse | 12 Women hoe'g the blind wheat Nfield

5. Seeds – contd sow'g Swear down Bonings corner & Lane Piece 3 Pecks of Mr Comptons Church bents 14li Mill'd hop 7 li Mr Rumbolls broad pr A | Soot, sow'g on Wheat 5 Cart loads Nf 2 Cart loads Ef 3 Cart loads Sf 3 Cart loads Sf, Ashes | Raftering Swear down lower fur: after Swedes being very grass'y, to Plough it athwart very thin next time, to harrow out the grass

6. Delightful mild morn, some heavy Showers afternoon Vegitation, most rapid Plough'g as yesterday & Rolling Sweardown.
7. BARKING, began Bushels gd & Dells Copse. Plough'g athwart Rafters Sweardown | Fine day
8. Fine grow'g morn, - very partial Thunder Showers afternoon Seeds sow'g in Wheat Westboro Larks lease & Hitchins 20 li Tail hop 3 li Broad 2 Pecks Ray } pr A, - the Machine about ½ hole. | Drag'g, - Harrow'g & rolling Sweardown lower furlong very tuffet'y, to rough for Barley
9. Sunday, - very warm fine day
10. Beautiful, fine day, the most pleasant this, Spring JL to MAS | Plough'g, Harrow'g & Rolling Sweardown, - too knobly & tuft'y to Sow Barley | Harrow'g Smiths ground rafters | Hoeing drill'd wheat Hitchens & sow'g Seeds, & harrow'g it in the Wheat. | Garden seeds all sorts sow'g all fail'd before. | Pig the last fat one Kill'd [In pencil] 11 S 8 li
11. Very cold benumb'g, - dull blight'g air | Plough'g Smiths ground rafters, for Oats. | Plough'g Burntmill Gd for Barley.
12. BEANS, 17 Women began hoe'g @ 1s pr day ea | Cold chilling air, with some blight'g rain, - almost too cold for Women in the field complate contrass wth the 10th | Plough'g & drag'g Smiths gd
13. Thick mizzle'g cold rain all day | Plough'g oldlay Nf & sow'g Vetches, make good work with the Rain
14. Steady soaking rain all night & all day * | Horses in the Stable * | Carters, lowering the earth in Barn, for a New hollow Floor * | Watermeads too wet for Sheep, in Pearces Piece all day, being grass'y after the Fold lay in Standing Pen with Barley Straw *
15. RAIN incessantly, much heavy'r than yesterday Night and day, to cause a flood, - greater than known before in many places * | Ewes & Lambs in Dells, and lay in standing Pen *
16. Sunday RAIN, all night, - nearly 3 days & 3 nights rain, fine day
17. Fine day, mizzle rain afternoon | The rain 14th & 15th, favorable for working the stiff land, - sow'd 8 Acres Barley chalk'y part of Swear down lower fur, - and 5 Acres Oats Lane Piece, and 8 Acres Turnips NfieldChurch furlong. | Plough'g Nfield oldlay for V [24]
18. Plough'g Nfield oldlay & sowed 8A Vetches – 3 Yrs Old 3 Bu pr A Turnips, - sow 8A Nfield very good work | Carting Tuftets off Barley land Fine day, mizzle rain afternoon | Lambs, cut 178 226 Ewe Lambs [Total] 404 cut more than 60 pr hour
19. Rain heavy in night, very fine day, - little rain evening. Plough'g & Sow'g Vetches Pearces piece, for Horses at Harvest. | Cartg, Tuftets off Barley land | Watermead East finish'd feed'g
20. Watermead Mrs Potters began feedg | Barley straw gave the Ewes & L's the last this morn'g in standing Pen, - fold'd Lawn Piece, - the Ewes began

24 V – Vetches.

Straw 1st Novr and continued to this day. * | Plough'g & Sow'g Barley behind the Public house, - waited to clean it of the Tuftets. | Fine day, - first fogg'y even'g

21. Very sharp FROST, the Potatoes and Ash & Oak leaves cut off | BARLEY, finish'd sowing, - Burntmill ground, and Swear down lower fur: | 2 Nag Horses, Roll'd Dean heath 25 Acres (1 day) | Carting Bricks from Kiln to the Culvert, Barn.
22. Finish'g off Barley land, and Plough'g Pearces Piece for Vetches | Carting fold &c to Wansdyke f | Fine day, some blight'g rain
23. Fine day, some blight'g rain | Dry sheep retd home from Sunday Littlecote Watermeads, began 30th April 4 Acres – 320 Sheep @ 45s | Ewes & Lambs, began feeding Burntmill mead, finish'd Mrs Potters mead. 4 days feed'g mead [...] Acres @ [...] and the fold last.
24. Fold Wansdyke furlong for Turnips. Barking, finish'd, 4 Men – 13 days each Trees, - small & large 217 Yards of Bark 3 Tons 4½ cwt @ 20s/Ton for throwing and Barking. | Draging & Plough'g Smiths ground for Oats | Little blight'g Rain
25. Fine day Plough'g Smiths ground & sow'd 4 Sacks Dutch Oats
26. Sow'd – 6 Sacks Dutch Oats & finish'd, - the land too wet to sow before this time | Plough'g in Vetches athwart La<ne>wn Piece, - 3 Bu: old pr A | Mare, - cover'd again | Very fine warm day | Hoe'g drill'd Pease & Vetchs
27. Very fine day, - the warmest and softest air, this Year The Wheat, Barley &c very much improved, since the last two days, - the Wheat turned from Yellow to Black. – and the first sowing Barley begin to Curl | Plough'g & Sow'g Vetches Lawn Piece, - Rolling & Harrowing Smith Ground
28. Smith Ground Harrowing | The heat oppressive, - very grow'g | Bark, - carry'g & faggots
29. Bark, - carry'g & faggots finish'd | Thwarting Lawn Piece for Swedes | Ewes & Lambs finish'd feeding Watermeads, Burnt mill Mead & began feeding Field grass Parlour bottom, - the Dry Sheep in Dells | Warm fine day
27. 3 – 3 Yr & 3 – 4 Yr old Oxen Welch from Chipp[enha]m
30. Sunday, - Fine grow'g Day, wth some, mizzle'y Blight'g rain
31. Very fine warm day. 8 Men 12 Women Waterfurrow'g Smiths Grod | Thwarting Lawn Piece for Swedes | Barley, last sow'g, appear'g | 4th Farrow of Pigs 9 5th Farrow of Pigs 11

June

1. Warm fine day | Thwarting great bottom for Swedes | Carting Stones from Pedlers Piece to Road | 1 Horse roll'd Smiths Grod 3½ hours [In pencil: 7½] Acres
2. 1 Team finish'd Cart'g Stones from Pedlers Piece | Very warm fine day, some mizzle'y rain even'g | Plough'g as yesterday. | Watermead, finish'd cuting

for Oxen & Horses behind the Mill, - began it the 30th Apl | Turnips, - sown 17th May shewing rough leaf attack'd by the Fly, very doubtful, their standing | Men thorough'g up of Dung in the Yard, till the Grass ready for mowing

3. Young sheep began feeding the Hop Clover Batts ground | Cold morn'g warm evening | Plough'g as yesterday, and Vetches, sow'd the last of Lawn Piece.
4. Fine warm day | Plough'g athwart Gt Botm
5. Plough'g finish'd Gt. Botm | 8 Old Oxen retd to Chip[penha]m | Very cold rough winds all day
6. Sunday, - beautiful & fine
7. Beautiful & fine Plough'g Pedlers Piece & Bricks from Kiln
8. Bricks from Kiln 5m | Very warm fine day | Beans hoe'g 2nd time, by the Men, - Women weed'g
9. Very warm fine day | Plough'g Cowlease ground athwart for Turnips
10. RAIN, steady from 6 OClock morn'g till, 1 OClock afternoon mizzle'y rain after, - very acceptable for late sown Barley and Oats & Vetches, wanted rain | Turnips, - sown 17th May now secure from the flea, - much injured and very doubtful before, - has a rough leaf size of half a Crown. | Horses in the Stable
11. Draging & Plough'g Wansdyke Furlong. | Vetches – Drill'd the last sow'g this Summer, - bottom of Pearces Piece for Horses – 4 Bu pr Acre, old | Dry day, - dull and growing
12. Fine Sunshine day, some mizzle'y rain in even'g | SWEDE, seed began sowing about 4 Acres fold'd, lawn Piece old Seed. And about 8 Acres Efieldbotm new Seed, not dung'd. | Plough'g & drag'g as yesterday | MOWING, - 8 Men began Field Grass Marls.
13. Sunday, - Fine all the morn'g rain all the even'g JACOB SPACKMAN TY[therto]N dd 62 [25] [This entry written within a thick black border]
14. RAIN all night and morn'g drying afternoon | Horses in Stable, Men turn'g dung in Morn'g Plough'g Snail hill, afternoon for Turnips | WHEAT EARS, Forebridge Piece the first seen burst'd, out [A historic record of ear-bursting and reaping dates for Wheat then follows – see Appendix C]
15. RAIN heavy all night, and morning till 10 OClock, dry afternoon, Horses in Stable morn'g, carry'g Dung in Waggons fm Pearces Piece to Efieldbottom afternoon
16. Very fine growing day. TURNIP=HOE'G, began Nfield Church furlong lower sow'd 17th – 18th May [A historic record of dates for the start of hoeing then follows – see Appendix C] Bean-rick, - small one put in Barn | 7 Ploughs, - Snail hill for Turnips | Mowers stop'd 14th & 15th

---

25 Jacob Spackman of Bremhill was buried on 18 June 1824 at Bremhill, aged 62, *Wiltshire, England, Church of England Deaths and Burials, 1813-1916* [accessed via Ancestry, 17 July 2020].

17. Rain in morn'g, fine day Jno Large [...] Bake} Lyneham Marrd, Jna Henley, - Died [26] | Plough'g Kite hill
18. Plough'g Kite hill | Very clear warm fine day BATH
19. Dry morn'g – RAIN 10 OClock & at times all day, - Horses retd from Plough
20. Sunday, - dry morn'g RAIN all the afternoon
21. Plough'g finish'd Kite hill | Harrowing Pedlers Piece and Cowlease for Plough'g & Sow'g | Fine day
22. Fine day | Plough'g & Sow'g Turnips Pedlers Piece. | 9 Women hoeing Turnips N field, - ready before | Hay Turned 1st Time
23. SHEEPSHEAR, - wash'd 18th, 808 Sheep, - 9 Shearers 1 Colt | Dry night & morning till 6 OClock, when mizzle'y rain began, continued all day by Storms till 3 OClock, when prospect of fine, turned the Sheep out of the Barn, and at 5 OClock, a heavy Thunder Storms, & rain'd all the even'g | Plough'g Pedlers Piece & Sowing Turnips, morn'g, Horses retd to the Stable, and Cart'g Dung after'n
24. RAIN, all night & all morn'g clear'd up, about mid day, and dry evening. Sheep <u>shearing</u> stopd | Horses, in the Stable morn'g, - Carting afternoon
25. <u>Sheep shearing</u>, resumed | Fine day | Ploughing Pedlers Piece, and finish'd sowing it to Turnips
26. Fine drying Day | Plough'g Cowlease, 3rd time | Plough'g Great bottom fur: n side & sow'd about 6 Acres <u>to SWEDES</u> about 8 li pr A, - own last Years seed, imperfect | VETCHES, - hoe'd, in Pearces Piece after Dung heap | 2 Sheep died after shearing 2 Sheep Kill'd for Sheep Shearing 60 li 1 Sheep Kill'd – (ill) after Shearing | 1 Porker 4 S 14 li Kill'd 2 Sheep 1 S 10 li Kill'd } S – Shear
27. Sunday, - some mizzle'y blight'g rain
28. Fine dry'g, day | HAY, - Carried the first from Marls, in good condition | Swedes, - Plough'd & Sowd about 4 A – great bottom by <u>the side of the Lane, - own</u> seed | Mr Potters, new house began prepareing, - Bricks from <u>Kiln, & removeing Gravel</u>
29. RAIN, mizzle in morning, | & heavy storms afternoon | Swedes, Plough'g & Sow'g Lawn Piece, Mr Scrivens Seed, fail'd own oldSeed | Cart'g Dung to Lawn Piece
30. Cart'g Dung to Lawn Piece and Sowing Swedes as yesty | RAIN, flying storms | Hay, - carrd remainer of Marls and part of Copes gds after 5 OClock

---

26 John Large and Ann Beak(e) were married on 17 June 1824 at Lyneham, *Wiltshire, England, Church of England Marriages and Banns, 1754-1916* [accessed via Ancestry, 17 July 2020].
John Henley of Goatacre was buried on 21 June 1824 at Lyneham, aged 37, *Wiltshire, England, Church of England Deaths and Burials, 1813-1916* [accessed via Ancestry, 17 July 2020].

even'g | BARLEY, ears, - appearing Nf | WELL, - diging Mr Potters [...] [in pencil] yds @ [...]

July

1. [In pencil] Steaning – diging Mr Potters [27] | Carting Dung Efield bottom where Swedes, - fail'd destroy'd by Flea Plough'g & Sow'g again Swedes Mr Scrivens Seed | Hay, - carr'd remainer of Copes grounds & Parlour field lower furlong | Very fine morn'g & frost Cloudy, but dry day
2. RAIN, heavy in the night and storms frequent in day | Carting Dung, Plough'g & Sow'g as yesterday | Mr Potters House, began building
1. Mangel Wurzel, - sow'd
3. RAIN, - frequent storms, - | Plough'g & sow'g Swedes as yesy | Carting Dung to Snail hill for Turnips
[4. There was no entry for this date, apart from "Cab gie" in the margin and "Sunday" in the main body – both written in pencil]
5. Ewes Fold, - removed to Wansdyke Furlong. | Dry Sheeps Fold, removed to Shepherds ground – ea for Wheat [28] | LAMB'S, - waned, - and put on Vetches N field sowd 18th Mar | Carting Dung Snail hill | Plough'g & Sow'g Swedes E field bottom | Very fine dry'g day | Hay – carrd Parlour field, and the Jockey ground & finishd at the Jockey. | WHEATRICK, - put into Barn (3 now in the Yard)
6. RAIN in the night, and some heavy storms in day, | Carting dung as yesterday, and Ploughing Kite hill
7. RAIN, - storms all day fine eve'g | Carting dung as yesterday. | Plough'g Wansdyke furlong
8. RAIN, - in morn'g fine after'n | Carting dung Kite hill | Plough'g in dung Snail hill | Swedes, - sow'd the last Gt bottom | Hay, cocking from Swath not turned - the 18 Acres &c nearly lost in the young grass finish'd mow'g it 28th June neither time or weather to attend to it before. *
9. Dry morn'g, Hay attempt'd carrying about 3 OClock when RAIN, - heavy thunder storm stop'd beginning | Cart'g dung & Plough'g as yest'y
10. Fine day | Hay, - carr'd, Pearces Piece 10A 30 Loads after 5 OClock even'g, till 12 OClock night | Carting dung Cowlease, - and Plough'g Snail hill
11. Sunday – fine day
12. Very fine drying day | Hay, - began carry'g the 25A 11 OClock morn'g, finish'd the Piece 10 OClock night. dble hand'd 1 sett, - 40 Tons [29] | Carting dung Cowlease | 5 Ploughs thwart'g Kite hill.
13. 5 Ploughs thwart'g Kite hill. | Carting dung as yesterday | Hay, - carr'd

---

27 Stean - To stean a well is to line its sides with stone.

28 ea – each, meaning both in this context.

29 dble hand'd – that is, double handed – in order to collect (carry) the hay quickly, Pinniger used twice the number of hands (workers) than usual. See also entry for 21 June 1830.

Batts ground, the the last of the field grass and top the Rick. | The most Sultry day this Sum'r

14. Exceeding Sultry dry day | Carting dung as yesterday till 10 OClock when the Element became over cast and dark with very heavy Thunder, & light'g, - follow'd with very heavy rain, - and large Hail * | Thunder, light'g & storms remainer of day. * | Lambs wash'd for Shearing | Plough'g Shepherd Gd athwart for Wheat afternoon

15. Plough'g as yesterday, - & Timber geting home. | Some mizzle'y Storms, and cool'r

16. Fine warm day | Dry meadows, - mowing | 12 Score Lambs, - the Chilvers and and Rams sheared by 4 Men | Carting dung Cowlease, and Plough'g it in, - and sowing Turnip Seed, - green rounds

17. Carting dung as yesterday, and Sowing Turnips Seed, Cowlease and finish'd. | Harrow'd down Snail hill and sow'd to green round Turnips | Warm fine day

1824 July 16th, - Dear Ellen unwell in the morning, - and in the evening 8 OClock, - expected would breathe her last, with an inflammation on the Chest, - after a very sorrowful night, appear'd a little better next day Saturday 17th, - with a doubtful hope of her recovery, - in the evening again appear'd Death struck, - yet her Pulse not so violent affording some little hope, but to our great sorrow her time was appointed, and expired at 5 O Clock in the morning 18th July without the least motion, - Her meek and affectionate Heart in health, and her unexampled patience in her pain and sickness, never betraying one sigh is a matter of surprise to us, and have engraven the shadow of her Angelic Soul on our Hearts, ever will be dearly cherished, and never, never erased. Sunday morn. Born 5th April 1822, Aged 2 years, and 15 Weeks. 3 feet high.[30] [Diary turned landscape for this entry, occupying a double page within a thick black border]

18. Sunday, - very fine day Glass rose unusually high to F in fair.

19. Very fine day, - all the dry meads mow'd, - and mow'g Thistles &c. | Plough'g and sow'g Turnips after Vetches fed off Nfield | Thwarting Shepherds ground for Wheat | Meadow Hay – carried.

20. Meadow Hay – carried and finish'd | never carried so soon & good | Very warm fine day | Ploughing & Harrow'g Kite hill | Watermeads, - began mow'g for Hay

21. Plough'g &c as yesterday. | Warm fine day

22. Warm fine day | Plough'g and Harrow'g as yesterday, and finish'd | Carted all the Dung from the Jockey yard, to Kite hill for Turnips. | Cow to Bull

23. Swedes, began hoeing 14 Women Lawn Piece and great bottom | Harrow'g

---

30 Ellen Pinniger of Little Bedwyn was buried at on 21 July 1824 at Lyneham, aged two years and 15 weeks, *Wiltshire, England, Church of England Deaths and Burials, 1813-1916* [accessed via Ancestry, 17 July 2020].

Swedes, & Pedlers Turnips | Turnips 2nd drift Plough'g and sow'g after Vetches Nfield | Oxen Plough'g Wansdyke fur: | Very fine day.

24. Very fine day. | Oxen Plough'g as yesterday | 14 Horses, geting the grass from East Watermead to Stockwells, dry mead, and finish'd, - to make it hay | Lambs removed from Vetch's N f: to Vetches Mushings
25. Sunday, - fine day
26. Turnips, - sow'd the 3rd drift after Vetches – Nfield | Oxen Plough'g Wansdyke Fu: | Swedes grow'g very fast, remarkably clean, & strong Plants nearly 40 Acres – will all soon want hoeing, - hands taken off to make Watermead Hay, and Wheat ripening fast | Watermead, Burnt Mill, mowd by 5 Men | Fine day, - unusual hollow winds, - succeed'd by rain, - not heavy, - in evening
27. Very fine drying day, | 17 Horses, Carting dung to Kite Hill for Turnips, if can get it to work, - if not for Wheat | Oxen Plough'g as yesterday | Pedlers Turnips began hoe'g
28. Very warm day, - C Lewes left | Hay watermead began make'g a Rick | Carting Dung as yesterday | Fold the Ewes, removed from Wansdyke Fur: to Marls, on the Clover for Wheat. | Oxen Plough'g as yesterday | Lambs fold'd Shepherds Gd have'n broken tie on Vetches
29. Very warm fine day | Oat Rick, - Wansdyke put into Barn | Hay – carrd fm Burntmill mead, - to Stockwells Mead | Oxen Plough'g as yesterday
30. Morn'g dull, expectd rain fine warm day | Hay, - watermead finish'd carrying. | Carting dung Kite hill | Oxen Plough'g athwart the Clover lay, Parlour bottom for Wheat
31. Fine morn'g, - thin rain even'g | Plough'g as yesterday, and Carting dung as yesterday | Half pass 10 O Clock evening Mary [...]

August

1. Sunday fine day
2. Fine day RAIN, - storm even'g | Oxen Plough'g as yesterday | Carting dung finishd Kite hill
3. Carting dung to finish Wansdyke Furlong. | Oxen Plough'g as yesterday | Fine morn'g, - repeated Thunder, - in the evening with heavy RAIN
4. Fine day RAIN, - evening | All Teams, - Ploughing in Dung Kite hill, road too wet for Dung Cart'g | Lambs & Young Sheep removed fm Mushings Vetches to Nf Vetches in Kid, - drill'd 30th Apl
5. Storms in day, heavy RAIN in the evening | Plough'g Marls after Fold and Plough'g Parlour Botm | Carting Dung Parlour botm
6. Carting Dung Parlour botm | Plough'g as yesterday | Frequent slight storms RAIN with Thunder afternoon
7. Plough'g & Cart'g Dung as yestery | Damp morn'g fine day
8. Storms of mizzle'y rain | Sunday
9. 3 Loads of Bricks from Hop grass | Carting dung & Plough'g as 7th | Rain

in night, fine day

10. Plough'g Kite hill, and 40 Qr Oats delivd at Marlbro 20 Qr Stale 7 Sco 4 li Sack Nt 20 Qr Fresh 7 Sco 14 li Sack Nt @ 28s Qr Blk Siberians | Fine dry'g day RAIN evening
11. RAIN, - heavy in night, and storms in morn'g, very fine afternoon. | Plough'g Wansdyke athwart this the 5th Plough'g. | Plough'g Kite hill.
12. Plough'g Kite hill Oxen finish'd | Horses Carting Dung to Parlour bottom & Thrash'g Oats, continued. | RAIN, heavy storm morn'g fine afternoon, Market brisk in consequence sold Wheat 12 Sco Nt @ 31s 6d Devizes.
13. Plough'g, Cart'g Dung &c as yesy | RAIN, heavy storm afternoon
14. RAIN, - light in afternoon Fine day | Carting Dung, finished | Plough'g &c as yesterday | VETCHES, – in Kid, the Horses began, - sow'd | Old Clover, Horses finish'd 7th Inst and began the 2nd cut the 7th now have Vetches & Clover for a change.
15. Sunday very heavy RAIN afternoon
16. Damp morn'g, fine afternoon | REAPING, began 40 reapers Wheat turn'g very fast, very little cut, this neighbourhood | Plough'g Marls, & Wansdyke.
17. Plough'g finish'd Wansdyke, fur: | Bricks from Kiln, Hop grass | Raftering – part of Parlour bottom, - very mild, dung'd on the Clover lay. | Horses thrash'g Oats | Fine day RAIN, in even'g and night.
18. RAIN, very heavy storms reapers do but little | Raftering as yesterday. | Very rough winds
19. RAIN, heavy in the night fine day | Plough'g Nether side of Marls Clover lay
20. RAIN in night and morn'g reaping afternoon | Ploughing as yesterday
21. RAIN, nearly all night, and RAIN, from 10 to 12 O=Clock at noon in Torrents, - the Wheat in greatest danger of being spoil'd not yet injured, - about 40 Acres cut * | Plough'g Marls as yesterday
22. Sunday RAIN, very heavy <heavy> 10 OClock, very fine afternoon.
23. Frost, - morn'g, fine day. Glass, rise'g gradually. | Plough'g Marls, - after Fold | Plough'g Mushins Vetches fed for Oats – next Year
24. Frost, - and warm fine day Wheat injured very little with the Rain, - still wet in Bond | Oat Rick, - put into Barn | Ploughing Mushings | 70 Acres of Wheat cut | [In pencil] Ewes [...] Rams [...]
25. Very fine harvest day The Wheat very damp in bond from the rain fall'n – Saturday 21st dry'd much to day, and load'd the Waggons, this even'g to begin a Rick tomorrow. | Plough'g Mushings finish'd | Plough'g Nfield athwart Vetches fed off. | Rolling Wansdyke furlong ready for Plough'g fr Wheat | Horses thrash'g Oats.
26. Horses thrash'g Oats. | Plough'g Shepherds again athwart | Plough'g Wansdyke fur: for seed | Lambs removed from Vetches Nf swell on them, - put on Sainfoin | Fine night, RAIN, began 5 OClock morn'g, contd till 10 OClock, - to prevent Wheat carry'g, fine after.

27. WHEAT, began making the first Rick. | Fine day | Ox Teams Plough'g Shepherds Gd
28. Ox Teams Plough'g Shepherds Gd | Very fine day | 4 days Thrash'g Oat Rick, and finish'd Kite hill. | Carr:g Wheat.
29. Sunday, - Very thick fog & very warm day
30. Very thick fog & very warm day Thunder, - all afternoon, rain at a distance. | Plough'g as 28th
31. Thick fog, - and very fine day

September

1. Thick fog, - and very fine day & Wheat Cart
2. Thick fog, - and very fine day & Wheat Cart
3. Thick fog, - and several showers of mizzle'g rain afternoon | Wheat Harvest finish'd [Particulars of the Wheat Harvest followed – see Appendix D]
2. Pease, small grays began hack'g | Barley, - Nf: began mow'g.
4. RAIN, - frequent storms & and soon dry again | Oxen Plough'g Shepherds Grod all the week. | Horses, Plough'g Wansdyke fur: | Turnips, - hoe'g Snail, - grown too large, whilst reaping
5. Sunday, - fine day
6. RAIN, all the Morn'g fine afternoon | Vetches, all hands, cuting | Plough'g as the 4th
7. Plough'g as the 4th | Some rain, - Barley turn'g for Carry'g tomorrow.
8. RAIN, - storms all day, - very heavy in evening | Thunder | Ploughing as yesterday, and finish'd Wansdyke fur:
9. Some light storms, - Glass, gradual'y riseing | Oxen Plough'g Shepherds grod for seed, - and Horses Plough'g Marls one earth after fold for Seed | Barley Nfield all moved, and in great danger of sprouting unless the weather change to fine, - it being very wet.
10. Fine fogg'y morn'g, Glass rise'g, every prospect of a fine day Put 26 Reapers to cut & tie up about 7 Acres of strong Oats, with intention of carry'g, - began carry'g at 3 OClock, - took up 4 loads when stop'd, - with heavy RAIN all the evening. ⋆ | Bad prospect of carrying Pease, - Vetches, Barley & Oats now, cut. | Plough'g as yesterday, and Oxen finish'd up: fur: Shephs
11. Oxen Plough'g Vetch Stubble Nfield | Horses Plough'g Marls | Dry after the morning,
12. Sunday, rain, in night and all the morn'g.
13. WHEAT=SOWING, began, - with 5½ Sacks old wheat, up'r part of Shepherds ground 2¼ Bu pr A | Drying morn'g, some flying storms Pease carr'd in even'g Straw dry – 10 loads – 2A 1R 33P | Oats, load'd the Waggons Nf | Oxen, Plough'g Kite hill for Seed
14. Oxen, Plough'g Kite hill for Seed Oats, - began Cart'g in morn'g and finish'd reaping & carry'g all the Oats Nfield | Drying night, some misle

rain frequent Mr Farebrother, - and Mr Smith[31]

15. Thick rain morn'g, - very fine sun shine afternoon. | No Corn, - dry enough to Carry | Ploughing Kite=hill, and sow'g WHEAT, - 4½ Sacks Old from Nether side to two half lands beyond corner of Hitchens.
16. Plough'g, & Sowing Wheat as yesterday | Very foggy, morn, and very fine after 1 OClock, when began Cart'g BARLEY the first from Nfield HYTY abu[...]e,
17. Fogg, - on till 2 OClock, - began Barley – Carting 3 OClock, - dark at 8 | RAMS turn'd 2 of Hilliers to the Stock Ewes, - 1 Hampshire (Mr Paines) - & 2 – two tooth Kennets to follow, - Sale Ewes, nearly all ram'd | 20 Wether Sheep & 1 Ram to Chip[penha]m
18. BEAN'S, began cuting | Plough'g, & Sow'g Kite=hill, yesterday, & finish'd this even'g with 6 Sacks of New Wheat, 16 Acres & 4½ Sacks of Old Wheat [Total] 10½ [Sacks] | Fogg'y – morn'g till 1 OClock, when very fine, and began Carting of Barley Nfield, till dark, when 5 Loads remain to carry,
19. Sunday, - dull yet dry day.
20. Rain, - in night and morn'g too wet for mow'g Oats | Wheat, began sow'g Wansdyke furlong, - & finish'd Plough'g fur: end of Marls. | Press rolling & tread'g Kite=hill finish'd. (wheat)
21. Wheat finish'd sowing Wansdyke furlong, and began sow'g Marls | Drying morn, RAIN, even'g | Barley, carr'd remain'r of Nfield.
22. RAIN in the morn'g to stop the Mowers, very drying after. | Wheat, finish'd sowing Marls further side. | Plough'g Parlour Bottom for Wheat
23. RAIN, thick all day, - and heavy in the evening | Plough'g Parlour bottom for sow'g
24. Plough'g – Parlour bottom and sow'g Wheat beginning lower end | Rain, - thin at times all the morning, - dry afternoon | Mowers stop'd, - the Barley in Swath = sprouting
25. Dry morn'g, - thin rain, even'g | Sheaf Oats, - carrd – 2nd time. | Plough'g & Sow'g Wheat, - Marls and Parlour bottom | Wheap up green, - Shepherds Ground and Kite=hill [32]
26. Sunday, - very fine drying day prospect of fine tomorrow
27. RAIN, close from 4 OClock morning to mid:day. | Ploughing, & sow'g Parlour bottom & Marls, - after the morning, - good work [Note in margin] Colt to Cowitch
28. Frost, sharp white morning fine day Oats, - Lane Piece began Carry'g 11 O Clock continued till Moon set | Plough'g & Sow'g as yesterday Mr Payne.
29. Frost, white as Snow, the Corn very wet till 11 OClock | Vetches, - Carried [...] A | Barley – Carried Burnt mill | Barley, - Carrd Sweardown lower fur: two Setts, till Moon down
30. Some little RAIN, - early in morn'g hollow wind, - expecting rain all day

---

31 Misle – mizzle.

32 Wheap – should be Wheat.

| Oats, - began Carry'g first in the morn'g from Bonings Corner, - finish'd 1 OClock, & Barley, - began Carry'g with two Setts - & stop'd, - by RAIN at 4 OClock | Oxen, - Plough'g Vetch stubble Nfield | TURNIPS, the forward ones Nf began feeding 26th, - sowd 17th May and fit to feed the last 8 Weeks now too old, - some run to seed having plenty of keep. | 1 Turnip weigh 10 li.

October

1. RAIN, heavy, from 12 OClock night, to 2 OClock afternoon 24 Acres Barley Dean heath 8 Acres Barley Swear down 12 Acres Oats – Swear down & Lane Piece 10 Acres Beans – Clay fur: [Total] 54 Acres now to Harvest. 20 Acres Clover seed – in the 6 & 18 Acres not cut Horses in Stable all day & Carters, thrash'g wth Machine The Sheep on the Pastures the Plough'd land very wet
2. Storms, - all day. | Plough'g, fallowg Batts Gd too wet for Wheat Sowing
3. Sunday, - Fine day Vetches, - Lawn Piece, finishd feed'g
4. Dry, - but dull day. | No Corn dry enough to carry | Barley, - Dean heath, - Cock'd part of, - to carry tomorrow if fine | Plough'g & Sowing Fold drift Marls & Parlour bottom, 3 Ploughs | 7 Horses, – resting, and turned out to Grass, - being nearly worn out, – no rest the last 12 Years | Oxen, - Plough'g vetch Stubble Nfield | Nearly 80 Acres Wheat sown.
5. RAIN, heavy all the morning and small warm rain all day Ploughing afternoon.
6. RAIN, Storm pouring afternoon | Horses, Plough'g Fold drift Marls | Oxen Plough'g, finish'd vetch stubble Nfield | Too, - wet to sow Wheat or feed Turnips
7. RAIN, in morning, warm close afternoon, - Turning the Swath's of Barley – Dean heath, - grown together like a mat, - under the Hedge's, and &c the middle of the cocks of Barley dry and uninjured * | Horses, finish'd Plough'g Batts ground | Oxen began began Plough'g Wheat stubble Nfield
8. Oxen, Plough'g as yesterday | Horses, - began Plough'g Wheat stubble, - Efield, - forebridge. | RAIN, in the night, dry morn'g very close and warm, - the Oats and Barley, - going very fast to ruin, - growing in Swath, and Cock, - all hands out in afternoon, opening the Cocks of Barley, & Oats into weaks, - prospect of being entirely destroyed, either in Cock, or, opened never knew so destructive a prospect, - except the Wheat 1821.[33] * | RAIN again in the evening, the Glass, fix'd, the last 4 days, below M in much rain, - riseing slowly this evening *
9. Glass, - gradually riseing, fine morn'g, - Men and Women turn'g Barley & Oats, by 6 OClock very drying, in hope of Carry'g RAIN, Storm drove

---

33 Weaks – Pinniger probably meant wakes (see Glossary). The oats that were lying in the swathe were raked up into lines (rows) to allow the air to blow through and dry them.

them home 4 OClock eve'g | Plough'g Wheat Stubble Nf & Forebridge Piece, too wet for Wheat sowing | Weyhill fair, - very good Highest Price for Ewes 44s Good Ewes @ 28s Highest Lambs 31s Good Lambs 23s Sunday 10th Octr, Glass, rose yesterday to R in rain, - now sunk to V in very dry, - dry morn'g, - RAIN all the day after. Prospect of all the Corn out being totally lost, now lay aspread in weak and growing, - the Clover seed grown standing, - the land cover'd with water, - the Chalk Pits full of water, - and the headlands of Wheat Kite hill & Parlour bottom flood'd, - no such time the last 11 years. Monday, – The Glass fix'd at V, -, The Horses &c at Plough for half an hour, & drove home, RAIN heavy till 2 OClock, dry evening [Diary turned landscape for this entry]

11. The Horses finish'd the Vetches & Clover Pearces Piece and began Watermead grass | Fold finish'd Marls, and removed to Nfield Vetch stubble for Wheat, all the Sheep on Turnips | Men out of Employ, for want of work in the dry.
12. RAIN, heavy in the morn'g and thin rain all day. | Plough'g Stubbles Nfield
13. Plough'g as yesterday | Frost morn'g, fine dry'g day | Barley, turn'd twice Southfield began carry'g 3 OClock, finished by 8 OC___, - in damp condition & dark [A table of wool weights for 1823 and 1824 follows – see Appendix E]
14. Very sharp white frost, Oats, - South field, Lane Piece & began Carting by 9 OClock & finish by: [...] about 12 Acres, 2 or 3 Sacks pr A britted, - in good condition.[34] | Oats Smiths ground began carry'g in evening | Oxen Plough'g as yesterday
15. Oxen Plough'g as yesterday | Frost & fine morning RAIN then after 10 OClock to 3 | BEANS, carried sound & good | Glass, - gradually rise'g
16. Glass, - gradually rise'g to nearly C in change | Very sharp frost & very fine | Top'd the Oat Rick & Thatch'd carr'd 14th, and untopd, and put about the ground 2 Oat Ricks Carr'd 28th Sepr, grown <u>and matted together</u>. | Barley, Dean heath, & Oats the remainer of Smiths ground too damp to Carry. | Plough'g as yesterday, and Wool, - delivered to Beckhampton
17. Sunday, - Fine morn'g, thin rain afternoon
18. Frost & fogg'y morng | Barley Dean heath began carry'g 3 OClock afternoon | Plough'g Stubbles Nfield
19. No frost, - Barley Dean heath began carry'g 7 OClock morng and finish'd, - Oats Smiths ground, and Vetches Nf | Ploughing Breach for Vetches
20. Ploughing Breach for Vetches | Fine morn'g, and fine day Cloverseed, the 18A & Pearces Piece Carried | HARVEST, finish'd.
21. Very fine warm day, the Roads full of dust, as mid=summer. | Wheat Rick old Dean heath put into Barn. | 6 Teams, - Breach as yes'y
22. Vetches, - sow'd 4 Sacks Breach | Fine day
23. Fine day Plough'g Breach

---

34 Britted - of crops, where the over-ripe grain has dropped out of the husk.

24. Fine day Sunday, - rain in evening, - Ty[...]n [35]
25. Rain morn'g Storms interval Oats 40 Qr delivd Marlbro | Plough'g Shepherds ground
26. Rain, - very heavy morn'g, and storms all day & evening | Plough'g Clay fur: too wet to Plough for Wheat
27. Plough'g Clay fur: too wet to Plough for Wheat | Fine day, some thin rain.
28. Plough'g & Sowing Sandy corner Marls | Fine morn'g & eve'g, storm afternoon
29. Rain mid day, to prevent sow'g | Fallow'g Clay fur: | Oats, - deliv'd to Ramsbury & Soot from Hungerford
30. Very cold drying Winds | Sowing finish'd Marls, good work | 5 Ploughs, - fallow'g Clay fur:
31. Sunday, - RAIN, all day

November

1. Thin rain, morn'g, fine after | Carting dung began Pearces Piece for Wheat | Ox Teams, - Plough'g Pearces Piece for Wheat
2. Ox Teams, - Plough'g Pearces Piece for Wheat and began Sowing Wheat Pearces Piece | Carting Dung as yesterday | Some small rain after noon.
3. Very fine drying day Carting dung, Plough'g & Sow'g as yesterday | GRASS, the Watermeads finish'd this eve'g, cuting for Horses and Oxen
4. Very fine, cold drying winds | 150 Ewes good mouth Appleshaw fair Mr Finch Chelmsford @ 28s 6d [36] | Ploughing & Sow'g, finish'd Shepherds ground | Carting dung as yesterday
5. Carting dung as yesterday finish'd | Plough'g and sow'g Pearces Piece
6. Plough'g and sow'g Pearces Piece

8, 9 & 10. Plough'g and sow'g Pearces Piece and finish'd

11. Plough'g Clay furlong, - being too wet for Wheat sowing
5. Fine day to Winchester
6. Fine day to Isle Wight
7. Fine day Sunday to Rue Street [37]
8. Some thin rain, Cross'd ye Island
9. Beautiful fine day – Portsmouth
10. East Cows & back to South'ton Fine day
11. Thin rain nearly all day Retd from Southampton
12. Plough'g & Sow'g Nfield began, Wheat Church fur: | Fine day
13. Fine morning, Plough'g, and sow'g as yesterday, - driv'g rain afternoon & fallow'g Stubbles

---

35 Ty[...]n – probably Tytherton.

36 Good mouth – sheep with a full mouth of teeth (a complete set of 8 permanent teeth present at 42-48 months old – also known as whole mouth) or younger. Appleshaw is a village in Hampshire, between Ludgershall and Andover.

37 Rew Street is a village on the Isle of Wight, two miles to the south-west of Cowes.

14. Sunday, - Thick rain all day
15. Fine day – one hail storm | Plough'g & Sowing as 13th | HAY, Stock Sheep began.
16. Fine day | Plough'g & Sow'g Wheat Nf
17. Plough'g & Sow'g Wheat Nf | Very rough Winds.
8. Sheep finish'd feeding Turnips N Field, - and began Pedlers
18. Very heavy rain after the morn'g dry evening | Plough'g & Sow'g in morn'g, - too wet after, and Horses in Stable
19. Mild morn'g & dry, - rain after noon | Plough'g & Sow'g as yesterday
20. Dry night & morn'g – thick driving after, - Teams at Plough only an hour.
21. Sunday, - Very heavy rain morn'g & evening, - dry day
22. Dry day, rain even'g Fallow'g Wheat Stubbles Nfield | Sheep, - on Pastures, too wet for feeding Turnips.
23. Boisterous Night & morn'g heavy rain & Tempestuous till noon, - dry eve'g | Horses & Men in Stable all day | Marlbro fair.
24. Fallow'g Nfield | Dry day
25. Dry day with partial Storms | Plough'g & Sow'g Wheat Nf: & finish'd, - sow'g.
26. Very sharp frost, and fine day | Bean Rick, old, - put into Barn | Fallow'g Clay, furlong.
27. Fallow'g Clay, furlong. | Frost & very fine day | Wheat Rick old, put into Barn
28. Sunday, - damp air.
29. Some heavy storms | Plough'g Clay furlong | Ewes, - discontinued fold'g and Pen'd in Stockwells Yard with Barley Straw evenings & Hay mornings and Turnips, Snail hill | Lambs, Pen'd at Cowlease Barn with Hay, - and Turnips Pedlers Piece | Oxen on Wheat Straw only.
30. Horses in the Stable | Storms all day Fauntleroys Ex [38]

December

1. Very cold, - and dry till evening when very heavy storm of Snow | Plough'g Clay furlong.
2. Barley Rick, put into Cowlease Barn. | Dry frosty morn'g, storms after and unusual heavy rain in the night, to cause a flood | Plough'g as yesterday Mr Pain

---

38 Henry Fauntleroy, partner in a leading London bank, forged securities valued at £170,000 (worth around £15m at 2019 prices, Measuring Worth, https://www.measuringworth.com/calculators/ukcompare/relativevalue.php. [accessed 18 July 2020]. He was executed at Newgate on 30 November 1824, British Executions, *Henry Fauntleroy* http://www.britishexecutions.co.uk/execution-content.php?key=2364&termRef=Henry%20Fauntleroy [accessed 18 July 2020]. The case and subsequent execution were reported in Devizes & Wiltshire Gazette, 2 December 1824 and in many other local and national papers.

3. Plough'g as yesterday Mr Pain | Frost'y morn'g fine day
4. Frost, morn'g, - cold Snow mid day heavy rain afternoon | Fallow'g, finish'd Clay fur &c.
5. Sunday, - extreme cold winds Sheep Hay twice a day, - the young the week pass'd – Hay twice a day
6. Very sharp, hard frost, to bear up fine day, and heavy rain even'g | Fallow'g Wheat Stubbles Nfield
7. Fallow'g Wheat Stubbles Nfield | Very cold west winds, all the Ewes, swell'd on young Turnips Snail hill, in great danger of dieing
8. Fine morning rain'y even'g | Plough'g as yesterday
9. Plough'g as yesterday | Dry day, - very heavy rain in the evening
10. Frost in the morn'g, fine day | Plough'g, as yesterday
11. Very sharp frost in the night thaw in the morn'g damp day | Plough'g as yesterday | 6 Men triming 450 Ewes before valued | Mr Pains, first Goods, sent
12. Sunday, - mild day
13. 2 Teams only fallow'g stubbles Nf The old Horses resting, - the Harness cleaning for valueing | Mild damp day, - the Glass unusually high to F in fair | 5 Men triming young Sheep 300 Rams &c for value'g
14. Very mild damp day, the Glass rose a letter | Plough'g as yesterday | Messrs Wooldridge & Rogers began valueing the Timber on Little Bedwin Estate for Messrs Guy & Pain [There follows a list of sheep marked to be valued – see Appendix B]
15. Very mild dull day | Ploughing as yesterday
16. Ploughing as yesterday finish'd Nfield | Mild, fine dry day
17. Dry morning rain after | Plough'g Fold drift Whitley fur
18. Plough'g Fold drift Whitley fur & finish'd Horses, work no more till they are Mr Pains | Fine day
19. Fine day – Sunday
20. Storms Men & Boys, in Stable cleaning the Horses prior to Valuation
21. Valuation, by Mr Wm R: Brown of Broad Hinton Manor Farm and Mr [...] Lewes [...] of the Live & Dead Stock on Lt Bedwin Farm, from Mr Guy to Mr Pain | Very Tempestuous, morning thick driving rain all day
22. Very Tempestuous, all day. | Valuation finish'd [Note in pencil] by W R Brown & Anthony Lewes of Chilton Candover [39]
[Stand-alone note on single page in pencil] 26th May, Mare Cover'd last time
23. Fine clear frosty day. | 3 Teams, - fallowing Whitley fur: | Barley, Dean heath finish'd thrashing 20th Inst
24. Rain in the night and morning dry afternoon | Horses in the Stable too wet to Plough
25. 1 – 6 Tooth Sheep of Mr Pains stole'n & kill'd – last night | Damp morn'g,

39 Chilton Candover is a village in Hampshire, 10 miles north-east of Winchester.

rough winds afternoon
26. Sunday, - Fine day.
27. Damp morn'g – fine day. | Ploughing Whitley furlong.
28. Ploughing Whitley furlong. Damp morn'g, heavy rain after to, stop Ploughing.
27. Mr Pains, - Ewes, taken, from the Standing Pen, to lay back on the Turnips. Snail hill.
28. Ewes, - retd to Standing Pen, being too wet after the rain. | Henry Tuckey, left the Jockey put there for a few days only 24th June 1821
29. Oat Rick on the ground put into water Machine Barn | Very fine day | Ploughing as yesterday [40]
30. Ploughing as yesterday | Rain morn'g, fine after
31. Damp day. | Ploughing as yesterday.

1825

January

1. Fine day, - high winds in night. | Plough'g Whitley Fur:
2. Sunday, - fine day, cold air
3. Fine day Mary, Christianed | Plough'g Mushings
4. Plough'g Mushings | Fine day, - thick rain evening | Ewes, - swell at Turnips Snail hill, removed, and began the Swedes.
5. Sharp frost morning, fine day | Ploughing Whitley furlong
6. Ploughing Whitley furlong Sharp frost fine day. | WHEAT RICK. the last old one put into the Barn. | Porker, the first 67 li Porker for Mr Miller 62 li
7. Plough'g as yesterday | Fine mild day, little rain Glass unusually high beyond fair
8. Glass to I in set fair | Ploughing as yesterday
9. Sunday, - fine day sharp frost in the night, thaw morn'g
10. Mild day | Plough'g as the 8th
11. Plough'g as the 8th | Fine day
12. Fine day Plough'g Whitley fur: finish'd | Barley Rick, the small fm Nfield put into Barn
7. Ewes fold & muckled forebridge Piece and Swedes
13. Very fine day, - like spring | Oat rick sheaf put into Barn | Carting Muckle & road earth to Forebridge Piece
14. Carting Wood to Labourers | Fine day
15. Fine day | Carting wood & road earth [The remainder of the diary page was filled with a summary of the sale of Little Bedwyn – see Appendix F]
16. Sunday, fine day, rain eve
17. Fine day, Plough'g, Westboro, for Vetches
18. Rain, - nearly, all day

---

40 water Machine Barn – the barn adjacent to the water-powered threshing machine. As opposed to Horse Machine Barn – see entry for 29 January 1825.

19. Some Snow & Sharp frost in night | Dry cold day | Plough'g as 17th
20. Plough'g as 17th finish'd | Cold dry winds
21. Little Snow & damp air Ploughing Cowlease, fr Swedes
22. Ploughing Cowlease, fr Swedes Fine day
23. Sunday, fine day
24. Fine mild day | Plough'g Cowlease & stiff part of Snail hill | 10 Ewes, Cast their Lambs, on Swedes, eat too many, and bad hay, eat too little, probably the cause [41] ★
25. Ploughing as yesterday finishd | Fine day | Sheep finish'd the Hay Straps
26. Plough'g the 18 Acres, old Clover lay for Pease | Frost sharp, morn'g, Snow and rain after
27. Rain morn'g dry after | Ploughing as yesterday
28. Ploughing as yesterday | Very fine mild calm day Glass rose very high | Wheat Rick, the first new put into Barn
29. Sharp frost, fine day Clover seed Rick, put into Horse Machine Barn. | Plough'g as yesterday
30. Sunday, - fine day
31. Foggy mild day Rabson Farm, look'd, over [42] | Plough'g, finish'd, half of the 18 Acres.

February

1. Misly, rain | Plough'g after the fold Ef
   January 31. Young Sheep finish'd Turnips Snail hill, and began Swedes before the Ewes, to eat the green probability of the Green the cause of their casting their Lambs, - 20 cast.
2. Fine day | Plough'g Pedlers Piece and sow'g Wheat
3. Misly rain & rough winds morn'g, fine day, - Chip[penha]m | Too wet for sow'g, - Carting Chalk to Road South field
4. Snow & frost in night, fine day | Work as yesterday | Porker 76 li
5. Sharp frost, morning, heavy Snow storm, mid:day | Barley=rick, the last Nf put into Barn | Carting Stones from Clover
6. Sunday, - frost & fine
7. Sharp frost, damp day heavy rain evening | Wheat Rick 2nd New put into Barn. | Carting Chalk to road
8. Ploughing & Sowing Wheat Pedlers Piece | Fine day, little rain evening
9. Sharp frost, beautiful'y fine | Vetch Rick = put into Barn Horse Machine | Ploughing & Sowing Pedlers Piece to Wheat finish'd
7. Ewes, brot from Swedes to Standing Pen Stockwells Yd on Hay & Water: Straw mid=day

41 In this case, cast refers to the ewes aborting their lambs.

42 See Appendix G. It appears that Pinniger was on the lookout for a farm to buy or rent: his notes on other farms are written at the end of Volume 4 of the diaries – others dated 1825 are also in Appendix G.

10. Very mild spring like day, Glass rise'g very high | Westboro, sow'd to Winter Vetches
11. Turning furrows Nf: for Turnips | Beautiful fine day | 50 Ewes, - cast their Lambs 3 - Lambs alive 4 - Sheep died, since Mr Pains Sk [43] *
12. Sharp frost, morn'g, very fine | Ploughing Snail hill for Barley
13. Sunday, - fine day
14. Fine day, damp air | Plough'g Snail hill
15. Plough'g Snail hill | Damp raw air
16. Fine day | Ploughing finish'd Snail hill
17. Ploughing Nfield, turn'g the furrow for Turnips | Damp mild day
18. Thick rain all the morn'g Horses retd from Plough'g N f
19. Beautiful warm fine day Carting dung to Nfield for Turnips
20. Sunday, - Fine day 100 Ewes cast their Lambs 50 Ewes with live Lambs
21. Fine day | Thwarting Nfield
22. Thwarting Nfield | Beautiful Summers day | MR PAIN & FAMILY
23. Ploughing as yesterday. | Wheat rick – 3rd new put into Barn, - 6 now out. | Cold dry day, - slight frost
24. Cold dry day, - slight frost | Carting dung to Nfield
21. Pease, - sow'd in the 18 Acres
25. Carting Stones Southfield | Very fine day
26. Cold with a little Snow | Carting Dung to Batts Gd for Swedes
27. Sunday, - fine day
28. Frost and fine day | Carting dung to Batts

March

1. Frost & fine morning very heavy rain even'g | Oat Rick put'g in Barn till stop'd with rain | Ashes began sowing
2. Fine after the morning | Carting dung to Nfield
3. Rain heavy in the night follow'd with Snow – fine day Carting dung as yesterday
4. Carting dung as yesterday | Dry day
5. Dry day Sharpest frost, this Winter | Carting dung – as yesterday
6. Sunday, - dry morn'g, rain afternoon
7. Thin rain all the morn'g | Carting dung as the 5th
8. Carting dung as the 5th | Very fine warm spring day
9. Damp air fine day | Carting dung & [...]
10. Carting dung & [...] | Warm growing day
11. Warm damp day | Ploughing the 18 Acres for Oats
12. Ploughing the 18 Acres for Oats | Fine warm day | Seed Clover remaining part of Rick, put into B[44]

---

43 Sk – Stock.

44 B – Barn.

13. Mid:lent Sunday Thin rain at times all day, warm & grow'g [45]
14. Exceeding cold winds | Plough'g as 12th
15. Plough'g as 12th | Sharp frost, some little Snow till mid day, fine after | Oat Rick put into Water=Machine Barn
16. Very sharp frost night and day in the shade. | Plough'g Clay fur: athwart
17. Plough'g Clay fur: athwart | Very sharp frost the Roads dusty | Market improveg, sold at D Wheat 11 S 15 li @ 36s Hard Barley @ 45s Barley in the Market from 28s to 51s [46]
18. Frost equally sharp wth yesterday fine day | Ploughing as yesterday.
19. Ploughing as yesterday. | Frost sharp, and warm mild Sun
20. Sunday, Frost & fine day
21. Frost, & warm fine day the roads dusty and very good the land work well. | Bean Planting general on heavy land | Watermeads, - Mr Wentworth began, very good, - Turnips all fed | Mr Pain, Oats began sow'g Clay furlong
22. Frost mild, & very fine | Soot sow'd on thin Wheat | Lambs Tails cut 242 Lambs 40 Barren Ewes 168 Ewes, lost & Cast lambs [Total] 450 | Pig – Kill'd 6 S 18 li
23. Frost and very fine day | Ashes, 2nd load sowing | Wheat=Rick, put into Barn 5 now out
24. Frost and very fine day | Barley market improving, sold the last 17 Qr Thin 14 Qr Nfield Tail } @ 44s
25. Frost and very fine clear day | Sow farrow'd one wth 3 one wth 7}
26. Warm fine day frosty morng
27. Warm fine day frosty morng Sunday
28. Warm fine day frosty morng | Sow farrow'd with 4
29. Frosty morn'g fine day
30. Dull damp day, very mild
31. Very fine clear day

April

1. Very fine clear day | Fat Pig Kill'd 8 Sco 12 li | Wheat Rick, put into Barn Corres
2. Frosty morn'g, - warm fine day
3. Easter Sunday The weather as warm as the middle of June, no Person remember so fine a March *

4, 5, 6 & 7. Frosty mornings, fine warm days Sowing Barley, Lawn Piece one earth. after Plough

8 & 9. X Frosty morning, dry'g and warm days

10. Sunday warm, as midsummer X Barley, - Drill'd Snail hill stale earth

---

45 Easter Day 1825 fell on 3 April 1825, 13 March was the third of five Sundays in Lent.

46 D – Devizes.

11, 12. Less frost, mornings, - and warm days
13. Less frost, mornings, - and warm days thin blight'g rain morn'g | Wheat Rick, - put into Barn 3 Wheat Ricks now out 2 Sows farrow'd – 10 each
14, 15, 16. Mild morn'gs, very blowing harsh Winds 2 Sows farrow'd 3 & 12
17. Sunday, - very fine warm day
18. Sharp frost, morn'g very fine warm day Wheat Rick, put into Barn.
19. Dry warm day
20. Dry warm day
21. Dry warm day | RAIN, began very mild 9 OClock evening, - continued all night
22. Rain continued, very mild all day, - the first rain since 13th March The Land too dry for sow'g many Farms neither Oats or Barley began. The Wheat improved rapidly | Pig 7 S 16 li, - the last Kill'd
23. Rain in morning, - very growing day.
24. Sunday, - very heavy Showers with hail.
25. Heavy Showers wth Thunder
26. Showers in morn'g very fine remainer of day CUCKOO
27. Showers all the morn'g dry eve | Swedes, - finish'd feeding
28. Some showers in the morn'g
29. Some showers afternoon fine morn'g | Wheat Rick put into Barn one rick only out
30. Showers in morn'g – dry after
27. MARE. Kitty foal'd evening Horse Colt.

May

1. Sunday, - Showers MARE, Kitty, - covered by Marquis of Ailesbury's [...]
2. Showers, - frequent, &c heavy
3. Storms, - frequent, &c heavy
4. Fine and dry day
5. Rain in morn'g, sultry and dry after, - Thunder lighten'g and rain in the last night.
6. Thunder, lighten'g, and rain in night, very fine day | Mare, - covered again
7. Thunder &c, - and heavy weather in the night, fine day
8. Sunday, - very heavy storms
9. WHEAT=RICK, - the last, put into the Barn | Fine day, - except one storm.
10. Fine day
11. Fine day
12. Rain nearly all day, and very heavy in the evening
13. Rain mid=day
14. Very fine day 1st Load of Goods, sent to Chippenham
15. Sunday, fine day.
16. Fine day.

17. Mare, (Kitty) with a Colt and a Yearling Colt sent sent to Bremhill Grove. (Mr Rawlings) | Fine day
18. Fine day | Oat Rick put into Barn | Swedes, sow'd Batts Gd | Turnip sow'd Burnt mill Ground 16th
19. Very fine day
20. Very fine day | Watermeads Sheep finishd | Wheat thrash'd, and all Winnowed [Details of the outcome (make out) of the 1824 wheat harvest then followed – see Appendix D]
21. Mr Edwd Potter, died, 2 OClock morn'g, Aged 75 years [...] [47] | Very fine day Oat Rick, the last put into Barn,
22. Very warm fine day | Sunday
23. Very warm fine day
24. Very warm fine day high winds
25. Very warm fine day high winds some Showers
26. Very fine day
27. Very fine day
28. Heavy Showers} Very acceptable
29. Heavy Showers} Very acceptable the Barley turng Yellow
30. Fine day, - Mowing fieldgrass began
31. Fine day, - Wheat nearly bursting the Ear, found one half out, of the first sowing

June

1. Some little rain. Turnip [...] began sowing Nfield
2. Cold winds
3. Heavy showers of Rain and Hail
4. Heavy rain all the morning
5. Sunday, some Showers
6. Very fine day | BEAN Rick, - the last Rick put into the Barn. | Sheep Wash'd | Oats accusacion[?] [Written in pencil]
7. Very fine day
8. Very fine day
9. Partial showers
10. Warm fine day | Sheep shear
11. Sheep shear
12. Sunday Fine day

13, 14, 15, 16. Fine days

17. Dull blight g day
18. Fine day
19. Fine day Sunday

---

47 Mr Edward Potter senior, late of Chisbury Farm buried on 26 May 1825 at Little Bedwyn, aged 75, *Wiltshire, England, Church of England Deaths and Burials, 1813-1916 for Edward Potter* [accessed via Ancestry, 20 July 2020].

20. Lambs waned & put on Vetches, Westboro | Ewes, - began Vetches Breach | Reads, Hay, - carried | Showers, - Mr Pain finish'd field Hay | 2nd Load Goods to Chippenham
21. Showers | Swedes, - Batts Ground Turnips Burntmill } fit to Hoe
22. Fine day
23. Fine day The last lot of Bedwin Wheat sold, to Mr Taylor Baldown Mills 118 Sacks 12 Sco 3 li to 12 Sco to 6 li 10 Sacks Tail 11 Sco 11 @ 35s
24. Fine day
25. Showers
26. Sunday, fine day, - heavy rain evening
27. Heavy Showers after the morning | 3rd load Goods to Chip[penha]m
28. Rain all day at times after the morning
29. Wednesday, Fine day - 4th & last load Goods to Chippenham and left Bedwin | Fine day
30. Thursday, - Storms | Entered Mr Guys Farm House Chippenham

July

1. Fine morn'g, heavy Storms afternoon
2. Fine day
3. Sunday, - very fine day
4. Fine morning, - heavy storm afternoon
5. Fine day
6. Fine morn'g, storm after'n

7, 8, 9. Fine day, but dull

10. Sunday, - fine day, very heavy storms. partially

11, 12, 13, 14, 15, 16. Hot days

17. Sunday Hot day
18. Hot day

17, 18 & 19. The heat over powering, - many People & Horses died [Diary turned landscape and written from top of page at end of entries for 11th to 18th]

19. Warmer than the East Horses, many Kill'd in the Coaches, - and Men discontinued working
20. Rather cooler
21. Rather cooler

22, 23. Mid day warm, mornings and evenings cooler.

24 Sunday to 30. Warm with some cooling Air

31. Sunday nearly as warm as the 19th
[Note in margin] 19th Saml Hardings House and Factorys, Burn't by Lightg.

August

1. Very warm, – The Wheat turn'd off - the midst of Reap'g generally | Thin rain all the morning quite lost in afternoon
2. Warm day

3. Some Storms
4. Rain nearly all day
5. Fine with partial Storms
6. RAIN, from 7 to 11 OClock morn'g, alternate Storms and very drying afternoon | Wheat harvest finish'd round the Plain All nearly cut, andbut little carried Avebury Parish
7. Sunday fine dry day
8. Rain morn'g fine day
9. Fine till 3 OClock when heavy storm.
10. Hail storm, mid day, & other Storms.
11. Glass rose, fine harvest day. BEANS in many places cut before the Wheat, very unusual
12. Fine Harvest day
13. Thin rain after morng
14. Sunday, - Storm afternoon
15. Some thin rain
16. The Glass rise'g very fine Harvest day.
17, 18. The Glass rise'g very fine Harvest day. The Wheat Harvest nearly ended
19. Thin rain nearly all day
20. Fine day,
21. Sunday, - as warm as some of the hot days in July
22. Warm fine day
23, 24, 25, 26. Very warm fine days, great want of Water in the fields for Cattle, - and the Grass dried up *
27. RAIN steady all day, to soak to soak the land, fill the Ponds & Brooks, - and give new life to all Vegetation *
28. Sunday – very fine day the Glass remained high. and very steady *
29. The Glass still rising, and steady rain all the morn'g fine after *
30. Very fine and very warm day, A new face on the Earth all vegetation growing rapidly *
31. Very warm fine day

September
1. Nearly as warm, as any day this Summer some partial rain in evening | HARVEST, - with few exceptions finish'd
2. Fine day and warm
3. Fine day and warm
4. Sunday, - warm fine day
5. Fine day, - much cooler morn'g Frosty, mornings
6. Fine day, - Frosty, mornings
7, 8, 9. Fine day, - Frosty, mornings
10. Fine morn'g, rain evening
11. Sunday very heavy storm, morn'g and evening

12. Heavy Storm, morn'g, dry after
13. Storms
14. Fine day
15. Fine day very warm
16. Storms even'g warm
17. Rain morn'g fine after
18. Storms heavy, Church time
19. Fine and warm
20. Very fine & warm
21. Heavy storms
22. Fine day
23. Fine day
24. Fine day some thin rain
25. Sunday some thin rain
26. Fine day, damp morn'g
27. Very fine, - fresh morning, & fine day.
28. Fog'y morning, warm fine day.
29. Fog'y morning, the Glass very high
30. Glass, sink'g, rain in evening

October

1. Fine morning, rain evening
2. Sunday, - fine day, heavy rain evening
3. Rain heavy in morning, and rain in evening
4. Storms, - fine mid:day
5. Thin rain at times very mild
6. Rain all night fine day

7, 8. Fine days

9. Sunday, - fine day

10, 11. Glass beyond R in fair, unusual fine weather

12, 13. Foggy mornings clear evenings fine calm, mild days.

14. Very mild fine day, storm in the evening
15. Fog'y morn'g, very fine day
16. Sunday, very fine day
17. Thin rain after morn'g
18. Rain all day, after morn'g
19. Rain heavy, afternoon
20. Very cold drying Winds
21. Pierce'g cold drying Winds
22. Cold dry day
23. Fine dry day
24. Very mild fine morning dull after.
25. Very mild fine day

26. Exceeding sharp frost, fine morn'g, rain evening
27. Fine day, - dull
28. Dry mild day
29. Very fine morn'g, Storm afternoon
30. Sunday fine day
31. Fine mild day

November

1. Damp, heavy day some rain
2. Fine day
3. Very boisterous and rainy
4. Very fine drying day
5. Sharp frost, thin rain at times
6. Rain, in night & morning drying day
7. Frost, rain afternoon.
8. Thin rain all day
9. Fine drying day | Gigg Cart weighd 6¼ Cwt Gigg Cart weighed with 4 Passengers 11½ Cwt
10. Boisterous winds and rain all day
11. Snow in night, Flood in the morning & fine day.
12. Fine day
13. Sunday, - Sharp, blk frost
14. Fine morning, and mild thin rain afternoon | The Mare & Colt, and the Yearling Colt, removed from Bremhill Grove 26 Weeks @ 5s 6d to Mr Vines @ 7s 6d per Week good grass, - and Hay, if necessary
15. Sharp frost and fine day
16. Frost and some thin rain
17. Fine morning, rain evening
18. Thin rain all day

    19th Novr 1825 Christopher Pinniger of Langley Expired half pass Nine O Clock in the morning (Saturday) Aged 50 Years 21st Septr last Intered at Tockenham on Thursday 24th Novr [48]

    Wednesday 23rd Novr was Married at Compton Catherine Pinniger of Cowitch and Thos Bethell of Trowbridge [49]

    Thursday 1st Decr was born at Chalfield Catherine Spackman[50] [Diary

48 Christopher Pinniger born 21 September 1775, WSA: 4381/2/1, *Pinniger genealogical memoranda book, c.1820-1943,* pp. 20, 151-152.

49 Compton – Compton Bassett, *Wiltshire, Church of England, Marriages & Banns, 1754-1916 for Thomas Bethell* [accessed via Ancestry, 20 July 2020]. Catharine, eldest child of John & Catharine Pinniger of Cowage [Cowitch] married Thomas Bethell and had three children, WSA: 4381/2/1, *Pinniger genealogical memoranda book, c.1820-1943,* p.27.

50 Catherine Spackman, ninth child of William Spackman and Susannah Spackman

turned landscape for the previous three entries, the first of which is written within a thick black border]

19. Storms
20. Fine morning, some rain
21. Storms
22. Fine drying day
23. Very fine till evening, thin rain after
24. Thin driving rain all day
25. Very fine all day
26. Storms
27. Sunday, - fine till even'g when Rain
28. Heavy Storms
29. Some Storms
30. Fine clear day, frosty even'g

December

1. Fine day
2. Rain morning, fine after Sa[...]h & Goods removed from Langley [51]
3. Fine, and dull day
4. Fine day, very cold air | Sunday
5. Snow, a little in the morning Rain nearly all day, and till midnight, - great flood
6. Rain heavy in morning, dry after
7. Fine day
8. Fine day
9. Fine day
10. Fine day
11. Fine day Thick cold fog – in evening Sunday
12. A Shower, in morning, fine mild day & unusual calm eve'g
13. Rain nearly all day
14. Rain heavy, in morning dry after
15. Fine mild day, cold morn'g
16. Dry day
17. Dry day
18. Sunday, damp day, heavy rain in evening
19. Rain'y morning, fine after.
20. Fine mild day
21. Thin rain and damp day

(neé Pinniger), WSA: 4381/2/1, *Pinniger genealogical memoranda book, c.1820-1943,* p.24.

51 This is Sarah Pinniger, sister of Thomas Pinniger, and wife of the late Christopher Pinniger of Langley who died 19 November 1825 (see entry above), WSA: 4381/2/1, *Pinniger genealogical memoranda book, c.1820-1943,* p. 20.

22. Fine Sun shine day
23. Frost and damp day
24. Frost morn'g, fine mild sunshine day
25. Fine mild, sun shine day
26. Fine morn'g storms after
27. Fine morn'g heavy SNOW, for an hour afternoon, wth Frost eve'g
28. Sharp frost & fine day
29. Sharp frost & fine day
30. Sharp frost Thaw & some snow
31. Sharp frost Fine day.
26. Colts removed from Cocklebury to Newlease @ 2s pr Week

1826

January

1. Sunday, - Storms of rain in evening
2. Dry day, - Littlecote, and Little Bedwin
3. Frost morn'g, fine day Gt Bedwin & Forsbury [52]
4. Excessive cold winds, and Frost, - Forsbury
5. Sharp frost morn'g, Snow & Rain all day after Forsbury & Newbury
6. Rain all day, - Little=cote
7. Fine day, - Kennet & home
8. Sunday, - dry day
9. Sharp, frost
10. Very sharp, frost
11. Snow in night to cover the Land and sharp frost
12, 13, 14. Clear fine day, - exceeding sharp & quick frosts.
15. Sunday sharp & quick frosts.
16, 17, 18. Sharp & quick frosts.
19, 20. Very mild dry Thaw
21. Mild like May morn, some little rain
22. Raw air, cold, - frosty even'g
23. Fine day, and dry
24. Fine day, and dry
25. Fine day, and dry Foggy morn'g
26. Frost morn'g – Thaw evening
27, 28. 1st HAY, bot Thaw fine days
29. Sunday bot Thaw fine days
30. Thin rain all morn'g, fine after
31. Mild fine day

February

1. Rain morn'g mild calm fine day

52 Forsbury - probably Fosbury.

2. Mild calm fine day | Abbot L[...] born [53]
3. Thick rain nearly all day
4. Fine day
5. Sunday, - fine mild morn'g rain afternoon
6. Thick air all day, high winds
7. Very fine drying day
8, 9, 10, 11. Very fine drying day
8. Colts removed from New lease to Corston @ 2s 6d
12. Sunday, - rain in evening andnight
13. Very fine even'g cold fog'y morn
14. Beautiful fine day, heavy rain in night
15. Thin rain at times all day
16. Some little thin rain
17. Very fine day
18. Cold hail Storms.
19. Sunday, - dry morn'g rain'y eve
20. Very drying fine day
21. Frost & fine morn'g rain even'g
22. Some thin rain at times
23. Rain, morn'g and evening
24. Sharp frost morn'g, fine day rain even'g
25. Fine day, Potatoe, Planting
26. Sunday, - fine day
27. Fine day, rain evening
28. Fine day SEED all sorts planted in Garden

March

1. Very fine day
2. Rain all day
3. Beautiful fine day
4. Cold air, dry day, COLTS and Mare, returned from Corston, starved
5. Sunday, - cold dry day.
7. Damp air morn'g, rain all the eving
6. Rain thin nearly all day
8. Rain morn'g, warm, and fine after
9. Beautiful, mild fine day | Colts, - taken to Cottles Sale and return'd to Chalfield
10, 11. fine continue
12. Sunday fine continue
13. Cold dry day
14. Rain in night fine day
15. Rain in night Summers day | MARE, taken to Cleveancey

53 Abbot Large.

16, 17. Frosty mornings, fine days
18. Frosty mornings, fine days
19. Sunday, cold dry winds
20. Cold thin Storms
21. Cold drying winds, the roads very dusty
22, 23. Cold drying winds, the roads very dusty
24. Good Friday Pierce'g cold winds, some little Snow in the night & morn'g
25. Cold drying winds
26. Sunday, - some very coldstorms of Hail.
27. Unusual sharp frost, and cold cuting winds all day
28. Fine & much milder day, some little rain
29. Fine day, - keen air.
30. Sharp frost, & very fine day
31. Sharp frost, & very fine day

April
1. Sharp white frost fine day
2. Some little rain in night. | Sunday = fine day
3. Warm Sun shine, Summers day
4. Fine mild day
5. Fine mild day
6, 7, 8. Fine mild day
9. Sunday, - Showers, - warm and growing
10. Fine day
11. Storms
12. Rain all night, high winds and Storms in day
13. Fine morn'g Storms after
14. Fine day
15. Fine day
16. Sunday Rain in morn'g fine after
17. Fine day
18. Fine day very sharp frost Potatoes cut up, - Surveying Monckton Farm [54]
19. Fine day
20. The warmest day, this spring sharp frost. | The Mare Kitty, taken from Cleveancey to Cockleberry at Grass and Hay - @ 3s 6d pr W [55]
21. Warm grow'g day some Showers.
22. Thin rain all the afternoon
23. Sunday, - Fine day CUCKOO
24. Very sharp frost, fine day The Mare, retd from Cockleberry, - not in foal
25. Fine day, - little rain in evening.
24. HAY, - 11¾ Cwt @ 3s from Cockleberry The last Hay 22¼ € @ 4s 6d is

54 See Appendix G.
55 W – Week.

5£ kept the Horse 10 Weeks @ 10s is 5£ [56]
26. The Old Mare, taken to Tytherton to be shot, & buryed, Mr Miller allow 5£ | Fine day, Nightingale (27th)
27. Storms of Hail and rain in the evening
28. Snow Storms, early in morn'g, cold winds all day
29. Very sharp F with Ice, keen air all day [57]
30. Sunday, - very keen air all day

May

1. Very sharp Icy Frost, - harsh winds, - The Potatoes, Pease, Cabbage Gooseberrys, - Ash &\& &c cut up
2. Sharp frost, mild'r day Kidney Beans - planted
3. Fine day, cold evening
4. Very cold, cuting winds
5. Very cold, cuting winds
6. Very cold, cuting winds
7. Very Sunday cuting winds
8. Frosty morn'g, warm fine day
9. Frosty morn'g, warm fine day 2 Colts Keynsham
10. Frosty morn'g, warm fine day
11. Frosty morn'g, rain in evening
12. Growing morning and fine day
13. Very sharp frost, the Potatoes cut off 2nd time
14. Sunday, - warm mild day
15. Warm fine day
16. Warm fine day frosty morn'g
17. Warm fine day
18. Warm fine day
19. Warm fine day
20. Thin RAIN, all the morn'g
21. Warm fine day, the rain quite lost | Sunday, Caucomber cut
22. Very warm day
23. Fine day, little rain, eve'g | Horse, put on Mr Guys Island.
24. Heavy shower in evening
25. Fine RAIN, - warm thin and quited for 11 OClock morn'g to 10 OClock even'g
26. Some showers, very growing
27. Sticking, Kidney Beans. | Rain, - heavy afternoon
28. Sunday, - fine day
29. Steady RAIN, all day – from 10 O-Clock
30. Steady RAIN, nearly all day.

56 C – Cwt, hundredweight.

57 F – Frost.

31. Dull, very, growing day.

June

1. Fine morn'g rain, evening WOOL: 8 Tons, - arrived at Chip[penha]m from Mr Gye of London, Candidate for the Boro: to employ the Poor | Peter Audry Obt. | X WHEAT=EARS, bursting, in forward pieces (Nr Salisbury)
2. Warm growing, day, sultry, and thunder
3. Warm growing, day, sultry, and thunder
4. Sunday Warm growing, day, sultry, and thunder X Wheat fully out at Allington Wm Laws, the 1st of June, being forwar'd'r than Salisbury
5. Warm fine day
6. Warm fine day
7. Warm fine day
8. Warm fine day
9. Warm windy
10. Very warm
11. Sunday warm
12. Chippenham Election Gye & Maitland very warm, - bordering on the heat of last July [58]
13. Warm fine day
14. Cool'r fine day
15. Cool'r fine day
16. Warm fine day
17. Warm fine day 5 O_Clock, morn'g
18. Sunday, - fine day very warm | 10 O=Clock, morn g ANN Baptized 21st June, by the Revd Mr. Short | Wheat, - full in Ear, - nearly every piece | Oats – full in Ear, - Barn bridge [59]
19. Very warm fine day
20. Very warm fine day | Colts returned from Keynsham 6 Weeks @ 6s pr W Ea | Colts remain at Chalfield, | Hay making, general the last week.
21. Fine warm day

---

58 Following the General Election, the members returned to serve the Borough of Chippenham were Frederick Gye of Woodgreen, Middlesex and Ebenezer Fuller Maitland of Shinfield Park, Berkshire (taken from the London Gazette of 30 June 1826), Dorset County Chronicle, 6 July 1826.

59 Revd Mr Short is probably William Short, vicar of Chippenham with Tytherton Lucas from 1823 to 1830, Clergy of the Church of England Database, *Chippenham with Tytherton Lucas* https://theclergydatabase.org.uk/jsp/locations/index.jsp?locKey=1705 [accessed 20 July 2020]. However, there is no record of the christening in the parish records: given that Ann was christened only three days after her birth, it may appear that she was christened at home, suggesting that she was not well at this time.

22. Fine warm day Westmead, began Mowing, & Mr Guy, began mowing The Grass wasted, very much the last fortnight, on all dry soils and the Barley now appearing sickly, - on the flint & Chalk doing well, - Cattle, dull sale for want of keep
23. Very warm day
24. Very warm day
25. Sunday warm day
26. Sultry warm, - Thunder storm afternoon
27. Thunder storm, morning, dry and very warm after
28. The heat oppressive
29. Dull fine day, The Two Colts at Chalfield, reduced to Deaths door, by the Strangles, - now just recovering
30. Warm day

July

1. Very warm
2. Sunday warm
3. Very warm, - the Grass on all thin soils, dry'd and wasted away same time under the hill. at Corton &c in a growing state, & the Crops good
4. Very warm, - the Drought exceed last Year, - and considered to exceed any year since 1785 Accounts in the News Papers of the Mountains in Scotland being ignited with the Rays of the Sun and on fire for 20 Miles in length, | All the Spring Crops drying up and the Corn advancing in price, - the Wheat, doing well and heavy sale.
5. Thunder, - lightening and Storm in the night, - very warm day Partial Thunder storms about the Country.[60]
6. Warm day WHEAT at Sheldon nearly ripe Barley & all spring Crops dri'd up prematurely
7. Not so warm, active wind cool pleasant evening.
8. Some blighting rain, morn'g warm day.
9. Sunday, - fine warm day
10. Thin blighting rain in the morning, warm after | REAPING, - last week in the neighbourhood of Box. | PEASE, blighted, - and made Hay off, commonly.
11. Heavy Partial Storms
12. Warm drying day the rain lost
13. LITTLE=COTE, - Thin rain morn'g very x storm after x heavy
14. FORSBURY, - fine warm day, - | Wheat, ready for the hook, and | OAT'S, at Little cote fit to Mow
15. KENNET, - fine day, - Haymaking finish'd, - watermeadows
16. Sunday, - fine warm day

60 The term 'Country' was used to refer to the local area at this time, not specifically referring to England.

17. REAPING, began in general in Chippenham Parish, - began at Sheldon 13th, - full ripe. | Fine day
18. Fine day Wheat, - turned off the last, day or two, very fast, fit to cut, at Cleveancey & backward soils, - Breadenstoke
19. Fine warm day
20. Fine warm day
21. Rain heavy storms, - evening | Beans, - cuting at Tockenham
22. Rain, partial Storms
23. Sunday, - warm day
24, 25, 26, 27, 28, 29. Warm days, - Wheat harvest Universal this week, finishd in forward, soils, - 3 Sacks of New Wheat in Devizes Market @ 30s Sack
29. FIRES, - Lyneham Court broke out 10 O Clock last night Xn Malford Mr Reeves last night Minety, - Mr Gales – 4 OClock this afternoon, Saturday [61]
30. Sunday, - very warm
31. BROADSTOCK, look'd over The Wheat nearly all cut, and carried, - the Pease carried Beans and Oats fit to cut and the Barley nearly fit. | Heavy Thunder & light'g and some rain in evening | very sultry, warmest day [62]

August

1. Very warm, some rain
2. Rain, little in morning, very warm day
3. Sultry, hot day, continued Thunder & lightening, from 7 to 8 O Clock evening, with heavy RAIN, - the Water rise'n in Brooks & Ponds, before quite dry
4. Very warm day, - in the evening Thunder light'g, and partial heavy rain
5. Very warm day
6. Sunday warm day
7. St Anns Hills Fair Lambs from 8s to 21s Ewes from 13s to 18s Very pleasant fine day, the Wheat, nearly all carried round the Hill, - Oats – Barley and Beans cuting
8. Very fine
9. Very fine Mr West Inter'd
10. Very fine Tockenham, cool eve'g
11. Rain morn'g, very fine after
12. Fine
13. Sunday, Fine Calne Church
14. Thin Showers, - morn'g
15. Fine day
16. Thin Showers, morn'g
17. Thin Showers, morn'g | Visitation
18. Confermation

---

61 Xn Malford - Christian Malford.

62 Broadstock – Bradenstoke.

17, 18, 19. Warm as any part of the Summer
20. Sunday, - warm day
21. Warm day HARVEST, finish'd at Cowitch, & Bremhill & in general, - others a week and more.
22. Warm fine day
23. Thin rain at times all day
24. Warm fine day
25. BROADSTOCK, look'd over Rain in morning, - dry day, the heat oppressive in the evening with Thunder & lightening.
26. Fine day
27. Sunday, warm fine day
28. Damp morning, warm fine day Bradford leigh Fair
29. Thin rain morn'g, dry after'n
30. Heavy Thunder and lightening with very heavy Storm, for an hour to mid:day, - and heavy storm again at 6 OClock even'g BROADSTOCK, - look'd over, the last time, and gave up
31. Heavy partial Storms

September
1. Partial Storms, fogg'y morn'g The Land yet, very dry, no grass, many Dairymen, and Graziers, giving their Cattle Hay, No drought, equal to that of the present Year, in this Country since 1762, - in which Year was no rain from the 3rd of May to the 19th of July
2. Thin Rain in evening
3. Sunday, - cool dry day
4. Rain thin in morn'g, warm and dry after
5. Dry day
6. RAIN, commenced: 4 O Clock morn'g, continued all day with heavy Thunder & Light'g
7. RAIN, all night, and morn'g together, - more than 24 Hours dry the remainer of the day.
8. Gentle rain, - at times all day the face of the earth change'd, from a brown, barren hue, to the most delightful, and fruitful green, - the Grass growing rapidly, - as well the stubble Turnips, - but few Turnips in the Country, before the stubbles almost a frightful prospect for the Farmer, - having but little Hay also, - a prospect of an abundance of Keep now.[63]
9. Cool evenings - strong dews, & fine day
10. Sunday fine day
11. Strong dew, - warm fine day
12. Strong dew, - warm fine day
13. Strong dew, - warm fine day
14. Thin rain after the morn'g

63 The intended insertion of "Grass" has been made.

15. Beautiful fine day
16. Fine warm day
17. Sunday warm day
[Pinniger then presents Wilton fair prices for 1825 and 1826 and similarly wheat and barley prices at Devizes Market - see Appendix H]
18. RAIN, heavy, nearly all day fm 11 O Clock to 11 at night. Yesterday particularly fine, the earth covered, with web. See RAIN after the Web 27 Sepr 1823 [64]
19. Very fine day
20. Storm, mid day
21. Frosty mornings, very fine day
22. Frosty mornings, very fine day
23. Frosty mornings, very fine day
24. Sunday, - heavy storms.
25. Storms, nearly all day, warm
26. Very warm fine day
27. Very heavy rain in night dry day
28. Very warm dry day, nearly as warm as any part of Summer.
29. Calne Fair, - warm fine day Lambs from 4s to 20s
30. Chippenham, new burial ground, Consecrated, by Bishop, (Burgess) of Salisbury The Service of the Church, except the Communion performed Proper Lessons, & Psalms 23rd Chap: of Genesis 5th Chap of John, begin 21st Verse 39 50 Psalms 100 Psalm, Old ver: sung [65] ⋆
Saturday morn 5 O Clock died Mrs C Hulbert (Seager) Aged 39 Yrs
Saturday eve – 9 O Clock died Nichs Thos Skull Ponting Aged 42 Yrs [This and the previous entry written within thick black borders]

October

1. Sunday, storm, mid day.
2. Very fine day
3. Very fine day
4. Storms heavy morn'g very fine after.
5. Very fine day WHEAT SOWING, began at Cowitch Corton, and two Cleveanceys
6. First very sharp frost, Potatoes & Kidney Beans, cut off | Very fine day
7. Very fine day
8. Sunday, - rain in morn'g & evening
9. Heavy Storms
[Pinniger then included a list of names and monetary values entitled "Royal

---

64 The entry for 27 September 1823 (see above) noted that the web was a token of fine weather.

65 This was additional land for the churchyard at Chippenham St Andrew's church, WSA: D1/60/2/7, *Chippenham. Addition to churchyard, 1826.*

Allowances (Annual Acct for 1825" taking up 6½ diary pages. He also included a list of years, monarchs and monetary values entitled "Revenue of England since the Conquest" which takes up a single diary page. Neither have been transcribed]

10. Rain heavy, all the morn'g
11. Damp air all day
12. Damp all morn'g dry after
13. Damp all morn'g very fine after
14. Very fine great'r part of day
15. Sunday, fine dry'g day
14. HORSE – sent to Kellaways in the evening
16. Dull day
17. Very warm sunshine morn'g
18. Fine day
19. Fine day
20. Warm, murk y weather
21. Warm, murk y weather
22. Sunday, - fine day
23. Thin rain at times, - heavy 10 OClock. (Trowbridge)
24. Very drying day
25. Heavy rain in night and storms in the day
26. Fine drying day
    [Pinniger then includes a description of prices etc. at Weyhill Fair on 10th October – see Appendix H]
27. Rain all the morn'g dry after
28. Fine dry'g day, cold air
29. Sunday cold air
30. Fine day
31. Drying day, little rain even'g

November

1. Thin storms, - frequent
2. Cold dry, healthy winds
3. Cold dry, healthy winds
4. Dry morn'g thin rain after [Note in pencil] Crook born
5. Sunday, damp day, rain even'g
6. Thin rain mid day COLTS, removed from Chalfield to Cockleberry, from Thurday 9th March to Thursday 2nd Novr is 34 Weeks 6 Weeks at Kainsham 28 Weeks at 2s ea
7. FROST, very hard, the first this season, - very fine day | Colt, Golden duns, began Breakg
8. Very sharp frost, fine day

9. Very sharp frost, fine day | Cattle generally fothered [66]
10. Sharp frost, - rain light after the morning
11. Damp day, rain in evening
12. Sunday, - fine day, rain even'g
13. Frosty morning, heavy rain afternoon
14. Frost in morn'g, Glass very low to M in much rain, - yet a clear fine day.
15. Fine day
16. Very sharp frost, and clear fine day, - rain in evening
17. Damp day.
18. Dry day
19. Sunday Dry day rain evening
20. Dry day
21. Dry day
22. Dry day
23. Beautiful, fine like May
24. Damp morn'g, very warm Sun clear afternoon
25. Very sharp frost morning SNOW, - heavy mid day to cover the ground
26. Sunday, - <u>FROST</u>, <u>unusually</u> sharp, - in the evening <u>intense</u>
27. Sharp frost, in morning
28. Frost in night, - thin rain all day – Swindon
29. Clear morning, - afternoon over cast and some rain
30. Damp day

December

1. Frosty morn, - heavy rain mid day
2. Storms and damp day
3. Sunday, - fine morn'g, heavy storm by mid day, - damp after
4. Dry day
5. Frost, morn'g, thin rain all day after
6. Storms mid day
7. Rain nearly all day

6, 7. Floods

8. Rain heavy before 9 OClock dry after the morn'g
9. Rain heavy in the evening
10. Sunday, damp day rain evening
11. Fine day wth exceptions of thin storms at intervals

12, 13, 14. Dry mild days

[Pinniger then describes (with diagrams) the eclipse of the moon on 14th November 1826 on one diary page and the eclipse of the sun on 29th November 1826 on the next diary page. These have not been reproduced in this volume]

15. Dry mild day, rain even'g

---

66 Fother - Alternative form of fodder; fothered – foddered.

16. Dry mild day, rain even'g
17. Sunday - dry day
18. dry day
19. Thin rain all day
20. Clear morning damp after
21. Very clear light morning fine day, very sharp frost in the evening
22. Very hard frost, morning quick thaw, some rain
23. Very mild, - fogg'y day
24. Sunday, - dry, dull, day – Glass very high beyond fair
25. Xtmas day Dry, warm calm day The Cattle still on the grounds requiring less Hay, - Rain, thin in the evening and heavy'r in the night GLASS, - as high as great part of the Summer, - halfway to sett fair
26. dry warm fine day
27. dry warm fine day Glass at F in set fair.
28. Glass at F in set fair. colder air clear fine day
29. Glass at F in set fair fine clear mild day
30. A Spring day, - the roads dry'd, - and wax'y, like a dry March
31. Sunday, fine mild day, clear Sun. Glass sinking
    [Between the diary entries for 1826 and 1827, Pinniger presents an Account of the Prices of Wheat from 1646 to 1826, taken from the Register kept in the Audit Books of Eton College.[67] These occupy three double-page spreads in the diary and have not been transcribed]

1827

January

1. Damp, flying Clouds, rain in evening and Tempestuous Storms in the night, - Glass sunk half way to rain
2. Fine morn'g, Snow storms in the Day, and exceeding hard Frost, in the evening & Night
3. Intense Frost, - light Snow storms
4. Intense Frost, - light Snow storms
5. Intense Frost, - light Snow storms
6. Intense Frost, - mild day, - Thaw, eve'g
    DUKE OF YORK, Died last eve 20 Minutes after 9 O Clock [This entry written within a thick black border]
    Lying in State 18th and 19th Jany Buryed at Windsor 20th Jany in his 64th Year
7. Sunday, fine mild day
8. fine mild day
9. fine mild day
10. Thick light rain all day
11. Heavy Snow storm, morn'g dry evening, and very Boisterous night

67 Pinniger spelt the college name as Eaton.

12. Freeze'g air all day sharp frost at night
13. Thin rain all morn'g, dry after
14. Sunday, - Very Boisterous dry day
15. Fine sunshine, - all day
16. Dry mild day, damp eve'g | Died Frederick Augustus Maillard Aged 33 Yrs [This entry written with a thick black line above and a double narrow line below]
17. Very mild fine day
18. Damp morn'g - dry after
19. Dry, day, - cold evening
    DIED
    1821 May 17. Roger Spackman, Chalfield, Aged 67
    1827 Jany 19th Ann Spackman, Widow of the above in her 79th Year [This entry and the previous entry written within a thick black border]
20. Keen air, very fine day, frost, morn'g
21. Sunday, - some Snow in night, & sharp frost, - sharp cuting frost all day
22. Some Snow morn'g, - sharp winds and frost all day
23. Snow little in morning, sharp frost
24, 25. Frost and little Snow
26. Frost and Rime, - very clear day
27. Very, sharp frost, fine day
28. Frost in night, fine mild thaw Sunday
29. Fine day, mild thaw.
30. Fine mild day
31. Fine mild day

February

1. Fine mild, spring like morn'g damp evening
2. Snow, - little, in morng, cold dry day
3. Sharp frost, fine day
4. Sunday Sharp frost, fine day sunshine
5. Sharp frost, fine day sunshine
6. Thaw, - damp air, in morn'g
7. Mild thaw, fine day, freez'g eve
8. Sharp frost fine day
9. Sharp cuting, freez'g winds
10. Sharp cuting, freez'g winds
11. Sunday, - very cold freez'g winds
12. Snow, in morn'g, mild thaw
13. Sharp frost, clear fine day
14. Frost and fine day
15. Frost and some Snow, in morn'g mild thaw after
16. Very cold sharp frost, freezing all day

17. Sharpest Frost this winter, and freezing all day
18. Sunday, - exceeding cold freezing winds
19, 20, 21, 22. Fine days, - sharp frost night and days, skate'g on the River
22. Mr [...] Tanner Miss Sarh Maskeline Married.[68]
23. Frost and fine sunshine, all day.
24. Frost and fine day
25. Sunday fine day rain even'g The first RAIN, in any quantity since 1st Jany, - the DUST scrope from the ROADS, great part of this month, some Farms short of Water now
26. Rain in morning, - beautiful mild spring day
27. Fine mild morn'g, rain all afternoon
28. Thick rain nearly all day.
[The next three double-page spreads in the diary relate to propositions made for the Corn Laws - these are transcribed as Appendix I]

March

1. Rain thick all morn'g, dry after'n
2. SNOW, unusually thick for an hour from 9 to 10 OClock morn'g. Blossoms as broad as Crown Pieces, Ancle deep melted by evening, cause a flood fine dry afternoon.
3. Frequent storms of Rain, flood rise'g
4. Sunday, rain morn'g, storms with drying winds
5. Sharp frost morn'g, tempestuous winds and rain evening.
6. Hail storm, heavy 8 OClock morn'g fine till 12 OClock, - when violent Hail storm, and very forked light'g with heavy Thunder, and very tempestuous winds and rain all eve'g
7. Calm fine morn'g, - coldstorms after
8. Cold drying winds, some storms
9, 10. Sharp frosts, & fine days.
11. Sunday, rain morning & evening Damp day
12. Rain heavy in night, fine mild day.
13. Fine mild morn, storms after.
14. Rain in night very drying day.
15. Heavy storm morn'g, storms of hail & rain after
16. Frost and fine morn'g, rain after Oats. Lewes, - Monkton, sowing Wheat. spring, - T Knight - sowing
17. Boisterous night, winds, - storms of rain and hail in day
18. Sunday. - fine day
19. Snow, in morn'g, dry mild, afternoon
20. Thin rain morn'g, dry mild day after

68 John Tanner (a widower) of Yatesbury and Sarah Rumboll Maskelyne (spinster, age 21) married on 22 February 1827 at Clyffe Pypard, *Wiltshire, England, Church of England Marriages and Banns, 1754-1916* [accessed via Ancestry, 20 July 2020].

21. Fine mild spring grow'g day.
22. Very fine morn'g drying day.
23. Very fine morn'g Gardening, Seeds
24. Very fine morn'g
25. Mid=lent=Sunday. Storm of Rain at 11 O Clock, - fine before & after
26. Beautiful, warm fine day
27. Rain thin from 9 to 2 OClock fine after
28. Fine morn, high winds after heavy hail & rain even'g
29. Cold hail storms & winds morn'g fine after
30. Oats, - sowing J____ P____ of Forsby finish'd began 20th - 50A, - Fine day [69]
31. Fine day

April

1. Warm misly rain all day Sunday
2. Fine warm growing day
3. Fine warm growing day SALE OF SHEEP at Clatford Farm Viz [Pinniger recorded the prices reached - see Appendix H]
4. [It appears that Pinniger visited Silbury Hill and the White Horse at Cherhill - perhaps on his return to Chippenham from the sale at Clatford, near Marlborough - see Appendix J which includes measurements of both features] Very fine growing day
5. Very fine growing day
6. Warm growing, showers
7. Warm fine day
8. Sunday fine day little thin, rain, even'g
9. Fine day, Mr Smith, Bremhill turned his Cows out to good Grass and, others, for want of Hay
10. Fine day to London
11. Very dark, foggy morn'g fine afr
12. Fine morn, very wet after
13. Good Friday, - Regent Park beautiful fine day.
14. Trench Farm, Kent Beautiful fine day, Pole'g the Hops, - began,
15. Sunday, - Rain little morn'g fine day
16. Fine day
17. Hail heavy morn, and heavy rain eve'g at Greenwich
18. Thin rain all day
19. Fine warm day, Windsor
20. CUCKOO, Fogg'y morn'g very warm, fine day, Dorney, and home from London. | DEVIZES Fair, very good Tegs expected would pay nothing for wintering, those bought in at 9s 6d, sold 21s and 12s for 24s &c.
    [Pinniger added the following to his diary at this point]

---

69 John Pinniger was the occupier of Upper Fosbury Farm in 1827, WSA: A1/345/392.

Mrs Pinnigers, observations on the Weather at Chipp[enha]m, see, and compare with the Weather in London.

10. Fine grow'g day, warm rain after noon
11. Beautiful, fine warm grow'g day
12. Rain in morning, fine after warm & grow'g
13. Fine clear day
14. Frosty morn'g fine clear day
15. Fine day
16. Fine day
17. Foggy morn'g fine day
18. Cool air, dull day, Glass sinking
19. Cold dull morn'g, fine after
    [Pinniger resumes his own diary entries from this point]
21. Fine morn'g, rain eve'g & night
22. Sunday, - very cold dry day
23. Cold dry air
24. Cold Hail & Snow Storms
25. Sharp, frost, Ice, - cold dry day
26. Sharp, frost, Icecles, - 6 In long, & large as a finger, - from steam of Cucumber frame, dry day and warmer
27. Fine warm day
28. Very fine day
29. Sunday, - very warm Sun and sultry
30. Very warm, - like Mid=sumr | Wm Jenner 48 Miss Jones 36 Married [70]

May

1. Fine warm growing day, | Scarlet Beans, - planting, 2nd lot
2. Beautiful warm growing day
3. Beautiful warm growing day
4. Beautiful warm growing day some blighting Rain
5. Thin rain great part of the day to do good, - wanted
6. Sunday, - Thin rain nearly all day, very growing
7. RAIN, in night and morning, fine day, - most providential the land being dry and warm prospect of a fruitful time. *
8. Frost and Ice in the morning fine day
9. Dull dry day
10. Fine warm day, some heat drops
11. Warm growing day
12. Warm growing day
13. Sunday, - Fine day, cold air | Church not lighted, - first time

---

70 William Jenner of Bremhill and Anne Jones of Stanton St Quinton married on 30 April 1827 at Stanton St Quinton, *Wiltshire, England, Church of England Marriages and Banns, 1754-1916* [accessed via Ancestry, 28 July 2020].

14. Very warm little rain eve'g
15. Very warm fine day
16. Fine morn'g rain enough in the evening to do good
17. Chippenham Fair, - Rain from 9 to 12 OClock, dry after
18. Very warm growing day
19. Very warm growing day X The Horse & Two Colts, sent to Cleveancey for Swindon Fair, - 21st
20. Sunday, - warm fine day
21. Very growing warm fine day X Sold Mr Summerset Barton 32£
22. Thin rain, frequent morning | Sticking Kidney Beans
23. Fine morn'g, storms afternoon
24. Fine dry day, - Mr Gye's treat
25. Fine dry day, very growing.
26. Fine dry day, very growing.
27. Sunday, - Rain morn'g & Eve
28. Rain in night & morning very fruitful time
29. Rain in morning, dry winds after
30. Littlecote Fine growing day, the Watermeads, mowing, now and ten days since, - some near Hay
31. Forsbury Very wind'y some thin rain Forward Turnips geting in rough leaf. | Mr Rum__ll, Plough'g for first Turnips

June

1. Cold rain, all morning, dry after, - no Wheat Ears near bursting, see last Year [71]
2. Thick rain all the morning dry after, home from Littlecote
3. Sunday, - cold showers
4. Warm fine day
5. Storms of thin rain morn'g dry after | WHEAT EARS, many quite out at Hardenhuish, and the whole field bursting the ear. | Revd Mr Short, Married to Miss Audry, Notton | Hay, making, at Cocklebury & Mowing, many places [72]
6. Very cold Storms & very frequent all day, - blighting | Swedes, sowing again at Corton old Seed not grow
7. Dry cold winds, very unlike June
8. Great change in Weather, very warm soft air, & very grow'g
9. warm soft air, & very grow'g

---

71 On 1 June 1826, Pinniger recorded "Wheat Ears bursting in forward pieces (near Salisbury)".

72 Revd Mr Short is probably William Short, vicar of Chippenham (see footnote to entry for 18 June 1826). William Short and Jane Awdry married on 5 June 1827 at Lacock, *Wiltshire, England, Church of England Marriages and Banns, 1754-1916* [accessed via Ancestry, 21 July 2020].

10. Sunday, - warm fine day
11. Warm grow'g day | SWEDE, - hoeing at Lowden. | MANGEL=WURZEL hoeing 2nd time at Lowden
12. Very growing warm day
13. Very growing Mr Guy, began mowing | Kidney Beans, grow rapidly 4 feet high
14. Blight'g morn'g, fine day
15. No Sun shine all day, blight'g air, and at times damp.
16. No Haymaking, damp all day, - some showers
17. Sunday, - warm fine day
18. Dull morn'g, very fine after
19. Rain morn'g, & evening, some Hay carrd bad between
20. Damp morn'g, warm fine afternoon
21. HAIL storm, & frequent storms Hay caught up at dry intervals | Wheat, fully in ear, and Barley and Oats appearing some pieces in ear
22. Shower, before noon, dry after
23. Very fine Haymaking day, the first since the 14th
24. Sunday, - very fine day, the Glass high & fixed for dry | Mr Walter CROOK died aged 69 yrs
25. Fine haymaking day
26. Fine haymaking day
27. Rain a little, morn'g & eve, dry between
28. RAIN, in night, till noon, wanted for Barley, Turnips, and very acceptable, for all vegitation
29. Warm fine day | Colt the oldest taken from Laycock Abbey sale, to Studley to Keep, Mr Mansell
30. Frequent Showers, to prevent Haymaking

July

1. Sunday, - Rain, heavy night and morn'g, very dry'g after
2. Rain all the morning dry after
3. Rain early in morn'g, very warm & drying after
4. No meadow Hay carryed at Kennet (WP) yet, very dry'g morn began carry'g mid day, nearly spoiled by the late Rains.[73] | Turnips, - hoeing 2nd time | Swedes, old seed sowed 6 weeks, not up | Swedes, other old seed just fit to hoe | Rain in the evening,
5. Fine hay making day
6. Fine hay making day very warm
7. Fine hay making day the heat oppressive | Surveying Grange, and Claverton Farms
8. Sunday, - very warm fine day
9. Very warm, day | Young Colt, returned from Cocklebury to be broke

73 WP – William Pinniger, his brother.

10. Fine day, not so warm
11. Fine day, and very warm
12. Fine day, and very warm
13. Fine day, and very warm Freegrove &c.
14. Fine day, and very warm Freegrove &c.
15. Sunday Fine day and very warm
16. Warm fine day
17. Warm fine day
18. Damp morning, - fine day | Pease, - cuting, at Chalfield | Wheat & Barley ripe in 10 days at Chalfield | Haymaking, finish'd at Chalfield
19. Thin rain frequent through the day
20. Rain, heavy in the night, very drying day
21. Warm fine day. [Pinniger then records the sale of three farms sold by auction at Malmesbury - see Appendix K]
22. Sunday, - rain in the morning, and damp evening
23. Warm fine day
24. Warm fine day BPS[...]s liberty [74]
25. Warm fine day 19th REAPING at Andover | Oats, mowing at Cockleberry
26. Very fine morn'g, RAIN, began thin at 1 O Clock, heavy in the eve'g, Miss A L retd to Cleveancey, after noon.[75]
27. Little rain, mid day, fine before & aft
28. Fine - warm day
29. Sunday, very sultry, - at 12 OClock night commenced, unusual quick lightening, Thunder, and very heavy Storm
30. Reaping, commenced very generally, Chippenham & Bremhill Parishes Very warm fine day
31. Very warm fine day

August

1. Fine dry harvest weather
2. Fine dry harvest weather
3. Fine Rain in morning
4. Fine

---

74 This is probably Broome Pinniger Smith, second son of Richard Sadler Smith and Elizabeth Smith (nee Pinniger), Thomas Pinniger's eldest sister, WSA: 4381/2/1, *Pinniger genealogical memoranda book, c.1820-1943,* p.14. The prisoners to be heard at the Insolvent Debtors Court, London on 19 March 1827 included Broome Pinniger Smith of Great Queen St, surgeon, Public Ledger and Daily Advertiser, 27 February 1827. It appears that he was freed from prison on 24 July. The death of Broome Pinniger Smith, surgeon of St Leonards, Hastings, second son of Mr. R.S. Smith of Bremhill occurred on 28 November 1847, Devizes & Wiltshire Gazette, 9 December 1847.

75 Miss A L – This is Miss Ann Large of Clevancy, Pinniger's sister-in-law.

5. Sunday, - fine day
6. Blight'g morn'g, fine day | WP, began reaping Oats | Very little Wheat cut in neighbourhood | WS Chalfield, making Wheat Rick [76] [Pinniger records prices at Tan Hill Fair - see Appendix H]
7. Very fine Harvest day
8. Very fine Wednesday - died Mr Canning, Premier of England aged [...] [77]
9. Fine day
10. Rain
11. Rain heavy, night & morn'g fine after | Two Yr old Colt, to Jno Hughes Tythern to be broke to Harness
12. Sunday, - dry cool air all day
13. Rain morn'g and even'g | Malmesbury, survey ye land again
14. Rain morn'g & Eve'g, damp day Tytherton
15. Sultry and Storms, Tytherton, and Tockenham
16. Sultry Thunder, lightening, and heavy rain after 4 OClock returned from Tock[enha]m by Tytherton
17. Thunder & partial Storms | This the last day, saw my, ever lamented Father alive, Tytherton | Took two Stocks Bees - 14 li Honey
18. Glass rising, light storms, no Wheat, carried Beckhampton to Marlbro', - some vegitated
19. Sunday, fine day, remarkable calm evening
[Diary turned landscape for the entry of the death of his father, Christopher, which is written within a thick black border]
Sacred, to the memory of my ever lamented Father who resigned his pious Soul to his Heavenly Father who gave it. Sunday at midnight, or the morning of Monday the 20th August, being just pass'd 12 O Clock 1827, in the 80th Year of his Age, - Born 10th March 1748
I am thine, save me; for I have sought thy precepts 119th Psalm 94 Verse
Sunday 19th Augst 1827. My lamented Father walked round the Orchard and Yards in his usual health about 12 O Clock, then dress'd himself for dinner, shaved himself remarkably clean, altho: his sight very impaired, and wash'd himself very nicely, and before dinner said grace by asking a blessing very devoutly made a very hearty dinner, - and some little time after Dinner drop'd to sleep, as was his custom, and about 3, O Clock awoke, and made an effort to rise and walk out, but found himself unable, when he asked for his Stick, and making a second attempt to rise found had lost the use of his left side, but with assistance rose from his Chair on on his feet, but unable to

76 WS – this is probably William Spackman, husband of Pinniger's sister Susannah.

77 George Canning [1770-1827] was Prime Minister for the final four months of his life, thereby holding the record for having served as Prime Minister for the shortest period of time: he died aged 57. UK Government, *History, Past Prime Ministers, George Canning* https://www.gov.uk/government/history/past-prime-ministers/george-canning [accessed 12 May 2020].

stand, drop'd on the floor never to rise again in this world, - was immediately bled by Mr Edwards very profusly but to no good effect, his eyes closed and breathing like one in a sweet sleep, continued so, his feet and hands rather cold till 9 O Clock, when revived a little, hands and feet became warmer and Pulse stronger, when sent for Mr Edwards who Cup'd him in three places on the Neck which bled freely, the Glass's reapply'd, and bled freely again, without the least reviving, continuing in a lethargy, occasionally interupted by a long breath and pause, then the breathing restored, a visible change soon took place for the the approach of Death, - the breathing interupted by a pause as before, and at intervals a cough, which finaly stopd the breathing directly after midnight, when all who believe in the atonement of a merciful Redeemer in whom my dear Father most devoutly trusted for mercy, will believe as a just and good Man, his Soul was received into the Mansions of the Blessed.

[Diary turned portrait for the entry of the funeral of his father, Christopher, which is written within a thick black border on three sides]

1827 Augst 24th My Father's mortal remains deposited in his Vault, Bremhill, Church Yard, by the Revd Mr Bowles, at 2 O,Clock Friday

Friends, at the Funeral

The Revd Mr Edwards, The Revd Mr West of Tytherton

Pall Bearers

Willm Pinniger, Thos Spackman, Nichs Ponting, first Cousins, Robt Stiles, Josh Edwards, Saml Viveash

Mourners

4 Sons, Mr Smith, Mr Spackman, Mr Buxston & Fred: Hatherell Servt (Mr Crook too ill, in Gout to attend [78]

[Pinniger then lists all of his father's family who survived him, including his grandchildren. This spans eight pages of the diary and has not been transcribed: however, they have been used in preparing the genealogical charts in the Introduction.]

20. Heavy partial Storms, the Harvest retard'd, by the Showery days
21. Very Fine morn'g, ThunderStorm in evening
22. Fine Harvest day
23. Very Harvest day
24. Very Harvest day | Funeral [This entry written within a black border] | Wm Spackman finish'd Harvest
25. Very fine Harvest weather
26. Sunday Harvest weather
27. fine Harvest weather

---

78 It appears that as well as his four sons, John, William, Thomas and Broome, the mourners included his three sons-in-law: Richard Sadler Smith, William Spackman and Henry Crook. Frederick Hatherell attended to represent Henry Crook and Mr Buxston is probably the Revd Mr Buxston (see entry below for 10 October).

28. fine Harvest weather | Wheat Harvest finish d at Corton & Cleveancey
29. Warm, fine Harvest Weather
30. Warm, fine Harvest Weather
31. Warm, fine Harvest Weather

September

1. Very, fine Harvest day, - nearly all the the Corn, carried, except Beans, and those, nearly cut, - & some carried.
2. Sunday Funeral SERMON, for our dear Father, by the Revd Mr Clifton 23rd Psalm 4th Verse Yea, though I walk through the Valley of the shadow of Death, I will fear no evil: for thou art with me; thy rod and thy staff they comfort me. The Psalms sung 121st and [in pencil: 56] New V Bremhill | Very fine warm day
3. Very fine warm day
4. Very fine warm day
5. Very fine warm day
6. Very fine warm day
7. Very fine warm day
8. Very fine warm day
9. Sunday, Rain heavy all afternoon
10. Appraising the Effects at Tytherton frequent Showers [79]
11. Damp air at times
12. Very heavy storms Cleveancey
13. Glass riseing, fine nearly all day
14. Warm fine day
15. Very close warm day, very growing
16. Sunday warm day, very growing
17. Mr Crook, finish'd Harvest, very growing
18. Mr Jas Large, finish'd Harvest rain afternoon
19. Cool, - Air changed, - very fine for finish g Harvest
20. Partial Storms
21. Rain heavy afternoon, Mr Jn L finish'd Harvest [80]
22. Heavy Storms
23. Sunday, - fine day
24. Warm grow'g fine day Wheat Sow'g began Sow'g at Avebury.
25. Warm fine day
26. Warm fine day Sultry & Thunder storm, eve'g
27. Rain night & morn'g fine after, Sultry
28. Sultry, - very fine day
29. Very heavy rain in morn'g fine after
30. Sunday, - very fine warm day

---

79 The effects at Tytherton were probably those of his late father.

80 Jn L - Probably John Large.

October

1. Calne Fair, held, Beautiful fine day
2. Rain in morning, very fine after
3. Thick fog till mid day, - very fine after | Salthrup, - Carting dung for Wheat too late | Rule, here, finish Wheat Sow'g Devizes Fair the 20th Octr & Barley sow'g Devizes F. 20th Apl
4. Thick fogg, - very fine day
5. Thick fogg, - very fine day
6. Thick fogg, - Beautiful fine day [Pinniger recorded sheep prices at Kington Fair - see Appendix H]
7. Sunday, fine day, Ground covered wth Spiders Webb, sign of Rain
8. Fine morning, rain all evening
9. Rain, heavy, at intervals all day
10. Tockenham TP ill | Fine morning, heavy rain all evening | The Revd Mr Buxston Married Rose, Daughter of Chas Shephard Esqr of Cobham, Surry *
11. Heavy storms, in night and Day
12. Rain in night, drying Day.
13. Fine drying day [Pinniger then recorded prices from Weyhill Sheep Fair held on 10 October 1827 - see Appendix H]
14. Sunday, Fine morn'g, damp eve | This morning at 9 OClock Died in the 58th Year of her age, Eleanor Wife of Anthony Guy Esqr of Chipp[enha]m A Lady universally respected and esteemed and in whom the Poor have lost a liberal benefactress
15. Fine morning, damp after
16. Damp morn'g, fine after
17. Fine all day
18. Bath wth Mr Edwards Junr, Rain morn'g, Beautiful fine after
    This morning 11 OClock Died Alice Spackman, Widow of the late Jacob Spackman of Tytherton in the 68 Year of her age.
    Sunday 21st Octr Died Thomas Lavington, Grocer Marlborough, Aged 37 Years Augst last. [This entry and the previous entry written within separate thick black borders]
19. Dull dry day
20. Very fine day, Mrs Guy's Funeral
21. Sunday, - fine morn'g, Storm after, n
22. Storms, in morning
23. Rain in night and all day
24. Very fine day, Mrs Spackmans Fu:[81]
25. Rain nearly, all day
26. Fine dry day

---

81 Fu: - Funeral.

27. Colt, returned from Jno Hughes much reduced, having, worked 11 Weeks, Wheat sowing &c.
28. Sunday, Rain nearly all day, to cause a flood
29. Very fine drying day
30. Very fine drying day Tock[enha]m
31. Very fine drying day Highworth and M Mill

November

1. Sharp Ice'y frost, fine dry'g Day
2. Storm, morn'g, very fine after
3. Very fine, Colt, returned from Mr Mansell very poor, - wth a Chill, very ill, 18W frm 29th Jun
4. Sunday, fine day
5. Beautiful, mild fine day

6, 7, 8. Beautiful, mild fine day Acct Settle'<g>d wth AG Glass unusually high [82]

9, 10. Warm fine days

11. Sunday, fine day
12. Very fine, roads dusty
13. Very fine, warm, clear Sun
14. Rain afternoon
15. Dry day, rain in evening and heavy in night | Acct settled wth Mr G
16. Rain heavy in morn, fine after
17. Dry day, high flood
18. Sunday, - damp fog'y air

19, 20. Mild fog'y days

21. Cold clear fresh air frosty eve'g
22. Sharp frost, Ice. 1 Inch thick, damp eve'g, freezing after
23. Marlbro: Fair, - very sharp frost & Snow, from 12 to 5, OClo
24. Very sharp frost & Slippery
25. Sunday, - Thaw, mild dry day
26. Very, fine clear day, Queenfield & Beazells Farms, - look'd over
27. Very thick, fog, all the morning Colt, to Cockleberry
28. Mild rain at times all day.
29. Very, heavy rain in night, fine sunshine day
30. Thin rain, at times all day

December

1. Boisterous Winds, some storms
2. Sunday, - damp eve'g
3. Beautiful day - Bath : Su & Sa P[inni]ger
4. Fine day
5. Dry day, and warm

---

82 AG - Anthony Guy.

6. Dry warm day
7. Sharp white frost, dry day
8. Fine clear day, very mild
9. Sunday, - sharp frost and thaw
10. Thin rain frequent
11. Fine dry morn'g, raid mid=day and eve'g
12. White frost, morn'g, damp eve'g
13. Little frost, fine day, & clear
14. Close rain
15. Rain morn, dry eve'g
16. Sunday, heavy rain afternoon
17. Fine dry day
18. Very mild pleasant morning rain afternoon
19. Rain all day
20. Fine day, FLOOD, - flood, highest since 2½ Yrs Past
    20th Thos Pinniger of Down ampney fell of his Horse in a Pond by his own Door, 10 OClock, eve, Drownd Aged 59 [This entry written within a thick black border]
21. Fine morn'g thin rain all afternoon
22. Dry day
23. Sunday, - clear dry mild day
24. Rain all day, fine eve, Flood
25. Fine drying night, & fine clear mild day
26. Fine clear mild day | Carrots, began feeding my Nag 13½ ⋲ [83] @ 48s pr Ton
27. Very murkey warm day
28. Beautiful clearmild day
29. Thick fog all day mild
30. Frost in morning, cold air
31. Rain all afternoon

1828

January

1. Heavy rain all day, and the highest Flood at Calne since Year 1707 to the ground floor Windows, Butcher row – , very mild and Daiseys in full bloom, as thick as in May [84]
2. Rain all afternoon
3. Fine day
4. Rain after the morn'g
5. Frost in morning

---

83 ⋲ - cwt or hundredweight.

84 This may be a reference to floods following the great storm of 1703, or a flood in 1725 referred to in A. E. W. Marsh, *A history of the Borough and Town of Calne* (Calne and London: Robert S. Heath, 1903), p.7.

6. Sunday, damp morn'g, dry after
7. Cold freezing winds
8. Snow at times all morn'g, fine eve
9. Freez'g all day, very cold
10. Exceeding cold freez'g all day little Snow. Poulshot
11. Much Snow in night, frost & thaw alternate all day.
12. Freezing and thaw alternate and rain in evening
13. Sunday, damp day high Flood Carts and Boat used in Street (Chipp[enha]m) for Passengers, - within Two Inches of flooding our House.
14. Fine day, rain in evening
15. Very cold dry morn'g, heavy Snow in evening and night
16. Damp day, rain in evening
17. Rain all night, snow gone, Flood nearly as high as Sunday.
18. Dry mild day.
19. Very fine Sun shine day
20. Sunday, very fine mild day
21. very fine mild day
22. Damp morn'g fine mild day
23. Damp morn'g fine like April
24. Beautiful fine mild day
25. Dry morn'g & eve'g, rain mid=day.
26. Very fine mild day
27. Sunday fine mild day
28. fine mild day admin[...]
29. Til[...] Cours[...] fine mild day
30. Littlecott Farm, Enford, for sale [85] | Beautiful May like day
31. Rain night & morn, damp day

February

1. Fine mild day
2. Fine mild day
3. Sunday, very fine mild day
4. very fine mild day
5. very fine mild day
6. Damp morn'g dry eve
7. Rain morn'g, dry after
8. Cold dry day
9. Littlecott Farm, very cold dry day
10. Sunday very cold dry day
11. Deep Snow in night to stop Oxford Coach
12. Freezing & Snow

---

85 Pinniger wrote notes on the value of Littlecott Farm at the end of Volume 4 - see Appendix L.

13. Sharp Frost, morn'g, cold winds all day (High Pen Sale)
14. Very heavy Snow all morning, and rain afternoon
15. Sharp frost & fine day
16. Sharp frost & fine day
17. Sunday, fine day
18. fine day
19. Warmer, fine day Devizes to Meet Mr Moore to Purchase Littlecott
20. Frequent Cold storms
21. Fine mild day
22. Heavy rain in morn'g, dry after
23. Rain morn'g dry after
24. Sunday - damp at times
25. Beckhampton Farm, look'd over with Mr Guy, to purchase [86] | Heavy rain at times till eve'g when fine
26. Damp morn'g dry after
27. Mortgage, took, very mild fine day | Beckhampton purchased
28. Very mild May like day
29. Fine mild day, Grains, Horse, began

March

1. Thin rain, all morn'g, mild dry after
2. Sunday, - fine day
3. Very fine day, dust ready fly | Gardening, began, | Norton Sale, The Farm 276A
4. Fine day Cleveancey
5. Snow & hail Storms, very cold | Norton & Halcomb
6. Very cold frosty Winds
7. Sharp freezing Winds all morn'g very seasonable, to check Turnips some in bloom, damp even'g.
8. Fine mild day
9. Sunday mild day
10. Fine mild day Bean plant'g general Cleveancey.
11. Very mild, and fine
12. Very warm & mild like May
13. Particulary warm and growing
14. Particulary warm and Cleveancey
15. Particulary warm and Cleveancy
13. HAY, finish'd 8½ ₠wt, began 27th Decr 11 Weeks @ 4s pr Cwt
14. HAY, rec'd 2₠ 3qr 12li @ 4s 6d is 13s finishd 18th Apl
16. Sunday, warm growing day
17. Pd Mr S[...] Legacy warm growing Cleveancey

---

86 Pinniger wrote notes on the value of Beckhampton Farm at the end of Volume 4 - see Appendix L.

18. [?...] Miss A L[...] warm growing Cleveancey
19. Rain in night the first for, since 1st March Very cold winds, Cleveancey
20. Thin partial Showers Trowbridge
21. Fine day
22. Heavy rain in night, and heavy SNOW from 9 to 10 O Clock, morn'g, cold storms after
23. Sunday, fine day [This is the final diary entry in Volume 4]
Opening Volume 5 of the diary, Pinniger recorded the heights of various members of his family: these have been transcribed under **Garden, family and health** in the BECKHAMPTON section above. He then noted certain details of the Beckhampton Enclosure - see Appendix M, followed by a list of questions which Miss Forty of Chippenham proposed asking Anthony Guy about his insolvency - see Appendix N. His diary entries commence with entries for 19th, 25th and 27th February regarding Littlecott and Beckhampton Farms. The next entry is for March 28th, followed by daily entries from April 1st. There are no diary entries for 24th, 25th, 26th, 27th, 29th, 30th, 31st of March 1828.]
February 19. Declined Purchasing Littlecott Farm
February 25. Beckhampton, look'd over, with Mr Guy to Purchase
February 27. Beckhampton Farm, Purchased of Mr Guy
March 28. Beckhampton, look'd over first time since Purchase'd

April

1. Sharp Frost fine day
2. Mild morn, fine day
3. Keen air, fine day | SEED'S, bought for Beckhampton
4. Good Friday, - very keen winds
5. Milder and lowering, Wm Philpot and Mr Guy, made Terms for quit'g Beckhampton
6. Sunday, Rain in night
7. Rain heavy in night, grow'g Day
8. Cold dry day
9. Rain in night, partial Showers in Day | Agreement signed wth Wm Philpot at Beckhampton, for Quiting his Farm | Rolling and Harrow'g the thwart for Barley, Drilling
10. Showers, and Thunder Storm, eve'g | Barley, Wm Philpot, began Drilling bottom of the 19 Acre Piece, and my Seeds, sowing after, heal'd with one tine of the Harrows [Written in pencil: No tine too wet] 2s pr Acre, Drilling, Wm Rumming 4½ Bushl Barley 2 Bushl Hop & Ray @ 32s pr Qr 6 ℔ Broad Clover @ 5d pr li } pr A
11. Dry'g morn'g heavy rain all even'g
12. Rain nearly all day, beat off sow'g | Tythes, took of Mr Ward, at Marlbro | Beckhampton Farm

13. Sunday, heavy Showers
14. Fine day, Partial Showers
15. Rain nearly all Day
12. Barley, - Wm Pin[...]r, began Sowing Mr Pumphry's Barley, shew'g green sow'd [...] March, when the Land work'd well [87]
16. Drying day, storms after
17. Rain morn'g, very heavy all eve'g Norton Sale
18. Very heavy storms | HAY 2€ 3Qr 12li began 14th March, now finish'd being 5 Weeks at 4s 6d pr € is 13s | Hay 13s Bran 1½ cwt 9s Grains 8 Bu 4s Cuting Chaff 3s 5 Weeks [Total] £1 9s 0d is 6s pr Week
19. Heavy storms, - dry nights
20. Sunday Storms
21. Dry morn'g, rain in evening | Devizes Fair, Sheep, Prices, about the same as last year [See Appendix H] | Barley Sowing impeded, the last week with frequent Showers
22. Mild grow'g day
23. Very drying & mild till eve'g, then heavy rain
24. Beckhampton, walk'd over the Farm Very drying winds all day, roads dusty
25. Very heavy rain afternoon, at times all day
26. Very fine dry g day Glass rising - CUCKOO, appear'd
27. Sunday, fine dry'g day
28. Beautiful, fine day. Malmsbury Fair
29. Beautiful, fine day. Cows, turn'd to Grass, generally, and grass good. | Vetches, began feeding at Chalfield | Clover & Vetches, very good in general from the mild winter.
30. Dull morn'g fine day

May

1. Beautifull, fine grow'g day.
2. Beautifull, fine Thundery weather in the evening, and little rain
3. Thunder & hail Storm at midnight | Cloudy day and partial storms.
4. Sunday Cloudy day and partial storms.
5. Warm showers
6. Warm showers some Thunder, very grow'g
7. Very fine day | Kidney Beans, planted, & 2nd Potatoes
8. Cold morn'g, some blight'g rain, fine after
9. Warm fine day
10. Some thin blight'g rain
11. Sunday, cold morn'g warm after
12. Warm fine day
13. Warm fine day look'd over Beckhampton, wth Mrs Pin | Wm Phillpot, finish'd, feeding the Marsh part of Silbury Mead, wth 200 Ewes and 100

87 Pin[...]r – Pinniger.

Tegs, Kept all a fortnight wth part of the other Mead. | W Philpot, began the Field grass 7 Acres expect keep the 300 Sheep & the Lambs till June when the Down will be good, which is now unusually good, white with Daisy's in places The Downs will after the Field grass, keep the Sheep without any Artificials till the after grass is ready | Wm Phillpot, finish'd drilling BARLEY on the Down land, - Robt P not finishd and the Land, much too dry and rough.

14. Wm Pinniger Kennet, finish'd Barley. sowing
2. Wm Pinniger, finish'd Swedes, and began 9 Acres of Grass, which will Keep all the Sheep till after the 21st of May, when the Down will be unusually good Very warm, and fine
15. Very warm fine day
16. Thin Showers, Thunder & light'g, night
17. Showers in morn'g, and afternoon Chipp[enha]m Fair growing day
18. Sunday, rain'y morn'g, fine after
19. Very fine, and warm
20. Very fine, and warm
21. Fine morn'g, Thunder & light'g after afternoon and heavy Thunderstorm
22. Heavy rain afternoon, the Rivers fill'd rain enough for all kind of land
23. Fine morn'g, thin rain in evening.
24. Fine morn'g, thin rain afternoon & eve'g
25. Sunday, fine day
26. Dry morn'g & eve, rain, mid day.
27. Fine morn'g, Rain after the mid, day the Rivers fill'd again, the Clover grass a heavy Crop, & fit to mow, - and the dry meadows nearly fit.
28. Fine morn'g rain all evening
29. RAIN, nearly all day
30. Very fine warm day Beckhampton | Mowing Field grass, in general, Kennet & Mowing Meadows at Kennett 27th Turnips, up, sowed 19th at Kennett Sheep on the Down, at Kennett week pass'd | Wm Philpots Sheep, on New field, make not bad work, intend keeping on the same another week, before stock the Down. | Turnips Wm Philpot, began sowing, to feed for my Wheat.
31. Fine warm day | WHEAT=EARS, the first bursting, half the Ear seen, - (at Cockelbury) | Hay 4 Cwt @ 3s 6d is 14s, Bran 1 Cwt 5s, Grains 4 Bu 2s, Chaff cuting 3s Kept the Nag 4 Weeks [Total] 24s

June

1. Sunday, fine day
2. Thin showers, to check haymaking | Nag first, turned out, Mr Guys field [Diary turned landscape for this entry]
3. Heavy showers, evening and night
4. Heavy showers, morn'g, drying after and following night, a Whirlwind, that

twisted off the tops of Oak, Elm, and Willow Trees by Turnpike road, near Stanly Lane.

5. Continual, heavy, cold Storms
6. Glass rising, drying day, wth some fly'g Showers
7. Glass rising, drying day, very cold morning
8. Sunday, fine morning, some thin blighting showers, in the evening
9. Very fine warm, Haymaking day, and carrying in many places
10. Very fine warm, began carry'g Sainfoin Hay and Shearing Sheep, Chalfield
11. Very fine warm
12. Very fine warm, Hay began carry'g at Tockenham and Freegrove, and Meadow Hay carrying universally
13. Very fine Sunshine day, finer Haymaking day | Mr Guy began Mowing | Wheat at Tockenham and Freegrove not yet Bursted, and Barley at Freegrove very short and yellow, as well in many other places, where sown late, chilled with the continued rain, begining of the Month, The Wheat at this time at Cockelberry, out even at Head, & the stalk 12 In long above the flag & the Barley 2 feet high, shewing the Beard of the Ear, sow'd, about 20th Apl | Mr Smiths of Bremhill, last sowing Barley after Swedes 20th May, growing rapidly, prospect of good Crop.
14. Very warm, Haymaking day, much Hay carried, this week in good condition
15. Sunday, very heavy Thunder, light'g and very heavy storm before day light, & continued till 5 O Clock, very dry'g after
16. Heavy storms, and very hot
17. Very violent and Awful Thunder, and and lightening, with Torrents of Rain very general Hay, wash'd away with the flood at the Whitleys, and other places. A Tree shattered, and a Cow kill'd under it, of Mr Butlers of Fowlswick, and A Shepherd, his Dog, and 5 Sheep kill'd at Collingbourne, and A Sheep Shearer, struck senseless in Mr John Larges Barn of Cleveancey whilst Shearing, Trees torn up, and Barns blown down Floods on the dry lands, greater than any time in the Winter.
18. Beckhampton Inn, Purchased of Mr Guy | Heavy storms, drying between them
19. Frequent flying storms, dry'g evening Hay, begin to receive injury, with heat and wet.
20. Small rain, morning, very fine dry'g day, - Glass rising since the 18th
21. Very heavy Storms, morning, and evening | RAIN, every day, this week, no Hay carried in good condition, a great quantity very much injured
22. Sunday, - dry morn'g, continual Thunder, after noon, for 2 hours with some rain, dry eve'g, Glass rise'g
23. Glass rise'g, very fine Hay make'g day, and a good prospect of continue'g
24. Beautiful fine day, Hay, Mr Guy began carry'g, mowed 13th, much stained and very much recovered

25. Very fine Hay making day
26. Very fine Hay making day Wilson & Son, Dentist, 13 Paragon Buildings [88]
27. Very fine exceeding warm | Wm Philpot, finish'd, carrying Hay The Dry mead, and 4 A: of Watermead injured, being much wash'd wth the Rain | 1 Rick of the Clover Hay, injured, much wash'd wth the rain | The Clover Rick, by the Wheat carried well, except about 3 Acres at the bottom from the Piece next the Firs | The other, Clover Rick, carried well this, afternoon, but all finish'd mowing Saturday 21st June, before the weather took up, to be fine, 23rd June, Monday | HAY, made by Wm Philpot at Beckn Viz [...] The Dry mead 4, Watermead about [Total][...] A 2 or 3, - of New field for my Wheat 17 of New field for Wheat, Year after 3 of New field for my Wheat next the Firs [Total][...] Deduct, what has been cut & is still left for the Horses [89] [Pencil note in margin] See Hay Aug - 6 & 18th & 27 June and Septr 26th & 15th | The remainer of the Watermead finish'd feeding 13th May, now a good crop except the marsh'y part, about 5 Acres the whole Mead being 9A | Wm Philpots Turnips, sowed 30th May on my Wheat Land, quite fit to Hoe. | Barley at Beckhampton, began sow'g 10th April, now 2½ feet high & much in Ear | Sheep keep at Beckhampton, very plenty, the Down good, and the second Clover good, and soon ready | Fold'd nearly 2 Drifts on Clover fed next the Barley for my Wheat Wheat at Beckhampton, very good none better to appearance
28. Very warm fine day | HAY, W P, K finish'd Carting, began mowing 27th May, much wash'd [90] | Turnips, hoe'd at Kennet, will spread 12 In sowd 19th May
29. Sunday, Very warm, fine day
30. Very warm, fine day | HAY, Mr Guy, finish'd making N.B. 6th June Weather began clearing up, to 15th very fine, & great quantity of good Hay carry'd. 15th Sunday Morn'g, very heavy RAIN and rain every day to 22nd 22nd Sunday, weather cleared up, and all Hay carr'd up to the Mowers | Mr Guy, and Mr Lock, commend Part | Mr Smith, Bremhill & B-P, of Tythn finish'd Haymak'g, unusually ear [91]

July

1. Shower, early in the morn'g, fine after Turnip hoeing, Corton & Cleveancey
2. Shower in morn'g, fine after, blighting.
3. Blighting dull morn'g, fine day
4. Lowering morn'g, partial Showers, fine day | Park, Town Farm, sow'g Swedes first sow'g Turnips, but just up for Wheat too late | See, the heavy

---

88 Mr Willson, dentist to his R.H. the Duke of Clarence, 13 Paragon Buildings, Bath, H. Gye, *Gye's Bath Directory, corrected to 1819* (Bath: H. Gye, 1819), p.7.

89 Beckn - Beckhampton.

90 W P, K - William Pinniger, Kennet.

91 ear - early.

rain 22nd & 27 May, stop'd sow'g a fortnight, at Park T-F, the Barley & Oats sow'd since, too backward, but look well | B & O - sow'd before the Rain in full ear

5. Thunder Storms, in eve'g fine day.
6. Sunday, - fine drying day
7. Shower in morning, drying day | Turnips, hoe'g at Freegrove, very good and Swedes good, want hoeing, and Barley at Freegrove, much improved.
3. Ploughing a Fold drift at Beckhampn and Fold put on it, two drifts next the Barley, fold'd before ploughed
8. Fine morn'g, - continual low Thunder all afternoon, with rain, and about 7 OClock, - one dreadful, clap of Thunder and Lightening, same time at Bath. RAIN in Torrents, such not known, by any Person living continued from 7 to 12 O Clock night 7 Inch's of Water fell in the time, the River overflowed, suddenly & several Persons drowned in their Beds
9. Fine morn'g, rain all the afternoon Glass sinking
10. Very fine drying day Bath
11. Fine morn'g, rain'y evening
12. Rain heavy in night, shower in day
13. Sunday, - Storms and Sunshine
14. Frequent Showers, Hay spoiling
15. Fine morning, showers after
16. Very drying day, much stain'd Hay, carried
17. Heavy Showers in morn'g, dry'g day Potatoes planted, after 1st Crop
18. Storms all morn'g, dry after, storms eve'g
14. REAPING, began near Salisbury
19. Very x storm 11 O Clock morn'g, very dry'g after, at 3 O Clock, dust fly x heavy
20. Sunday, very heavy rain nearly all day after 11 O Clock
21. Very drying, after the morn'g Hay got together in a bad state, and Rain very heavy in the evening | COLT removed from Cockleberry to Corton, lent to Mr Crook to ride
17. Mr James Large, finish'd Hay=mak'g
21. Mr John Large Mr Crook & others } not finish'd mowing
22. Heavy storms in morn'g fine after
23. The Meadow Hay Rick and Two Clover Ricks not thatch'd at Beckhampton, expect injured having been made a month | Very fine morn, Glass rising, heavy Storms 11 O Clock at Beckhampton | Lambs on the last Tie of Clover by Furze Down, to be fallow'd for Wheat | A weeks keep of Clover left for the Lambs on the 19 Acre Piece for Wheat All the Wheat Land fallow'd, except about 2 Acres of 19 A piece & 3 A of Furze Down Piece | Ewes on the Down & follow Lambs on the Clover do well | About 5 Acres of the 19A Wheat field sow'd to Turnips - just shewing | The 2¼ A: of Turnips Furze Piece, good, and heading | The 17 A: of New field Clover

still left for Sheep | Wheat ripening a good colour at Beckhampton, and Barley very good Crop | Reaping at Allington, Chalfield &c.

24. Rain in the Night & storms in the Day
25. Heavy storms in morn'g, Thunder & rain afternoon
26. Dry night, heavy storms mid=day Awful Thunder and light'g, at 3 O Clock after noon with heavy rain
27. Sunday, rain morn'g, dry after Glass rising
28. Very fine harvest day, Rain heavy 6 O Clock eveening | 3 or 4 days to finish Hay making at Chalfield
29. Flying Storms, afternoon, Haymaking finish'd at Cowitch & many places
30. Fine day, Probably partial Showers
31. Very fine Harvest day, Probably partial Showers | 20 Minutes after 4 O Clock afternoon Thomas, my Son born.

August

1. Rain in morn'g and even'g - drying between | Reaping began at Corton and Cleve'y [92] | WHEAT RICK, made at Whitley, and many other places | The continual rain and high Floods the last three week have cause'd irreparable loss's, Hay wash'd away, Corn injured.
2. Rain all the morn'g, dry afternoon
3. Sunday, - heavy rain Thunder & light'g before mid=day, dry eve, Flood in place<s>'s
4. Very fine harvest day, and very heavy rain in eve'g, most awful Thunder and light'g in night wth heavy rain
5. Rain all morn'g, fine afternoon
6. St Anns hill Fair Lambs about 2s pr head, higher than last Year 10s to 22s 6d | Dry morn'g till 9 O Clock, rain all day after, except at short intervals the wetest fair, the last 12 Yrs | Reaping Avebury Parish, about a third part, none carried, the Two Mr Philpots, not began | Wm Philpots Mead by Silbury Hill, now in grass Cock, 3 Loads carried (the part fed begining of May,) ready to cut a Month since, wait'd for fine weather | Wm Philpots, Hay ricks not yet thatched, made by middle of June
7. Warm fine morning, very heavy storms wth Thunder, WHEAT sprouting, lodged, and standing, advanced this day at Devizes 2s pr Sack, highest price 36s pr Sack |This is now the most uncertain point whether the Wheat will be saved or lost by immediate fine weather or a continuance of rain Three fine dry day's only in July, and probably rain on those days in some places
8. Very stormy all day, dry eve'g
9. Very heavy rain in night, and very boisterous winds, and heavy storms all morn'g, dry eve, Glass sink'g.
10. Sunday, very heavy storms before Church, and stormy afternoon
11. Glass rising, dry morning, heavy storm after 12 O Clock, and frequent

92 Cleve'y - Clevancy.

Storms till 4 O Clock, from this time Rain pouring all night for 12 hours

12. FLOOD this morning to cover Mr Guy's Island, frequent Storms till 6 O Clock even'g, when fine, a fog rising. Stars very thick, and milk'y way seen, portend'g fine weather, WHEAT not very much sprouted, if dry from this time
13. Fine fogg'y morn'g, dry, Glass steady Rain, all afternoon
14. Rain in night, and thin steady rain all day | FLOOD, again on the Island | Wheat advanced to 40s pr Sack.
13. Mrs Sarah Maskeline, Bupton. Died, suddenly, yet very ill. 13th August 1828 Aged [...] [This entry written within a thick black border]
15. Glass, rise'g, dull dry day, wth exception of a few drops of rain after noon.
16. [Written in pencil] UNLUCKY SATURDAY, carry'g Corn | Fine day, not very drying, some thin Rain afternoon Many people tempted to carry Wheat, Mr R S S of Bremhill and others still very wet in bond, but fearful of more rain, The Wheat sustained no very serious injury yet by growing, but unfit for Market, for 12 Months to come.[93]
17. Sunday, Fine morn'g, heavy storm of Rain abt Noon, and heavy storms aftr
12. The Flood, carried away the Hay in Mr Philpots Mead by Silbury Hill and other Meads nearer Marlbro
18. RAIN, a Storm in the morning Fine day, except Storms at a distance | Reaping finish'd at Beckhampton No Wheat carried in Avebury Parish, - wet in the bonds of Sheafs and sprouted much, the outside ears faded Many People carrying this afternoon, tempted too soon, Glass rising. | Wm Philpot, - Pot dunging my Wheat Land, the 19 Acres, will be Fold'd over and Pot dung over about half a dress'ing by the 23rd The 9 Acre Down Piece, to Fold and Pot dung, which will be in done in good time | Turnips hoeing, just up on the 23rd July a good Plant and clean The Meadow Hay Rick not thatched waiting to put the Hay from Silbury hill mead on it
19. The weather Glass nearly to F at fair best prospect of fine weather yet Wheat carry'g universally. Very fine day
20. Beautiful warm clear, harvest day Wheat Ricks made 16th & 18th, smoking and heating, some Ricks, taken apart *
21. Beautiful fine Harvest day
22. Wheat Harvest, finish'd at Kennett and Cowitch &c. Beckn | Marlbro Fair Some thin flying Showers, mid-day yet much Wheat carrying Ewes 20s to 34s Lambs [...] to 24s Wethers two tooth 25s to 29s 6d
23. Oats nearly finish'd, Harvest Barley, began carry,'g, and generally mowing, except Philpots, quite fit | Very Warm fine drying day | The Mr Browns finish'd Wheat Harvest in excellent condition, - those Farmers who carry'd before Tuesday 19th, carry'd very bad, and have turned their Ricks and others, purpose taking back to the Field to dry them, being wet in bond and Corn soft

---

93 Mr R S S of Bremhill - Richard Sadler Smith of Bremhill, Pinniger's brother-in-law.

24. Sunday, very warm, ripening day (Mrs S Rum[...]ll distress) [94]
25. Bradford Leigh Fair, Very warm ripen'g day Lambs to 22s & 4 Tooth Ewes 23s Fat Wethers Mr Spackman of Ch: [...] li pr Qr @ 7d li [95]
26. Very warm fine day
27. Very warm fine day (Mrs Rum[...]ll, liber'd)
28. Very warm fine day WHEAT RICKS at Bre=ll turning 2nd time, and many other Places [96] | BEANS, cuting, Harvest soon get in, some Farmers, have already finish'd | Barley carry'g at Cowitch Oats finish'd
29. Very warm fine day
30. Very warm fine day
31. Sunday, - fine day, not so warm Thomas my Son, Baptized, at home by the Revd Mr Short[97]

September

1. Dry, dull day - seed mowing BEANS, Carry'g generally, and Clover
2. Very warm, clear fine harvest day
3. Very warm, clear fine harvest day
4. RAIN, heavy storms partially, Kennett &
5. Warm fine day, RAIN partially
6. Warm fine day
7. Sunday, fine day
8. Warm Sun and wind | Harvest Tockenham near end by Beans nearly carryed | Barley, began carry'g Freegrove
   Mr John Mitchell Chippm Obit 80th Year [This entry written within a three-sided thick black border]
9. Harvest, finish'd Mr Smith Beaversbrook | Very warm fine day.
10. RAIN, heavy in night, and unusual heavy Storms in the Day. | Wm Philpot, agreed wth Mr Guy on quiting Farm
11. Very heavy rain in night, and heavy Storms in day.
12. RAIN, heavy mid day, dry afternoon with Thunder
13. RAIN, heavy in night, dry till evening when rain again
14. Sunday RAIN all the afternoon 25th Augst TURNIPS W P__ Kennett, began feedg
11. TURNIPS, Wm Philpot, began feeding wth Ewes, will hold all this Month, when all the Wheat land, will be Pot dung'd and fold'd
13. Ploughing the Wheat Land 2nd time Finished
15. Glass rising to R in fair, very fine day Wm Philpot of Beckampton, about 6A of Barley to carry, and made 4 Ricks Harrowing the Wheat Land | HAY,

---

94 Rum[..]ll – probably Rumboll. This entry and that for 27 August suggest that a legal process for debt was commenced and resolved.

95 Ch: – Chalfield.

96 Bre=ll - Bremhill.

97 William Short, vicar of Chippenham.

Wm Philpots Silbury Mead 2nd Crop, much stained, in swath considered not safe for Sheep, HAY, 2nd Clover in swath, stained too old for Sheep | Barley, much not carry'd at Kennett

16. Glass to T in set fair, very fine day | Goosberry cutings, and Rasberrys planting, - proper season
17. Glass sinking, yet very fine day | Mr Crook, Corton, finish'd Harvest with Beans | Mr Jas Large, finish'd Harvest with Barley | Mr Jno Large 2½ A of Barley to cut
18. Fine warm day & Bristol
19. Fine very thick fog Downend
20. Fine day, rain in eve'g
21. Rain in night, fine day
22. Some little rain fine day Kingsdown Fair
23. Very warm and fine | RAM LAMBS, W P__ of K: turned 8 to 402 Sale Ewes, 4 at a time, Rt Lines at 23£
24. W Philpot, began Ploughing for Seed | Warm and fine
25. Warm and fine Didmarton Wheat sowing & up fine at Didmarton
24. John Beams of Abingdon Obit 24th Septr Aged 33 Years [This entry written within a thick black border]
26. Warm fine day Rain in evening | Hay W Philpots (see 15th) made on the old field and Rick top'd
27. Rain in Night, dry day
28. Sunday, light Storms | FOLD, finish'd all the Wheat Land at Beckhampton 28A, - and turned back - to fold the top of Furze down Piece a second time, & Pot dung'd also
29. Calne Fair, - Rain all morn'g, dry eve Mr Smith 100 Wethers 27s cost 17s Ewes 21s to 25s Lambs 16s 6d to 21s
30. Showers in the morn'g dry after

October

1. Showers in the morn'g dry after Beckhampton, 1st Turnips finish'd
2. Fine dry day MARKETS advancing, a bad yield at Devizes 4s pr Sack, highest 47s 6d Wheat Barley 45s Qr
3. Fog'y morning Warm day
4. Showers, afternoon
5. Sunday, Showers, morning & evening
6. Beckhampton, Very heavy rain in Night to cause a Flood, and cover the roads, Showers in day | Ploughing Furze down Piece for Seed (2nd) Young Turnips, began
7. Some light Showers, Fine day, FLOOD whole'y over the meadows
8. Very heavy rain in night, fine day. | Fold removed from Wheat Land to the Barley Stubble adjoining
9. Very fine day

10. Very fine Wheat Sowing, began at Corton and Cleveanceys
11. Very fine Glass rising to R in fair
12. Sunday, Beautiful fine day
13. Beautiful fine day
14. Beautiful fine day Young Horse from Corton to Cowitch to help Wheat Sowing
15. Beautiful fine day [Pinniger then recorded sheep prices from Weyhill Sheep Fair on 10th September - see Appendix H[98]]
13. WHEAT, began driling, Beckham[pto]n by W Rumming 3 Bu pr Acre Mr Crooks Seed @ 37s pr Sack, Furze down Piece, 5A pr day @ 2s pr A | Wheat at Kennet shewing green sowed the 3rd
15. Turnips, last gave the Sheep at Beckhampton, Hay and old field and Down only for the Sheep
16. 20 Acres drill'd, 8A, to Plough and finish driling | Very fine, like Summer
17. Very fine day Devizes Market decline 4s Qr lower, 46s Sack, highest, - last week 48s to 50s S
18. London Market 10s Qr lower | Very fine day
19. Very fine day Sunday
20. Very fine day like a May day [Pinniger then records prices at Devizes Fair - see Appendix H]
21. Very fine day
22. Little rain in night, fine day, very heavy Rain from 7 to 9 O Clock eve'g, wth heavy Thunder & Lightening
23. Fine drying day, little rain in morn'g | (Fine rain for Wheat sowing, too dry before.
24. Very fine day
25. Little rain, very fine day
26. Sunday, Frost and fine day
27. Damp dull day.
28. Cold dry day Rudloe Sale
29. Chipp[enha]m Fair, - Beautiful, fine day
30. Beautiful, fine day
31. Very fine day | WHEAT DRILL'G, finish'd 28 Acres at Beckhampton @ 2s pr A wth 21 Sacks end of the 19A Piece, next the Down | W Philpots 200 Stock Ewes & 87 Ewe La neither Grass or Turnips, have had a little bad Meadow Hay, will not ate it Fold on Barley Stubble [99]

November

1. Spring like day, dry, mild & clear

---

98 The date of 10 September is an error – Weyhill Sheep Fair was held on 10 October in the year 1828, Salisbury & Winchester Journal, 15 September 1828.

99 La - Lambs.

2. Spring like Sunday mild & clear
3. Fine day, little rain in evening
4. Fine day Littlecote
5. Very fine day
6. Dry morn'g rain all afternoon, acceptable for late sown Wheat
7. Dry day, Cloudy and warm
8. Exceeding cold dry winds, & freezing
9. Sunday, very sharp frost, thick ICE
10. sharp frost, thick ICE and thick fog all day <nearly>
11. Very sharp frost, and very thick Fog all the morning
12. Very sharp frost and fog'y morning mild afternoon
13. Frost in morning and fog'y, mild day some little rain mid day
14. Rain all afternoon
15. Damp morn'g fine after, rain'y even'g
16. Sunday, rain in night, and at times all day
17. Fine dry night and fine mild day
18. Frost little in night, very fine mild day
13. The FOG, in London so dense last night and this morn'g & even'g, - occasioned many accidents, - The Coaches lost 4 & 6 hours Mail arrived at Chippenham 10 O Clock instead of 6 O Clock, this morning
19. Fine mild Day
20. Fine mild Day Hazleberry
21. Fine mild like May
22. Rain heavy in morn'g damp day
23. Sunday, - Dry mild day
24. Marlbro Fair, - quickest Trade of the last 7 Years And the greatest ROTING Year of the last 20 year, produced by the wet Summer. [Pinniger then reported sheep prices from the fair - see Appendix H; he then made observations on the importation of foreign wheat - see Appendix I] | Very mild, more like Spring than Winter
25. Very mild and dry Wheat last sowing Beckhampn up, - the first sowing up strong. Meads by Silbury hill fed by Wm Philpots Sheep, suppose'd not to Rot after the Frost. | All Wm Philpots, Sheep now sold and off the Farm, Ewe Tegs at Keep
26. Rain in morn'g, dry after, and very heavy heavy rain in even'g
27. Glass rising and fine day
28. Very mild fine day Everly Justice Meeting [100]
29. Forsbury, - very mild fine day
30. Sunday very mild fine day

December

1. Very keen, cold dry'g winds, The Roads as dry, and as good travelling as

100 Petty sessional division of Everleigh.

Sumr
2. Frost and Ice, in morn'g, very cold dry air
3. Very fine mild night and day
4. Damp mild day
5. Fine mild day, last sowing Wheat at Beckhampton, after Turnips, too thin no visible cause
6. Very fine mild day
7. Sunday, damp morn'g, very heavy Storm afternoon
8. Heavy rain all morn'g with Thunder
9. Damp wth rain in morn'g, dry eve'g.
10. Dry day, except a light shower
11. Very wet morn'g dry after
12. Dry and very mild
13. Dry and very mild
14. Sunday, fine mild Day
15. Showers, and very mild
8. Fallowing, W Philpot, began for me the Wheat Stubbles at Beckhampton
15. Mary Alice Pinniger of Tytherton Died 15th Decr 1828, Born 1st Octr 1818 Daughter of Broome & Elizh [This entry written within a thick black border]
16. Some Showers, and mild
17. Very heavy rain in night, windy fine day and Flood
18. Very heavy rain in night, winds, and storms in the Day - Flood
19. Rain in night & high winds, dry after the morning
20. Very mild dry day
21. Sunday dry day calm like May partial Storms.
22. Very mild dry day
23. Very mild dry day
24. Very mild damp day
25. Rain very heavy in night, Flood over the Island at Chipp[enha]m in the morn'g clear keen Air [101]
26. Very sharp Frost by morn'g, thaw and fine day
Elizah Pullen of Lyneham Died 1 O Clock morn'g 26th Decr 1828 Aged 35 [This entry written within a thick black border]
27. Fine mild day
28. Sunday, heavy Storm mid day, fine eve.
29. Very sharp frost, fine clear day & Thaw.
The late Miss Ann Large of Cleveancey Gift of Linen, distributed to the Poor of Lyneham & Hillmarton £100, 5000 Yds [102] [This entry written

---

101 The island is the island of Rea in the River Avon at Chippenham, near the town bridge, on the west side.

102 Ann of Cleeve (Clevancy), daughter of Richard and Jane Large, baptised 5 August 1804 aged one year, *Wiltshire, England, Church of England Baptisms, Marriages and Burials 1538-1812*. Ann Large of Cliffansty buried on 24 March 1828 at Lyneham,

within a thick black border]

30. Frost and fine day. Thaw
31. Mild day, dry morn'g & eve, rain mid=day | CHALKERS, left their work, pretend'd no Chalk to be found.

1829

January

1. Miss Pullens, Funeral.[103] | Keen air some rain
   Miss Ann Large of Melksham Died instantly after taking her lunch by rupturing a Blood vessel 2nd Jany 1829 – Aged 45 Years [This entry written within a thick black border]
2. Beautiful – fine mild clear day
3. Danl Hughes, Wick died aged 62 Yrs | Thin rain all the morn'g, dry mild eve.
4. Sunday, heavy storms morn'g
5. Very cold winds, frosty eve'g
6. Very sharp frost, fine clear day
7. Snow at intervals, and cold air.
8. Cold air, dry day.
9. Very cold dry day Fallowing Wm Philpot finish'd the Wheat Stubbles at Beckhampton 28 Acres for me. | Miss A Large, Bury'd at Melksham
10. Very cold dry day CHALKING, Wm Philpot began for me at Beckhampton, the 2nd Piece by furze Down 9 Acres @ 52s pr Acre with 3 Carts, 6 Horses, 6 Men & 2 Boys If done by Men from Wells with Barrows at 42s pr Acre
11. Sunday Very cold dry day. frost. morning and evening
12. Thaw and very cold winds | 1st Cart Colt, to Kellaways T Seager to keep on Grass at 2s 6d pr Week
    Mary Daughter of Wm & Su: Spackman of Chalfield, Died 12th Jany 1829 Born [...] [This entry written within a thick black border]
13. Frost morn'g, thaw, and very cold
14. Sharp Frost, and very cold day
15. Very cold day, damp eve & Thaw | 10 Pieces of Baltic Dantzic Timber

---

age 31, *Wiltshire, Church of England Deaths and Burials 1813-1916.*
Will of Ann Large, Spinster, proved at Chippenham 11 July 1828. Executors: John Large (brother) and Thomas Pinniger (Brother in law). Her many bequests included: to her sister Mary Pinniger "my pearl brooch with my father's hair in it for the use of her daughter Mary Pinniger"; £50 to the poor of the parish of Hilmarton to be divided amongst them within 12 calendar months of her death either in money, bread, meat, clothes or wearing apparel, as her Executors may think proper; £50 likewise to the poor of the parish of Lyneham. WSA: P3/1828/22, *Will of Ann Large, formerly of Lyneham.* Given that £100 bought 5000 yards of linen, then the cost per yard was just over 4¾d.

103 The funeral is that of Miss Elizah Pullen of Lyneham: see entry for 26 December 1828.

bought

16. Cold dry day
17. Frost, the sharpest, this winter, freez'g all day | M S Buryed at Cliffe [104]
18. Sunday, - Snow all morn'g, not cover the grod and sharp frost
19. Very sharp frost, clear sun shine day | Chalking wth Carts, cover 3 of 4 Pole squares which is 48 Pole @ 52s pr A, - pr day = 10 Carts Loads of 15 Buckets ea: 150 Buckets to 16 Poles or a 4 Pole square, (9 Buckets pr Pole is 144 Buckets to 16 Pole.
20. Very sharp cold Frost, Sunshine day
21. Very sharp cold Frost some Snow
22. Very intense cold Frost
23. Very intense cold Frost some Snow
    [Note in margin, double underlined] 23rd The coldest, cutings Winds and freezing Day, since 28th Decr 1820
24. Very intense cold Frost Butchers, have the greatest difficulty, with Carpenters Hand Saw of Saw'g Beef, apart
25. Sunday, Very keen cold sharp Frost the Roads as dusty, as dry summer Thaw, Snow, rain & frost in night
26. Thaw, and very mild clear Sun
27. Frost morn'g – damp afternoon
28. Frost morn'g, mild warm Sun mid=day | 2nd Horse, bot of Thos Hillier 30£ | (Cart'g Stones at Tytherton) finish'd
29. Mild fog,'y day, Frost, morn'g
30. Frost morn'g, - fine clear day. & Thaw
31. Frost morn'g – Thaw after

February

1. Sunday, very sharp Frost and freez'g all day
2. Very cold sharp Frost, and warm clear Sun at mid day, Chalk'g continued
3. Sharp Frost, mild'r day STONES, - Brother W P_ Kennett began Carrying for me from Calne to Beckhampton, for Building
4. Sharp Frost, mild Thaw all Day. Glass at T in Set fair
5. Very mild Thaw, clear fine day
6. Very mild Thaw, Thick fog all day
7. Very mild Thaw, clear fine and dry'g
    Benjn Pegler of Foxham Died 1st Feby 1829 Aged 64 Years. [This entry written within a thick black border]
8. Sunday, - Cold dry day
9. Fine day, Cold air STONES, Brother J P_ Cowitch, began carry'g for me from Calne
10. Beautiful mild Spring day
11. Beautiful mild, clear fine day Sale at Mr Rummings Notton 19 Cows, (only

---

104 Mary Spackman.

2 Calves, from £15 to £23 10s Average 19£ ea & 3£ odd over. Rick of good HAY £75, put at 36 Tons [105] | CHALKING, W_ Philpot, finish'd for me, began the 10th Jany

12. Thin mizly rain all day.
13. Very fine mild day
14. Very fine mild day Thin rain in even'g | Brother J P_ carry'g Stones for me all the week – 6 Loads, (18 Yards)
13. 2 Horses, to keep Kellaways TS [106]
15. Sunday, - Rain morn'g, fine after
16. Very fine day
17. Fog in morn'g, Beautiful fine cheering day
18. Fog'y damp morn'g, fine after Bristol | CHALK, began diging, for Building Feby 12. Willm Pike of Gt Bedwin Died in his 59th Year

    14th Revd Bartholomew Buckerfield of Marlborough Died in his 63 Year[107] [This entry and the previous entry written within adjacent thick black borders]
18. Nathl Atherton & Miss Atchley, Married
19. Fine dry'g day
20. Very fine mild day
21. Thin rain at times all day
22. Sunday, rain heavy in night fine day
23. Rain in night, fine dry'g day, very little Snow | Lambing, Mr Kemms time out Willm time out 16th now 80 Lambs
24. Fallowing Old field began Rain in morn'g damp day
25. Close thin rain after the morn'g
26. Rain, incessant after the morn'g
27. FLOOD, beautiful, fine day, Hay Mr Rich
28. Timber, landed, & took to Yard. HORSES, brot from Kellaways | Very fine day

March

1. Sunday, Very sharp frost, thick Ice | TEAM started first time 10 O Clock at night, for New Waggons, load with Slates from Bristol
2. Very sharp frost & very cold dry'g day | Sawing, began for Building
3. Very cold harsh drying winds | TEAM, returned 4 O Clock, morn'g having injured the Waggon, in Bristol

---

105 3£ odd over - that is, on calculating the average, there was between £3 and £4 remainder, total sale achieved between £364 and £365.

106 T Seager – see entry for 12 January 1829.

107 Revd Bartholomew Buckerfield was rector of Marlborough St Peter and St Paul from 1796 until his death, Clergy of the Church of England Database, *Buckerfield, Bartholomew* https://theclergydatabase.org.uk/jsp/search/index.jsp [accessed 22 July 2020].

4. 1st Waggon to Beckhampton 4 Ton, 14 € | Fine Sun shine, harsh Winds, some thin rain
5. 2nd Waggon to Beckhampton 4 Ton 2 € wth Slates | Stable making, from part of the Skilling | Jno Freegard, the Carter left | Very fine day, Cold winds | Thos Millerd engaged as Carter at 9s pr W and 1s ex: pr Week, if behave well.[108]
6. Fine dry'g day some little rain | Timber remainer brot to Yard
7. 3rd Waggon from Bristol wth Slates | Fine day
8. Sunday, fine day, very little rain in morn'g
9. 3rd Waggon to Beckhampton, and Stable finish'd, - Horses remain to begin Work | Keen air very fine day
10. Keen air very fine day with warm Sun | Horses commenced Work, by Carting Chalk Stone, from the Old road for Building [...] Loads wth 3 Carts & 6 Horses | Tarter Oats & small gray Pease, Mr Brown Sowing in his Whitelands | Oats. Mr Wentworth sowing
11. Carting Chalk as yesterday | Very keen dry'g winds
12. Very keen dry'g Winds Ice, Wroughton (Poles) | 2 Loads of Chalk to Calne for Lime | 2 Loads of Stone back
13. 1 Load of Chalk to Calne | 2 Load of Lime 8 Qr ea Waggon back | Very keen dry'g winds | Nettleton wth Wm Philpot, to see Farm
14. Team to Box, for 150 feet of Ashler | Very keen & cold, little Snow. | <u>Willm Philpot took that Farm</u>
15. Sunday, mild morn, very keen cold eve'g
16. Ice'y sharp frost | POTATOES, planted at Beckhampton | 2 Load Stones from Calne
17. 2 Load Stones from Calne | Cold frosty morn'g, mild eve'g
18. Rain at times during the day | 2 Load, Stones Crop fm Calne
19. 2 Load Lime Calne | Mild, & fine warm morn'g | Seeds bot of Mr Line 7 Qr Hop & Ray @ 23s 1 Sack of Ray @ 30s 112 li Broad @ 6½ li } for 28 Acres | 2 Ploughs @ £4 1 Sowing Machine £3 3s od } Tasker
20. Beautiful mild fine day | 2 Loads best Stones from Calne (or 1 Ld Lime)
21. Beautiful mild fine day | Poles from Derry hill
22. Sunday, mild dry day, cold eve'g
23. Board from Cowitch | BEES removed to Beckhampton wth Yoke | Rain in morn'g very dry'g day
24. Cold morn'g – clear fine afternoon | Trowbridge | Upper furlong Tanhill road & Whitelands thwarted by Wm Philpot for Barley
25. Very sharp Icey Frost, very fine day | 1 Load of Stone & 1 Load of Lime | <u>SALE</u> <u>at</u> <u>Beckhampton</u>
24. 1 Load Stone & 1 Load of Lime
25. Garden at Chippenham done up and Seeds sown
26. Frost and very fine day, warm & clear [...] Load [...] | Frameing, began,

108 ex: - extra.

Roof of Brewhouse &c.

27. Very mild warm fine day | 2 Load Crop | Barley Wm Philpots, last put into Barn | 2 Wheat Ricks of Philpots still out | Plants Cabbage, the first planted at Beckhampton | Cart & Shandy took to Beckhampton
28. Sharp frost & very fine day [...] Load [...]
29. Sunday, Mid:Lent | Fine rain all last night, misly morn'g, fine after
30. Misly rain at times, mild | PLOUGHING, commenced, with 6 Horses, low'r furlong Tan hill road 7 Acres left to Thwart
28. Mr Philpot, Harrowing, the 9 Acres of Wheat by Down, 1 Tine, prepare'g to Hoe'g it
31. Mr Philpot finish'd Ploughing. for me in thwart'g Tan hill road Piece | Mr W R Brown & Mr Jno Stratton Valueing the Plough'g Hay &c &c, for Thos Pinniger & Wm Philpot at Beckhampn | Dry morn'g Morn'g rain afternoon

April

1. Fine morning, Cold afternoon with a light Snow Storm | Plough'g finish'd thwarting the Barley land 28 Acres | Sale at Compton, Jos Rogers bought 3 Rick Staddles
2. Heavy Hail Storm & Cold | Ploughing the Chalk'd Land
3. Wm Philpot quited the House at Beckhampton | Fine day, Cold | Mr Brown sow g Coal Ashes' on Clor [109] | 2 Waggon for Staddles fm Compton
4. 2 Waggon for Staddles fm Compton | Fine morn'g Rain all afternoon | 1st Load of Goods, Brother Willm return Waggon took to Beckn
5. Sunday, Rain, very heavy in Night and showers in day
6. 2nd Cart & Shandy wth Chairs to Beckhampton by Mr Rich's 2 Horses | Frequent Showers | Plough'g Chalk'd Land & Cart to Compton for remainer of Staddles
7. Waggon Load of Furniture, to Beckhampn own Horses | Marter earth, Carting | Frost and fine day
8. Very white and Ice'y frost, very warm Sun and Showers at intervals | Remainer of Goods pack'd at Chippenham
9. Thursday, <u>All Family</u> removed from <u>Chippenham to Beckhampton</u> | Dry morning, Rain before the 2nd & last load Waggon could be load'd, intervals of Rain, and dry till 6 O Clock, Waggon arrived at Beckhampton, half pass 12 O Clo at Night
10. Dryer Day, with some Showers
11. 1st Load of Timber on Carriage fm Chipp[enha]m | Cold storms, and very boisterous and heavy rain all night | Men grubing Banks in Meadow. | Planting Privet cuts
12. Sunday. - heavy Storms
13. Storms at times all day | 2nd Load of Timber on Carriage

109 Clor - Clover.

14. Dry morn'g, very stormy & boisterous after
15. Frequent cold storms, Glass at Y in stormy [110] | Fallowing Chalk'd Land, and grubing Banks,
16. 3rd Load of Timber, on Carriage | Brother Wm 3 Horses, rolling Clover with new Iron Roller | Pump put to work by Mr Slade | Rain all the morn'g dry after
17. Good Friday. Dry morn'g Glass rising, Hail storms in succession after 10 O Clock. - and frequent storms all day. | 40 Bushs of Ashes sown on Wheat | Rolling Grass, - Horses put into Stable to wet to work
18. 4th and last Load of Timber on Carriage from Chippenham | Brother Wm finish'd Rolling my Grass 30 Acres in 2 days, and 2 hours | Fine day, Old hedge wood brot from Kennet
19. Cold Storms of Hail & Rain, Sunday.
20. Devizes Fair, fine day, some light cold storms | Carrying Dung from Rick Yard to Down in Store heap for Swedes
21. Hedge round the Meadow, cuting and laying by Henry Hendon of Compton | Drying day, some rain, Dung & Chalk Cart'g - Wheat Rick took in
22. Rain all the morn'g & damp all day Jno & Wm DOBSON at home & charged the day. | Cart'g Dung as yesterday & rolling Mead | Cuting Hedge as yesterday
23. Rain all the morn'g very drying afternoon | Carting Dung as yesterday | Thatch took off Stable
24. Stable took down | Dung Cartg as yesterday | Rain heavy in Night & in the morn'g to keep, the Horses in Stable
25. Pierceing cold Storms & Winds | Marter earth Carting
26. Sunday, White Frost, Fine day cold winds some rain
27. Cold storms of hail & rain | Team to Puthall, for a Ld of Bushes | Brother Willm's Team Draging my Barley Land | Mr Wentworths Sheep began Silbury Mead | Willm Philpots, last Years HAY-COCKS got out of Silbury, mead, the Grass kill'd
28. Very Tempestuous, Ricks uncovered &c. none in this Parish, remember, the like | Horses, draging and harrowing Tan hill: Fur: | Heavy storm beat off, afternoon. | Mr Wentworths 7 Women gathering Docks for me Tan hill Piece CUCKOO, hear'd in the Vale | Dobsons, hedging Whitelands | BUILDING began Brewhouse &c.
29. Draging &c as yesterday | Range &c. put up | Women, work as yesterday | Very cold high winds & cold Storms
30. Rolling Tan hill Piece, Drag'g White lands & Potatoe ground. | Ice, over all the Water in the morn'g | very cold winds, & storms of Snow & rain

<u>RAIN 27 Days in April</u>

---

110 Glass at Y in stormy. This indicates that the needle on the barometer was pointing towards the letter Y in the word "STORMY" on the dial, indicating a very low atmospheric pressure – see Figure 2.

May

1. Some rain in morn'g, & cold winds mild'r after the morn. | Harrow'g Tan hill Piece | Rolling, Plough'g & Drag'g White lands | DOBSONS, struck | [...]
2. Cold harsh Winds morn'g, mild'r after | Rolling & harrowing all the foul parts of the Barley land & Ploughing the foul headlands &c. all harrowed down, ready to begin Ploughing for Seed | Richd Chivers's Sheep 100 Tegs sent to Kennet, on W Ps Swedes, want'd gone for Barley
3. Sunday, Rain in morn'g dry'g winds after
4. OBIT, Edd Crook, half pass 3'O-Clock in the morning Aged 21 Yrs 27th July 1828 [This entry written within a thick black border]
4. BARLEY SOW'G, BEGAN, | Fine mild calm day, the first since we came to Beckhampton
5. The Revd Mr Palmer, of Fordington Formerly, Curate of Bremhill Obit Aged 84 Years [111] [This entry written within a thick black border]
   Driving thin rain in the morning dry'g after very windy. | Wheat hoe'g John Amor began next the Down, | Plough'g & Sow'g as yesterday
6. Rain, heavy Showers in the morn'g. very fine day & Shower in eve'g. | Plough'g & Sow'g as yesterday. | Brother Wm Ps 5 Horse & 2 Ploughs helping Barley Sowing. | Rt Philpots SWEDE'S finish d, and his Flock on the Down. | Mr Wentworths Flock finish'd SILBURY MEAD, - and began Field Grass. his Tegs on the Down. | Kidney=Beans - Planted, and 2nd Crop of Potatoes - Planted.
7. Plough'g & Sowing, - as yesterday about 10 Acres, - finish'd | Cold Showers, intervals of Sunshine & dry'g winds.
8. 4 Ploughs, Mr Wentworths 2 Ploughs, Brother W P_ 2 Ploughs, Our Own } Plough'g & Sow'g Barley | 19 Sacks Barley & 28 Bu: of mix'd Seed } from the bottom to 12 lands above the lower Furlong Tan hill road
9. The first May like day, being calm very mild & dry and growing. | 8 Teams - as yesterday | BLED the Horses in the eve'g, about 2 Quarts of Blood from each.
   Edwd Crook, - Buryed at West Tytherton [This entry written between two thick black lines]
10. Sunday, very pleasant, warm & calm & very growing.
11. 1 Plough of Mr Wentworths & ours, finish'd Sowing &c. Tan hill Piece, with (23½ A) 29½ Sacks of Barley 6½ Qur of Hop & Ray & 4 li of Broad Clover pr A | Potatoes Planting in Meadow.
12. 3 Teams of Mr Wentworths - and 1 Team of our own - Plough'd Whitelands

---

111 Rev John Palmer was vicar of Fordington from 1799 until his death, having been curate of Bremhill with Foxham and Highway from 1789, Clergy of the Church of England Database, *Palmer, John (1799-1829)* https://theclergydatabase.org.uk/jsp/search/index.jsp [accessed 22 July 2020].

except headlands & Sow'd with Barley | Rolling & Harrow'g to finish Tan hill Piece | Mr Tanner, finish'd feeding Swedes for Barley

13. Barley Sow'g FINISHED, the Headlands of White lands 28 Acres, with 34½ Sacks of Barley 16 Sacks of Hop & Ray @ 23s Qr 1 Sack of Clean Ray @ 30s Qr 112 li of Broad Clover 6½ d pr li | Chalk, diging, Isaac Maslen, and Aaron Andrews, began | Rolling Tan hill Piece | 1 Team of 3 Horses, to Mr Wth half a Day [112]
14. 2 Teams of 6 Horses to Mr Wth all Day | Very warm fine day
15. 1 Team - 3 Horses to Mr Wentth all Day | 2 Horses Roll'g & 2 Horses, moving Stones & to Calne for Trussels [113] | SWEDE'S Brother Wm finish'd feed'g | Barley 1st sow'g up green, The Chalk digers, Robing on the Turnpike road & absconded. | Fogg'y morn'g, very warm fine day
16. Very fine warm Day, Didmarton | Mr Tanner & Brother W P_ finish d Barley. Sowing | 2 Teams to Mr Wentworths
17. Sunday, very fine day.
18. Very warm fine day | Thwart'g Wheat land for Potatoes, and forward Turnips | Women turning Dung heap for Swedes | Hoeing Wheat 19 A _ Piece.
19. Warm fine Day | Plough'g as yesterday, 2 Teams | [...]
20. Ilsley 3rd Market, - Sheep up 2s pr head C_ Pinniger bought 82 Wether Tegs @ 26s 6d 100 Wether Tegs @ 27s Tegs generally asked 28s to 34s 2 & 4 Tooth Wethers 28s to 48s Small Chilver Tegs allow'd me @ 26s Mr Hewett | Wheat hoeing finished | Barley last sowing up green over the ground | Very warm fine Day.
21. 1 Team at Plough as yesterday | 24 S_ Wheat to Devizes for Wm Philpot | 2 Load Bricks returned 16 Hd | Wheat Rick put into Barn | Warm fine day
22. Warm fine day Vetches 1 Sack Sow'd
23. Warm fine day Vetches 3 Bush sow'd | Mr Wentworth, Sow'g 1 Bu: Vetches 7s and Sow'g 4 li of Rape @ 4d } pr A
24. Sunday, Very gentle rain from 10 O-Clo to 2 O-Clock, very much wanted for the Barley & Grass, and some spots of yellow Wheat.
23. Watts's Mare, Carried in Shoulder at Plough
25. Chalk, - geting home with 4 Carts | Dry, harsh, high winds, - the Rain yesterday quite lost.
26. Very cold benumbing Winds in morn'g warmer afternoon | Harrowing & rolling for Potatoes, and Carting Dung for Potatoes
27. 12 Doz Hurdles from Ramsbury Park | 2 Horse Plough'd in 5 Sacks Potatoes in half an Acre, every other furrow = 5 Women = droping the Potatoes | Roll'd after to close the furrow & keep the moisture in. | Cold windy morn'g, calm & mild after

---

112 Wth - Wentworth.

113 Trussels - this could be either trusses or trestles, in connection with the building work at Beckhampton.

28. Warm fine day. | Ploughing & Sow'g a 2nd Drift of Vetches on Wheat land. | Roof began puting on BREWHOUSE &c.
29. Lambs W P_ of K: cut | 2 Ploughs for Vetches, Wheat land, to take the Sun - and Sow when have rain | Windy morn'g, warm fine afternoon
30. 12 Doz Hurdles from Ramsbury Park | 2 Horses, Harrow'g & Rolling Vetches sowd 28th | Cold morn'g - warm & fine after
31. Sunday, - fine day.
RAIN 6 Days in May

June

1. Grass began cuting for Horses. | Plough'g for Vetches. | Very fine Day.
2. Journey to Cholsey, Very fine [114] | Plough'g as yesterday finish'd, and Plough'g Drag'g & harrow'g - headlands of Chalkd Grod
3. Very fine Day | Roll'g & Harrow'g for Turnips | Ilsley 4th Market Mr John Hewett, - Mapledurham Bought 100 Ewe Tegs @ 28s Bought 130 Ewe Tegs @ 29s Mr Geoe Castle - Laten Down [115] | Mr Washbourne began Mow'g (Sainfoin) | Wm P_ K_ began Mowing Field grass [116]
2. Rt Philpot, began Mowing Silbury Mead
3. WHEAT EARS, shewn at Ilsley
4. Very fine day | Mowing began Field Grass, by Geoe Brewer, borrowed of Mr Wentworth not having a Mower. | Ploughing Chalk'd land athwart for Swedes | Slateing - Brewhouse, began. | W P_ K Sow'g Turnips - and Mr G_ Brown - Plough'g in Swede Seed
5. Plough'g as Yesterday | WHEAT-EARS, Mr Browns out | Field Grass growing up rapidly, want cuting. | Rt Philpot carrying his Meadow Hay yesterday and to day.
6. 229 Ewe Tegs, from Mr Budds Winterbourne fold'd for Swedes - the 2nd Down Piece | Plough'g as yesterday | Very fine day Rt Philpot, finish'd his Hay'g, Silbury mead. | Brewhouse, finish'd Slateing
7. Sunday, fine day Sharp frost, Kidney Beans & Potatoes cut.
8. Plough'g for Swedes as the 6th | Geoe King & Geoe Brewer began mow'g Field grass - Down Piece | 229 tegs wash'd at Cherhill by Jno Eley & Robt Dyke @ 5d pr Score & 3d pr Score in lieu of Beer & Victuals | Very fine day, Swathes turned
9. WHEAT EARS, - our first seen Plough'g as yesterday & Harrow'g after | Very fine day, BEES, Swarmed.
10. Plough'g as yesterday | Hay - Ray began carry'g - the first 3 Loads | 5 Men

---

114 Cholsey - village 2 miles south of Wallingford, originally in Berkshire, now in Oxfordshire.

115 Laten – probably Latton, Wiltshire. George Castle was listed as a sheep dealer in 1821 at the Hungerford Sheep and Lamb Fairs, Salisbury and Winchester Journal, 18 June 1821.

116 Wm P_K_ - William Pinniger of Kennet.

Shear'd 11½ Score Tegs @ 1s pr S 1 Winder 2s Allowed over 1s | From 8 O Clock in morn'g to 8 in eve'g | Gammon for Breakfast, Fillet Veal & Bacon & Plumb Pudding for Dinner | Very fine Day

11. Very fine Day | Plough'g athwart finish'd the Chalk'd Land for Swedes, Roll'g & harrow'g Chalk'd | Hay Ray, carr'd remainer below road about 6 Acres | 800 Bricks W P_ K = fm Devizes
12. Very fine Day | Mowers removed from Clover not over ready to Dry Meadow, ripe & dried up | Carting Chalk all Day, no Hay ready | Barley improving altho very dry, the Roots of the Barley got to the Cool Chalk
13. Harrow'g & Rolling Chalk'd Land, & began Carting Hay next Piece above the Road at 10-O Clo morn'g, intending to finish the 11 Acres, a Thunder Shower beat us off at 4 O-Clock.
14. Sunday, Very warm fine day
15. Damp early in the morning, Glass sinking from fair below change, dry windy Day. | Hay, finish'd carry'g Furze Down Piece & carryed the home Meadow Hay after 6 O,Clock | Clover finish'd Mow'g ¾ Acre left for Horses expect to serve them a Month. | Rolling & Harrow'g the Chalk'd Land & finish'd
16. RAIN, commenced 5 O-Clock, morn'g & Storms till 9 O Clock - dry'g after | Carting Dung, began for Forwd Turnips on Wheat Land | [...] | Silbury Mead, began Mow'g, No - Hay mak'g to Day
17. Carting large Sarsan Stones till 11 O-Cl Carting Dung, till 2 OClock Carting Hay - till 4 O-Clock when stop'd with Thunder, Lightening & Rain | WAGGON HOUSE, began Building, by the Maslens | Fine & Warm till 4 O-Clock, when rain & all the eve'g by Storms, very heavy rain partially
18. Fine morn'g - very heavy rain partially No Hay - carryed | Carting Dung as yesterday | Grass mowing Finish'd
19. Carting Dung from Yard finish'd & began Cart'g Dung from Down on the Chalk'd Land for Swedes | Hay, carry'g from 3 to 5 OClock, when RAIN stop'd, fine morn'g | Mr Kemm, Turning hoe'g
20. Cart'g Dung as yesterday | Plough'g in Dung on Wheat Land for Turnips | Very Tempestuous winds after the morning | Hay fit to Carry, but cannot Cock or Carry, RAIN, in evening | Mr Amer - Sow'g Turnips
21. Sunday, heavy Storms morn'g - Dry eve'g
22. Rain at times all day | Cart'g Dung on Chalk'd Land
23. Dry'g Day with partial Showers, very heavy Storm in eve'g, - the Hay much injured in Silbury Mead by it, all in hack [117] | 45 Sacks Wheat to Devizes for Wm Philpot & 2m Bricks fm Devizes
24. Little rain in morn'g, very warm drying afternoon | Plough'g for Turnips | Hay, finish'd Carry'g Clover &c by Tan hill Road | Barley, mending much

117 To hack is to loosen the earth round potatoes, preparatory to earthing them up. Here, Pinniger appears to be saying that the hay is injured as a result of the rain beating the earth up into the hay, thereby discolouring it.

since the rain

25. Delightful, fine, calm, & mild morn'g the Glass rising, every prospect of a fine Day, Threw all Silbury Mead into Weak, - & all of it Hay, near fit to carry by 11 O-Clock, when a few heat drops, - from that time began rain'g very close & continued till 5-O-Clock in eve'g [118] ★ | Plough'g for Turnips | New Brewhouse, Paved & Complated
26. Baked in Brewhouse & prepared for Brew'g | TURNIPS, began sow'g Tankards 2 holes every other place & small Cog Wheel | Drying Day after the morn'g, the Hay Silbury Mead Cock'd
27. Fold removed from Swede land to Wheat, land adjoining, whole ground. | RAIN in the Night & all Day with a few short intervals of dry weather | Turnips finish'd Plough'g & Sow'g the Piece began yesterday | Plough'g began the Chalk'd land for Swedes
28. Sunday, RAIN nearly all Night & Day
29. RAIN, all Night & Day till 12 O-Clock, Horses & Men in the Stable, dry after and Plough'g Chalk'd land for Swedes | Well=House &c. took down, preparatory for Building Cellar: Dairy &c. | Vetches & Barley lain in ground since Seed time now appearing
30. RAIN in Night & morn'g - dry from 8 to 11 O-Clock, heavy Storms all Day after | Carting remainer of Dung on Chalk'd land & Ploughed Chalk'd land, Rain prevented sowing Swedes
RAIN 16 DAYS IN JUNE

July

1. RAIN in the Night & heavy driving rain all Day. Ricks covered with Straw only much damaged, & the Hay not carried quite spoiled Horses in the Stable, Men at Home the Sheep unsettled on the Down & the Shepherd soaked with wet Glass to C in much rain ★ | 2 Masons, make'g 7½ day to week @ 2s 10d 1 Masons, make'g 7½ day to week @ 2s 8d 1 Server @ 2s 3 Carpenters And 3d pr Day ea for Beer
2. Dry morn'g, - heavy Storms after Thunder & heavy hail Storm, Glass ris'g | Chalk, Carting for Building & before New Nag Stable, | Too-wet to Plough for Swedes | New Cellar &c. began Building
3. Rain in Night & succession of Storms all Day. | Plough'g for Swedes, - Chalk'd Land | Marter Earth - 1 Horse Carting | Wheat, side of the Hill, ledge'd quite flat | Barley - shewing the beard | Potatoes, all shew'g in the Field | Beams - put on Waggon House | Glass sunk to M in much rain
4. Very boisterous night, - Tree blown on Brewers Cottage | Dry till 10 O Clock morn'g, frequent heavy Storms after through the Day | Horses return'd from Plough'g | This evening closed a very uncomfortable wet week, the Hay not touched since the 25th June, now like Dung ★ | 10 Carpenters & Masons loseing half their time No chance of sowing Swedes

118 Weak – Pinniger probably meant wake (see Glossary).

Women weeding Vetches in morn'g

5. Sunday, Storms in morn'g dry after

6. Dry windy Day, Glass rise'g to nearly Change | Hay, or rather Dung, moved the Cocks Silbury Mead, not moved since 26 June | SWEDES, began sowing Chalk'd Land made good work. | Plough'g for Swedes | Iron Gates, Jas Yeo, - fix'g in Meadow

7. Glass sinking, dry night, RAIN from 6 O-Clock to 1-O Clock & frequent Storms remainer of Day | Horses in the Stable, | The Clover grass reserved for the Horses nearly gone, - Mow'g Grass, edge of the Down for the Horses by the Wheat and baiting the Horses evenings

8. Swedes, finish'd Plough'g & Sow'g | Storms in morn'g - dry'g after

9. Mr Winkworth. - Carr: 800 Bricks | Dry morn'g, Hay Silbury Mead threw heavy Storm 11-O Clock & light Storms after | [...] | Sheep giddy Kill'd 36 li for Masons & Carpenters Dinner - for put'g Beams on Waggon House | Plough'g for Turnips

10. Plough'g - and Sow'g Tankard Turnips, on the Wheat land, intend'd for Vetches, the dry weather prevented sowing | Dry Night & morn'g, Hay Silbury Mead dry except Cocks, hoped to carry, Skuds of rain in succession after & heavy rain afternoon | Mr Kemm, carry'd Clover all the morn'g quite spoiled | Mr P_ K_ carry'd Ray Seed early in the morn'g - dry

11. Dry morn'g, heavy Storms after | Plough'g as yesterday & finish'd Turnip Sowg

12. Sunday, Very heavy Storms morn'g drying after till eve'g, when Storm again

13. Storms in morn'g & evening | Fallow'g a Fold drift, Clover Stubbe Down Piece | Jas Robbins, - came to work, having no Steady Man in the Parish to work for me

14. Storms heavy very frequent, to drive Carpenters into Barn & idle | [...] | Fallow'g the 1st Fold Drift (on whole grod | 1 Team to Chippenham for Posts = viz & Pumps. (=37 Posts - 5 of them Dble)

15. Chalk, Carted 25 loads | Drying afternoon took up 1½ Load of Hay Silbury Mead when at 5 O-Clock a Very heavy Storm

16. Rain in morning & at 12 O-Clock very drying after, many People carrd Hay | Marter earth Cart'g for Building | 1 Waggon Load of faced Stone & 1 Cart Load of Lime fm Calne | 115 Wether Sheep from Tockenham to Keep

17. Drying morn'g till 6 O-Clock, heavy rain with few intermissions after | Straw removing from New Waggon House by Mr Jno Larges 2 men, - (Whales)

18. Horses in the Stable, Rain all the morning, - Storms at intervals | Roofing Waggon House, Force Pump erected

19. Sunday - Dry Day - with exception of light Storms

20. Fine Day, except a light Frisk, Storms partially [119] | HAY, Silbury mead carried 4 Large Loads | Potatoes, Hoeing, by Cleveancey Men | Chalk, -

---

119 Frisk – probably flisk (see Glossary).

Carted 23 Load for New Stable.

21. Rafters from Calne for Waggon House | HAY - Silbury mead, - finish'd carry'g | Fine Day, - Cart'g Marter Earth. | Clover grass finish'd yesterday for Horses and Mow'g Silbury Mead, before the Hay off
22. Very warm fine day, Wheat ripening very fast at Haslebury, & begin Reap'g next Monday 27th - began last year 17th | Isaac Blackman & Rt Church, fallow'g before Fold - Clover stubble Down piece | James Robbins &c. finish'd hoe'g Potatoes and toping Ricks | Raftering the Waggon House, and the Masons (Maslens) finish'd
23. Fallowing before Fold | Harrowing Turnips - Wheat Land | New Bay Horse 5 Yrs Old from Thos Hillier | Warmest Day this Summer | Toping Hay Ricks
24. Toping Hay Ricks | A frisk of Rain 1 O-Clock, fine after Cooler Day | 1 Team all Day - for Mr Wentworth | 1 Team after 12-O-Clock - for Mr Wentworth | Harrow'g 1st Turnips finish'd | 2 Turnip Hoers began (fm Ogbourne)
25. Close warm Day, - large drops of rain in evening for a short time | 1 - Team all Day for Mr Wentworth | 1 - Team to Calne for Board for Lathes | Hurdles removed to young Vetches to feed off being choaked with Charlock
26. Sunday, - A fine Day, frisk of Rain in eve'g
27. Very fine day - no rain Thatch'g - Hay Ricks began, thatch'd two and 4 Women, made Elms[120] | 1 Team, for Mr Wentworth | Hurdles removed from Clover to Vetches & Charlock | Roll'g & Harrow'g before the Fold | Swedes - shewing rough leaf
28. Fine & dry all Day | Chalk, - Carted 21 loads | Vetches & Charlock - began feeding
29. Rain heavy in Night & morning and Storms all Day, - Barley - beat down | Chalk, Carted 21 load | New Waggon House, began Slateing
30. 1 - Team for Mr Wentworth | 1m Bricks by Mr Wentworth Waggon | Cart'g Road earth to raise Waggon House | Very heavy Storms, partially
31. Some Frisks of rain | Turnips, - hoeing 2nd time | Carting Road earth to Waggon House | Cart to Calne for Lime | 3 Rick Staddles put up in the 3 last Days

RAIN 25 Days in JULY

August

1. Chalk, Carting to Waggon House Very fine Day
2. Sunday, fine warm ripening Day Glass very high, prospect of fine weather, Reaping commenced many Places tomorrow, and generally in the week
3. Fine dry morn'g, RAIN incessant from 9 O-Clock morn'g to 4 O-Clock eve'g | Wm Philpots - last Wheat delivered to Devizs & 1800 Bricks retd |

120 Elm - (Also Helm or Yelm) Small bundles of fresh straw, damped and laid out straight for the thatcher's use. Dartnell & Goddard, *Wiltshire Words,* pp. 51-52.

Journey to Bristol for Slates

4. Showers in morning dry'g afternoon | Hurdles removed to protect the Barley by Tan hill road Fair Day | REAPING Wm P_ K_ & Mr Wentworth began, - My own Wheat want cuting, no Reapers on the Farm | Last Rick Staddle put up | 2 Carpenters & 2 Tilers finish'd for Present & left, - the Young Carpenter left - the 1st Inst

5. Some frisks of Rain in morn'g, dry'g after | Chalk - Cart'g to Rick Yard, by Waggon House | REAPING began 4 Taskers in evening | Old Brewhouse Tiles took off

6. Dry morn'g - Dizzle'y rain after 10 O-Clock at intervals all Day | 1800 Bricks from Devizes

7. Warm close Day, some dizzle y rain eve'g [121] | 5 other Reapers put, 2 of the first started | Hurdles return'd from Barley to Vetches | Rolling & harrow'g Swedes

8. Very fine clear Harvest Day 8 Reapers | 2 Waggon Load of Slates from Devizes

9. Sunday, Rain in night & Storms afternoon

10. 14 Taskers, Reapers & 2 Women, from Avebury | Rain heavy in night, dry & fine after 8 O Clock morn'g | Mr Kemm, began feeding Clover & Tankard Turnips with Lambs hoe'd 19th June 1500 Slates from Devizes & 6 Qr Coal

11. Marter earth, Cart'g for New Stable | Plough'g Fold drift, Vetch ground | 4 Tockenham Reapers left & 4 Rowde Men put on, Furze Down | Very fine Harvest Day

12. Very fine Harvest Day | Blind Horse ret'd to Cleveancey | Rolling & Harrow'g Swedes | Lime & Coolers from Calne [122]

13. Turnips finish'd Harrow'g | Chalk, Cart'g to Mud hole in Yard | Rain heavy 4 O-Clock morn'g, - dry till 12 O-Clock, rain all day after, stopd Reapers | 2 Ogbourne Reapers Paid off, a few Grips to tie | Clover 2nd Crop, the 2 Whites from Lineham began Mowing

14. Rain all day, and heavy at times, Reapers in the Barn discontented | 6 Horses & 2 Waggons to Devizes for 1600 Bricks and 300 Ladies-Slates [123] | Wm Philpot Obit 50 Years [124] [This entry written within a thick black border]

---

121 dizzle y - drizzley.

122 Cooler - A large open tub. J. O. Halliwell, *A Dictionary of Archaic and Provincial Words: Obsolete Phrases, Proverbs, and Ancient Customs, from the Fourteenth Century, Volumes 1-2* (London: J. R. Smith, 1865), p.269.

123 Ladies Slates - these are slates which measure 8 inches by 16 inches. See entry between entries for 19 and 20 October 1832 below.

124 William Philpot of Box buried at Avebury on 21 August 1829 age 50, *Wiltshire, England, Church of England Deaths and Burials, 1813-1916 for William Philpot* [accessed via Ancestry, 22 July 2020].

15. Rain at times all Day The Marter from old Building, Carting on Fallows for Wheat_ Down Piece | Old Bricks & Stones Cart'g away fm Brewhouse | The young Turnips & Swedes want hoe'g | 3 Reapers from Devizes & 1 from Lavington Paid off, tired of waiting for fine weather their remaining work, made over to another
16. Sunday, Fine, and drying
17. Fine and very drying, Glass returned to N in change | The Wheat quite dry, Load'd 3 Waggons in evening to make a Rick tomorrow | Mr Wentworth put his Talavera Wheat into 2 Ricks in Field | Chalk, Carting to Waggon House | The Wheat that is not cut is dead ripe, not much cut at Yatesbury Winterbourne &c.
18. Dry night, Rain began 5 O-Clock morn'g, Storms & very drying alternately, Reapers at work at intervals | Foundation of Cart Stable began | 2 Horses Cart'g Materials for Masons the others in the Stable | The two Mowers (Whites) Struck for 2s & Beer, instead of 1s 6d for Mowing the 11 Acres of second Clover, - and began it | The two Mowers finish'd the remnants of Reaping
19. Heavy Storms & drying Winds alternately Reaping & Mowing at intervals | Unusual heavy Storm 6 O-Clock eve'g | The Charlock & Vetches, finish'd feeding, and the Hurdles removed to Silbury Mead for the Sheep to feed it, - 87 Hurdles required for me to divide the Mead. | Ploughing a Fold drift Vetch land for Wheat | The Horses at this time & few Weeks pass'd less than 2 Bu: of Oats pr Week - to bait mornings only.
20. Rain in Night & storms till after mid day, dry after, Glass sunk to M much rain yesterday, now rise'g to R, in rain | Wheat sprout'g in Sheaves, on the ground. | Garden Wall, Carting for New Cart Stable | Cellar & Offices, Slated in
21. Waggon House finish'd Slateing and Slaters helping at Cart horse Stable 8 Hands in all | Fine dry'g Day Wheat got in good condition by 1-O-Clock when began carrying, made 1 Rick & part of another | Ploughing and harrow'g Fold drift in morning | Sheep began Silbury Mead | <u>Wm Philpot - Buried</u> <u>at Avebury</u>
22. Dry Night Marlbro Fair Wheat began Carting half pass 5 O-Clock morning, RAIN very little began at 7-O-Cl and continued at intervals, very drying winds between till 3-O-Clo when a Storm too heavy to continue carrying, finish d the 2nd Rick | Finish'd the 2nd Wheat Rick & Thatch'd them and made half the 3rd Rick in very satisfactory condition | <u>Very heavy Rain</u> 7 O-Clock evening and continued for two hours, the Rick Yard in all low places covered with Water much more rain fell in the time, than in the same time at any period since have been at Beckhampton
23. Sunday, Heavy Rain in Night and frequent Storms in the Day.
24. Heavy rain in Night & frequent heavy Storms in the Day | Wheat sprouting in Sheaves on the ground | 2 Waggons to Calne for Lime & Stone, and

Hurdles to Silbury Mead

25. Fine dry Day & remarkable calm pleasant evening, the Sun set red, and all appearance of fine tomorrow. | The Wheat very wet in bond & growing and the Sheaves very wet one side of the Tything, Mr Wentworth & many others carrying all Day | Our Wheat Rick, half made on Saturday 22nd the Rain gone through the middle of it, and the Wheat sprout'g took it apart & put about the Meadow the out side of the Rick quite Dry. | 2 Waggons to Calne for Lime & Stone | Jos Bromham & Danl Dyke, Hoe'g Turnips 1st time - the second Sowing | The two Whites finish'd mow'g Clover & returned home | 8 Masons & Carpenters began Sleep'g over new Nag Stable

26. Dry morning Glass rather sinking | Wheat began carry at 9 O-Clock when the dew off. Rain soon after a little and by 12 O-Clock heavy'r finish'd the 20 Acre Piece by 3 O-Clock made a Rick & Thatch'd it, & raised the one made Saturday to the eves lay'g | Very high winds & Rain in evening. | Brother Wm P_ Kennett, began carry'g Wheat & made his first Rick, and | Mr Geoe Brown - Avebury began carry'g | The Wheat lost its colour outside the Sheaves by the the Rain since the 22nd | The Wheat carried last Saturday 22nd, is good in colour, all admit now, those were fortunate that carried Saturday. | 2 Little Pigs from Kennett

27. Plough'g finish'd the Vetch land fed off | Heavy Storms & drying Winds alternately the Wheat at 3 O-Clock as dry as yesterday | Rain & very high winds in evening

28. Rain in Night dry morning, heavy Showers about 2 O-Clock, Stop'd many People carry'g Wheat for half an hour and Showers after till 4 O-Clock, very drying remainer of Day | Load'd 3 Waggons from Furze down Piece to top the 4th Rick, the Wheat dry with the exception of a few Sheaves wet in bond | Mr Budd of Winterbourne not carried any Wheat, and Mr Stratton of Rabson not finishd reaping | Carting Dung to 2nd Furze down Piece

29. Carting Dung to 2nd Furze down Piece | Top'd the 4th Wheat Rick & Thatchd it | Very fine Day, with exception of a Shower 5 O-Clock eve'g

30. Sunday - Dry day, except a Shower in evening

31. <u>WHEAT=HARVEST</u> finish'd by carrying remainer of Furze down Piece to Barn, some few of the Sheaves yet wet in the bond - and green with grown Wheat | Fine and drying all Day, the best Harvest day yet | Hay - the 2nd Clover carried 3 Loads from Nether Down Piece | Turnips the 2nd Sowing for Wheat, finish'd hoeing the first time, and began Hoeing Swedes the first time on the 2nd Down Piece | Draging & harrow'g Swedes | Pease W P_ of K, carrying - quite spoiled

<u>RAIN 22 DAYS IN AUGUST</u>

September

1. Carting Dung as the 29th | Glass yesterday at G in change, with every

appearance of fine weather | Thick rime all Day at times, & in the eve'g Rain to prevent carry'g Corn | Straw to the INN sent the first for 14 Coach Horses of Mr Wm Laws of Chipp[enha]m came yesterday

2. 2 Waggons to Calne for Lime & Stone | Carting Marter earth. | Very. fine Harvest day, little rain afternoon | Many Places round Calne, Mowing, & making dry Meadow Hay, grown since the rain.

3. Very fine Harvest Day | Hay - the 2nd Clover, carried in good condition | Chalk - Carting to Yard | W P_ of K_ finish'd Wheat Harvest in good condition

4. Barley, began Mowing Jos Brom[ha]m &c. | Swedes, not finish'd hoe'g 1st time | 2 Waggons to Calne for Lime & Stone | Harrow'g Turnips hoe'd 1st time | Harrowing before the Fold | Front Wall of New Stable up | Straw Waggon load to Inn | Mr Geoe Brown & Mr Wentworth finish d Wheat Harvest in good condition Very fine day (Friday)

5. Damp air all Day, some rain, cold winds Glass lower | Plough'g Fold Drift, where Hay carried from Down Piece

6. Sunday, - Rain heavy in night, dry morng Storms all day after Thunder and heavy rain partially
Willm Chivers - Obit 55 Years - pass'd [This entry written within a thick black border]

7. Dry morning heavy Storm evening | Ploughing Clover lay for Fold &c

8. Ploughing Clover lay for Fold &c | Rain heavy in night & storms in morn'g afternoon & Eve'g - Thunder & lightening and heavy Storms | Walls of Cart Horse Stable finish d, and Roof began puting on | Privy began building

9. Carting Dung on Fold drift Furze down Plough'd Yesterday | Drying Day, rain in evening | Wall end of Skilling, building by Stable

10. Much rain in Night & morning = Horses in the Stable till 10 O-Clock = Chalk Cart'g to Yard remainer of day = Dry'g Winds till 4 O-Clock rain all eve'g

11. Glass rising - dry'g Winds till 4 O-Clo: when commenced raining all the eveing & very violent from 8 to 9 O-Clock | Carting Dung & finish'd the 9 Acres Down Piece | The Women spread'g the Dung | Wm Hampshire of Connock, began Mow'g Wheat Stubbles | Barley sprout'g with continual rain

12. Swedes finish'd hoe'g 1st time, very hard time for it, the Charlock and Chick we<t>ed set, as fast as hoed | Cart'g Dung for Wheat, Clover lay 19A Pie [125] | Rain in Morn'g, Showers at times all day

13. Sunday, - Dry morn'g, heavy rain after noon.

14. Carting Dung as the 12th and finish'd from the Yard, all the Wheat land Dung'd except 2 Acres | Turnip Hoers - seconding - the last sowing Turnips | Tileing, began, Cart Horse Stable | Storms at times after the morn'g, Thunder Lightening & hail storm afternoon

125 Pie - Piece.

15. Fine drying Day, wth exception of Partial Storm | Ploughing remainer of Clover Stubble for Wheat
16. Rain heavy in Night & all Day, except intervals of dry | The 6 Horses & 2 Waggons to Jno Days of Pickwick for 4 load of Stone Tile | The Barley in a bad state, Swolne ready to burst, but not much sprouted, except under the Hedges. The Weather being Cold & frosty Nights
17. 3 Rams of Mr Wentworth - 2 of 6 Tooth & 1 of 8 @ 50s ea | Sharp Frost fine till 4 O-Clock, moved the Ears of the Barley Swaths off the ground the Grass grown through the Swath, yet not much Sprouted | Chalk - Carting to Rick Yard, the Land too wet to Plough for Wheat
18. too wet to Plough for Wheat Rain a pelting Storm & frequent lighter Storms
19. Rain - frequent Storms | Cart'g Chalk to Yard & Grist for Pitching New Cart Stable | 2 Sides & 1 end of Cart Stable Tiled, the other end waiting for Tiles from the end of Dweling House | Silbury Meadow, the last tie gave the Sheep, began 19th Augst 228 Tegs - our own 115 Tegs - Messrs A & R Larges [Total] 343
20. Sunday - Fine day - Dry all Day
21. Rain in Night, very drying Day | Chalk, Carting to Yard | Began Keeping 100 Pick'd Ewes for Mr Wentworth, put to a good Ram of Mr Wm Neats, having plenty of of keep my Down very good (10 Ram'd this Day | Pitching Cart Horse Stable & laying Floor | Wheat Sowing, Mr Wentworth began
22. Carting Chalk, as yesterday, & road earth to Stable | Hurdles removed from Silbury Mead to Turnips | Fine dry'g Day, many People carry'g Barley | Barley moved the Ears off the ground not moved before since Mowed the 4th Inst and under Ears of the Swath as bright as when standing, - this plan generally approved, - the Barley often moved by some People is become black [The following note is written in the margin] Note this is a good Rule
23. Very fine morning, partial Storms at a distance | Turning Barley, Cocking & Raking | Barley began mowing again, having some hopes of carrying | Plough'g for Wheat Sowing, began Down Piece next to the Swedes | Masons Pitching before New Cart Stable Carpenters, fiting up, Rack, Manger &c.
24. Masons finish'd Pitch'g before Stable and Messrs Strong & Ruttiford planing front of House [126] | Plough'g as yesterday in the morn'g and began BARLEY, carrying at 12-O-Clock | Rain a little at 2 O-Clock - with this exception a very fine Harvest Day, Fog in bottoms, Morn'g & Eve'g | Sun r[...]
25. FOUNDATION of House began | Frost the sharpest this season, very fine drying Harvest Day | Barley - carried the remainer of that mowed the 4th Inst about 8 Acres put in lower Rick and 4 loads the remainer on Staddle by

126 Ruttiford is possibly Thomas Rutherford, stone mason of Quarr, Calne, J. Pigot, 'Commercial Directory 1830' in WRS47, p.66.

the Barn.

26. New Cart Stable, Horses first occupied | Cloudy dull morn'g, little rain in eve'g Cobweb over the ground yesterday eve is often a sign of rain soon | 1 Team to Pickwick for Tiles to finish the Stable & 1 Team Plough'g for Wheat | No Barley fit to carry, being full of grass

27. Sunday. Rain heavy in night & rain'g morning, finer after wth partial Storms | TURNIPS began feeding a very good piece with a tie of old field grass - adjoin'g & feed on Wheat Stubbe before going to fold

28. Heavy partial Storms, other places fine Two Teams Plough'g for Mr Wentworth to finish sow'g 30 Acres of Wheat, myself but 2 Acres Plough'd for Seed | Tileing the end of the Stable to finish it | Painting began Windows of the Inn | Plint began laying for New House [127] | Carpenters, enclosing Pump &c.

29. Fine day, Storms at a distance | 2 Waggons to Honey St Wharf for 2 Logs of Memel Timber for the Joist of New House

30. Carting Wheat Stubble & Ploughing till 1-O Clock | Barley, - the fresh mowed began carry'g in very fair condition being grass'y | Very fine Day | Sawyers - cuting Joist for House | Barley, finish'd mowing | Ewes - 3 Ram'd, part of the 11 - put to Mr Wentworths Ram | RAIN 24 Days in Septr

October

1. Blowing Dry'g Winds | Plough'g in the morning till 11 O-Clock, & Carry'g Barley after, - top'd the 2nd Rick by the Barn

2. Dry windy night, Barley began carrying by 8-O Clock, stop'd with rain mid day | Ewes. - 6 ram'd by Mr Wentworths Ram

3. Rain heavy forenoon, to drive Teams from Plough & frequent Storms after

4. Sunday, - Rain mizley afternoon

5. Rain heavy Storms, kept Masons at home | Plough'g for Wheat, Vetch land fold'd

6. Very drying fine Day, yet some little rain | Barley, - began carrying remainer of Tan hill Piece at 12 O-Clock, & finish d at 8 O-Clock, in best condition of any yet

7. Rain began 8-O-Clock morn'g, & continued Storms all day | Ploughing for Wheat | End of the House took out

8. 1 Team Plough'g for Wheat | 2 - Load Stone from Calne | Pierce'g cold morn'g & very sharp frost thick Ice, excessive cold cuting, dry'g winds all the morn'g mild'r afternoon | Barley, fit to carry by 4 O-Clock afternoon, when began & carryed Whitelands 12 or 14 Loads | Sent a Letter of complaint to Mr G | Harvest finish'd

10. Plough'g Down Piece for Wheat Geoe King the Carter & 3 Plough Boys left

---

127 Plint - probably plinth.

9. Plough'g Down Piece for Wheat | Harvest Supper | Frost & Ice morning, fine, but dull Day
10. RAM, turn d the first to the Ewes 22 Ram'd | Glass rose to I_ in fair, thick air in morning fine day, best prospect of fine weather yet, Weyhill Fair, Sheep fallen 8s pr Head from Wilton Fair
11. Sunday, some mizley rain mid=day
12. Damp morn'g, very fine day | Rams, turn'd the other 2 to Ewes, 40 Ramd | 1 of Mr Larges Wether Sheep, Died on its back in a furrow in my Fold, suppose'd to be very full with Turnips, & could not get on its legs, - the furrows not harrowed down, - which ought to be | Clover seed Mowed, - but little seed, purpose salting it for the Horses | W: P_ of K_ finish d Harvest | Women drawing Charlock from Swedes at 1s pr Acre instead of hoe'g second time
13. Rain in morn'g, damp air at times in Day | WHEAT SOWING began Nether Down Piece 3 Bushels pr Acre. Press'd & 5 tines with the Harrows made good work
14. Press'd and finish'd the Piece 9 Acres with 26 Bu: of Wheat, - from Mr Jno Large Cleveancey @ 27s 6d pr Sack | Rain heavy in morn'g & light cold storms during the Day
15. Frost & very cold morn'g fine Day | 2 Waggon Load of Stone from Calne and 1 Wag: Load of Sand from the Low and 1 Cart Load of Sand from the Low to the Hail Farm | New House got up to Window=top of Parlour, - and the Foundation of Garden Wall in.
16. Rain morn & eve, and damp Day | Plough'g athwart 2 foul Lands for Wheat next Down Piece | Treading the Wheat with the Sheep make good firm work | Iron Gates, fixing by Rick Yard and Paint'g Iron Gates
17. Iron Gates, fixing by Rick Yard and Paint'g Iron Gates and Plough'g as yesterday | New House ready for Joist of 1st Floor
18. Sunday, Very fine Day
19. Rain in morn'g and damp air at times but very fine afternoon | Clover seed Hay carried | Thwarting the two Couchey Lands & pick'g it off, - and the first Waggon Load of Dung from the Inn put on it, and the Fold follow | Trim'd & mark'd Mr Abm Larges 114 Wethers for Devizes Fair | The Joist's laid in New House
20. Devizes Fair, Rain at times from 12 to 3 O Clock Ewes 21s 6d - the highest, Mr Cowards Lambs, but few sold | Devizes Fair, Bought of Mr Coward of Roundway 37 Two tooth Ewes @ 16s ea | 2nd Waggon load of Dung from Inn | Drag'g - Harrow'g & Picking Couch from Wheat Land | Picking Tuffets from Wheat up | 3 Teams of Mr Wentworths, Ploughing for Wheat for me.
21. 3 Teams of Mr Wentworths (10 Horses) Plough'g Press'g & harrow'g for Wheat, (Turnip Land) | Wheat sow'd 21 Bushls from Turnips to last Fold drift to Furze Down Wheat about 7 Acres | Dry morning, very driving rain

before 1-O-Clock, Pouring rain by 2-O-Clock and Teams drove home, Wheat only broke in | 2 Load of Stones from Calne

22. 6 Horses to Honey Street for 3 Logs of Timber on Clarke & Robbins's Carriage | Some Storms, mid=Day
23. Sharp frost & fine warm Day | 1 Team of Mr Wentworths half a Day | 6 Horses to Honey Street, for 3 other Logs of Memel Timber | Potatoes out of ground Frozen, and 4 Men diging Potatoes | Cart load Dung from Inn [Pinniger then recorded sheep prices from Weyhill fair on 10th October - see Appendix H]
24. Potatoes, Diging in Field 4 Men and 4 Women 20 Sacks, fill'd | 2 Horses took Timber Carre to Honey St 4 Horses finish'd harrow'g the 7 Acres of Wheat sow'd 21st [128] | 3 Waggon load Dung fm Inn | Mr Wentworths 100 Ewes returned home came 21st Septr | Sharp Frost, very fine warm Day
Danl Edwards, of Calne Obit 23rd October Aged 81 Years, in June last [This entry written within a thick black border]
25. Sunday, Very fine calm mild Day
26. Fine mild clear warm & Summer like | Potatoes finishd Diging in Field 22 S 24th 20 Garden 11 [Total] 53 | Plough'g & Sowd 2 Drifts of Turnip Land the 1st next the Hay Ricks wth 7 Bu Wheat 2 of Mr Wentworths Teams half a Day | 195 Lambs 50 Sale Ewes of Brother Wm began feed'g my Turnips, & brot Hay 2 Truss's at Night | 1 Load Dung from Inn, & finish'd Fold drift foul Land
27. Very foggy all the morn'g, clear afternoon | Load of Lime & faced Stone from Calne | 2 Horses, harrow'g Potatoe Land & Plough'g it | Christr Pinniger & Emma Smith, United [129]
28. Close rain in morn'g, fine dry'g Day after | Load of rough & faced Stone | 2 Horses, harrow'g & Plough'g Potatoe Land for Fold
29. Fold removed to Potatoe Land, muckled from the Inn | Plough'g a Fold Drift for Wheat | The first sow'g Wheat appear'g & the Women picking Tuffets of Grass off
30. Fine mild morn'g, Storm mid=day, fine after | Plough'g & Sow'd the last Fold Drift next Down Piece
31. Finish'd Harrow'g the Wheat Sowd yester'y | 1 Team Plough'g half a Day at Kennett Brother Wm | Rain 10 O-Clock, fine remain'r of day | 48 Bushls Grains. Horses began this morn | Plants 300 set in Garden

November

1. Sunday, Very sharp Icey frost cold dry Day
2. Frost in night, cold dry Day | 4 Carpenters began frame'g, the Roof for the

---

128 Carre - Carriage.

129 Christopher Pinniger of Box and Emma Smith married on 27 October 1829 at Bremhill, *Wiltshire, Church of England, Marriages & Banns, 1754-1916 for Christian Pinnger* [accessed via Ancestry, 22 July 2020].

House 2 Sawyers at Roof for the House 6 Masons 3 Servers to Masons at the House. & the Garden Wall | Both Teams Plough'g for Brother Wm Wheat Sow'g

3. Load of rough & faced Stone fm Calne | Fold removed from Potatoe Land to Wheat <lands> Stubble Furze down Piece, and Muckled from the Inn, for Swedes | 265 Ewe Teg 3 Rams 114 Wether Tegs [Total] 382 | Fold 9 by 21 Hurdles [is] 189 [times] 2 [is] 378 - 2 Sheep to a Coop | Fine mild Day

4. Rain in night & damp all morn'g, and close rain from mid day | Horses in the Stable | Masons &c Stop'd, working | Chalk, Cart'g in morning

5. Frost and Fine Day | Wheat Rick, - put the first into Barn | 19 Sacks Wheat to Devizes & 2T 2€ 0Qr Coal home | 2 Horses Plough'g Potatoe Land for Wheat

6. Fine mild drying Day | Wheat sow'd the Potatoe Drift & a Turnip Drift | Window heads put on the new Chamber Windows | Jacob Pinnigers 6 Pigs brot to Stubbleing

7. Very fine Sun shine Day, little Frost, morn'g | 2 Horse's - Cart'g Chalk for Wall & muckle for Fold | 4 Horses to Tytherton for Cider

8. Sunday, Fine but cold dry Day

9. Very fine Sun Shine Day | 3 Horses finish'd harrow'g Wheat sow'd the 6th | 3 Horses Cart'g Dung before Fold | Up'r String course put on the new part of House | Foundation, of back Garden Wall put in

10. Tiles took of part of Old House 2m of Bricks from Cannings 3 Horses and 3 Horses, Cart'g Chalk | Damp day, close rain in eve'g - and heavy before mid=night

11. Foggy morn'g, very mild day & pleasant till 1-O Clock, when Storms till eve'g | The Roof, took off, end of the House | Load of Lime & rough Stone from Calne and Load of Road earth to Mr O_ Viveash | 2m of Bricks fm Cannings, 1 of Brother Wms Horses to help

12. 2m of Bricks fm Cannings and 1 of Brother Wms Horses to help | 12 hd of Bricks from Devizes, by Brother Wms & Mr Wentworths Waggons | Hay, cut for Horses & Paid Wms borrowd from his Sheep | Load of Lime & rough Stone fm Calne, and a Load of road Earth to Mr O Viveash | Rain in Night and thin storms in Day | Building began on the Walls of the Old House

13. Dry dull day very mild | 2m of Bricks from Cannings with one of Brother Wm Horses | 2 Load of Rough Stone from Calne

14. Mizly rain at times all the morn'g | 4 Horses to Honey Street Wharf, Piece of Timbr 12 Dales & 4 Pipe Staves [130] | 3 Horses (1 of W Ps) Cart'g Muckle fm Inn, and 40 Bu: Grains from Kennett.

15. Sunday, little rain, morn'g & eve'g fine - day

---

130 Dales – probably Deals. Pipe staves - Each of the staves which are hooped together to make a cask or barrel. Lexico, *Pipe stave* https://www.lexico.com/definition/pipe_stave [accessed 1 May 2020].

16. Frost in morn'g, fine cold Day | 4 Horses, Cart'g Chalk for Build'g House | 2 Horses took Timber Carr: home | Beams & Joist began puting in New House for the Atticks | Sheep Kill'd one Giddy, (spot on Should'r
17. 4 Horses, Cart'g Chalk as yesterday | 2 Horses, Cart'g Muckle from Yard & Sheep Cages from Avebury to Fold | Unusual Sharp frost Ice to bear cold thick air clear evening
18. Frost morn'g fine day GRASS, Horses finish'd | Wheat Stubbles, brot remain'r home | Cart'g Muckle from Yard to Fold | HAY, the black Meadow began giving our own sheep & Mr Larges | Rafters began put'g on part of New House | RAMS, turned 2 of Mr Wentworths to 265 Stock Ewes
19. Frost exceeding sharp, Masons dress'g Stones till 10 O-Clock, Marter not work till then | 4 Horses to Devizes wth 20 S Wheat & 1m Bricks home | 2 Horses to Quemd Mill [131] | Brother W Ps Sheep home from my Turnips to be trim'd for Marlbro Fair
20. Unusual & exceeding sharp Frost, fine mild Day, thaw, and Glass at A, in fair | Half the House Rafter'd & kept the House Raring [132] | Chalk - Cart'g for House & Rick Yard
21. Plough'g Turnip land for Wheat | Very sharp Frost in morn'g thaw after and pierce'g cold winds
22. Sunday, frequent Storms & mild'r
23. Cold dry'g Day | Cart'g Muckle, before Fold | Mr Large's 114 Wethers sold at Marlbro Fair @ 22s
24. Very cold dry'g Winds | Plough'g Turnip Land for Wheat all the old Turnips fed off, and one drift of the Young Turnips | All the Masons employ'd yesterday & part of this Day on Garden Wall for want of Stone from Box. Load of Stone came at 10-O-Clock

    Mr GUY of Chippenham declared, absented from his Crediters & the Docket struck by Mr Bush of Trowbridge. Myself & Brothers met at Mr Guys Office to consult with Messrs Goldneys &c. on the best means to be taken, till 10-O-Clock at Night [This entry written within a thick black border]
24. Bitter cold Snow'y night, much drift'd
25. Bitter cold Snowing morning, the land quite covered wth Snow & drifted 6 feet deep by the side of the Road | 2 Load of faced Stone from Calne | The Masons & Sawyers, gone home cannot work in the weather | Snow a little all Day
26. Snow a little in the Night, Men diging a road through the Snow on the Down, - Dry & mild'r Day | 6 Horses to Calne for a load of R & F_ Stone[133] | Sheep on the Down, Turnips cover'd wth Snow

---

131 Quemd - Quemerford.

132 Raring - possibly rearing - that is, going up, continuing towards completion.

133 R & F - rough & faced.

27. Cart'g Muckle, before Fold | Dry mild Day
28. Thick fogg'y air all Day | 3 Pigs, put up to fatten for Pork | WHEAT STUBBLE, began fallow'g 1 Team | Load Lime from Calne, - and 8 S Wheat to Chilvester Hill | House, - the Masons, finish'd & ready for the remainer of Roof, and the Sawyers Paid off.
29. Sunday, - Dry, but thick cold air
30. Wheat sow'd 3 Sacks on Turnip Land with the headlands, up to the Hurdles Two-drifts more of Turnips to feed off & Sow | 4 Carpenters came this morn'g, to put the Roof on the remainer of the House | 9 Masons, at, the House, Chimneys & Wall | SOW = brim'd, - Mr Hitchcocks Boar

December

1. Cold drying Day | Horses, finish'd harrow'g off the Wheat Sow'd yesterday
2. Plough'g Wheat Stubbles, & 9 S. Wheat to Devizs | Very thick, damp fogg'y Day | Garden Wall, finish'd (both Sides)
3. Very thick foggy Day | Cart'g Dung before Fold, 16 Load frm Inn and 2 from home | Slate=ing, - began the House by 5 Masons by 3 Servers to Masons, by 4 Carpenters
4. Mizly rain at times, mild & fine for the Season | No Corn in the Barn, 4 Men cleaning up the refuse from Garden Wall & House | One end & front of the House covered in. | Plough'g & Sowing a Turnip Drift to Wheat - about 1½ Acre to feed off, and finish Sowing
5. Fallow'g Wheat Stubbles | Stubbles, carried the remainer | Foggy morn'g - clearer by 11 O-Clock | Barly Rick, began puting into Barn the Straw as damp as when harvestd
   <December 20. Sunday, Snow in morn'g, little frost dry_day>
6. Sunday, - Foggy - and fine Day
7. Frosty morn'g, very thick fog | Fallow'g - Wheat Stubbles
8. House, complately Slated, and lead'd the hips & Ridge &c. The Slaters, returned home, - Jno & W Robbins and Boy remain to finish Chimney &c. and 1 Carpenter | Fallow'g Stubbles | Frost in morn'g, thick foggy Day
9. Frost in morn'g, thick foggy Day | Cart'g Dung from Yard, before Fold Obit Robt Ashe Esqr of Langley House Aged 82 Yrs [This entry written within a thick black border]
10. Fallowing Stubbles, | Frosty morn'g - clear Day | Cress, puting on Cellar [134] | Carpenters making - Yard Doors
11. Sharp Frost & dense Fog, all Day | Fallow'g Stubbles
12. Carting Muckle, before Fold Mizly Rain & Damp air | Masons, finish'd Cress'g Waggon House
13. Sunday, - Dry cold air
14. Very fine, clear Sun & mild | Fallow'g Stubbles
15. Fallow'g Stubbles | Very cold frosty morn'g, thick fog | CHALKERS,

134 Cress - probably crest, or ridge of roof.

commenced & sunk a Pit 8 feet came to Chalk | Masons, the 2 last finish'd, and left the Building till Spring

16. Fine Day, thick air afternoon | Fallow'g as yesterday
17. Fallow'g as yesterday | Very thick fog, & freezing afternoon
18. Porker Kill'd 80 li
17. JNO PINNIGER OF COWITCH Struck the Docket against ANTHY GUY OF CHIPPM
18. Snow in night & all forenoon | Cart'g Dung before Fold | 2 Men diging Post holes for Yard Doors being very damp for thrash'g Barley | HAY, Sheep twice a Day, & on the Down
19. Frost & little Snow, thaw & frost again eve'g | Cart'g Muckle before Fold | Diging out Watercourse, in Meadow, and Post holes for Yard Doors
20. Sunday, Snow in morn'g, little frost, dry day
21. Snow heavy, large blossoms till 11 O-Clock thaw & freezing after, very slippery | Jas Robbins & Isaac Blackman, requested to find work at home, having little for them, but could not. & returned to me again this morn'g at reduced Wages fm 9s to 7s | Very Slippery, Sharp frost | Waggon & 4 Horses went to Box, for Freestone, - returned to Studley=brook the Horses unable to draw it home altho the Thiller rough'd - and the others half rough'd, Road very slippery [135] | 2 Horses, Cart'g Marter earth for Plastering next summer
22. Very heavy Snow, all the morn'g and nearly all day, the weather too bad for the Horses to go to Studley for the Waggon, although all the Horses roughed | Hay Carted to Sheep & Grains fm Kennet
23. Very cold winds & heavy Snow at times all Day, the Snow deep on the land The Coaches impeded some hours | The Sheep on Hay only, & hardly keep up their condition with the bad Meadow Hay, began the two old Clover Ricks of Wm Philpots for them, - Sheep require Sheep require Water with the Salted Hay altho the Snow deep on the land | 6 Horses to Studley for the Waggon 2 Men thrashing Barley, & 3 Men truss'g Hay & puting Timber in Waggon House Loft but little to employ the Men | Sow, brim'd | Jno Pinniger of Cowitch Godwin of Beaversbrook Rumming of Draycot Salter of Langley } attend d a Meeting of Commissioners at Bath, to prove & declare Anthy Guy of Chippenham a Money Scrivener & BANKRUPT
24. A Guy, returned home to Chippenham | Snow at times all day, & excessive cold freez'g Winds, - the coldest Xtmas remembered | Cart'g Hay to Sheep & 2 load Muckle before Fold | Hay, the Sheep began the worst old Rick of Wm Philpots mornings, 5 or 6 Tons of bad Meadow hay, left for Nights

---

135 Thiller - The shaft-horse of a team. Dartnell & Goddard, *Wiltshire Words*, p.166; Rough - To make rough, applied to horses' shoes when they are made rough to prevent them slipping in frosty weather. J. O. Halliwell, *A Dictionary of Archaic and Provincial Words: Obsolete Phrases, Proverbs, and Ancient Customs, from the Fourteenth Century, Volumes 1-2*, (London: J R Smith, 1865), p.693.

| The Carpenters, having finish'd the Yard Doors & Pew for Church, have returned home till the Days are longer | James Robbins & Isaac Blackman return'd home till the weather become open & mild
Obit Ralph Hale Gaby of Chippenham Aged 80 years [This entry written within a thick black border]
25. Very cold frosty, but little Snow
26. Snow large blossoms, short time morn'g sharp frost, - Men diging a Road through the Snow on the Down. | 4 Waggon load Muckle before the Fold
27. Sunday, Very cold frosty air all day
28. ANTHY GUY OF CHIPPENHAM GAZETTED A BANKRUPT [This entry written between two thick black lines]
28. Excessive cold chilling & cuting frosty Winds and hoar frost | 5 Horses & Waggon to Box for Copeing and Portaco.[136]
29. Horses taking Hay, only to Sheep &c. | Fattening a Porker with Wheat Meal | Very cold frosty morn'g & Snow afternoon
30. Very cold freezing Day | Waggon load of Straw, borrowed of Brother W P_ for Coach Horses | Hay to Sheep
31. Very sharp frost & keen frosty air all day | 11 Qr Barley to Devizes & 34 Bu Coal for Poor
RAIN IN OCTOBER 18 days
RAIN IN NOVEMBER 9 days
RAIN IN DECEMBER 12 days

1830

January

1. Sharp Frost, Fine day, the Snow on the ground, blown over the Road on the Down a foot deep in a Night and day, Men diging it out | Muckle to Fold, & 2 Horses to Mill for W P_ K
2. Waggon to Devizes for Coal 2T 1C 2Qr | 3000 fm D_ applied 9th Decr reckoned to 11th and to 2nd Jany is 22 Dys @ [...] | Heavy Frosty day | Sow – brim'd the 2nd
3. Sunday, very cold & thaw
4. Freezing in Night & morning very slippery, and a thick mizly damp air all day | Farringdon, to meet W L__J | Muckle to Fold
5. Very thick mizly morn, and clear Sun shine after, very gentle thaw, & beautiful clear stil, light & mild eve'g | Flints, Carting from Beckhampton Furze to_ home, Road
6. Flints, Carting 20 Cart loads of 20 Bushl each in the two Days - for 10 Waggon Loads | Sharp Frost, in Morn'g, cold thaw after The Workmens Xtmas Supper
7. Rain, heavy in night, Snow disappear except in deep spots | Chalkers, began, again | Fallowing Stubbles, began again | 1st Load of Muckle fm W

136 Copeing and Portaco - Coping and Portico.

Ps Straw | Cart'g Earth for Bank White lands, fm Mead

8. 2 Teams Plough'g as yesterday | Bank, White=lands, began, making | Very sharp keen frost

9. Very sharp keen frost Attempt'd Plough'g to hard, returned home to Cart'g Chalk to widen the Road to the House

10. Sunday, Sharp Frost, little Snow in morn'g fine day | Fold, finish'd, Furze down Piece

11. Sharp Frost, some Snow | Chalk, Cart'g as the 9th | James & Isaac returned to work | The Fold return'd over the ground again the last 10 night, the Snow prevented the Muckle treading

12. Sharp Frost, heavy Snow Storms | Chalk, - Cart'g as yesterday | Barley all in Barn thrash'd

13. Hay, the bad Meadow, the Sheep finish'd | Chalk, Cart'g as yesterday | Very sharp cold frost, heavy Snow, after'n

14. Very sharp cold frost, no Snow | Cart'g Marter earth

15. Cart'g Marter earth | Sharp Frost, & very cold Snow Storms

16. Barley, Rick, from White lands put into the Barn | Dull, heavy day, some very small rain at times | Horses, in the Stable, the Carters, helping at the Rick

17. Sunday, Intensely cold & Sharp freezing

18. Intensely cold & Sharp freezing within doors | Carting Marter earth

19. Carting Marter earth to Rick yard

20. Exceeding boisterous Night, Snow drifting all Night, and all Day | Neither Thrashing or any other work the Men can do, except all hands giving the Sheep Hay, - the Fold partly fill'd with Snow as high as the Hurdles | After=noon, the Sheep brot home to Yard a prospect of their being buryed in the Fold | The Coach's the Mail turn'd over whilst lead'g the Horses (the Passengers walking) between Beckhampton Inn, - and the White House on the Down, - the Snow being the Snow being from 2 to 3 Feet deep over the Road, - the Coach drawn up with Mr Wentworths Cart Horses. - 5 Hours behind being 10-O-Clock, instead of 5 | The Monarch, Coach stop'd at the same Place 11 O-Clock, instead of 5 The Coachman & Horses returned to Beckhampton, - the Coach. drag'd to the White House by Mr Ws Horses both Coach's proceed'd to Cherhill abt 1-O-Clo | The 4 Day Coach's came to Cherhill & returned to Calne, being unable to proceed to Beckhampton &c | 2 Devizes Day Coaches came to Beckhampton & obliged to remain | Mr Guy's, 2nd Day, appearing before the Commissioners of his Bankruptcy, and to prove Debts no possibility of attending myself the Roads being impassable. [Note in margin] The Weather Glass gone back, the parts round the Dial to between, Stormy & very dry [137]

21. The Mail through Devizes the only Coach up the Road last night, and that

137 The barometric pressure was therefore likely to have been below 28.2in Hg or 955mm Hg.

at 1-O-Clock instead of half pass 9 - in eve'g and stop'd this side Silbury hill, return'd to Beckhampton Inn | Two hundred Men employ'd diging a Road in various places between Marlbro: & Cherhill, - being in many places level with the Banks 3-4 & 5 ft deep | By 3-O-Clock afternoon the Road made just passable, - and enlivened by nearly all the Coach's Night and Day, detained at Marlbro & Calne last night & this morning with 6 Horses each. pass g here from 3 to 5 | The Winds continued blow'g with little Snow till 10-O-Clock last evening, - calm Night and beautiful clear sunny mild Day SEE- Remarks on exact such weather on the 20th & 21st Jany 1814[138] Horses, moving Fold, and Hay to Sheep in the Yard The Men, helping on the Road. | STRAW Barley, the Sheep began mid=day, Hay, Night & morn'g & in the Yard all day, the Snow too deep to Fold them.

22. Rain last eve'g & Night with boisterous Winds & damp Day, the Snow much melted | Nothing for the Horses to do, & in the Stable all day 2 - Carters & 2 other Men - working for the Surveyer, making a Road through the Snow to Monckton [139]
23. Frost in morning, very slippery | Snow a little, great part of the Day | Carting the earth round the Chalk to the Moss'y part of the Down | Turned, the 2nd Sow & Brim'd again
24. Sunday, - Thick Fog, Freez'g & Thaw'g The first Lamb - one of Mr Cowards Ewes
25. Mild'r Frost'y Night, thaw'g Day | Carting Earth from Chalk Pit on Down as the 24th
26. Carting Earth from Chalk Pit on Down as the 24th | Thaw all day, & mild'r
27. Thaw all day, some Snow in the morn'g | Carting as yesterday on the Down
28. Carting Road Earth to Rick Yard | 14 Qr Barley to Devizes | Frost in morn'g - Dry day & gentle thaw
29. Frost in morn'g - Dry day & mild thaw
    Carting Road Earth to Rick yard till 10-O-Clock, when all hands call'd off, to assist in geting Wm Paradise out of the Chalk Pit, being buryed in the Earth fallen in upon him when filling the Bucket in chalking the land. Jno Chivers of Avebury & others got him out in very, great danger of their own lives at about half pass 12 O-Clock. [This entry written within a partial thick black border]
30. Snow in the Night & a Snow storm in the morn'g. Thaw & fine after | 7 Waggon load of Dung - from the Inn from Brother W Ps Straw to the Right hand of Tan hill Road and Fold on, for Turnips | The Sheep in the Yard the

---

138 On 20 January 1814, Pinniger recorded "SNOW drifed, the roads impassable - SHEEP DUGOUT of the Snow, Ewes fold in Gate Close Horses in the Stable" and on 21 January 1814 " Fine sun shine'y day, mild & wind still & yesterday the coldest of days".

139 Surveyer – the surveyor of highways responsible for overseeing the upkeep of local roads, one of the parish officers.

last 10 Days the land till now too much covered with Snow to Fold them

31. Sunday, - Damp & mild yesterday eve this morn, exceeding hard Frost & very cold freezing winds all Day

February

1. Very sharp hard Frost, the ground not hard'r all the Winter. Snow to cover the land in morn'g | 2 Lambs fell unexpect'd in the Fold in the Night, one frozen to death | Cart'g Earth to Rick Yard, for a Lambing Pen. | The Chalking, being dangerous work discontinued it, 4¾ Acres done out of 9 A
2. Exceeding cold & sharp Frost The Snow in the night much drifted, & all the morning, - and so intense to prevent the Labourers work'g out of doors. The Horses in the Stable & Men in Barn | Articles this day, freeze by the Fire & the Washing Copper lid, frozen on whilst a good fire under & the Water Milk warm | The afternoon became clearer, when drew 4 Waggon load of Dung from the Inn, and - 4 Men helping dig a Road through the Snow on the Down being fill'd again | The Sheep returned from the Fold again (being there only 4 nights) filld with Snow & too cold, to the Straw Yard. | STRAW discontinued serving the INN having none to spare
3. Pierce'g cold driving Snow all day and in the night. | The Horses in the Stable, the Carters and all the Men, assisting in diging the Snow - to make a Road for the Coaches, - which blow in as fast as it is dug out, - and the Men unable to support themselves in the cold & left off
   Mr Christr Broome of Brompton Intered at Cliffe Died 26th Jany 1830 in his 51st Yr [This entry written within a thick black border]
   The cold driving Snow, almost insupportable, & the Roads fill'd with Snow almost impassable, - the Hearse & mourning Coach nearly stop at Winterbourne, - 20 Men with Scoopes preceded the Procession from thence to Cliffe.[140]
4. The Roads, yet impassable, - all our Men & Carters, assist'g - the Horses in Stable | Very Sharp Frost & Cold day
5. The Frost last night so very sharp, have prevented us doing any thing with the Horses this morning sent the Men & Carts to the Chalk Pit on the Down, which they could not enter with Pick axes & returned home then tryed the bank of Road earth, which is equally hard, returned home & put the Horses in the Stable, - Men loading old Thatch &c. for a standing Sheep Pen.
6. Frost again if possible sharper than before, very cold all night, & all water within doors frozen up if in small quantity s and the Water frozen in Buckets a difficulty in breaking. The Horses in the Stable, - the Men removeing Faggot Pile &c. | The Frost stoping all out door labour am under the necessity of requesting 2 Men Jas Robbins & Isaac Blackman to remain at home.

---

140 Scoop(e)s - shovels.

7. Sunday, Very Boisterous Snowy night, the Snow much drifed the road from Beckhampton to Avebury fill'd & impassable | Thaw & mizly rain at times all day
8. Very heavy storm in morn'g mild thaw & dry after Carting earth from ditch to the Iron Fence

Nicholas Ponting, of Langley Died last Night Aged 82 Years [141]

Mrs Hart, late Sarah Chivers Died this Morning Aged 45 [142] [This entry and the previous entry written within individual thick black borders]
9. Rain in the night, & thin rain till 10-O-Clock, fine after | Carting Earth, before the Front of House | The Sheep taken from the Yard to the Fold, - the Snow gone & mild, the first time since 1st Feby | WELL, in the Rick Yard, Lawrence Chivers & his Son, sunk, in less than 2 Days & Steaned the top 11 feet @ 1s
10. Sharp Frost again, - & fine mild Day | The 6 Horses & 2 Waggons to Derry hill for 60 Ash Poles & [...] Oak Posts | One old Clover Rick of Wm Philpots, finishd by the Sheep, see 24th Decr began it | Mr GUYS, first Sale, at the Farm | Rt Philpot. WHEAT SOW'G, his Turnip land Plough'd two Months, Frost & Snow prevented sow'g before [Note in margin] RED=WHEAT.
11. Frost in morning, dry day, the Sun bright & Warm | Fallowing Wheat Stubbles again only 2 days Ploughing since 17th Decr being 8 WEEKS, - Frost yet in the ground | BARLEY, at Devizes Market no sale for mine, being damp, yet bright, aske 25s Qr no biding put it by in the Mow for Seed others sell from 20s to 40s pr Qr
12. Plough'g as yesterday | Cold dry Day
13. Damp dull Day | Ploughing as yesterday | COW & Calf Alderny from Mr Line
14. Sunday, dry Day
15. Foggy damp Day | Fallow'g finish'd, the 20 Acres Wh: Stubbles & began fallow'g the Chalkd Land
13. 12 Ewes, Yean'd 9 Lambs alive 1 Ewe died
25. 20 Ewes, Yean'd 16 Lambs alive
15. Robt. Bailey, Wool Stapler of Calne Died this morn'g Aged 51 Years [This entry written within a thick black border]
16. Foggy Day | Ploughing as Yesterday | Turnips, The Ewes, put on again to finish them for Wheat sowing having waited since 18th Decr by the Snow

---

141 Mr. Nicholas Ponting of Langley Burrell was buried on 15 February 1830 at Langley Burrell age 81, *Wiltshire, England, Church of England Deaths and Burials, 1813-1916 for Nicholas Ponting* [accessed via Ancestry, 23 July 2020].

142 Sarah Hart of Seend (wife of Robert Hart) was buried on 13 February 1830 at Hilmarton age 45, *Wiltshire, England, Church of England Deaths and Burials, 1813-1916 for Sarah Hart* [accessed via Ancestry, 23 July 2020].

& Frost | Hay, began the New Rick, on the Down, for the Ewes & Lambs & Cow

17. Very thick Fog & Sharp Frost, Highway Sale | Ploughing as yesterday
18. Ploughing as yesterday | Sharp Frost, Cold wind, fine Day
19. Very sharp Frost & Cold Snow Storm in the morning, fine afternoon | Ploughing as yesterday
20. Very sharp Frost, and fine clear Day Attempted Plough'g the Chalk'd Land frozen too hard, & tried the Fallows too hard also The Turnips finish'd feeding, took the Hurdles home & Carted Muckle before the Fold
21. Sunday, Sharp Frost & fine Sunny morng, heavy rain afternoon
22. Very sharp frost, Snow Storms in the day | Thwarting began the Fallows for Barley very mellow after the Frost | Labourers, began taking their Meals twice a day
23. Rain in Night & morn'g, mild & fine after | Ploughing the Turnip Land for Talevera Whe
24. Damp morn'g, mild & dry afternoon Mr Smiths Sale at Beaversbrook [...] Cows with [...] Calves, worth [...] | Fallowing, finish'd the Chalk'd Land for Swedes, The Chak'd Pits, fill'g in, in the upper Pit is buryed a Bucket cost 15s 1 Pickax & 2 Scoopes 13s 6d, 2 Waistcoats and a Frock
25. Thwarting, for Barley, the Wheat Stubbles | Fine mild dry'g day | The Men Banking, Barn empty | Mr Wentworths Ewes & strongest Lambs taken to the Swedes, a standing Pen adjoining, - his Hay, little left | Mr Geoe Browns Ewes time up, and taken to the Turnips & Swedes, a stand'g Pen adjoining for Lambing
26. Rain heavy in the Night, Dry day rain in evening. | WHEAT Talavera, sowe'd 2 Sacks @ 35s waiting since 18 Decr to finish sowg | The last sow'g Wheat just appearing sow'd 4th Decr 12 Weeks
27. Fat Pig kill'd 9 S 11 li | Fine dry Day, The Men continue Banking Whitelands, no Corn in Barn | 4 Horses to Pickwick for 80 yards of Border Stone
28. Sunday, Very dry windy day

March

1. Calm mild Day, - The Wheat brown before with the Frost, becoming quite green | Thwarting for Barley
2. Damp morn'g, very mild fine Day | 2 Teams Ploughing for Brother W, in the Down Penning for Talevera Wheat LAPWING Birds _ appear | Mr Wentworth Drill'g Pease
3. Dry mild & fine Day | BARLEY Rick, next the Barn, put into Barn | 2 Teams Ploughing for William as Yesterdy
4. The Ewes, put into Standing Pen in Rick Yard - their time out tomorrow | Sharp white Frost, fine drying Day | Thwarting for Barley
5. Thwarting for Barley | Fine mild calm growing Day

6. Fine mild calm growing Day | W & 4 Horses to Box for Coping [143] | 2 Horses Plough'g as Yesterday
7. Sunday, Sharp Frost, thick Ice fine Mild Day
8. Sharp Frost, very Cold Winds | Ploughing the old Lay, muckled & fold'd for Vetches [Pinniger then described the planting of trees and shrubs in the garden of the new house - see Appendix O]
9. Ploughing as yesterday | Rain in the Night & thin rain in the day | Freeth - Planting in Whitelands | Lambs coming - 13 in a night
10. Lambs 100 & 4 Ewes died | Ploughing as yesterday | Mild fine morn'g, cold winds after
11. Fine mild Day | VETCHES, Spring, sowed 2 Acres 4 Bu: pr A, old ones | Thwarting for Barley | Women, Picking Docks of Wheat
12. Ploughing as yesterday | Dry day, very boisterous winds after noon | Posts & Rails, Geoe Smith, puting round the Meadow
13. W & 4 Horses to Pickwick for 87 Yds of Bordering Stone | Coal Ashes, sow'g wth 2 Horses | Fine dry day | 168 Lambs - in 8 Day - & 5 Ewes Died
14. Sunday, very cold drying winds, after the morn'g
15. 1 Load of face'd Stone & a load of lime fm Calne | 2 Horses thwart'g for Barley | Cold boisterous Winds & heavy Hail storm | Flints diging in Garden Walks | FARM, Mr STILL, - began Measuring
16. Plough'g as yesterday - 1 Team | 10 Qr Barley to Devizes, & 45€ Coal home | Cold Storm & Cold Winds | CARPENTER 2 - and MASONS 3, commenced Building again | 10 Bushls Malt, Brewed 1 Hogshd Strong Beer 1 Hogshd Ale Beer 2 Hogshd Table Beer
17. Cold dry morn'g, thick Ice, warm eve 4 Horses to Honey St for Timber 2 Horses Cart'g Earth to Garden Walks &c | BARLEY, W Neate Overton, began Sow'g
18. Cold windy morn'g, warmer after 4 Horses to Tytherton for 6 Qr Wick Lime & 3 Black Thorn Bushes @ 3d ea [144] | 2 Horses - Plough'g | Market improving, by Duty taken of Beer
19. Very fine mild Day Mort [...] 3--- Transfered | 2 Teams Thwart'g for Barley
20. 1 Teams Thwart'g for Barley | W & 4 Horses to Box, for Portico | Very fine mild'r day, little rain in the evening | The Talevera Wheat sow'd 26th Feby just shewing
21. Sunday - Mid:Lent, fine & mild
22. Some showers of rain, - and mild | 1 Team to Derry hill, poles & Rods | 1 Team finish d thwart'g for Barley and Hurdles carried to Swedes The Swede greens, growing fast, and the Lambs want more food than Milk yet fear Swedes will be gone before have Meadow Grass
23. Dry cold Winds | Harrowing the Barley Land | 220 Ewes & Lambs, began Swedes Down Piece 20 Not Lamb'd 20 No Lambs 5 Died, since began

143 W - waggon.

144 Wick lime - probably quick lime.

Lamb'g [Total] 265 - The Stock before XtMas

24. Harrow'g the Barley Land as yester'y | Fine mild Day | Potatoes, the first, Plant'd in Garden
25. 3 Horses to Devizes with 12 Qr Barley and 1000 Bricks home | 3 Horses, Harrow'g for Barley | Very fine mild Day
26. Very sharp Frost & thick Ice, very warm mild & first Summer like day | Rolling & harrow'g, Barley land work very mild & kind | Ewes & Lambs 17 - and 10 Ewes too Lamb took from Standing Pen to Swedes | The Ewes do well on Swedes and drink Water freely
27. Delightful mild, calm, clear Summers Day. Frost in morning | Rolling & Harrow'g - the Barley land finish'd, fine time for it & work well | Wall for Stair Case above the Window | The Attic floor laid, and ready for Lath & Plastering
28. Sunday. Frost & warm fine Day
29. Sharp frost & very warm fine Day more like JUNE than March | W & 4 Horses to Pickwick for 94 Yards of Bordering Stone | 2 Horses, Carting Road Eearth, for filling Walks in Garden.
30. Sharp Frost, warm fine day | Rolling rough part of Barley land | Wheat, harrowing the Down Piece, and the Piece adjoining 2 Tines across FAT PIG Kill'd 11 S 4 li, see the fellow to it Kill'd 18 Dec'r 4 Score, gained in 14 Weeks 7 Score is 10 li pr Week at 7s pr Score @ 3s 6d pr W
31. Beautiful fine Day, (Highway Sale) the Ground chop'd - with the Draught Very little rain in the whole Month, and at times like Mid=Summer | Thwart'g for Swedes, and Thwarting the Swede Land for Turnips & Wheat

April

1. SNOW, before Day light to cover the Land & continued till 11 O-Clock thin rain at times, remainer of the Day | 10 Qr Barley to Devizes, & Cart'g Earth to the Flint holes in the Meadow
2. SNOW, to cover the ground again before morning, and continued, large Blossoms wth hail & rain at intervals all Day, and keen blows the wind and piercing is the cold - more Snow fell this Day, than any one day this Winter & drifted 3 feet Deep | 6 Horses to Honey Street, for a Piece of Memel Timber 34 feet long & 30 deals 21 feet long on a Carriage | Ewes & Lambs, Standing Pen made comfortable on the Down wth plenty of Hay before this would eat no Hay mornings
3. Frost morn'g, SNOW, large Blossoms and driving all the morn'g dry after | 4 Horses & Waggon to Honey Street for Oak Sleepers & Sills | 2 Horses, Cart'g Muckle to Standing Sheep Pen.
4. Sunday, Very sharp Frost, the Force Pump stop'd with the Frost fine clear day.
5. Very sharp Frost, thick Ice, Force Pump stopd again, the Country covered with Snow, fine warm clear day, by the evening the Snow Melted | W & 4

Horses to Box for Portico &c | 2 Horses to Kennett wth a load of Hurdles and a load of Hay for my Ewes & Lambs to feed of Swedes for Brother Wm | 2 Billet Men throwing up Dung in Rick Yard Standing Pen.

6. Very sharp Frost, to stop the Pump thick Ice, Fine Day | SWEDES, finish'd Feeding, began 23rd March, a fortnight 9 Acres | Ploughing for Barley, began
7. Ploughing as Yesterday | Beautiful mild morn'g & Day, rain in the evening, Alexr Pinniger App: to Mr Belcher [145] | My Ewes & Lambs, taken to Kennett to feed Bro: Willms Swedes
8. The 3rd Fat Pig - 9 Sco 5 li | Plough'g as Yesterday | Very Fine Warm Sun, mid=day
9. Ploughing as yesterday | Very Fine Day X | Good Friday, All, Labourers wth Masons, Carpenters & Sawyers attend d at Church, Avebury | X Rain heavy, all the evening
10. Frequent heavy Showers of Rain & Hail | Ploughing Swede Land for Turnips Barley land too wet | Barley finish'd thrashing the 3rd Rick & 1 to thrash, Men cleaning Hedge by Meadow
11. Sunday, frequent Showers all morn'g dry afternoon
12. Shower in morning - dry & warm after | ROLLING Grass, & Ploughing Swede & Turnip Land
13. Rolling finish'd 28 Acres & Plough'g as yesterday Very fine Day, all the Men leveling the Garden & diging Gravel in it [Note in margin] Dr R[...]g last Visit
12. SOW. farrowed 7
14. Very fine drying day | Wheat Rick the first carried put into Barn | BARLEY SOWING began, sow'd 6 Acres 5 Bu: Barley @ 28s pr Qr 5 Pecks good Ray @ [...] 6 li Broad Clover @ [...] 6 li Mill'd Hop @ [...]
15. Thin rain, warm & growing | Ploughing for Barley | 3 Men Wheeling Earth into Garden & 2 Men cleaning Hedge
16. Much rain last Night, & frequent Showers in the Day | Ploughing Swede & Turnip Land too wet, to Plough for Barley | Hay taken to Kennett for Sheep and 48 Bushl Grains returned | 5 Men wheel'g earth into Garden &c and Masons puting Coping on Wall
17. Masons puting Coping on Wall | Ploughing as yesterday | Showers in morning, fine after
18. Sunday, very fine day
19. Heavy Showers of Hail & Rain in morning - dry after | Waggon & 4 Horses to Box, for Stone for Portico 2 Horses Rolling & Harrowing off Barley sow'd 14th
20. Ploughing for Barley | Heavy Showers in morning, Horses returned from Plough, dry afternoon, & very boisterous Winds blew of the top of my 2

---

145 Mr Belcher – possibly Edward Belcher, maltster, Timber Street, Chippenham, J. Pigot, 'Commercial Directory 1830' in WRS47, p.68.

Stacks of Hay

21. Damp morning - cold dry winds after | Barley, sow'd - 8 Sacks | Finish'd wheeling earth into Garden and sow'd all small Seeds & Pease not prepared for it before | Mr Brown, finish'd feeding Swedes the 14th & began - water meadows & old=field [Note in margin] CUCKOO
22. Showers in morn'g & rain afternoon | Ploughing for Barley | Bank by Cottage Garden, began makg by the billet Men & others
23. Much rain in night & heavy Storms in day to prevent Ploughing & Men Working | W & 3 Horses to Tytherton for Iron Gates | 3 Horses, Cart'g Stones for Yard Wall &c.
24. Ploughing for Swedes & Turnips, too wet for Barley | Much rain in Night & morn'g, drying boisterous winds & Storms. | Ewes & Lambs do very bad in the dirt on the Swedes yesterday, at Kennett.
25. Sunday, fine growing Day | Mr Wentworths Ewes & Lambs began Silbury Mead, & lay back at night his Tegs on the Down, the last week
26. Fine mild day, Plough'g for Barley
27. Fine mild day, Plough'g for Barley | Brother Wm Ewes & Lambs, removed from his Swedes to his Watermeads his dry Sheep & my Ewes & Lambs remain to finish the Swedes | Front Garden Planted with Fruit Trees &c. in full leaf
28. Mr Philpots Swedes, finish'd, and began his Watermeads 8½ Doz HURDLES required for him & me each to divide the Meadow, and 30, for him, me & Mr Wentworth ea | Ploughing for Barley | Warm fine day
29. Warm fine day | Ploughing for Barley, finished | BARLEY, Finish d Sowing 19A 3R 33P by Mr Wentworths & the Down | 25½ Sacks of Barley [...] Qr of Ray [...] of Broad Clover [...] of White Clover [...] of Mill'd Hop [...] of Foreign Clover [...] Total [...] | The 7 Acres next the Down sown with 5 Pecks of Ray 3 li Foreign Broad Clover 4 li White Clover 6 li Mill'd Hop | Hay 20 Truss's took to the Sheep at Kennett for 6 Nights & 8 Doz'n of Hurdles returned from K_ to Silbury Mead to divide half the lenght of the Mead between Rt Philpot & me. The 4th & last Pig Kill'd 9 S 14 Li 9 S 5 li 11 S 4 li 9 S 11 li [Total] 39 S 14 li
30. Very fine Day | Rolling & Harrow'g off Barley | Kidney Beans Plant'd | Odd Men, Diging Flints & Gravel for Garden | Odd Men, Diging Flints & Gravel for Garden

May

1. Fine Day, - some thin Showers | Finish'd harrow'g off the Barley and Plough'g Swede Land
2. Sunday, Very fine Day
3. Sharp Frost in morn'g, Beautiful fine Day | Sheep removed from Swedes at Kennett, to my Water=Mead the Marsh part, the other part reserved for the Horses & Hay | The Old field geting very good to take the Sheep & want to be fed to fallow it for Turnips | Waggon & 4 Horses to Grove for 12 Doz

Hurdles, and 2 Horses geting Hurdles from Kennett to Silbury Mead | The first sow'g Barley, up and green and the second shewing green | Kidney Beans = Plant'd

4. Sharp Frost & fine day | Harrow'g single'y the Wheat, much fill'd with Chick weed = do good | Rolling the Barley | Hay - the last of Down Rick, brot home for Horses | Wall building up in Yard, Windows, put in Kitchen & Room over
5. Beautiful fine Day Tailing the Sheep, 2 Men, from 8 O-Cl to half pass 10 - wth a Man - a Boy & Self | 1 - Ewe died in eve'g wth Murrin [146] | 2 - Windows put in front of House | 2 - Ploughs, for Mr Wentworth (Barley)
6. 2-Ploughs for Mr Wentworth Vetch Sow'g | Very warm, much light'g in Night & Thunder 5 O-Clock morn'g, wth heavy Storms of Hail & Rain, partially
7. 9 Qr Barley to Devizes & 20 Post's home | Rolling Barley finish'd | Warm Thunder'y Day, little rain
8. WOOL, weigh'd, Mr Geoe Bailey [Details of wool are to be found in Appendix E] | Heavy Thunder & lightening 12 O-Clock and heavy rain a short time | Ploughing Swede Land & harrow'g Wheat | Wool - delivered
9. Sunday, Rain in Night and Steady rain all the morn'g till 12 O-Clock very acceptable for late sown Barley
10. 3 Horse to Box for Stone, last Journey | 3 Horse Plough'g Swede land | Rain in Night some little in the Day
11. 4 Horses to Poulton, for 12 Doz Hurdles | 2 Horses Plough'g as yesterday | Potatoes, Planting in Meadow | Hay remainer of W Philpots, from Wheat field | Turnips Mr Wth began sow'g after Swedes | Very benuming cold winds
12. Ploughing for Swedes | GRASS, The Horses began Silbury Mead with Hay & 1½ Bu of Corn pr Horse the next Fortnight, till the grass older | Mr Philpots Ewes & Lambs, left the Mead to Down & Newfield
13. Plough'g for Swedes | Damp Air, in morn'g, dry & Cold afternoon
14. Potatoes, Planting in Field, Rick botm | Very sharp FROST, to nip the Potatoes and very warm day | Thwarting the Swede Land finish'd & draging & harrow'g it | Ewes & Lambs, removed from Silbury mead, having fed the Marsh & Bank only, began the 3rd would have kept them a fortnight if had Hurdles to lay back | now fold'd on the Oldfield, the Lambs having a Fold before the Ewes and the Old field tie'd out | Posts & the Iron Gates puting up before the House
15. Drag'g & Harrow'g as yesterday & Hurdles removed from Silbury Mead | Warm fine Day
17. BAITING the Horses began on the Old field, before the Sheep
15. FLOOR, the first Chamber, laid down
16. Sunday, Warm, fine day

146 Murrin - murrain.

17. Warm, fine day | Drag'g & harrow'g the Swede land
18. Rolling & harrow'g the Couchey Swede land | Very warm Day | WATER taken to the Sheep in a Barrel in much want; on old Field | 3 Men Thrash'g Wheat, the odd work for Building, being nearly done
19. Very fine warm Day | Fallowing the old field, just fed off for Turnips | Mr Wentworths Women, helping Couch'g the Swede Land | Carpenters, put'g down Chamber Floors Masons, preparing ground Floors
20. Very fine warm Day | 3 Horses Plough'g as yesterday, and Fold removed on Fallows | 15 S_ Wheat to Devizes & 9¼ Hd Bricks retd | The 3rd WHEAT RICK, put into Barn | Charlock, in full bloom draw'g from the Wheat
21. LAMBS, - Cut & Tail'd in 2: Hours 114 Ewe Lambs 112 Ram Lambs, 3 Kept for Rams [Total] 226 got by Mr Wentworths Sussex Ram Leaving 5 Sco 9 Rams @ 1s 3d is 6s 10d Leaving 226 tails @ 1s Hd 2s 6d [Total] 9s 4d | Some Rain in the Night, dry'g day | Plough'g as yesterday
22. Ploughing athwart the Couchey part for Swedes | Rain heavy in Night, fine grow'g day
23. Continual heavy Storms with Thunder & Light'g from 10 to 2-O-Clock very acceptable, for late Barley & the Grass, Gardens &c, dry after | Sunday
24. Rain with Thunder, from 4-O-Clock to 10 in the morn'g, Dry after | Fallowing the Old field, just fed with the Sheep for Turnips & Wheat, the Swede Land too wet. | PORCH, by back Door, erecting
25. Fallowing as yesterday, & 16 Sacks of Wheat to Devizes, 800 Bricks & 10 Posts returned from Devizes | Frequent Showers of Rain
23. The HORSES BLED, 4 Qts each
26. Very cold boisterous winds with Storms | Carting road earth to fill the Gravel Pit being too wet for Ploughing
27. Heavy Storms | Thwarting Couchey spots of Swede land | Calf Sold 15 Weeks old 40 li pr Qr @ 6d pr li
28. POTATOES, hoeing in, in the Field 3½ Sacks of small, planted 70 Pole 10 & 11 rows to a Pole
27. WHEAT EARS, at Durrington near Amesbury, half out
28. Draging, Harrow'g & Ploughing for sow'g Swedes | Storms & finer day
29. Sharp Frost ICE, in the Cabbage & the Kidney Beans cut Fine day, rain in even'g | SWEDES, began sow'g Down Piece 2 holes & blank is 2 li pr Acre | Charlock in full bloom, hide the Wheat, drawing them
30. Sunday, much rain in Night very drying day, rain in eve'g
31. Damp cold day, Ray grass fit to cut, but the weather bad | Ploughing & harrow'g a drift for Turnips - Down Piece

June

1. Carting Dung, from Down standing Pen, & Plough'g it in for forward Turnips | Fine morn'g heavy Storm after

2. Fine day Glass gone up to change | Carting Dung from home to Down Piece for Turnips & Plough'g it in | Mr Jno Canning, hoe'g Turnips & Mr Davis - Turnips hoe'g, hoe'g the last week
3. Mowing, began the Ray grass, just begin'g to bloom White Lands | Fine fogg'y morng till 8 O-Clock, when very heavy Thunder & Lightening & Rain in Torrents, & continued heavy rain till 3 O-Clock & thin rain all the eveg to cause a Flood | No Market at Devizes, or the smallest remembered, not twenty Farmers present, no return made
4. FLOOD in Silbury Mead, cover my Mowing Grass, not half so great a flood since we have been at Beckhampn
3. 16 S_ Wheat to Devizes & 1000 Bricks returned | Cart to Box for Chimney Pieces
4. Rain & blow'g winds in night fine dry day | 16 Dry Sacks sent to Devizes to change the Wheat weted yesterday | Turnip land too wet to Plough Carting Earth to Garden, front of House
5. Frost & fogg'y morn'g, very warm fine day, Glass rising | TURNIPS, began sowing, Down Piece, for Wheat after | Swedes, sow'g Down Piece through the Pits | Giddy Sheep kill'd for Labourers & a Lamb died
6. Sunday, Fine day, rain in eve'g
7. Much rain in the Night and frequent Storms in the Day | Cart'g road earth to front of House to wet to sow Swedes | Willm Pearce, removed fro Monkton to my Cottage, | Horses, baiting on Down, not yet stocked, - the old field finished
8. Some rain in the morn'g, fine after | Ploughing, & sow'g Swedes | Mowers stop'd yesterday, began again
9. Ploughing for Swedes | Rain thin nearly all day, too wet to Plough afternoon | Sheep finish'd feeding the Old field, & began it the 14th May, - now began stocking the Down, & folding for Wheat after Swedes the middle Down Piece
10. Heavy Storms Swedes, finished Sowing part of the Piece. sow'd 2 holes & part 3 holes every other place in the Machine for Arial the land being rough in places, but dung'd well
4. Clover, the Horses, began, the Flood being on Silbury Mead
10. Clog weeds, diging out of the Barley, very strong [147] | WHEAT EARS, some of Farmer Philpots quite out, no appearance of mine
11. Drying at times till eve'g, when a little rain Turn'd the Swathes a second time of the first cuting grass | Thwart'g as yesterday
12. Much rain in Night, & frequent rain till till noon dry after | Women turning Dung rick yard | Thwart'g as yesterday | Rt Ph_ t - sowd his Swedes | The Portico erecting
13. Sunday; light Storms & cold dry winds

---

147 Clog-weed - Hog weed, also known as cow parsnip in North Wiltshire, *Heracleum sphondylium*. Dartnell & Goddard, *Wiltshire Words,* p.80.

14. The COW, bull'd by Mr Peplers, Hail [148] | Drag'g the thwart'd land for Turnips & Wheat | Dry windy day rain in eve'g | Field grass, finish'd mow'g, none carrd | Watts & Philpot, carry, Hay
15. Dry windy day, damp air at times & close rain in eve'g | Rolling & harrow'g the last sow'g Swedes the first sow'g up | Mr GUY'S, Sale of Furniture commenced Mr Wentws Turnips harrow'd Mr Wentws Lambs waned on Vetches | HAY, carr'd Whitelands well from 12 to 5-O-Cl before the rain | Jos Bromham, left his place & engagement | Turnips W Ps at Kennett, want hoe'g
16. Rain heavy in night and dizly morn'g, dry afternoon, Glass rising [149] | Thwart'g a Fold drift, Old Field | Turnips at Rockly, hoe'd fine & hoe'g & Swedes nearly fit to Hoe.
17. Rain in night & damp lowering day | Thwart'g as yesterday | Barley, finish'd weeding, & Couching the Turnip Land - Old field, no Haymaking
18. Glass, sinking, no Haymaking, damp windy day & rain in eve'g | Roll'g & harrow'g, Couchey land
19. Drag'g & harrow'g last Fold Drift, Old field till 12 O-Clock, when HAY, fit to carry clear'd 10 Acres by 9 O-Clock. in good order | Rain in night, dry windy day, heavy shower after 9 O-Clock, night
20. Some rain in Night, very fine day | Sunday WHEAT EARS, some of ours quite out & the Talavera bursting
21. Fog, in morn'g & very fine, shew'g a fine day, but Glass sink'g all day & Cobwebbs on the ground last night, both indicating, coming rain, the wind shifting to every quarter | The Swathes turned early, & very dry by 12 O-Clock, when began carry'g double hand'd, till 6-O-Clock when RAIN stop d with 4 load only left to finish Tan=hill P | Plough'g Couch'y land = 3rd time
22. Rain in the Night & morn'g, to keep the Horses in the Stable till 9-O-Clo: dry - after, no haymaking | Fallow'g the Old field
23. Fallow'g the Old field | Fine day, HAY, carried the remin'r of Tan hill Piece, in good condition, which finish the Clover hay
21. Dry Meadow, Jno Titcomb, began mow'g
23. Sheep WASH'D at Calston 259 & a loaf of Bread, Piece of Bacon & 5 Qt Bottle of Ale, | Warrant obtained for Jos Bromham
24. Fine day, Fallow'g as yesterday, and 18 Sacks Wheat to Devizes, 13 Deals, Laths & Coal returned | Markets rather improved, sold 50 S_ Wheat from 11 S 10 li to 11 S 13 li Nt @ 31s 6d | The 2 Hay Ricks top'd & well secured from the rain - RAIN in eve'g - 10-O-Clock | Dry meadow, finish'd Mow'g & began mow'g SILBURY Mead
25. Rain heavy in night & morn'g, & frequent heavy storms in day, very warm

---

148 Hail - This is probably Hayle Farm located between Quemerford and Cherhill, just north of the turnpike road (the modern A4) at NGR SU 022 699.

149 dizly - drizzley.

| Fallow'g as yesterday

26. JOS BROMHAM, commited to Marlbro: Bridwell, for leaving my Service | Fallow'g as yesterday | Hurdles, removed from Old field to Newfield to wean the Lambs | Dry fine day & very warm, rain in night
27. Sunday, rain in night & very heavy rain mid day
28. SHEEP SHEAR, 259 Viz 3 Rams 2 Wethers 254 Ewes [Total] 259 4 Men @ 1s pr Score 1 Boy - a Colt - beg'd 1s } began 8_ in morning finishd ¼ pass'd 7 - eve | Some thin Storms | Fallow'g Old field
29. Very fine day, | Wheat Rick, the last, put into Barn | Fallow'g as yesterday | Some Pieces of our Wheat in full Ear | The Down Wheat first sow'g not in Ear being thin & stock'd out, and the Wheat sow'd 4th Decr not in Ear, the Talevera Wheat sow'd 26th Feby, in Ear | Barley, shew'g the beard | Swedes, and Turnips, Down Piece in rough leaf | A nother Man & Boy & Mr Rutherford Plaster'g New House, in all 6 & 4 Carpenters [150]
30. Very fine Haymaking Day, carried the Hay, home Meadow, after 5 O-Clock | Fallow'g FINISHED the old field for Wheat | <u>RAIN IN JUNE</u> [...]

July

1. Rain early in morn'g, dry till 4' O-Clock when very heavy rain till 6-O-Clock eve | Mow'g finish'd Silbury mead | Waggon & 4 Horses to Etchelhampton for leg wood Posts
2. Rain in Night & heavy Storms in Day | Dung, Carting from Rick Yard to White=lands for Winter Turnips on the second Clover, - <u>The Land being to wet</u> to work the Couchey part for early Turnips | The <u>LAMBS, waned 225</u> - on the second Clover White=lands & fold'd there, the Vetches to wet for the Lambs
3. Very heavy rain in Night & morning & frequent heavy Storms, with very heavy Thunder & Lightening in the day | The <u>Land too wet</u>, for the Horses to do any kind of work on <u>the Farm</u> | Carting Chalk to Rick Yard
4. Sunday, Dry morning, storms afternoon.
5. Fine & very drying Day HAY, Carried 1 load the first from the Silbury Mead, mowed 10 Days | Chalk, Carting to Rick Yard
6. Dry morn'g till 10-O-Clock, frequent Storms remainer of the day | Carting Dung to Whitelands & the Women spreading it | <u>Vetches</u>, put into Cages & the Lambs began feeding them, morn'g & night, and on the second Clover, mid day, sowd the 11th of March, a good Piece
7. A Tempestuous Storm of rain, mid day & frequent storms in the day | Ploughing Whitelands for Turnips, one earth | The Women spread'g Dung
8. Fine after the morning, Rain in the eve'g | Hay carried one load & carried 3 Loads being the remain'r in the wet, to make it dry in home Meadow. Ploughing as yesterday | Rolling & Harrow'g the Turnips Down Piece

150 Mr Rutherford is probably Thomas Rutherford of Calne – see entry for 24 September 1829.

make very good work | TURNIP, hoe'g began, the above, by 2 Men from Bremhill John Ponting & Jas Townsend | 2 Masons, turning a new Arch over the road in the Meadow 4 Carpenters at [...] pr Day 3 Plasterers at [...] 2 Plasterers Servers at [...]

9. Fine day, some thin Showers, and appearance of rain | HAY. carried that brot from Silbury mead & finish'd Hay making | Ploughing as yesterday, & rolling &c as yes [151]
10. Fine day, except a Shower | Press'g Whitelands for Turnips
11. Sunday, Rain in morn'g & heavy Storm afternoon
12. Whitelands sow'd to Greenround Turnips | Ploughing athwart the Vetch Lands & the two adjoining Pieces | Horse the Bay, false & mang'y turned to grass for Sale
13. Fine Summers hay making day | Harrow'g - burning & cleaning, Couch'y land | Rolling Swedes, the Horse Rake tie'd behind make good work, Swedes too small to harrow | Toping the Hay Ricks
14. Toping the Hay Ricks Rolling & harrow'g White lds [152] | Horse raking the Swedes, after the Hoers making good work | Rolling & harrow'g, before Couchers, and Ploughing after | Very warm drying & windy day
15. Thin rain in morning, very fine after and windy | Plough'g as yesterday, and harrow'g and Horse raking Couch | Potatoes in the Field earthing
16. Potatoes in the Field earthing, Plough'g &c. as yesty Very fine Summers Day | To Downend, Stone Quarry, from Beckhampton 33 Miles Chippenham to Ford 6 Miles Ford to Marshfield 4 Miles Marshfield to Bridge yeate 7 Miles Bridge yeate to Mangotsfield 3 Miles Mangotsfield to Downend Quarry 1 Mile | Bridge Yeate to Hall lane Colliery ½ Mile Hall lane to Wellsbridge Turnpike (HILL) and Brokam hill 1 Mile return to Bitton 1 Mile, thence through Upton to Lands down 4 Miles
17. Ploughing the Couch'y land | 2 - Horses & Cart to Box for Covering to the the Portico | The Swedes, - finish'd hoe'g 1st time The Turnips finish'd hoe'g 2nd time - Down Piece * | Dry morn'g, rain mid day, & at times till night
18. Heavy rain in morn'g, and at times through the Day | Sunday Tockenham
19. Thwarting Couchey Land, till 1-O-Clock | Started 5 Horses Waggon & Cart at 8-O-Clock eve to Downend 33 Miles for for Paveing for the Hall
19. Fine day, partial Storms
20. Partial Storms | Horses on their Journey, to Downend
21. Horses returned home by 1-O-Clock | Fine & dry all day
22. Rain heavy in the morning, dry after | 2 - Horses Plough'g as last day, and 3 - Horses to Devizes wth 22-S-Wheat
23. Plough'g as yesterday, FOLD removd frm Down Piece | Very hot fine day rain in eve'g

---

151 yes - yesterday.

152 White lds – Whitelands.

24. Dizley rain in morn'g fine after | Drag'g - the rough Couch'y land | Mowing the Old Clover, too much left for | the Horses, now geting too old, and the young Clover ready.
25. Sunday, Very hot close Day
26. To Salisbury Assizes, - Very hot close Day | Rolling & Harrow'g - Wheat land | Mowing, began, Silbury=Mead
27. Plough'g & harrow'g as yesterday | Very hot ripening Day
28. Ploughing & drag'g Vetch land fed off | Very Warm fine day | SWEDES, began hoe'g - 2nd time
29. Thwarting the Fallows began & 10 S-Wh: to Devizes & 10 Deals retd | Very ripening fine day, Market lower | HAY, - Carr'd first from Silbury Mead.
30. Mow'g - finish'd [...] & Carrd Hay | Thwart'g as yesterday | Very warm & heavy Thunder Storm, partialy | Vetch's Lambs finish'd, about 2 Acres began 6th July Carpenters, puting up Sheds, back of House Masons. Paving, Hall, Court &c. } this week
31. Lambs, began uppr Down, very good, not stock'd before, fold'd Down Piece | Very fine ripening Day | Thwart'g as yesterday | HAY, finish'd, carry'g Silbury Mead well without rain | Second, Clover, - began Mow'g for Horses

August

1. Sunday, - Fine warm morn'g, rain at times all afternoon
2. CARPENTERS, 2 remain'd at home, out of the 4 - work geting nearer a close | 3 Masons & 2 Boys, - Paving Court & Plaster g | Some thin Showers | Thwart'g Fallows, for Wheat, very foul | Nag, borrow'd of Mr Rt Large
3. Ploughing as yesterday | Some flying Storms, dry'g afternoon
4. Very fine Day | Plough'g as yesterday
5. Plough'g as yesterday | Very fine day
6. Very fine day cool air | Thwarting finish'd the Piece by Tan hill road for Wheat
7. Fine day | Ploughing began for Wheat sowing the middle Down Piece | Reaping, Mr Wentworth began the first in this Parish my Wheat having Stock'd out is yet green not fill'd, & likely to blight *
8. Sunday Rain in morn'g, dry after
9. Very heavy Thunder Storm, mid=day & Storms after | Plough'g for Wheat as the 7th
10. Plough'g for Wheat as the 7th | Lambs, finish'd Fold'g - Down Piece & Fold removed & adjoined the Ewes Fold. Rt hand of Tan hill road | R Philpot, Sow'g Turnips for Wheat after Seed Ray, Muckle'd & Fold'd many other People feeding Turnips a Month Pass'd for Wheat | Some heavy Showers
11. Fine morn'g, Rain all the evening | Ploughing finish'd Down Piece for Wheat | Lambs on Down still good, and a tie of Clover evenings, Ewes follow next day | Reaping generally round Calne

12. Drag'g for Wheat Rt hd of Tan hill Road | Frequent Showers
13. Rain heavy till 11 O-Clock, dry after | Horses in the Stable, all morning, Cart'g Chalk in Rick Yard, afternoon
14. Dry morn'g, very heavy rain all the afternoon. Flood, expected at Chippm | Horses drag'g as before, stop'd at 2-O-C
15. Sunday, some thin Showers
16. This morning, People in general began began Reaping, that had not began before, ours still unripe & the Down Wheat blighted, being thin & stock'd out | Men preparing Rick Yard, for Harvest | Turnips, began hoeing Whitelands | Plough'g the Vetch land fed off athwart 2nd time, being foul | Fine morning, frequent heavy Showers afternoon
17. Some heavy Showers | Ploughing as yesterday, and other couchey spots
18. Ploughing the Couchey & Chalkey brow | Very fine ripening Day, very dry'g winds, the best day yet for carry'g Wheat, Mr Wenth began Carry'g | REAPING, began afternoon the last probably in the Country ripening very fast 3 Men & five Women - our own People & 4 Men from Calne, on Down Piece | Drag'g Turnips, Whitelands
19. Fine day. Wheat universally Ricking One Sample Sack of New Wheat only in Devizes Market (Fay) sold at 31s 6d Sold my last Wheat 11S 11li @ 32s every prospect of Wheat being cheaper Ploughing as yesterday 2 Masons finish'<g>d, dress'g off the Portico having been 5 days at that & put'g the top on, Carpenters, lay'g best Parlour Floor
20. Frost, in the morn'g, cut the Potatoes | very fine harvest'g day | Plough'g as yesterday | Lambs trim'd their Heads & Tails for Marlbro Fair | BEWARE of reaping Wheat TOO=GREEN since some Farmers have seriously injured their Crops, by so doing [Pencil note in margin] H C finishd
21. Very fine day Wheat ripe only in part, part of the Lands, remain to ripen | Plough'g the Couch'y Spots finishd & began drag'g them
22. Sunday, Fine morn'g thin rain all the evening
23. Frequent thin Showers in morn'g drying afternoon Drag'g as the 21st 2 Carpenters, now finish'g, Door Ways &c 2 Masons & 1 Boy, Plaster g eves &c, and cleaning front of House
24. 2 Men fixing the Spouting round the House | A shower early in morn'g fine after, about half our Wheat cut whilst Mr Wentworth nearly finish'd carry'g | Drag'g as yesterday
25. Fine morn'g, frequent heavy Storms after WHEAT began carry'g, 1 load & stop'd with the rain Draging Wheat land the middle Down piece, for the Press'r to follow Rt Philpot, Sow'g Turnips for Wheat after Grass Seeds
26. Draging as Yesterday, til 12 O-Clock when began Carting Wheat, made 1 Rick 15 Loads by 6 O-Clock, when a Thunder Storm, stop'd Carrying - Very drying morning - Rolling & harrow'g - Couch'y land [Note in margin] CALF GONE 44 li / 4 pr Qr
27. Summons from Messrs J T Gunning Thos M Crutwell & Robt Savage

Commissioners in Anthy Guys Bankruptcy | Fog'y morn'g, Wheat wet, fine till 3 O-Clock, rain after, & very heavy all the eve'g Draging finish'd Down Piece and began Press'g it for Wheat Finish'd mow'g 2nd Clover, and BARLEY, mow'g began, Jno Titcomb

28. Heavy, rain in morn'g, Reapers leaseing, - drying & heavy Storms alternately, through the Day, Glass - sunk to C, in much rain Plough'g Wheat land thin geting green homeward end. | Stair=case, first flight put up
29. Sunday, some rain in the eve'g
30. Frost, in morning. Glass return'd to A in change, fine day yet a thin Shower in eve'g, some Wheat Ricks half made Saturday, better now in the Field | Pressing, finish'd the Wheat Land Down Piece, - Harrow'g Turnips, for hoe'g 2nd time Whitelands
31. Plough'g, finish'd as 28th | Frost in morning, very fine day Glass rose to near F in fair | Reaping - finish'd

September

1. Frost in morn'g fine day Wheat Carting began by 6 O-Clock morn made the 2nd Rick & ¾ of the 3rd
2. Rolling & harrow'g Couchey Land till 10-O-Clock, when began Cart'g Wheat & FINISH'D, by 8 O-Clock, & began Thatch'g the Talevera Rick, when an ECLIPSE of the full Moon stop'd the Thatcher, but work'd on by Candle light & finish'd it by Moon light after 12 O-Clock | Very thick fog in morn'g & little rain at 11-O-Clock at night, very fine day
3. Very fine day, thin rain in eve g | Rolling & harrow'g, Wheat field | The Old Barley, Rick, put into Barn | COLT bought of Thos Martin Bitton COLT bought of [...] Pearce of Easton (23rd Augt
4. Very fine drying day | Rolling & Harrow'g in morn'g | BARLEY, began Carrying about 6 Acres
5. Sunday, Rain heavy in night fine day, thin rain in eve'g
6. Very windy morn'g, & Storms after | Carr'd a Load of Barley & topd the Rick & Thatch'd it | Hurdles removed from Clover to Silbury Mead | Cart & 2 Horses to the Low, for Sand Stucco'g the Rooms | Clover too old for Horses, mowed another piece for Hay, the Barley not quite ripe stop'd mowing
7. Barley mow'g again riper since 4th | Heavy Storms partially, with Thunder & Lightening, favorable with us, made the 2nd Barley Rick
8. Very fine drying Day | Clover, made a Rick of 2nd Crop, too old & too much for the Sheep & Horses | 3 Men & 1 Boy, began Stucco'g the House
9. Rain all the morn'g & all the even'g | Plough'g foul Spots of Wheat field | Examination [Pencil note in margin] Barley finishd Mow'g
10. WHEAT SOWING, began, the middle Down Piece part, about 6 A with 17 Bu = of own old Wheat Limed & Urine. - good work | Some light Storms & Sunshine | Lambs, began feeding Silbury Mead

11. Dry day, - Ewes tread'g the Wheat sow'd yesterday, good work | Plough'g - the white brow for Wheat
12. Sunday, - very heavy rain in the Night & heavy Storms in Day
13. Dry Night Ewes finish'd tread'g the Wheat 6 times on a Land 3 hours tread'g 3 Acres | 30 - Cul Ewes, drawn, and the 6 Rams uncoupled, put on TURNIPS, Down piece (began & good) | Plough'g - the Vetch land 3rd time for Wheat, - very Couch'y | 1 - Carpenter fiting up doer frames &c 2 - Masons, fixing Chimney Pieces, Grates 3 - Masons Stucco'g - Cornish'g & c. [153]
14. Rain heavy in Night - dry'g day, with partial Storms, some People, Carting Barley in a Cold State | Plough'g as yesterday
15. Plough'g as yesterday | Some heavy Storms, dry morn'g
16. Very heavy Storms | FINISHED Plough'g the Couch'y spots of the Wheat land, fine dry weather wanted for harrow'g & cleaning it Women continue gathering Docks from it all the Summer
17. Fine fog'y morn'g, mild calm & warm Glass rise'g gently, a better prospect of dry weather, some little rain Carting long Dung, before the Fold on the Chalky brow for Wheat & Spit Dung on light spots, after the Fold [154]
18. Heavy rain again in the Night dry morn'g, Storms after | Carting Dung above the white brow after the Fold for Wheat
19. Sunday, Dry morn'g, heavy rain all the eve'g Barley - Sprouting, in Swath Mrs [...] Seager [...] [155] [This entry written between two thick black lines]
20. Rain heavy in the night, dry'g morn very heavy storm after noon | The Wheat land too wet to harrow or Plough, | Cart'g road earth for Whitelands bank | Cart'g Chalk to Cowhouse
21. Cart'g Chalk to Pig sty & Skilling | Very heavy rain in night, dry morn & heavy storm afternoon, Glass below much rain | Willm Coleman, Dig'g Flints
22. Frost & very fog'y morn'g, very fine day Forsbury, much Barley cut | Grass, Silbury mead, began mow'g for the Horses, finish'd the Clover | Chalk, Cart'g to Rick Yard & road | Many People carry'g Barley, prospect of fine

---

153 Cornish'g - cornicing.

154 Long dung - Fresh unrotted dung with straw therein. Thomas Potts, *The British Farmer's Cyclopaedia: Dung* (London: Scatherd & Letterman, 1807). Spit dung - That sort of manure which has undergone complete fermentation or putrefaction and is reduced into a somewhat earthy state, so as to be dug or taken up by the spade or shovel in a sort of spit manner (Abraham Rees, *The Cyclopaedia, or Universal Dictionary of Arts, Science and Literature, Volume 33*. London, Longman, Hurst, Rees, Orme & Brown, 1819) under Spit-dung.

155 Mary Seager of Highway (wife of William Seager) was buried on 25 September 1830 at Highway age 71, *Wiltshire, England, Church of England Deaths and Burials, 1813-1916 for Mary Seager* [accessed via Ancestry, 23 July 2020].

tomorrow | Mr Wentworths Rams, to Ewes [Pencil note in margin] W Robbins Pit

23. Very heavy rain in Night & frequent heavy storms in the day | Cart g Chalk to the Road & to the INN for a Stable
24. Violent Storm of Hail, rain & Thunder & frequent heavy Storms | Cartg Dung before the Fold
25. Very dry'g day, some thin Showers many People, carry'g Barley | TOO wet all this week to work the land | 10 Qr Barley to Devizes 1000 Bricks to Inn & 16 Bu: Lime to Inn | Cart g Marter Earth to Inn - & Road | STRAW, the first to the Inn
26. Sunday, Dry day, thin rain eve'g
27. Glass raised to F in fair, fine day except a little thin rain 10-O-Clock | BARLEY, Carry'd remainer, and finish'd Harvest about 6½ A
28. 3 Horses & 2 Waggons, helping Mr Low Barley harvest | 4 Horses, help'g - Brother Wm harvest Draging in the morn'g | Rain bow, in fog & thin rain morn'g fine till 4 O-Clock, rain all eve'g
29. Plough'g, Tan hill Rt hand Piece began Mr Lows Team, helping } for Seed | Dry day
30. Sharp frost, fine day | Ploughing as yesterday till 11-O-Clock when 4 Horses to Kennett & 3 help'g Mr Low Barley Cart

October

1. Thin rain at times all the morn'g, dry aft | Ploughing for Wheat | Nag, returned to Mr Rt Large
2. Close rain in morn'g & Storms afternoon | WHEAT, began sow'g the Piece Rt of Tanhill road 4 Sacks on 5 Acres | Carting Dung before the Fold
3. Sunday, little rain, morn'g & afternoon
4. Very sharp frost, fine day | Ploughing a Fold drift on the Chalk'y brow for Wheat | 3 of my Ewes turned to Mr Wentworths best Ram | The Lambs finish'd the Marsh of Silbury Mead
5. Very sharp Frost, & very fine dry'g day | Nearly all the Farmers in the Parish have Barley, not harvest'd | Sow'd the Fold drift, Plough'd yesterday & Carting Dung, before the Fold | Painter, began Yard Doors &c | Wm & Jno Robbins, Tiling at the Inn, Privy
6. Frost & fine day, dul morning W P of K_ and Rt P_ of B_ finish'd Barley Harvest [156] | Plough'g for Wheat Sow'g
7. Plough'g for Wheat Sow'g & 10 Qr Barley to Devizes | Very fine harvest day
8. Very fine very thick fog in morn'g | Plough'g as yesterday
9. RAM'S, the 3 Lambs from Mr Wenth Stock, turn'd to the Ewes | Wheat sow'd 4 Sacks - on about 5 A | Very fine day | Hurdles, removed from Silbury Mead to Turnips - Down Wheat=land | The Masons, having nearly

---

156 Rt P_ of B_ - Robert Philpot of Beckhampton.

FINISHED the HOUSE, gave them all a finishing Supper [Pinniger then recorded prices at Weyhill Fair on 9 October and Wilton St Giles Great Sheep Fair on 13 September - see Appendix H]

10. Sunday, - fine dry day | Lambs, began the Down Turnips
11. Rt Butler & James Hillier, puting up the Bannisters | 5 Loads of Dung, from 5 Loads of Straw from the Inn, before the Fold | Remainer of Hurdles fm Silbury Mead | Fine day | Ploughing for Wheat
12. Ploughing for Wheat | Very fine day, 1 Load Gravel to Avey [157] | Mr Wentworth, lent of 2 of his Ram L [158] | Very fine day
13. Extraordinary, fine mild weather | Plough'g as yesterday, dig'g Gravel
14. Potatoes, dig'g in Field | Plough'g as yesterday & 13 Qr Bar'y to D [159] | Sharp Frost & very fine day
15. 2 Teams, Plough'd up to the Fold | Very fine day
16. Very fine day | Potatoes finish'd in Field dig'g 19 Sacks | 2 Teams Plough'g for Mr Wentworth
17. Sunday, fine day
18. 2 Teams Plough'g for Mr Wentworth now out of Debt & 1 day over | Warm clear sun shine, like midst of Summer, foggy morn'g | Fold finish'd Tan hill Piece & put on Down Potatoe Land
19. Rain a little in the morn'g, and a little in the Day, Glass sinking | Ploughing the last fold drift
20. All the GRASS, mow'd for the Horses in Silbury Mead, & the Ewes put on it | Plough'g finish'd Tan hill Piece | Beautiful fine Day 17½ Sacks Purple Potatoes from Down 6 Sacks White Apple Potatoes from Down 7 Sacks Mag Pie Potatoes by Wm Pierces 16 Sacks Prolifficks Potatoes by Wm Pierces 3 Sacks from Garden [Total] 49½ | W P of K, Sold Ewes 21s Lambs 13s | Estate at Avebury Sold to Mr Munday of Wedhampton
21. Very fine day, Glass rising | Wheat sow'd 3½ Sacks on last Fold drift The bottom of the Piece too dry to sow | Clover, the last cut for Horses, and began GRAINS
22. Fine morn'g, rain afternoon | Wheat Rick, the first put into the Barn, cover'd up in the storm | Press'd about 3 Acres of the lower end of Wheat land T_hill piece to dry to sow | Pew - finish'd 8 days for one Man | Painting Doors, Shutters &c 1st time
23. Fine day | FALLOW'G Wheat Stubbles began | Plasterers - FINISHD
24. Sunday, fine day
25. 5 more Horses to the Inn | Thin rain nearly all day, Horses in Stable | Wheat began thrashing
26. Rain heavy in the night, drying day except a storm mid'day | Wheat, finish'd sow'g Tan hill Piece = good work

---

157 Avey – Avebury.

158 Ram L - Ram Lambs.

159 D - Devizes.

27. Cart'g earth by the Road in the Meadow & fill'g the Gravil Pit in the Meadow | Plants - set 200 - for Spring Cabbage - Fine morn'g, Showers in the day | Carpenters, hanging Sashes & fastening
28. Rams, returned Mr Wentworths 2 La & own best old Ram put to Ewes [160] | Carting earth to Pit as yesterday | Fine cold day, rough winds
29. Fine cold day, rough winds | Carting earth as yesterday finish'd
30. Fine day | Wheat. sow'd Potatoe Land, Down | Sheep, fold'd on last sow'g Wheat not dung'd before (T-hill piece) [161] | CARPENTERS, finish'd the inside of the House
31. Sunday, dry cold day

November

1. Fine day, Fallow'g Stubbles
2. 20-Sacks Potatoes fm Calne, and 48 - Bu: of grains fm Kennett Fallow'g & fine day [Note in margin] 3 Small Pigs put to fatten
3. Fallow'g Rain, heavy in even'g too wet for the sheep in Fold on the Wheat, removed to the Turnips
4. Heavy Storm, mid day & after | Fallow'g & 11 Sacks Wheat to Devizes & Gate Timber return'd | Turnips - on the Down finish'd feed'g, began 13th Septr with 30-Cul-Ewes & the Rams & 10th Octr with 125 Lambs [...] Acres, some of the Turnips weigh'd 15¼ li
5. Fine day, rain in the evening | Plough'g Turnip Land for Wheat | HAY, the Stain'd Meadow, the Sheep began evenings only | The Painter, left | The Fold, began the Old Field, top of Tan hill Piece, left hand | 48 Bu: of Grains fm Kennett
6. Very boisterous day, & driving rain all day, very heavy in evening | Horses, in the Stable all day
7. Sunday, dry day FLOOD
8. Dry fine day | Wheat Sowing FINISHED, the Turnip land 4¾ Sacks Old Wheat 17 Sacks New Mrs Edwards of Wilton ½ Sacks Mr Crook, Corton [Total] 22¼ on 27 Acres
9. Fine day, sharp frost, thick Ice | Fallow'g Stubbles
10. Fallow'g Stubbles Very heavy Storms & very heavy rain in evening | Rt Philpot Sow'g Rye | Last sow'g Wheat, Tan hill piece appear'g
11. Fallow'g, - Heavy Storms
12. Fallow'g and Cheese & Wine fm Cleveancy | Fine dry'g day
13. Cold boisterous winds, in the morn'g heavy rain in the evening | Fallowing
14. Sunday, dry day - Ewe died
15. Fallow'g - fine dry'g day
16. Most tremendous Winds & rain all day, Horses in the Stable
17. Fine day, - Fallow'g

---

160 2 La – Two lambs, in this case, two ram lambs.

161 T-hill – Tanhill.

18. FROST sharp in the morning dry cold day | 1 - Ewe died Sunday last from a bite of the Sheep DOG, poisen'd & mortifyed in the Shoulder and another dieing ★ | Stock of EWES 253 Rams 3, Ram Lambs 3 Lambs 123 Wethers 2 [Total] 384 [162] ★ | Fallow'g Stubbles
19. Fallow'g Stubbles | Very sharp frost, long Icles, fine | Banking began the 17th - top of Whitelands | Carpenter making Gates 16th | Mason Colouring end of House 19th | First Pig Kill'd Porked - 100 li | Hay, Sheep, twice a day
20. Rain heavy, afternoon | Fallow'g
21. Sunday - dry morn'g, rain afternoon
22. Dry morn'g, Thunder & rain afternoon | Fallow'g Stubbles | An other Ewe died with a stopage
23. Fine day, Fallow'g
24. Fine Sharp Frost - & Fallow'g | 20 S_ Wheat to Devizes & 2T 2C 2Qr Coal retd | SOW, Brim'd, Mr Popes Boar
25. Fallow'g athwart the Down Stubbles [Pinniger then wrote a short article entitled "Alarming Times" - a local account of the so-called Swing riots - see Appendix P] Sharp Frost, fine day Small Market at Devizes in consequence of the Thrashing Machines being distroyed last week, and the Farmers with the Cavelry scouring the country for the Ring leaders of the Rioters
26. Assisted with 200 Horse Men & the Cavelry in taking 4 of the Rioters scoured the Country from Rockley to Winterbourne, Cliffe, Corton, Hillmarton, Lyneham & Wootton=basset | Fallowing as yesterday | Very hard frost, cold fine day The 3rd Cart horse sent to Tytherton for T Hillier, lost by them the 1st cost 30£ sold for 24£ [Loss £] 6 | the 2nd cost 32£ sold for 23£ [Loss £] 9 | the 3rd cost 32£ sold for 20£ [Loss £] 12 | [Total Loss] £27
27. Assisted as yesterday at Oare Pewsey, Milton, Easton &c. took 7 Prisoners | Very cold, dry day | Fallow'g as yesterday | Carpenters, Puting up Whitelands Gate
28. Sunday, very heavy Rain last last night & damp morn'g
29. Rain in the night, thick air dull day | Carting Muckle, before the Fold | 1 - Carpenter only come to finish the jobs - : Sheep Troughs &c | 4 other Pigs put to fatten
30. Storms all the morn'g - dry after | Fallowing the Down Stubbles

December

1. Fallow'g the Down Stubbles | Very cold thick air
2. Cold dry day | Rt Philpot sow'g Wheat after seed Vetches & Pease | Fallow'g as yesterday
3. Fallow'g as yesterday | Foggy, Day | FALLOW'G=FINISH'D - the Wheat

162 In counting his stock of sheep, Pinniger noted that 12 ewes had died out of an original total of 265 (see entry 23 March above).

Stubbles, no more Ploughing till Plough the old field in the Spring for Vetchs

4. Carting Dung, behind the Fold not Muckle'd before | Fine dry mild Day | James Hillier, the Carpenter having quite finish'd, left us, no Tradesmen from this day at work | The last sow'g wheat appearing, after Turnips on the Down
5. Sunday - Dry cold day
6. Very boisterous night with rain, Glass low, dry mild day | Carting Muckle from the Yard, before the Fold, - on Oldfield
7. Dizly rain, and damp Day | STONE, began bringing from Calne, 3 Waggon load wth 6 Horses, for Rick Yard Wall & 48 Bu: Grains from Kennett
8. Dry mild day | EVER=GREEN &c. Shrubs, planted | Carters & Boys, unload'g Stones &c the HORSES, resting, no field work to do
9. Very heavy rain in the night, have not seen so much water on the land rain all the morn'g, dry after | 3 Waggon loads of Stone fm Calne
10. Horses in the Stable, except fetch'g 48 Bu: Grains fm K | Dry mild day
11. Dry mild day | 3 Waggon loads of Stone fm Calne
12. Sunday, Piercing, cold storm of a little Snow in morn'g & Sharp Frost
13. Exceeding Sharp Frost, & fine Day | 3 Waggon load of Stone fm Calne
14. Very sharp Frost - & fine Day | Horses in the Stable 1st Wheat Rick, finish'd Thrashing
15. Frost in morn'g, rain all the eve'g | 3 Waggon load of Stone fm Calne
16. 29 Sacks Wheat to Devizes & 24 Bu: Coal retd for Labourers | Cold storm wth a little Snow in morn'g mild remainer of day & freezing eve'g
17. Sharp Frost, rain commenced 10 O-Cl & cold damp remainer of Day. | Barley, put a small cut of a Rick into the Barn, having no Thrashing to employ the Men & want the Straw for the Sheep | Carting Muckle from the Inn before the Fold
18. STRAW, the Barley, the Ewes began in the Yard, morn'g - the Down after and Hay evenings | The 2nd Porker, 7 Sco 2 li | 3 Waggon load of Stone fm Calne | Very sharp Frost
19. Sunday, Very sharp, rain in the evening
20. Slight Frost, in the morning dry mild Day | 3 Waggon load of Stone fm Calne
21. Very sharp Frost, fine day & thaw | Barley, Rick put into Barn the first | A=GUY'S, Estate this day Paid his Crediters 1s in the Pound which require Twelve Thousand Pound [163] | Horses, - resting in the Stable
22. 3 Waggon load of Stone fm Calne Roads exceeding heavy | thaw & very mild - rain in evening | Grubing - the 6 Elms, in Rick Yard
Mrs Jacob Large of Lyneham Died [This entry written beneath a thick black line]

---

163 See discussion of this point regarding Guy's creditors under **Events recorded beyond the farm** in THE BECKHAMPTON YEARS.

23. Frost & some Snow in the Night, & Snow nearly all the Market at Devs | Cart'g Muckle & Hay before the Fold | Old Hay, the 2nd mow'g the Sheep began
24. Exceeding Sharp Frost, freezing all day, & bitter cold Snow storm in the morning | 3 Wag'n load of Stone fm Calne | The Ewes in the Yard all day Straw twice & Hay in Fold Night & Morn'g The Lambs in Fold all day Hay thrice
25. Xtmas, Fine clear day, Snow cover the ground & Intensely cold
26. Sunday, The sharpest Frost this morn'g - the Pump, being frozen and not last year | Dry freezing Day
27. Little more Snow in the morn'g, and a little in the day | 3 Waggon load of Stone fm Calne
28. DEEP SNOW this morn'g, 6 Inch's on the level & heavy Snow in the day & freez'g, Horses in Stable
29. Some Snow in the Night, and Thaw after the morn'g, Horses in Stable
30. Thaw all day | 9½ Qr Barley to Devizes & 40 Bu: Coal back for the Poor
31. Tempestuous rain'y night, Snow all disappear'd by morn'g, dry windy day, Horses taking Hay to Sheep

1831

January

1. Very sharp Frost, fine clear mild Day | Carting Muckle - before the Fold
2. Sunday Thaw & dry day
3. Thaw & dry day | 3 Waggon load of Stone fm Calne
4. Horses taking Hay to Sheep | The Lambs, - began the Old Rick | The Ewes began the New Rick second cut | Hay the Sheep began 5th Novr the stain d Meadow, now finish'd
5. Dry Day
5. Dry Day and thick Fog | Horses in Stable, Men throw'g Trees
6. 9 Qr Barley to Devizes & 2T 2C 1Q Coal | Frost in morn'g fine Day
7. Frost very sharp & very fine Day | 3 Waggon load of Stone fm Calne
8. Very sharp quitet Frost, fine day | Carting Muckle before the Fold
9. Sunday - Sharp Frost, fine day
10. Thaw and damp day | The first 3 Waggon load of Stone to the Inn, for Garden Wall fm Calne *
11. Carting Muckle & Hay to the Fold | Sharp frost & fine day
12. Slight Frost & mild thaw, | 3 Waggon load of Stone to the Inn
13. Foggy damp day Horses in the Stable
14. 3 Waggon load of Stone to the Inn | Cold dry day | Threats & Fires, continue
15. Gentle Frost, fine Day | Horses in the Stable | The 6 Elm TREES, in the Rick Yard finish'd throwing | The 1st Barley Rick = finish'd Thrash'g the 13th Inst

16. Sunday, gentle Frost, some rain afternoon
17. Mild Thaw, damp Fog | 3 Waggon load of Stone to the Inn Kill'd the 3rd Pig 8 S 6 li} Kill'd the 4th Pig 8 S 8 li} sold 8s 6d Kill'd the 5th Pig 9 S 14 li [Total] 26 S 8 li
18. Dizly mild day | The 6 Trees got to Saw Pit, from the Rick Yard A second Lamb died in the Mur [164]
19. 3 Waggon load of Stone to the Inn | Damp & very mild Day
20. Dizly rain, all afternoon | 10 Qr Barley to Devizes & 50 Bu: Grains returned | Cart g Hay & Muckle to Sheep
21. 3 Waggon load of Stone to the Inn | Thick Dizly rain all day
22. Rain heavy in night and all day | the Sheep tuck and do ill, continually wet, the Barly Straw use'd, the Ewes must now go on the Down, till can put a Barley Rick in Barn. Carting Marter earth to the Inn and the Roots of Trees - drawn out
23. Sunday, Rain great part of Day, and Flood
24. Snow in night & heavy Snow in morng and at times after | 3 Waggon load of Stones to the Inn
25. Carting Hay and Muckle to Fold | Sharp Frost in night and freezing all Day
26. Very sharp Frost & very fine clear freezing Day | 2nd BARLEY Rick put into Barn | Horses in Stable
27. 3 Waggon load of Stone to the Inn the last for the present being 24 Lods for the Garden Wall | Very cold blow'g Snow and Rain
28. Waggon & 4 Horses to Cowitch for Black Thorn Bavins [165] | Very sharp Black Frost
29. Very sharp Black Frost | Carting Muckle & Hay to Fold
30. Very sharp Frost & very cold | Sunday
31. Sharp Frost, and Snow commenced 9-O-Clock, and in the eve'g very drifting & exceed'g Cold | Yesterday, a fourth Teg Sheep died with the Murrain, and this morn'g a Ewe died of the same know no remedy, but puting Tar about the Fold to smell [...] | 2 Waggons sent to Devizes for 120 Bu: of Coal Ash's @ 3d pr Bu Turnpike Free

February

1. Horses in the Stable all day SNOW this morn'g, so drifed the Lambs Fold near fill'd The Ewes, brot home to Yard and the Snow so drifed by the Evening - had the greatest difficulty in geting the Tegs to Mr Philpots Yard, impossible to get them home 2 COACH'S stop'd just beyond Beckhampton Inn & 1 at the Inn
2. SNOW very much drifted, the Horses unable to move the Coaches this morn, our Horses assisted in drawing them up the Hill, and afterwards took the Passengers of 2 Coaches from the Inn to the Coaches in our Waggon

---

164 Mur - probably Murrain.

165 Bavin - An untrimmed brushwood faggot. Dartnell & Goddard, *Wiltshire Words*, p.8.

(full) | Our Lambs brot from Mr Philpot Yard to ours | Snow heavy at times till even'g when sharp freezing

3. Frost morn'g, Snow heavy all the afternoon, thaw & rain eve g | Horses & Carters nothing to do except geting Hay to Sheep
4. Horses in the Stable, Carter & Boys sawing Wood, in Waggon house loft | Thaw & rain in the night, driving SNOW, heavy all morn'g, drying winds after | HAY, the Sheep consumes very fast The 1st Rick of Meadow} The 2nd cut of Clover} The Stack of old Clover Meadow &} The old Stack of W Philpots} by the 1st of Feby HAY, remaining the 1st Feby The last years Rick, Tan hill Piece The New Rick, - left - Tan hill Rd The Clover Rick at home Whitelands The Dry meadow Rick 380 Sheep find too many to winter | The Cow, leting dry, having calved 12 Months since, and Calve very soon | Great many Men diging a road through the Snow to Avebury 6 feet deep by Mr Browns
5. Very fine day and sharp frost | Sheep returned to the Fold, the Land still covered with Snow & Ice | Horses in the Stable, Carters winnow'g
6. Sunday, Frost in morn'g, Snow hail & rain after
7. Mild morn'g, damp after | 11 Qr Barley to Devizes & 100 Bushl Grains returnd in 2 Waggons
8. Damp rain'y day | 100 Bushls Coal Ashes fm Devizes in 2 Waggons 6 Horses
9. 100 Bushls in 2 Waggons 6 Horses | Rough winds in morn'g mild day
10. 100 Bushls Ashes fm Devizes | Beautiful mild fine day | Lamb the 1st from a Sale Ewe | Ewe another - died
11. Fine day, 2 Ploughs assisting Mr Wentworth, Wheat Sow'g Talevera
12. Fine mild day, 1 Plough all day for Mr Wentworth & 1 Plough from 10-O-Cl
13. Sunday - Fogg'y & damp day
14. Fogg'y - mild fine day | PLOUGHING, 1st time since 3rd Decr the old field, Muckled & For Vetchs | 3 Horses, Cart g Muckle before Fold
    Obit Willm Pinniger, of Cowitch at 6-O-Clock, morng Born 7th May 1812 [This entry written beneath a single thick black line]
15. Fine day - 8 Lambs | Ploughing as yesterday wth 4 Horses | 1 - Horse to Calne for 6 € of Manure Salt for the Down | 1 - Horse, Cart'g Flints for Surveyer
16. 1 - Horse, Cart'g and 1 - Cart'g Turf (bank) | 4 - Horse Ploughing as yesterday | Very fine mild Spring like day
17. Very fine cold drying Winds | 10 Qr Barley to Devizes, and 100 Bu: Ashes Retd
18. ASHES, began sowing 30 Bush: pr A | And Marter earth Cart'g | Ploughing as yesterday
19. Ploughing as yesterday | The 6th PIG kill'd 10 Sco 12 li | Cold storms & drying Winds
20. Sunday, Fine day [Note in margin] New House first occupied, Parlour

21. Very Sharp Frost, cold dry day | VETCHES, began sowing on the old field Muckled & fold d (5 Ac) 2½ Bushl of Winter & ½ Bu: Oats pr Acre | Carting Marter earth
22. Damp morn'g - cold dry afternoon | Harrow'g off the Vetches, and Ashes sow'g on Clover | 20 Lambs, from the sale Ewes
23. Harrow'g the Vetch land finish d | THWART'G, began the Down land for Oats | Fine mild day
24. Fine mild day rain in night & morn'g | 11 Qr Barley to Devizes, as 50 Bu: Ash's and 50 Bu: Grains retd
25. Very cold drying Winds all day rain in the night | Barley, put the last Rick into the Barn, stained | Thwarting for Oats
26. Rain in the Night and all the morn'g too wet to Plough, sow'd a load of Ashes, - and Cart'g earth and Muckle to Fold | [...] Fold removed from old field to home Meadow, the Ewes time for lambing, being out 5th March
27. Sunday, Rain heavy in morn'g drying after
28. Very fine drying day, except a cold hail storm | Thwarting for Oats as 23rd

March

1. Thwarting for Oats as 23rd | Mr Wentworth 2 Ploughs Thwarting for Oats as 23rd | Very sharp Frost & fine day | Cart'g Earth to Bank &c.
2. 2 Teams Plough'g as yesterday | Mr Wentworth 2 Ploughs yesterday | Very windy morning, & heavy rain all eve'g
3. Blowing winds & rain, nearly all day | Hay & Muckle to Standing Pen, Rick Yd | Ewes, put in stand'g Pen, too wet for the Lambs to fall in Meadow
4. Warm dul day | Carting Dung to dry Meadow [...] Bushls of Lime from Calne to Inn
5. Damp morning - dry mild eve'g | 2 Waggon load of Lime fm Calne | HAY, the new Rick, left of Tan hill road began | EWES, time out to begin Lambing 30 Lambs from Stock Ewes
6. Sunday, Wet morn'g dry after
7. Very fine mild day | Thwarting Couchey Fallows for Swedes | The Bank upper side of Whitelands plant'd wth Thorn 100 to a Pole
8. Ploughing an Acre of the old field for Potatoes | Carting, Chalk for the Rick Yd Wall | Turnips runing fast to Green put the Ewes - not lambed on, in Whitelands, or will soon spoil | Damp day, rain at times
9. 100 Lambs (no Ewe died) | Very fine mild day | Carting Chalk for Wall at Inn
10. Carting Chalk for Wall at Inn | 12 Qr Barley to Devizes & 8 Qr Oats ret'd [Pencil note] 16 B Coal | Very sharp Frost & fine day
11. Rain heavy mid day & storms at times | Thwarting for Swedes, till the Rain when Horses put into Stabl
12. Ploughing as yesterday Slight Frost in morn'g, very Tempestuous, afternoon (Wind & Rain) [Pinniger then left a blank space - alongside in the margin

in pencil, he wrote the word "Threat"]

13. Sunday, Most TEMPESTUOUS Night wth Lightening, Mr Neate of Overton, Chimney Stack of new house blown throug the Roof Mid Lent, blow'g rain nearly all day, to cause a Flood [Note in margin] Thunder & Lightening in the eve'g
14. WALL, commenced building by the INN Garden 3 Men 2 Boys | Blowing morn'g wth Snow & Rain dry afternoon | 2 Horses, Cartg Stones for Wall - the others in the Stable
15. 2 Horses, Cartg Stones for Wall - the others in the Stable | Rain heavy in Night, and great part of the Day, Calne flooded these three days, the Masons not working half their time | The Ewes & Lambs confined to the Pen, whilst the Turnips grow'g to green & spoiling & Sow'g &c hinder'd | 150 Lambs, in 10 Days 4 Ewes gone by, not keep a Lamb 4 Ewes want Lambs No Ewe died, no Lambs bought and only 4 doubles
16. Rain heavy in Night, blowing winds & some rain in the day | Cart & 2 Horses to Calne for Lime | Carting Sarson Stones to Wall, Inn | Vetches appearing, sow'd 21st Feby The Oats not appearing sow'd 21st Feby
17. Rain in Night, very Boisterous morn'g - impossible to load any Hay for the Sheep, calm eve'g | 2 Carts to Calne for Lime, Rick Yd Wall | Carting Sarson Stones, for Inn Wall
18. Carting Marter earth, the Commissioners Men fill'g the Carts | 2 Carts to Calne for Lime Rick Yd W | A change in the Weather, fine mild calm day, - very Tempestuous the last six days | SOW FARROW'D 7 | Plants Cabbage, set 200, and Potatoes, Pease & Beans planted in the Garden.
19. Frost in morning, fine warm day | Carting earth as yesterday | Thwarting for Swedes
20. Sunday, very fine mild day | Straw the Sheep discontinued
21. Beautiful calm mild day | Privet, planted top & bottom of Whitelands | Ploughing old field for second sowing of Vetches | Harrow'g the Thwarts for Oats 17 Barren & 6 - without Lambs } turnd wth young Sheep 30 - Ewes to yean | 2 Men dig'g Chalk for Rick Yd Wall [Note in margin] Bedroom first occupied New House
22. Cold air fine dry day | Ploughing as yesterday, and harrowing - Fallows for Beans & P | Chalk - 2 Men diging for Rick yard Wall
23. BEANS, kidwells - drilld 3 Sacks on 2A 3R 20P, on lower furlong | PEASE, marrow fats drill d 2 Sa on 1-3-0, adjoining Beans | VETCHES, winter sow'd the 2nd drift about 1½ Acre | Very fine day, very cold air | Young Sheep lay back on Swedes no Hay
24. COW CALVED, bull'd 14th June | Rain thin all morn'g, dry after | 1 - Ewe & 3 Lambs died | The last fat PIG KILL'D [For details of weights, see Appendix A] | Harrow'g Fallows & 13 Qr Barley to Devizes & 64 Bu: Grains returned
25. Snow to cover the Hills in morning & very cold drying winds all day |

Drag'g & harrow'g Down Piece for Oats | 6 Women draw'g Swedes, going fast to seed, very thick ICE this morn'g to check the greens | Mi[...] Thorne [...] | 4 Lambs died last night

26. Much rain in night, to prevent to prevent harrow'g Oat land | very boisterous after the morn'g Chalk, Carting for Rick Yard Wall | The Ewes, began Meadow Hay being exceeding poor & the Lambs appear all stunted, know no cause
27. Sunday, beautiful fine day | Turnips, Whitelands finish'd
28. SWEDES, the Ewes & Lambs began took all from Rick Yard | and lye back being very mild wth Hay | 9 Ewes only to yean | Frost'y morn'g very mild fine day | Plough'g began for Oats, 1 Team of Brother Wm helping
29. 2 Ploughs of Brother & 3 Ploughs of Mr Wentworths helping | Ploughing & harrow'g - Down Piece as yes [166] | Very cold drying Winds
30. Very cold drying Winds | Plough'g Harrow'g & Roll'g as yestery | 2 Ploughs Mr Wentworths & finish'd the Piece | SALT - 9 Bushls mix'd wth Earth, sowd over an Acre of sour Down | 2 Lambs died
31. 1 Lambs died | OATS, black Tarter<s>y Willm Rumming DRILLD, Nether Down Piece 8A 3R 6P at 2s pr A - with 13 Sacks 34li pr Bu at 25s 6d pr Qr | 3 Horses drawing the Drill, 1 Horse harrow'g after, and 2 Horses rolling and harrow'g before the Drill | WALL at the Inn, finish'd point'g in the evening | Very cold drying winds

April

1. GOOD FRIDAY, very cold dry g East Winds | Rolling & harrow'g, the Oats, drill'd yesterday | The Masons began dressing the Chalk for Rick yard Wall | The Swedes, nearly all drawn better had been drawn a fortnight being almost in blossom.
2. Rain & hail in the night, & in morn'g SNOW to cover the ground, fine day and milder | CHALK, Carted 24 loads for Rick yard Wall | The SPRINGS, continue riseing farther up the Down, & sinking at home the Marsh of Silbury mead still flood'd
3. Sunday, cold dry day
4. Beautiful mild fine day, the Lambs much recovered, in lyeing back on the Swedes the last week the Ewes also improved, eat very little hay N B Observe in future, in the begining of Feby draw the Swedes a little, by breaking the tap root, after which will keep good and not run to seed | 2 Ploughs for Mr Wentworth | Barley finish'd thrashing | Diging Chalk & Masons dress'g it
5. Masons dress'g it | 2nd WHEAT RICK put into the Barn two others remaining | Sharp frost in morn'g very fine day | 2 Ploughs for Mr Wentworth
6. 1 - Plough for Mr Wentworth and paid Debt | Plough'g Old field, for Vetches 3rd sow'g | Very fine mild day

166 yes - yesterday.

7. 13 Qr Barley to Devizes, the last to sell, [...] /26 Posts & Coal [...] retd | Ploughing Whitelands athwart for Turnips | Thin rain after the morn'g & heavy rain in the evening
6. WALL, Rick Yard Masons began building
8. Very heavy Storms | Carting Dung from the Inn
9. Carting Dung from the Inn and home Yard | Some showers of growing Rain | Dry Flock taken from Swedes to the DOWN, not having enough Swedes for Ewes & Lambs [Notes in margin] 9th Mr Kemm began Watermead E & Lambs 13th Mr Brown began Watermead E & Lambs
10. Sunday - Very heavy in morng dry afternoon
11. Foggy morn'g & very fine warm day | Plough'g Old field for Vetches & Hurdles removed to Kennett for 140 of my Dry Sheep to feed off SWEDES [Pinniger then records the weight of his wool clip - see Appendix E]
12. Very fine warm day till 4 O-Clock when heavy Thunder & lightening & tremendous HAIL storm & rain | VETCH'S, the 3rd sow'g on Old field sow'd about 2 Acres
13. Very heavy Thunder Storm in the morn'g - dry afternoon | CHALK, Carted 20 Loads after after 9-O-Clock
14. Fine dry day, very growing Wheat & Grass, improving | Chalk & Sarson Stones, Carting and Ploughing Whitelands
15. Very fine mild growing day | 10½ doz Hurdles from Putthall & 6 B_ rods | GRASS SEEDS, sow'd in the Wheat the middle Down Piece 8¾A 1½ Bushl Pearces' early Ray @ 35s Qr 4 li Broad Clover @ 10d li 4 li Trefoil (mild hop) @ 7½d li 3 li Dutch (white Clover) @ 8½d li Machine, half the hole | Harrowing the Seeds, in the Wheat 48 Bushls Grains from Kennett 9 Cwt Salt from Calne
16. Harrow'g finish'd the seeds in the Wheat 9 Acres - 3 tines | Plough'd for Oats, began, Swedes fed off further Down Piece | EWES &c - round tail'd | LAMBS, Tails seer'd 204 & Ear-markd the near (left) ear | Beans - Pease - & Oats Just appearing out of ground | Beautiful mild grow'g day [After this entry, within a blank space in the diary, Pinniger wrote the word "Fires" in pencil]
17. Sunday, mild growing day
18. Frost in morn, ICE, thick on water very fine warm day | Plough'g as the 16th
19. Plough'g as the 16th | Very fine mild Day
20. Beautiful fine warm day | 7 Qr Seed Oats from Marlbro, morn'g | Plough'g up to the Sheep as yestery with 3 Ploughs & 6 Horses
21. 12 Sacks Wheat to Devizes & 60 Bushls Grains return'd | Rolling & harrow'g after the Plough for the Drill'g Oats | 3 Horses from Kennett, harrow'g | Cold dry day, COW turnd to grass
22. DRILL'D by W Rumming 10 Sacks Georgian Oats 161 li pr Sack, after

Swedes, 3 Horses & 3 H_ roll'g & harrowg Warm & some Showers in the day | HIGH CONSTABLE, entered Office [167]

23. Dry morn'g, Thunder Lightening & storms of Hail & rain evening | Finish'd roll'g & harrow'g the Oats drill'd yesterday & sow'd the Seeds | The Masons, having used all the Chalk, and being too busy sowing to supply them, will discontinue building the Rick Yard Wall, for present
24. Sunday, Cold dry day
25. SWEDES, finishd the Down Piece by the Ewes & Lambs, feed'g them one month | Plough'd the remainer of Swede Piece & sow'd the Oats & Seeds | The Ewes & Lambs, removed from the Swede Land to dry Meadow, to give the Water mead 2 more days to dry being over shoes in water, & the Marsh still under water, & of no use. | CUCKOO | Oats, Georgians sow'd 12 Sa & 1 Bu on the further Down Piece 8¾A | Seeds, sown in the Wheat & Oat Pieces 24 Bushls Ray 86 li Broad Clover 80 li Mild Hop 54 li Dutch } on 17½ Acres is about 12 li small Seeds pr Acre | HAY, the Ewes, discontinued [Note in margin] 25th Jno Pullen Elizah Large } Married [168]
26. Rain, nearly all day | Hurdles, taken to Watermead in the morn'g, & Horses in the Stable | The Horses & 2 Ploughs to Kennett after noon, to assist Barley sow'g having finish'd our spring sow'g | Mr Kemm, finish'd feed'g his Watermedd & began his New field
27. WATER=MEAD, began feeding, the further side, the home side being too wet | Fine grow'g day, some little rain | Plough'g at Kennett as yesterday | Flour Seeds, planted in Garden | Women, began, weeding Wheat | Rt Philpot began, feeding Silbury mead & mowing Silbury mead for Horses
28. Heavy Storms, after noon, warm morng | Chalk finish'd dig'g for Rick Yard Wall | Plough'g at Kennett as yesterday
29. Plough'g at Kennett as yesterday | Some heavy Storms | POTATOES, began planting by Meadow
30. POTATOES, finish'd planting by Meadow | Fine warm growing day, some gentle warm Showers | My Brothers Land at Kennett being to wet for Barley sowing, our Horses rolled our Field grass 20 Acres & the dry Meadow

May

1. Sunday, Shower in morning fine mild dry day | Ewes & Lambs W Ps, began Water mead
2. Fine dry grow'g day, Shower in eve'g | Both Plough Teams at Kennett Barley sowing | Potatoes Plant'g, after Hay Ricks Wheat Field

---

167 Pinniger served as High Constable of the Hundred of Selkley for one year: see diary entry for 17 April 1832.

168 John Pullen of the parish of Lyneham and Elizabeth Large of the parish of Lyncombe married on 25 April 1831 at Lyneham, *Wiltshire, Church of England, Marriages & Banns, 1754-1916 for John Pullen* [accessed via Ancestry, 23 July 2020]. Lyncombe is now a district within the city of Bath.

3. Showers nearly all the morning drying afternoon | Both Ploughs at Kennett, and 48 Bu: Grains from Kennett | Mr Wentworths Sheep finish'd his Watermead, removed to old fied
4. Rain all the morning Horses in Stable showers of HAIL & rain afternoon | Chalk Carted 12 loads to Rick yard
5. Exceeding cold storms of hail & rain in morn'g & Showers after. | Sharp white FROST in the morn'g | 13 Sacks of Wheat to Devizes, 16 Bu: Coal 12 Post & 4 Plank's retd | 1 - Plough fallow'g Whitelands, being to wet for plough'g for Barley at Kenne [169]
6. FROST in the morning, pierceing cold storm of HAIL & SNOW, mid=day warmer at intervals | Both Teams at Kennet Barley sow'g
7. This morning, the most severe Frost, none equal to it remembered by any Person living at the period of year, - The Potatoes under cover, cut to the ground, the Beans, Clover, cut & black, as well as the Ivy creeping the wall & the evergreen Shrubs thick ICE and Windows cased with frost like the mid of Winter, great loss sustained by the Nurserymen. Cold dry East Winds, the Glass riseing. [Diary turned landscape for this entry] | 4 Horses to Box for Table'g for Barn | 2 Horses to Kennet Barley Sow'g | HOEING, Pease began, 3 Women & 1 Man, do well, too wet before
6. Yound Sheep removed from Swedes at Kennett to the Down, very bare began feeding them 11th April [170]
8. Sunday, - very cold harsh East winds, FROST MORN'G
9. FROST & dry harsh winds east Both Teams at Kennett, finishd Barley sowing | Pigs waned
10. Plough'g Whitelands for Turnips and Draging for Swedes | Pierceing Cold dry harsh winds | Soot & Ashes sowing on thin Wheat
11. Draging BEAN'S very clean, no time to hoe them, the OATS being choak'd with Charlock, 6 hoers on | Ploughing finish'd Whitelands and draging it | Very cold East Winds. Frosty morn'gs
12. Warm day, & mild Frosty morn'gs | Drag'g & harrow'g Whitelands | The Ewes & Lambs, finish d feed'g all Silbury mead, except the Marsh
13. Frost, very warm calm morning and very cold East Winds evening | Rolling & harrow'g Whitelands finishd | Rt Philpots Sheep finish'd Silbury md | SOW Brimb'd by the Rabson Boar
14. Frosty morn'g, warm Sun cold air | Ploughing the Couchey Swede land | A Man & 6 Women continue hoeing Charlock out of the Oats
11. Masons 3 Men & 2 Boys returned to building up the end of the Barn
15. Beautiful fine mild day, frost | Sunday
16. White Frost continue every morng very fine mild day | Ploughing finish'd the piece of foul land for Swedes
17. Very warm fine day | 3 Load of Hurdles removed from Silbury mead to

---

169 Kenne - Kennet.

170 Yound - young.

the Down Pieces, & 5 Waggon load of Sarson Stones from Avebury the Meadow Wall | Ewes & Lambs, removed from Silbury Mead to the Down, 20 days in the Mead & great part of the Marsh, still under Water, Fold'd on lower long Piece for Swedes

16. 3 Masons 2 Boys & 2 Sawyers came to work
18. A fourth Mason came to work | Chalk, Cart'g for the Meadow Wall | Fine warm Sun, and wind'y | Ewe's & Lambs on the upper little Down & the first tie of New field grass | A very Barren time, the fields cut up by the severe frost on the 7th have not grown since, no feed on the old field
19. Very tempestuous winds all day, and a few drops of rain at intervals | Carting Dung from Rick Yd to Whitelands for Turnips | LAMBS, cut by Abednego Blackman of Charlton 93 at 1s pr Score
20. Carting Dung from Yard to Whitelands and finish'd the piece, and 48 Bushels Grains from Kennett | Sultry warm with Thunder & some Showers and a very heavy Storm for 2 hours at midnight
21. Beautiful warm growing Day | Ploughing Whiteland, and began sowing <u>TURNIP'S</u>, Tankards, 2 holes and blank 2 li pr Acre Horses BLED, (too cold before) in the evening (Saturday) [...] Qts of Blood each and 2 oz of Salt Petre to each, drenched [Pencil note in margin] 16 Mr Brown 17 Mr Wenth | The Horses, continue at Hay & Corn Silbury Mead too wet to cut the grass Mr TANNER persisting in pening back the water
22. Sunday Fine mild day
23. Ploughing Whitelands & sow'g Turnips | The Tegs finish'd Fold'g Potatoe land and fold removed to Swede Land
24. Rolling & harrow'g Whitelands and Ploughing Potatoe land | Beans, began Hoeing, Charlock get'g up | Warm day & some Showers
23. Very fine warm day
25. Warm Sunshine day (Christening D) | Ploughing & draging Potatoe Land & Plough'g Couch'y Swede Land
26. Fallows began Thwart'g for Swedes the homeward Piece | 20 S- Wheat to Devizes & 44 € Coal retd | Very warm fine day
27. POTATOES, Plough'g in bottom of Tan hill Piece ¾ A [...] and [...] | 1000 Bricks from Totterdown | Thwart'g as yesterday | Very Fine warm day
28. Very Fine warm day | Thwart'g as yesterday & Rolling Oats and 48 Bu: Grains fm Kennett | Charlock mowing in the first sow'g Vetches | 3 Carpenters removed the Wood House & began framing for the Skilling by the Rickyard
29. Sunday, A most gracious gentle rain from 8 O=Clock in morn'g - to 8 - eveg
30. Beautiful fine dry grow'g day | Both Teams thwart'g for Swedes | GRASS, Watermead, began mow'g for the HORSES
31. <u>Wheat</u> Rick put into Barn full of <u>MICE</u> | <u>N B</u> <u>Dress</u> <u>the</u> <u>Ricks</u> after Harvest at any <u>COST</u> by Tops Successor | Thwarting as yesterday | Foggy morning,

very fine day

June

1. 1 Team Thwart'g as before | Rolling Oats, discontinued being too high | Harrow'g the Swede land | Very fine day Mr Wentworth began sow'g Swedes Mr Philpot began sow'g Swedes | Hoeing the Pease 2nd time full of Charlock & no time yet to weed the Oats
2. The most Summer like calm morn'g & very warm Day | Plough'g & harrow'g as yesterday & Carted 12 loads of Chalk | The New field grass too=old for the Sheep to eat it, discontinued, having fed about 2 Acres, - now on the Down alone, a few days, - the Old field now good & fit to feed.
3. Very warm Day | Thwart'g finish'd for Swedes & harrow'g Swedes
4. Plough'g the 4th time the Couchey part for Swedes & Harrowing | Very warm fine day
5. Sunday, fine morn'g Storm in even'g | Incendiary, Mr Neat of Monkton last night (Saturday) a large Wheat Rick, Bean & Oat Rick, with Three Barns, Sow & Pigs, & 40 Sacks Wheat destroy'd by Fire, between 10 & 11-O-Clock 3 Waggons, Gig &c & 3 Ricks of Straw [171]
6. Fine day, Pease hoed 2nd time | Plough'g & harrow'g for Swedes, next Wheat | Mr Brown & Mr Wentworth, mow'g Field grass
7. 3 Load of Stones from Calne | Very fine day
8. Very fine day Fold finish d further Piece Swede Land, & began Folding Old field for Turnips - muckled from the Inn, Lambs a Fold of grass before, | The Old Field, began feeding with the Ewes & Lambs, having fed 1½ Acres of New Field only with the Down, 22 Day | Hurdles removing from New to Old Field | Plough'g last Fold drift for Swedes | 5 Loads of small Chalk for build'g the Skilling - Carted, & hope the LAST | Roof, puting on Skilling
9. Ploughing as yesterday & Roll'g & harrow'g | 20 S: Wheat to Devizes | Very warm morn'g RAIN in the eve'g very acceptable
10. Heavy grow'g Showers | 2 Horses Plough'g & 4 to Tytherton for the Iron hurdle Fence, the Inn &c [Note in margin] 10th WHEAT EARS, right of Tan hill road appear

---

171 Mr Neat of Monkton is probably George Neate of Winterbourne Monkton. George Neate (transcribed as Wale on Ancestry) was a 60 year-old farmer at Winterbourne Monkton according to the 1841 Census, TNA: HO107/1185/15. *Census returns. 1841 census. Wiltshire Hundred: Selkley, including Parish: Monkton Winterbourne, ED25* [accessed via Ancestry, 23 July 2020]. George Neate of Monkton was buried on 3 September 1847 at Winterbourne Monkton age 69, *Wiltshire, England, Church of England Deaths and Burials, 1813-1916 for George Neate* [accessed via Ancestry, 23 July 2020]. See also his will, TNA: PROB11/2065/124, *Will of George Neate, Farmer of Winterbourne Monkton, Wiltshire, 12 November 1847*; WSA: 1409/14/13, *Copy will of George Neate, farmer of Winterbourne Monkton (proved PCC), 1847.*

11. Rain in night, Showers morn'g & eve'g | TURNIP, hoeing, began Whitelands | SWEDE, sowing began Whitelands | GRASS, mowing began Hop & Ray &c | Weeding finish'd Pease & Beans & began weeding Oats
12. Sunday, Rain in morn'g dry after
13. Rain early in morning fine growing day | Ploughing & sowing Swedes | COW, bull'd at the Hail Farm [172]
14. Very Warm fine day | Plough'g & sow'g Swedes as yesterday | Women, cuting Charlock, from Oats with Reaping Hooks
15. Ploughing & sow'g as yesterday | Heavy Showers, after the morning | Front Gates, James Hillier began puting up
16. Iron Fence, James Yeo began puting up at Cottages | Plough'g & Sow'g as yesterday | Heavy Showers
17. Heavy Showers & Plough'g &c as yesterday
18. Frequent Showers & very boisterous Winds | Swedes, finish'd sow'g about 13A
19. Sunday - frequent Showers in evening
20. Ploughing for last drift of Vetches | Carting earth to Iron Gates | Slateing Henhouse | <Front> Fine dry day | SHEEP=WASH'D, at Horton @ 4d Score, the best place, Usage, 2 Loaves, Bacon & Cheese, 5 Qt Ale & Beer, the Beer not enough, gave 1s 6d extra
21. Very warm fine day | HAY, field, began carrying | Ploughing as yesterday
22. Ploughing as yesterday | Very fine day & Carry'g Hay
23. Very warm day & finish'd Hay | Ploughing as yesterday
24. Summer VETCHES, sow the last drift | OAT'S in Jag [173] | Rain in morn'g, very dry windy Day | CALFE 40 li pr Qr @ 6½d, - Young Calfe 22s
25. Plough'g old field for Turnips, and rolling & harrow'g Vetch ground | Showers frequent, after the morn'g
26. Sunday, Showers in morn'g fine eve'g
27. Very fine morn'g, rain 11-O-Clock and rain pouring at 2-O-Clock, Carting dung to Fold & Plough'g for Turnips | SHEEP=SHEAR, 18 Score & 6 - at 1s pr Sco 6 Men & a Boy, - charged 1s for Boy as Colt began 8 in morn'g - finish d - half pass 6 - eve'g | Shear'd 8 each in 42 minutes, at finishing, too hurried, to make good work | 1 Man & 1 Shepherd in Barn, 3 Women to wind the Fleeces & stack away, quite enough their keep all day | Turnips, began hoe'g 2nd time | Mowing, home Meadow began
28. Showers at intervals & Thunder & light'g eveg | Plough'g & Carting Dung as yesterday
29. Ploughing & Carting Dung as yesterday | Fine day
30. LAMB'S - waned on Vetches, very good | Vetches, the Horses, began, | Ewe died, murrain | Pressing the Old field, for Turnips | Harrowing Old

---

172 See diary entry for 14 June 1830.

173 Jag - The awn and head of the oat. Oats are spoken of as 'well-jagged', 'having a good jag', 'coming out in jag'. Dartnell & Goddard, *Wiltshire Words*, p.86.

field & fallowing as yesterday | Very fine day

July

1. Very warm fine day | TURNIP Tankard, sow'd a drift next the Vetches, one earth old field | Turnip, Greenround, sow'd a drift next the Potatoes, old field | Fallowing, Old field | The Marsh of Silbury Mead, began mowing, too wet to feed or cut for Hay before
2. Fallowing as yesterday & Cart'g muckle | Beautiful fine day | WHEAT RICK, the 4th & last put in Barn
3. Sunday, little rain in morn'g, fine after
4. The most beautiful summer like Day | HAY, carried the dry meadow | Fallow'g as before & rolling & harrow'g last sown Turnip seed | Painting Gates &c, & Carpenter, finish g Poultry house &c.
5. Rain heavy in morn'g, very fine after | Fallow'g & Cart'g Dung before Fold | 3 Kennett Men, mowing Silbury Mead, great Crop in Marsh
6. Beautiful fine day, Plough'g &c as yesterday | Cart'g Marter earth from New gates | Swedes, began HOEING, (Brewer & Ash)
7. Plough'g Fold drift for Turnips The 2 Colts resting, the 1st time | Very fine day, heavy Thunder Storms partially
8. Pressing & harrow'g the Fold drift | Very warm fine day
9. Very warm fine day, Lamb died murrain | Ploughing &c as yesterday | HAY carried the marsh Silbury mead
10. Sunday, very warm morning, very heavy Thunder & lightening, and rain, with very large HAIL, in evening
11. Very fine day | Fallowing & Carting Muckle
12. Fallowing & Carting earth fm Wall by Meadow | Very fine day, Rain, heavy in evening

11. SALE EWES, drew, and put on Vetch's [The flow of diary entries was then interrupted by a 'stock-take' of sheep on the farm and records of sheep losses thereafter - see Appendix B]
12. Mowing, Watermead, fed off by the Sheep
13. Rain heavy, frequent | 4 Load of Fancy Flints, from the Down & Carting Rubble from the Inn Wall
14. 22 S Wheat to Devizes & Cart'g dung from Inn | Frequent Showers in morn'g dry after | Turnips sow'd the 2nd Fold drift
15. Frequent heavy showers | Fallow'g Wheat Field, finish'd except half a Fold drift
16. W & 4 Horses to Box - for Coping &c for Rick Yard Wall | Frequent heavy Showers
17. Sunday - Dry day
18. Rain in morn'g - fine remainer of day | Carting dung & removing Hurdles | 2 Masons & Boy, came to finish
19. Dull day, some Storms | Carting Stones for the Parish

20. Some rain, very drying winds at times | Harrow'g Fallows, & Carting Chalk from Machine to new Skilling
21. Heavy Storms all day, at times Market advanced 1s pr Sack | Fallowing Old field
22. Fallowing Old field FINISH'D | Swedes, finish'd hoeing 1st time, & began 2nd time | Some light Storms, - no hay made the last fortnight here | Tileing the Skilling Edmond's Cottage
23. Very drying morning, Hay turned & dry by mid day, when rain'd remainer of day | Carting Stones, for Parish 2nd day
24. Sunday, very warm fine day
25. Very fine drying day | HAY carryed Silbury mead, Sheep fed | Rolling & Harrow'g, the fallows
26. Harrow'g the fallows | Very warm fine day, the first thick foggy, morn
27. Very warm fine day | 4 Horses to Box for Coping [174]
28. 16 Sacks Wheat to Devizes & Ploughing began the Vetches fed off | Sultry Warm day
29. REAPING, Wheat began Enford Parish & 25th at Cullern, Marshfield &c | Ploughing as yesterday | Very Sultry morn'g, - Thunder, Lighten'g & rain from 3 to 8-O-Clock, at (Tytherton)
30. Very warm, some Thunder & Rain, afternoon | 5 Horses, Waggon & Cart to Box for Copeing for Inn & Tiles for Cottage the FINISHING JOURNEY
31. Very warm day

    Died Bryan Maskelyne Littlecote Aged […] [175] [This entry written within a thick black border on two sides]

August

1. Warm fine day, Fallow'g as before Swedes finish'd hoe'g 2nd time | Monckton Penning BURNT down 9-O-Clock - eve'g [176]
2. Rain much in Night & Showers in Day Carting Dung before Fold Masons came to finish off
3. Reapers prepared to begin mid:day, heavy Storm after to prevent | Cart'g Chalk to Gateway, by Cottage Gardens
4. Plough'g as before PEASE, began cuting | Warm Day, partial showers & Thunder
5. Fogg'y morn'g warm day, Plough'g as yes'y

---

174 Coping – this would have been coping stones.

175 Brian Maskelyne of Littlecot in the parish of Hilmarton was buried on 3 August 1831 at Clyffe Pypard age 32, *Wiltshire, England, Church of England Deaths and Burials, 1813-1916 for Brian Maskelyne* [accessed via Ancestry, 23 July 2020].

176 Monckton Penning – Monkton Penning is one mile to the east of Winterbourne Monkton, Kelly's Directories Ltd., *Kelly's Directory of Wiltshire* (London: Kelly's Directories Ltd., 1898).

6. Harrow'g Vetch land fallow'd | Very fine day | REAPING, began | Masons & Carpenters, finish'd and complated, the Building, by fixing the Cope'g & Tileing, Edmons's Cottage Shed &c *
7. Sunday Dry till 10-O-Clock, Thunder & heavy rain, nearly all day after
8. Fine dry day | Pease finish'd cuting by own 3 Men & 4 Won & began reaping with 15 Taskers [177] | Plough'g Vetch land fed off SHEEP, gave a Table Spoon full of TAR to each in hope, to prevent their dieing by the MURRAIN, & tar'd sticks in the Fold = 20 died since Shear day
9. Fine morning, rain 3 O-Clock & all the eve'g | Thwart'g behind the Fold
10. Thwart'g behind the Fold | Very fine reaping day
11. Very fine reaping day Plough'g as yester'y
12. Very fine reaping day Cart'g Dung before Fold | Wheat Rick Mr Wentworth made & many this side Marlbro | Dry new Wheat at Devizes Market @ 32s 6d
13. Very fine day, Thwarting & harrow'g | REAPING, finish d in the morning | Wheat harvest finish'd , Cowitch, Whitley &c | 2 Labourers, W Coleman & T Chandler, left their Service, can neither, Mow or hoe Turnips
14. Sunday - fine drying day
15. WHEAT, began carrying, made nearly 2 Ricks 12 Acres very dry Hands employ'd at the Rick and Load Silas Wiltshire, Jacob Minchin (Men Geoe Edmonds Boy & 2 Shepherd Boys
13. TURNIPS, Sale Sheep & Ewe Lambs began Whitelands & Vetches discontinued being too Old the Horses finish the Vetches
15. The finest harvest day this season
16. Very fine day, Thunder all the afternoon & little rain 7 O-Clock eve'g, to stop carry'g, heavy partially | Wheat finish'd 2nd Rick & Carried 4 load of Pease & made a Wheat Rick from the Down Piece 3A 3R 0P
17. Wheat harvest FINISH'D, made the 4th Rick & put 6 Acres into the Barn | Fine harvest day, heavy Thunder afternoon
18. OAT'S, began Mowing by Silas Wiltshire & Jacob Minchin, having no Men in the Parish to do it. | Very drying some little rain | Draging Turnips, Old field, & thwart'g Fallows | TURNIP'S, Old field, hoeing, spoild for want of hoeing before harvest, neither of our 4 Men could do it
19. Very heavy Storm in night, drying day | Cart'g Dung before the Fold, the whole of the Wheat field dung or muckled for next year | BEAN'S, 3 Taskers, began draw'g - & left it
20. High winds & heavy Storms Plough'g before the Fold, 1 Piece of Oats cut
21. Sunday, - Dry windy day
22. Very fine day, Plough'g Vetch land fed off Marlbro Fair, Prices higher [...] Lambs [...] [...] Ewes [...]
23. Very warm fine day, Plough'g as yesterday
24. Fine day, Rain in the evening | Plough'g the Pease land for Turnips | Oats

---

177 Won - women.

finish'd Mowing

25. OATS, black began carrying 12 O-Cl | Rain heavy storm in night, dry'g Winds | Ploughing as yesterday
26. Fine day, Plough'g as yesterday | Black Oats, finished carrying | Calf gone
27. No dew in morn'g, White Oats, began carryg stop'd carry'g with rain 10-O-Clock Plough'g as yesterday
28. Sunday, - fine dry day
29. Fine dry day White Oats finish'd carrying & Harvest finish'd except Beans
30. Beans, all hands (11) cuting & finish'd. | Ploughing the Pease land for Turnips & harrow'g Turnips | Very fine day
31. No dew in the morning, rain after 9 O-Clock till 12-O-Clock | BEANS, being dead ripe, began carry'g by 6,O-Clock & finish'd at 10-O-Clock and HARVEST, finished | Ploughing the Bean stubbles for Turnips

September

1. Ploughing the Bean stubbles for Turnips | Rain without intermission from 8-in morn'g to 8-O-Clock in eve'g | Vetches, the Horses discontinued being to old, - 1 drift left for seed
2. The second Clover, the Horses began | Ploughing the Bean Stubbles for Turnips | Willm Pierce, our only man & two Strangers from Cliffe, began Thrashing Wheat yesterday | Plants, Cabbage, seting for May Cabbage | Fine dry day, some Thunder & lightening
3. Very fine Harvest Day | TURNIP Green rounds, sow'd the Bean & Pease land, good work | Watermead grass, the Horses began the Clover being seed'y
4. Sunday, some rain in morn'g fine after [Note in margin] SOW farrow'd 9
5. Rain thin nearly all day after 8-O Clock | Plough'g Vetch land fed off & fold'd
6. Plough'g for Seed Wheat began | Partial Storms
7. Partial Storms & Plough'g as yesterday
8. King William the 4th & Queen Adelaide's CORONATION | Heavy partial Storms | Ploughing as yesterday for Seed
9. Rain heavy in morn'g, Horses returned from Plough, Carters weeding Hedge, dry after
10. Fine drying day | Harrow'g & Rolling, to clean Wheat land
11. Sunday, Foggy morning, & beautiful fine Summers day
12. Fine drying day, as any this harvest, like the commencement of another Summer. | Plough'g after the young Turnips fed off, not hoe'd being very foul
13. Drag'g the young Turnips & plough'g for Seed | VETCHE'S, seed, began cuting | Rain heavy storm in night, fine day
14. Rain heavy nearly all night, fine day | Ploughing for seed

15. Ploughing for seed & 23 S- Wheat to D= and fine day [178]
16. Ploughing for seed & very fine day
17. [Note in pencil] Rt Philpot sow'g Turnip Seed | Very fine day, nearly every one finish'd Harvest | Ploughing for Seed
18. Sunday, - Very warm fine day
19. Dry in the morning, very heavy Storm in the evening | Ploughing for Seed
15. Vetches, the last sow'g, the Sale Ewes & Chiver Lambs began [179] | Oats, began giving the Horses
20. Fine day, rain in evening and all night | Ploughing as yesterday | Ploughing as yesterday some storms in the Day
21. Ploughing as yesterday Fine day
22. Very fine day | Plough'g Whitelands for Seed
23. Plough'g Whitelands and very fine day
24. fine day | 6 Horses to Corton & Cleveancey with 23 Sacks Seed Wheat & 12 Sacks returned | 2 Men triming Stock Ewes | Vetches seed carried
25. Sunday - very fine day
26. very fine day and warm | Stock Sheep, finish'd triming 224 | Ploughing Seed Vetch land
27. Very fine morning, Thunder & Storm afternoon WHEAT=SOWING, began with the Tockenham Thick set 2½ Bu: pr Acre uper end of Tan hill Piece (left hand)
28. Sowing Wheat as yesterday | Very heavy rain in morn'g, dry afternoon
29. Dry morning, heavy storm eve'g | Plough'g for Seed & changing 6 Sacks seed Wheat with Mr Pike of Milton
30. Glass very low to much rain, yet a very warm drying Day, rain in evening very heavy | Harrowing off the Wheat sow'd

October

1. Ploughing for Seed, The Down Stubbles cut | Very fine day, altho Glass very low | Vetches, finish'd feed'g last sow'g
2. Rain heavy in night, dry morn'g, heavy rain eve'g
3. Fine Day, rain in evening | Wheat Mr Pikes, began sow'g above Turnips | RAMS, turned to Stock Ewes, the 3 Lambs
4. Wheat sow'g Mr Pikes | Very heavy rain in night, fine Day | Turnips finish'd Whitelands, began 13th Augst | Turnips Tan hill Piece began, feeding Silbury=mead, Sale Sheep & Ewe Lambs, began
5. Rain heavy in night fine Day, drying Thwarting the Wheat sow'd the last 2 days
6. Harrow'g off the Wheat & 2 Horses to Devizes for Coal | Thick air in morn'g, dry'g afternoon
7. Fallow'g, the Vetch Land fed off | Dry windy morn'g, rain afternoon

---

178 D= - Devizes.

179 Chiver - Chilver.

8. Thin rain, nearly all day Clover Seed, began mowing | Ploughing Whitelands for Seed
9. Sunday, rain in Night dry day & rain eve'g
11. Dry morn'g rain afternoon 15 S- Wheat to Cleveancey, removing Hurdles &c
10. Carting Muckle, from the Inn to Nether Down Piece (Oat Stubble) for Turnips & Wheat next year | Fine day rain in the evening | FOLD, finish'd the Wheat land Tan hill piece
11. FOLD removed to Nether Down piece
12. Rain in night, and frequent storms in in the Day | Very unfavorable wet time for the Sale Sheep do not settle on Turnips or Mead, put them on the young fresh Clover = do better | Finish'd Ploughing Whitelands and sowed it with 3 Sacks of Mr Crooks Wheat
Mr Nath[...] Atherton Senr of Calne drown'd himself in the Mill=pond Monday eve'g 10th Octr (Mind depress'd) [This entry written within a double line border on three sides]
13. Frequent Storms | Finish'd harrowing Whitelands, sow'd yesty
14. Continual heavy storms, dry evening | Carting Dung from Yard to Down Piece the land too wet for Wheat sowing
15. Carting Dung as yesterday | Fine Day, the first dry day since [...]
16. Sunday, fine dry day
17. Ploughing, Couchey Turnip Tan hill piece | Pease Rick, put into Barn, and | Oats, put a few into Barn for Horses | Very fine day
18. Very fine day Draging Couchey Wheat Land & carry'd Stubbles, on Down Piece
19. Marlbro Sessions & Beautiful fine day | Clover Seed - Carried & Stubbles
20. Very fine day | Thwarting Couchey Wheat Land
21. Thwarting Couchey and draging | Rain in Night, fine morn'g Storm after [Pinniger recorded the dates and weights of pigs killed during 1831-32 – see Appendix A: he then recorded his sheep stocks as at October 20th 1831 - see Appendix B]
22. Dry morn'g storms all afternoon | Carters & Boy's - picking Couch & Grass off the Wheat land, none clean to Plough
23. Sunday, rain nearly all day
24. HAY, the 80 Sale Ewes began, wth Turnips | Wheat Stubbles began Ploughing too wet to Plough for Wheat | Dry day
25. Ploughing again for Wheat, the Couchey part | Some Storms
26. Heavy Storms, Ploughing as yesterday
27. Ploughing as yesterday | Fine day, very heavy storm in evening
28. Fine day, Ploughing as yesterday
29. Finish'd Ploughing for Wheat except the Turnip & Potatoe Land
29. Fine day, except a Shower
30. Sunday, - Glass very high, fine morn'g thin rain all the evening

31. Very fine day Wheat sow'd about 6 Acres wth 4 Sacks about 4 Acres of Turnip & Potatoe Land to do, to finish sowing

November

1. Fine till mid=day, rain all afternoon | Harrowed athwart the Wheat sow'd yesterday in the morn'g, too wet to harrow it more
2. Dry fine day | Fallowing Wheat Stubbles
3. Rain all day, - Horses in the Stable
4. Dry all day, very ICEY frost in the morning (cut off the D[...] | Carting Dung from the Inn, to Down Piece | 2nd Fat Pig Kill'd 7 Sco 7 li
5. Thin rain at times all day | Fallowing Wheat Stubbles
[Diary turned landscape and the following written in the margin] Dreadful Riots in Bristol [followed by] Saturday 29th Octr [rest of page left blank] [180]
6. Sunday, thin rain at times & damp day
7. Lightening & rain in morning, and heavy storms of hail & rain continually | The Land too wet to fallow, Carting Dung to Down Piece, from Yard
8. The Land too wet to fallow, Carting Dung to Down Piece, from Yard | Dry cold Day
9. Dry cold Day sharp Icey Frost | Fallow'g Wheat Stubbles, (48 Bu: Grains) K
10. Fallow'g Wheat Stubbles, | Exceeding sharp Frost, rain in evening | Pease haum, began giving Stock Ewes
11. Very fine mild day | 22½ Sacks Pease sent to Devizes Sheep gates & Iron returned
12. WHEAT RICK, the 1st of the 4 put into the Barn, Carters helping, Horses in Stable | 7 Small Pigs & 3 Bu: Pease, sent to Cleveancey Cheese & Butter returned | Very fine calm day | POTATOES, finish d diging [...] Prolifficks in the Meadow [...] [...] Prolifficks in the Field [...] [...] Purples in the Field [...] [...] White apple in the Field [...] [Total] [...]
13. Sunday, very fine day
14. Frost sharp, fine day rain in evening | Draging Potatoe ground & Plough'g Stubbles
15. Plough'g Stubbles | Very heavy SNOW, & hail storms, mid=day dry evening
16. SNOW in the morning to cover the ground, rain eve g | HAY, began giving the Teg's (or Lambs) the old dry meadow | Fallowing the Stubbles
17. Fallowing the Stubbles Frost morn'g fine day | Young SOW=BRIM'D, Mr

180 There was widespread unrest across the country, following the defeat of the Second Reform Bill in Parliament on 8 October 1831, which stalled efforts at electoral reform aimed at extending the right to vote. However, it was three weeks later, when riots took place in Bristol (29 to 31 October). Susan Thomas argues that the defeat of the Reform Bill was not the primary reason behind the riots in Bristol, Susan Thomas, *The Bristol Riots* (Bristol: Bristol Branch of the Historical Association of the University of Bristol, 1974), pp. 1-2.

Popes Boar
18. Freezing all the morning, rain in evening | Fallowing as yesterday
19. 3 Waggon load of Sarson Stones from Avebury | Dry Day
20. Sunday - Cold dry day
21. Very boisterous, Stormy night, damp day | Fallow'g Stubbles | SWEDES, the Lambs or Tegs, began feeding fold'd with Hay, mornings, finish'd the Watermead yesterday | 2nd Fat Pig, Kill'd 8 Sco 8 li
22. Exceeding damp mizly day | Ploughing & Carting Hurdles fm Silbury Mead
23. Fallow'g Stubbles | Damp morn'g, finer day
24. Very fine mild, clear day | Fallowing as yesterday
25. Carted the Wheat Stubbles & Dung from the Inn to the Fold | Damp day, some Rain
26. Damp air mild day | Fallowing Stubbles
27. Fine dry day, Sunday
28. Sharp frost fine day | Ploughing the Potatoe Land & the Turnip Land finish'd feed'g yesterday for Wheat | Isaac Gilbert, began leveling the Bank in Silbury mead
29. Exceeding sharp frost & freezing all day with difficulty Plough, as yesterday
30. Sharp frost, & thaw Ploughing as yesterday
28. The Ewes, began SRAW, in the Yard & HAY the Old Clover Rick began, in the Fold evenings, the Down being very bare, but have done well on the Down hitherto, no Hay[181]

December

1. WHEAT sowing finish'd, the Turnip, and Potatoe land about 4½ Acres with 3½ Sacks 1 Team from Kennett, made good work | 4 Sacks, thick set from Tockenham 12 S 4 li Nt 8 Sacks Burrel from Corton 11 S 16 li Nt 6 Sacks Burrel from Mr Pike Milton 11 S 12 li ½ sacks Burrel own Down Wheat 11 S 18 li & 11 S 11 li [Total] 18½ [Sacks] sown on 28 Acres. | Damp morn'g fine day
2. Thrashing Clover used for Sheep Carting Dung to Down Piece | Extraordinary, mild fine day
3. Extraordinary, mild like spring morn'g | Fallowing Stubbles
4. Thin rain in morn'g mild dry eve'g | Sunday
5. Fine mild dry day, Cloverseed finish'd head g | Cart g earth to yard, for plant'g Trees | Fallowing Stubbles
6. Fallowing Stubbles | Fine day, rain in evening, pick g Stones from Wheat Stubbles in Carts
7. Very boisterous windy night with rain, winds & rain all morn'g, Horses in Stable
8. Very mild, light showers in the day, heavy rain in the evening | Fallowing

181 SRAW - straw.

Wheat Stubbles

9. Fallowing till 2 O-Clock, when very heavy rain Horses returned home
10. Fallowing & Hurdle Cart. from Swedes | Rain heavy, night & morn'g, mild & dry after | FLOOD, Silbury Mead, stop the Men, leveling the Bank [Pinniger then listed the trees and climbers planted in the Yard - see Appendix O]
11. Sunday, - Very boisterous stormy night and day
12. Very boisterous with rain all night & all day great FLOOD, all my Silbury mead under water except the Bank in the evening | The Horses in the Stable
13. Windy night, calm dry day | Water continue in the Mead, the leveling Men return'd home, to Grafton | Carting Dung to Down Piece & finish'd it
14. Fine dry day, Fallowing Stubbles | Last sowing Wheat just appearing
15. Rain in evening, Fallow'g as yesterday
16. Rain in evening, Fallow'g as yesterday
15. The Men returned to their work, Silbury mead
17. Very heavy rain in night Dry day | Finish'd Fallow'g the Wheat Stubbles | 3 Pig Killd 10 Score
16. Straw for Sheep all used the weather too wet to put more Oats into Barn for them they do well on the Down & Hay in fold Nights only The <u>Lambs on Swedes & Hay</u> Nights only
18. Sunday, Very heavy rain in Night the Wheat Tan hill Piece flood'd, and great Flood in Meads, fine day
19. Very mild fine day Fallowing, began Oat Stubbles Down Piece | Dung'd and Fold'd, finish'd this night, began 11th Octr is 69 Days or 10 Weeks with 230 Sheep and 107 Lambs [Total] 337 } 8A 3R 3P
20. Fine morning, rain afternoon, DUNG | Carting to Old field, and the FOLD removed to it, from Down Piece, for Vetches & Potatoes | Wheat the last sowing appearing
21. Very mild clear fine morn'g, rain eve'g | Fallowing the Down Piece, Dung'd for Turnips
22. Fallowing the Down Piece, Dung'd for Turnips | Cold windy morning, rain afternoon
23. Large blossoms of SNOW & rain in morning fine clear day | Fallowing as yesterday
24. Fallowing as yesterday | Glass, rising to near FAIR, fine mild clear more like <u>May</u> than <u>Xt mas</u>
25. Sunday & Xt mas Day, Sharp frost and fine day
26. Sharp frost and very fine clear day | 20 Sacks Wheat to Devizes & 2 T o € 3 Qr Coal retd | Fat Pigs, began 2 S- of Pease & Barley Meal with Potatoes
27. Damp day, Carting Dung before the Fold
28. Fine mild dry Day, damp evening | Fallowing the Oat Stubbles
29. Fallowing the Oat Stubbles [Note in pencil: W Davis Ill] Fine dry cold Day

30. Fine dry cold Day Plough'g as yesterday
31. Very sharp Frost, thaw & dry day | Carting Dung from Inn, before Fold

1832

January

1. Sunday Very sharp hoar Frost
2. Very sharp hoar Frost | The Oat Rick, Georgians, put into the Barn for the Ewes to have the Straw mornings, - doing well till now on the Down and 7 Truss's of field Hay nights (old) | The Tegs on Swedes & 2 Truss s of Meadow Hay nights
3. Sharp black Frost, yet exceeding calm & pleasant Day | The land too hard to fallow, the Horses in Stable no Dung Carting or any thing else to do | The Ewes in the Yard at Straw & Water mornings
4. 2A 2R 0P - of the Old field Muckled & Fold'd in 16 Days | The Frost continue very hard, and the Horses in the Stable
5. the Horses in the Stable Very fine mild Frost | The 3 Boys grubing FURZE on the Down having nothing else to do
6. The 3 Boys Thick Fog all day, thaw but little, The Horses in Stable
7. Thaw, thick Fog & some little rain | 2 Waggon load of Dung from Inn to Fold
8. Sunday, thick Fog'y Day
9. Heavy storms, afternoon | Carting Dung before Fold
10. Very mild, heavy Storm afternoon | Fallowing Oat Stubbles
11. Fallowing Oat Stubbles | Very mild fine morn'g, Storms afternoon
12. Very mild fine morn'g, Storms afternoon | Fallow'g Oat Stubbles
13. Fallow'g Oat Stubbles | Very mild dry day
12. SWEDES, The EWES, began feeding and reduced their HAY nights from 7 to 4 Truss's, the Hay probably not enough to keep the Sheep all the Spring, but the Swedes probably enough The Ewes do well with Oat Straw & Water mornings, Swedes after & Hay in Fold | The Tegs (104) Swedes all day & 2 Truss's of Meadow Hay in Fold
14. Fallowing finish'd the Oat Stubbe, Down = piece for Turnips & Wheat | Frost in morn'g - fine day, except'a - partial Snow Storm
15. Sunday, very fine mild day, like Spring
16. Very sharp frost, and very fine calm mild Day | WHEAT, Rick, put a second into the Barn two, remain in the Yard | Ploughing athwart, began the Bean & Pease for Barley
17. Ploughing athwart, began the Bean & Pease for Barley | Unusual mild calm day, no frost | The Sheep do well on the Swedes, (233)x the Ewes x & Rams, have a brea[..] of 20 by 2 hurdles daily only, which being large some weighing 13 li each, is plentiful | The 104 Lambs have 10 by 2 Hurdles of Swedes
18. Very fine mild Day | Plough'g as yesterday

19. 17 S- Wheat to Devizes & 40 Bu: Ash's retd | Very thick Fog & raw air all day
20. Fine mild day, the last sowing Wheat grow'g fast, geting green (1st Decr) | Fallow'g the Bean & Pease Stubbles for Barley
21. Fallow'g the Bean & Pease Stubbles | Very fine mild day
22. Sunday - mild damp day
23. Fine mild dry day | Fallowing as the 20th
24. Fallowing as the 20th | Fine dry day
25. Heavy rain in night beautiful fine day | Cart'd the Dung from the Inn (6 Waggn load)
26. Fallowing as the 24th | Sharp Frost in morn'g, heavy rain in eve'g
27. Rain from last eve'g till midnight, & Snow from that time till 3-O-Clock this afternoon | The Horses in the Stable all day | The Young Sheep Hay twice, the 1st time
28. Very sharp frost, fine day, Snow still cover the ground, the Sheep on Swedes again | Fallow'g as the 26th
29. Sunday, fine mild dry day
30. Very mild fine day, Glass high
31. Fallow'g finish'd the Bean & Pease land
31. 60 Bushs Ash's - 4 Horses from Devizes | Very fine clear mild day, the Snow laying about the Hills

February

1. HAY, finishd the Old this day about 16 Tons and of the new Meadow about 3 [...] [Total] [...]| New hay left to carry through the Winter the remainer of Meadow about 15 Tons The New field hay began this day about 18 Tons [Total] [...] | 6 Horses & 2 Waggons to Devizes for 90 Bu Ash's | Dry day, rain in even'g
2. Heavy rain in night, storms in day hail Sto=in eve'g | 15 S- Wheat to Devizes & 90 Bushls Ashes returned
3. Sharp frost in morning, dry day | 6 Waggon load of Muckle before the Fold | HAY, increased the Ewes allowance from 4 to 6 Trusss
4. Rain in night dry fine day | Waggon & 5 Horses to Oxford for 12 doz Hurdles &c
5. Sunday, Cold winds & Storms
6. Cold Storms, nearly all day with boisterous winds | 2 Waggons & 6 Horses to Devizes for 90, Bu: Ash's
7. 2 Waggons & 6 Horses to Devizes for 90, Bu: Ash's | Very fine Day
8. Beautiful fine, spring like Day | Ploughing began the old Field, muckled & Fold'd for VETCH'S
9. Ploughing began the old Field, muckled & Fold'd for VETCH'S | Mild rain all morn'g very fine after
10. Frost in morning, very fine mild day | Ploughing as yesterday

11. Ploughing as yesterday | Rain in night, Frost in morn'g, fine afternoon | Two tooth Ewe DIED, murrain, the first since the 14th Augst
12. Sunday, very cold dry day
13. Dull cloudy day & very cold
14. Very cold fine dry day | Ploughing as the 11th & 13th
15. Sharp Frost in morn'g, very fine day | Ploughing for Vetches | Rolling Field grass & dry medow = do well
16. Ploughing as yesterday | Sharp Frost & very fine day
17. Some rain in night, very fine Day | ASHES, began sowing, - sow'd the further down Piece,
18. Very sharp & white Frost | 2 Waggons to Devizes for 100 Bushls Ash's | Simkins of Littlecote 2 days Mole catch'g
19. Sunday - Very sharp frost & very fine day
20. Very sharp frost & very pleasant | Ashes finish'd sowing the 2 further Down Pieces 560 Bushls @ 3d is £7 on about 18 Acres is 31 Bu pr A
21. Very sharp Frost & very fine mild Sun shine day | Fallowing athwart began the Swede ground fed off athwart for Barley 2A 3R 28P Swedes began feeding 21st Novr wth 104 Lambs and the Ewes & Rams 12th Janr 233 [Total] 337 finish'd the Piece this day, and began the 10A 0R 22P piece [Total] 13A 0R 10P *
22. Sharp Frost & fine day | Ploughing as yesterday | Vetch Rick - put into Barn
23. Ploughing, finish'd the Swede ground fed off | Frost in morn'g, very fine drying day | The 1st LAMB, the Ewes time out 28th and no Ewe warp'd [182] | The Ewes put in Stand'g Pen, adjoining the Swedes, more Lambs expected fold'd till this eve'g, and Hay evenings only, as well the Tegs with Swedes
24. Frost & very fine day | Ploughing the fallows athwart, bottom of Tan hill Rt hand Piece for PEASE
25. Ploughing the fallows athwart bottom of Tan hill Rt hand Piece for PEASE | 120 Bushels of home burnt Ashes & Soot sow'd on home Meadows about 3 Acres & Cart'd Muckle on the same wth 2 Horses | Some little rain in night, very drying day
26. Sunday, fine day
27. Fog'y morning, fine day Ploughing for Pease as before Flints diging in Meadow @ 1s 6d Load of 42 Bu: | [Pencil note in margin] 20 Lambs
28. Dry cold day, Ploughing as yesterday | EWES, their time out for Lambing 28 Lambs, no Ewe warp'd, no dead Lamb
29. Little frost in morning, very fine day | Ploughing, finish'd the Pease land

March

1. 20-S-Wheat to Devizes & 2T 4C 0[Qr] Coal retd & 24 Shears Carting earth

182 Warp'd – this is taken in the sense of twisted, suggesting that no ewe has had difficulty lambing due to the lamb being mis-presented at birth.

to Meadow, fill up Flint holes Exceeding fine weather the last month NO RAIN since the 2nd FEBY worth notice, the land dry & mellow, & fine time for the Ewes, lamb'g

2. Very fine day | VETCHES, winter sow'd 3 Sacks on 5A 3R 24P own seed, on the Old field Muckled & fold'd
3. Very cold winds & little rain in evening | Carting Straw to Ewes & Lambs & Ploughing the | Down fallows athwart for Turnips SWEDES, have drew, about 4 Acres half up, to prevent runing to green
4. Sunday, rain in the night & boisterous winds & RAIN all day
5. Cold dry'g winds all day, Plough'g as the 3rd
6. Boisterous Winds & RAIN incessantly all day | Thrashing Machine, fix d by 1,=O-Clock, &
7. thrash by 1-O-Clock - the 7th by [...] Sainsbury of Marlbro
7. Frost in morning & heavy SNOW, for an hour in the morning, mild & fine after | Finish'd thrashing by 1-O-Clock & Cart'g Stubbles to the Sheep
8. Sharp frost in morning fine day | Plough'g as the 5th | 100 Lambs, in 10 Days No Ewe warp'd No - Ewe died & 4 Ewes only without Lambs
9. Sharp frost in morn'g, beautiful fine day Ploughing as yesterday
10. The Black Oat Rick put into the Barn | Very Sharp Frost, very fine day | The young SOW, farrow'd 11 - (time up 8th) brim'd 17th Novr is 16 Weeks to the 8th

    Emma Gaby Obit 10th March 1832 Born 22nd May 1805 [183] [This entry written within a thick black border]
11. Sunday, Frost in morn'g, very fine day
12. Thwart'g the Down Piece as the 9th | Thick fog in the morn'g beautiful fine mild day
13. beautiful fine mild day | Ploughing as yesterday | 70 Ewes to lamb
14. Ploughing finish'd thwarting the Down Piece for Turnips | Rain in night, dry till mid day, SNOW heavy till the evening, when 2 Lambs died with the Scour & a Ewe with stoppage | OAT=STRAW, of the 2nd Rick, began giving the Ewes, and to bed them, & may save some hay, being short of it
15. Rain in the night & bitter cold winds & heavy rain all the morn'g, mild dry evening. 3 Lambs died last=night, Carting Stubbles & muckle to Sheep too wet to Plough.
16. Frost in morning, rain all the evening | Carting Chalk to Gateway, by G-Brewers
17. Carting Sarson Stones from Cottage at Avebury & Muckle fm Inn - to the Meadow | Land too wet to Plough, Cold Storms | Emma Gaby's Funeral Mr Abram Henly Tockenham Obit 16th March 1832 Aged 81 [This entry written within a thick black border]

183 Mrs Emma Gaby of Calne buried on 17 March 1832 at Calne, aged 26, *Wiltshire, England, Church of England Deaths and Burials, 1813-1916 for Emma Gaby* [accessed via Ancestry, 24 July 2020].

18. Sunday, Cold storms
19. Mr Henly's Funeral, Cold dry day | 28 Sacks to Devizes, land too wet to Plough
20. 3 Horses to Derry hill for Poles & Rods | 3 Horses began thwarting land for Swedes | Cold storms
21. General FAST, Mild fine day | Ploughing as yesterday, before & after Church service
22. Fine mild day, Ploughing as yesterday John Offer, laying Messrs Brown & Kemms, hedge
23. COW, Calved 13th June, the 40 Weeks up 19th Mar | 3 Horses Plough'g & 3 to Mr Haywards Chirton with 10 Qr Oats | Cold Storms, - 28 Ewes to Lamb
24. Cold storms of hail & Snow & cuting winds | Harrow'g the land, for drilling Pease
25. Sunday, dry fine mild day
26. dry fine mild day | PEASE, marrow fats, began drilling bottom of Tan hill piece Rt hand | Potatoes & Pease, planting in Garden | Plough'g Swede Land, 1st Ewe, died, from lambing
27. Ploughing as yesterday, and Drilling Pease | Very fine mild day
28. Fog,y morn'g, very fine mild day, like midsummer | PEASE, Drill'd 6½ Acres, with 6 Sacks of Marrowfats, in 3 days, wth single Drill | Swedes, going to seed, drawing all. | Roll'g, Field grass, 2nd time
29. Frost & cold harsh winds | Ploughing old field, after the Fold for Vetch's
30. Ploughing old field & Carting earth to the Cottage Garden at Avebury. | Fine day harsh drying winds
31. Fine day harsh drying winds | Plough'g as yesterday [Pinniger then recorded the death of Richard Sadler Smith of Bremhill, along with verses written by Smith's son and the vicar of the parish, along with Pinniger's own observations on the man (this latter was written on an inserted sheet in the diary) - see Appendix Q]

April

1. Sunday, Frost & fine day | Straw discontinued giving the Ewes being very dry weather, do not tread into Dung The Swedes, all drawn, except those half drawn in Feby much check'd, good plan
2. Frost & fine day, Carting Earth to Garden Ave'y | Harrow'g Wheat | The COTTAGE - at Avebury taken down
3. Carting home the Timber & thatch of Cottage | Harrowing the Wheat | Frost & very warm
4. Frost & very warm almost oppressive | Sheep lay back on the Swedes 23 to yean & barren 12 no Lambs 3 Ewes died 188 with Lambs [Total] 226 | Harrow'g the Wheat all 3 tine, & drag'g a part very tuffy & hard 2 & 3 tine

4 & 5. Clearing & sorting, the Stones of the Cottage

5. Draging the Tuffy Wheat & harrow'g the young Wheat finish'd | Some little Frost & oppressively warm after the morning | Funeral, of the late lament'd Mr Smith
6. Thwarting for Swedes | Very cold morning fine day
7. Very cold morning fine day | Barley, began Ploughing for one earth after the Swedes fed of, some of our Neighbours, half done Barley sowing | GAS Tar,g the Granary, Iron Fence'g &c[184]
8. Sunday, Boisterous cold winds
9. Boisterous cold winds | Ploughing for Barley SWEDES=SPOLING two ways, either not drawn they run to stalk & Flower, or drawn they waste & dry up, to avoid both, have put them in heaps, about an Acre remaining half drawn to afford green for the Lambs
10. Ploughing for Barley after the Sheep | Cold drying winds
11. Cold drying winds Carting Dung from the Inn to the Old field, to get it good for Sheep after the Watermeads
12. Some cold rain in the morn'g, dry cold winds after Carting Dung from home yard to Old field CONTENTION in Devizes MARKET The Mealman, persist in buying Wheat by the Imperial Bushel only & the Farmers in selling by the Winchester Bushel only but little business done in consequence
13. Cold dry day, Carting Muckle fm standing Pen to old field | Ploughing last years Pease land for Barley
14. Ploughing last years Pease land for Barley | Carting Hurdles, from the Standing Pen, and the remainer of Muckle to Old field, and the whole covered. The whole of the WHEAT=LAND, DUNGD for next year & ⅔rd Fold'd | Dry day, and milder
15. Sunday, fine mild day, little rain even'g
16. Mild grow'g morn'g, very heavy Shower of RAIN & HAIL, afternoon | Ploughing as the 14th
14. The Muckle, raked of the Meadow, being good feed for a Cow under it, put on the 17th March
17. Very white Frost in the morn'g, warm fine day | Resigned my Office of HIGH CONSTABLE to James Looker of Monckton [185] | Ploughing for Barley
18. Ploughing and 7 Bushels of Winter VETCHES sow'd, the 2nd drift | Boisterous Winds in the morn'g - Cold rain after
19. Some rain & hail storm morn'g - very dry'g after | Ploughing for Barley. EWE DIED, very unexpectedly with swoln udder DEVIZES, Market, attend'd by Dealers from Newbury & Reading, in consequ of the contention
18. Mr Kemms Sheep in his Meadow

---

184 Whilst Pinniger wrote *Tar,g*, it is likely that he intended to write *Tar'g* – that is, Tarring.

185 Monckton – Winterbourne Monkton.

20. GOOD FRIDAY, - very boisterous stormy morning - dry afternoon, Plough'g for Barley, TEG DIED, no apperant cause
21. Fine drying day, Ploughing finish'd up to the Sheep for Barley & harrow'g for Turnips another EWE DIED, swoln udder
22. SWALLOWS, & CUCKOO. appear, Fine drying day
23. Dry, morn'g rain afternoon | BARLEY, began sowing, the Pease & Bean ld [186]
24. Rain heavy in night & very, cold rain incessantly from morn'g till even'g, a Lamb perished thereby. | Horses in the Stable & Carters cuting Chaff
25. Drying morn'g some showers in even'g | Barley sow'd up to the Sheep 9 Sacks
26. Seeds sow'd & the Barley harrow'd off, made very good work | Cold dry'g windy day, some light Storms
27. Fine day, warmer, in the evening | Ploughing, another Turnip drift, fed off | Potatoes, planting, Meadow garden.
28. Sharp white Frost, in morn'g, fine day heavy rain in the evening | Barley sow'd 3 Sacks, another drift to the Sheep | ROUND TAILING, the Sheep 330 and searing the LAMBS (188) tails off & Punching the Ewe lambs off___ Ear, by John Chivers, his son John, & 3 Boys & 1 Girl helping
29. Sunday, dry cold day, COW turn'd to Grass
30. Rain in night & morn'g, dry after, Horses in Stable, morn'g, Cart'g Dung from Inn & Hurdles taking to Silsbury [*sic*] mead & Plough'g after

May

1. Ploughing for Swedes in morning. Horses retd home, very heavy rain, Cart'g Dung afternoon for Potatoes
2. Carting Dung for Potatoes, 48 Bushls grains & 15 Sacks Potatoes from Kennett | Frequent heavy storms
3. Heavy rain in night & driving rain all day, as much rain on the down, as any time in winter | Carting Chalk to level the yard (Hay Rick thrown down)
4. Waggon & 4 Horses to Bremhill Grove for 48 Black thorn Bushes | Carting Hurdles to Watermead | Fine growing day

1. [...] Philpot began feeding Silbury Mead
2. Mr Kemm finish'd feeding his Mead

5. Thin driving rain, morn'g & eve'g | Ploughing the fallows for Swedes
6. Sunday, Rain in morn'g dry after
7. Very fine growing day after the rain | Ploughing, finish'd thwarting for Swedes | WATER=MEAD, the Ewes & Lambs began feeding & HAY, discontinued | Oats finish'd Thrash'g GEOE BAILEY, OBID, the Cryer of Calne
8. Very fine growing day | Plough'g dung in for Potatoes | The 10 young Pigs waned 2 months old

---

186 ld – land.

9. Very cold winds all day | The young Sheep Tegs, turn'd on DOWN | The SWEDES, finish'd, HAY, discontinued fold'd Nether Down Piece for Turnips | Ploughing for Barley | Pease began hoeing all hands
10. FROST, very sharp, very dry cold harsh winds | Ploughing & sowing Barley FINISHED
11. FROST, again very sharp, the Potatoes & Clover cut up Cold dry harsh winds | Barley finish'd Harrow'g off & sow'g seeds | 23½ Sacks of Kennett Barley 3 Qr of Devon Ray 2 Cwt of Own Red Clover ½ Cwt of Mill'd Hop ½ Cwt of White Dutch { sown on 18 A | Rt Philpot finish'd feeding his Mead
12. Rolling the Barley & 1 Team at Kennett finish'g Barley Sowing | Rain a little in morn'g, dry harsh winds
13. Sunday, very cold harsh winds, vegitation at a stand, or gone back (the grass)
14. Mild'r day, Roll'g, finish'd the Barley and draging the Potatoe Land | [Pencil note in margin] Coach Horses left
15. Rain, nearly all the night, cold winds & HAIL storms in the Day | Harrowing the Swede & Plough'g Potatoe land | LAMBS 91 - at 1s pr score cut by Abednego Blackman of Charlton [187]
16. Sharp white Frosts continue, harsh cold Winds | Ploughing & harrow'g Potatoe land 3rd time | SOW, brim'd Mr G Browns Boar
17. Sharp white frost in morn'g, cold rain at times | Ploughing a fold drift for Turnips Down Piece | Pease finish'd hoeing 1st time
18. Ploughing the couchey part of down piece for Turnips | Soot, Ashes & Hen dung mix'd sow'd on a land of weak yellow Wheat | Rain in the night, storms in the day, mild'r

Obit Mr Danl Janner, of Cricklade 12th May 1832, AGED 52 [This entry written between two thick black lines]
19. Sharp white Frost, warm fine day | WHEAT=RICK, the 3rd put into Barn one remaining in the Yard | Ploughing & harrow'g for Turnips
21. Ploughing & harrow'g & rolling for Turnips | POTATOES, Plough'd in an Acre, 2 Horses 5 Women droping - 7 Bushls Prolifficks 9 Bushls White Apple 6 Bushls Red Apple 9 Bushls Black [Total] 31 The half land home side Prolifficks The next half land is White Apple The next half-land is Red Apple The next half land is Black & the adjoin'g ¼-land is Black
20. Sunday, fine mild warm calm day
21. The wamest summer like day yet | The last sowing Barley appearing
22. Fine warm day | Ploughing, Harrow'g & Rolling the Turnip land | TURNIPS, Norfolk whites, sow'd the 1st drift, Nether Down Piece | Mr Wentworth finish'd feed'g Swedes by Tegs
23. Very fine warm day, Plough'g for Turnips The Delia's appear'g
24. Very fine warm day, Plough'g for Turnips the 2nd time finish'd the Down Piece | Silbury MEAD, finish'd feeding all except the marsh kept for the

187 Probably Charlton St. Peter, near Pewsey.

Horses 18 days (17th) Mr Wentworth mead finish'd | Rolling Potatoes & Vetches | Pease began hoe'g 2nd time

25. Very fine warm day | Turnips Norfolk Whites, sow'd - 2nd drift together 3 Acres, Ewes & Lambs removed from Mead to Down & put with the Tegs, and fold'd on the old field, the Lambs a fold before, the Clover good, the only feed for the Lambs till the Vetches are ready 5 or 6 We | 2 Tegs, brot dead Lambs, one of the Tegs died (the Lamb being very large in the night) | Seeds, Cabbage & Winter greens sow'n, -
    Obit Mr Robt Canning Whitefield 25th May 1832, Aged 64 [188] [This entry written within a thick black border]
26. Very fine day, Hurdles removed from Silbury mead & Plough'g Old field for Vetches
27. Sunday, fine warm day
28. fine warm day | 12 Doz Hurdles from Axford Coppice | The Swede land began Plough'g 3rd time | 10 Pease Hoers on, growing very fast, fine time for it, the last sow'g Barley up fine and the Wheat doing well
29. The 6 Horses make 3 Ploughs squinting the Swede land as yesterday [189] | Very fine warm day
30. Fine morn'g rain in evening | 3 Ploughs as yesterday
31. Rain in night & Showers morning dry after | 5 Horses to Honey Street for Garden Posts & Posts for Hurdle House

June

1. 3 Ploughs as before, for Swedes | Fine growing Showers in morn'g dry after
2. Fine growing day, dry | 3 Ploughs finished squinting the Swede land, the third Ploughing | GRASS, began mowing for Horses Silbury mead The Pease finish'd hoeing 2nd time | STRAW, began serving the White Hart Coach Horses at the Inn
3. Sunday, - Very fine warm day
4. Warm dull day | Harrowing the Swede land & Ploughing the old field, at the bottom of the Vetches | The Carter mowing Charlock in Vetches | MOWING, the Field grass began, further Down Piece chiefly Ray, early Pearce's quite ready
5. Dry fine day, Rolling & Harrowing the Swede Land, Weeding began the Barley
6. Rolling & harrowing the Swede & Vetch land | Dry warm day, Shower in the evening
7. Shower, morn'g & evening, dry mid=day | Rolling, Harrow'g &c. as

188 Robert Canning of Whitefield, Ogbourne St George buried on 31 May 1832 at Ogbourne St George, aged 64, *Wiltshire, England, Church of England Deaths and Burials, 1813-1916 for Robert Canning* [accessed via Ancestry, 24 July 2020].

189 Squinting - Ploughing land for a third time, working in a diagonal direction to the first and second ploughings.

yesterday

8. Fine dry day | VETCHES, Summer, DRILL'D, 3 Bushels on 1¼ Acre | Rolling & harrow'g the Swede land | The Sheep counted, now and in October in October Ewes 226 died since 6} 220 Tegs 104 died since 4 } 100 3 Rams two tooth & 4 tegs 7 [Original total] 337 [Total after deaths] 327
9. SWEDES, began sowing Rt of Tan hill road 2 holes & mis a place | Warm fine day
10. Sunday - very fine day
11. Witmanday - frequent showers | The 6 Horses & 2 Waggons to BOWOOD, for 1 Hd Fagots & 15 Timber tops | WHEAT=EARS - the first, seen quite out (by old shepherds shore) [190] | Wheat ours, the left of Tan hill road, the EARS, burst'd half out, and lodged in spots with the rain, being very heavy
12. Warm heavy showers | 3 two Horse Ploughs & sowing Swedes
13. Warm Showers, Plough'g for Swedes
14. Heavy Showers partially | Plough'g & sowing Swedes | The grass in the Old field, is become very heavy | the Sheep tread it under foot in the double fold | began tieing it out, before the Fold Sheep keep at present exceeding plenty
15. Dry morning heavy Showers after partially | Ploughing & Sowing Swedes
16. Fine dry day, Field Potatoes began hoeing HORSES, bled & 2 oz Salt Petre each. Webb, OBIT. | SWEDES, finish'd sowing.
17. Sunday, fine dry day
18. Home Meadow, began Mowing, fine dry day | 2 Horses to Burbage for Thrashing Machine | 4 Horses to Devizes for Timber & Coal | COW=BULL'D at the Hail Farm
19. Warm fine day, Roll'g & Harrow'g for Turnips 4 Horses, Thrashing Wheat | SHEEP WASH'D, at Horton 16 Score & 6 @ 4d pr Score, the usage as last year gave satisfaction, Jas & Stephen Davis only assisting
20. Fine warm day, Shower in the evening | 2 Horses Plough'd 1 Land Down Piece & sow'd to Norfolk White Turnips | 4 Horses finish'd Thrashing the Wheat at half pass'4 O-Clock, began 7 ea morn 2 days, 39 Sacks
21. Ploughing Down Piece & sowing Green round Turnip seed for Wheat | Very fine morn'g, Rain in the evening | HAY, the further Down Piece

---

190 Shore - A broken place or gap in a fence. Dartnell & Goddard, *Wiltshire Words,* p. SA510.There was a breach in the Wansdyke where the old Bath and London coach road passed through, the breach being known as Old Shepherd's Shore: today, ¼ mile to the south of this point, the Devizes to Swindon road (A361) passes through the Wansdyke - this point and the Wessex Water facility alongside (originally Devizes Waterworks, built in 1879) are known as Shepherd's Shore. See H. F. Chettle, W. R. Powell, P. A. Spalding and P. M. Tillott, 'Parishes: Bishop's Cannings' in *A History of the County of Wiltshire: Volume* 7 (1953), 187-197 http://www.british-history.ac.uk/vch/wilts/vol7/pp187-197 [accessed 7 September 2018] and related footnotes 11 and 24.

began carrying at 1-O-Clock, took up 5 Loads when the rain came | TURNIPS, the Norfolk Whites began hoeing

22. RAIN & WIND in the night, to beat down 7 Acres of excellent strong Wheat just in Ear to appearance 12 Sacks pr Acre, now in all probabillity not more than 5 Sacks, the upper side of Tan hill road, left hand | Storms in the morning dry after | Ploughing the Down Piece for Turnips
23. Ploughing the Down and sowing Green round Turnips | Boisterous drying winds
24. Sunday, very Tempestuous Winds all day, like begining of March & very cold to require the Fires lighted, unlike Midsummer.
25. Boisterouse drying winds, the Hay, Down Piece finish'd carrying, being injured, sow'd a Peck of SALT to every load
26. Very fine day | Ploughing & Sowing Turnips Down Piece | Mowing, part of the Marsh, Silbury Mead being too much for the Horses | Hay - the home Meadow carried
27. The CALF weigh'd 13½ Weeks old 50½ li pr Qr 6d | SHEEP=SHEAR, 16 Score & 7 @ 1s pr Sco by 6 Men, began at 8-O-Clock & finish'd @ 8 helpers as last year &c. | Turnips - finish'd sow'g Down Piece | The Grass brot from Silbury mead to make at home, hay | The finest calm mild day yet
28. Very fine & very warm day | 20 Sacks Wheat to Devizes and Ploughing for Swedes the 2nd time, the first destroyed by the Flea | Barley shewing the Beard | [Note in margin] the 'Bastard Calf' brot
29. Swedes, sow'd 2 lands 2nd time next the Pease Rolling the Turnips & Swedes to keep the moisture in the ground | Very warm fine day, FALLOWING the OLD FIELD | Geoe Smith began framing the Skilling for the further Meadow | [Note in margin] Rt Philpot mowing Silbury mead
30. Very warm fine day, My Haymakers at Kennet (having none to make) | The 6 Horses to Calne for 3 Load of Stones for a Stable at Avebury

July

1. Sunday - very warm fine day
2. very warm fine day | Fallowing the Old Field & rolling the Swedes before Plough'g & Sowing again
3. Fallowing & Carting Marter earth to Avebury for the Stable | Ray Seed began mowing, not quite ripe, to make it better Hay, weather fine | Turnips, began hoeing the 2nd time, about 3A Swedes, no hopes of any, being apperantly quite destroy'd by the Flea, and the land too dry, to have any hope of the Seed grow'g by sowing again | Very warm fine day
4. Very warm fine day & Ploughing as yesterday | Isaac Gilbert, began leveling the Bank in Silbury=mead | The last tie of Old Field, given to the Sheep | My Haymakers, assist'g at Kennett
5. Very warm fine day, hay, mak'g W P- of K: finsd | Fallow'g & 20 S- Wheat to Devizes | No prospect of any Pease, my piece for want of rain

6. The Ray seed finish'd mowing, & the Horses taken from Plough, to begin carrying the first cuting at 2 O-Clock, very dry, when RAIN the since the 22nd June | Fallowing as yesterday
7. Fallowing as yesterday, RAIN in the night & this morning, very drying after
8. Frequent thin showers of RAIN, Sunday
9. RAIN in the night & morning, very drying after with thin BLIGHTING RAIN at intervals, the SEED, Pearces early Ray began carrying, Fallowing as before
10. Thin frisk'y, rain in the morning very drying after, the Seed finish'd carry'g in dry condition, Fallowing as yesterday
7. Obit Rickd Box Chippenham Aged 86 [Underlined with a single thick black line]
9. The Masons, began building a STABLE at Avebury
10. The Lambs WANE'D 193 & began the VETCHS
11. VERY heavy Thunder Storm in the night & heavy rain in the morning, to wet the land deep, Plough'd up 2½ Acres SWEDES quite fail'd & sow'd again, rain in evening much wanted for the Pease to fill, & the Meadow Silbury & young Turnips
12. Fine drying morning, heavy Storms after | Plough'd & Sow'd 2½ Acres more of Swedes the 2nd time 14 li to the 5 Acres & 3 holes every other place in the Machine | The new Skilling finish'd Raftering
13. Plough'g up the Swedes fail'd &, sowing new green round seed at 1s pr li, 2 li pr Acre, 2 holes every other place | RAIN heavy in the night, to lodge the Wheat and Barley & Showers in the Day
14. Rain heavy in the night & morn'g dry after | Plough'g & Sow'g Turnips as yesterday.
15. Sunday - Warm fine day
16. Very fine drying day | Finish'd Ploughing & Sowing, the Swede land to Turnips
17. Fallowing the Old Field & harrowing it | Very warm fine day | Wheat Rick, the last of the 4 put in Barn
18. The WHITE=HART, Coach=, horses left the Inn | Very warm fine day, Fallowing
19. Very warm fine day, Fallowing
20. Very warm fine day, Lime & Tiles fm Calne, Fallowing | Mr Wentworth, began mow'g Silbury Mead | Haymaking, finish'd, Cleveancy &c.
21. Fallowing FINISHED, the old field for Whet | Ash's sow'd on 2 Acres of Young Swedes made on the Old field | Isaac Gilbert, finish'd the leveling of Silbury Mead | Cold air fine Day
22. Sunday - Fine dry Day
23. The Thrashing Machine, brot from Marlbro | Plough'g the rough fallow'd Land by Vetches | Warm fine day | Mr Wentworths 4 Men mow'd SILBURY=MEAD, a part 4¼ Acres

24. 4 Horses thrash'g Wheat, and 2- harrow'g the Down Turnips | Warm fine day
25. Some blighting RAIN, early in morn'g, dry day | Thrashing Wheat & harrow'g Fallows
26. Thrashing Wheat FINISHD the Rick in 3 Days | Cart'g Stones, Bricks & from the Stable Aveby | Very warm fine day
27. Very warm fine day HAY, carried Silbury Mead 4½ Acres (6 loads) | Fallowing the Vetch land fed off, & finish'd Carting the Stones from Avebury, the STABLE finished building | John Chivers began Thatch'g Hurdle House
28. Fallowing as yesterday | BEAN=RICK, put into the Barn | Very fine warm ripening day
29. Sunday, very fine day
30. very fine day | Carting earth, back of Hurdle house | Reaping, partially, the Wheat not ripe at Hazleberry &c
31. Very fine warm day | Carting Stones at Avebury

August

1. Carting Stones at Avebury | Fine till 11 O-Clock RAIN remainer of the day | The Carpenters finish'd the HURDLE=HOUSE | The Swedes & Turnips partially distroyed by the Flea, my sowing a second time | TURNIPS Down Piece 9 Acres hoe'd once some large enough to boil
2. Thrash'd, the BEAN RICK 28¼ Sacks with the Machine | Fine dry day, the headland of Swedes hoe'd, the only part of the Piece not plough'd up
3. Plough'g & harrow'g Vetch land fed off | Some little rain, afternoon
4. Fine day & warm, some little rain in morn'g | Plough'g as yesterday | Mr Kemm, began Reaping, the first in our Parish
5. Sunday, some heavy rain in night fine day | The new STABLE, at Avebury, first used
6. Draging & harrow'g the Vetch land. | Beautiful fine day, some rain evening | REAPING, began in the evening, by 4 Taskers, the lodge'd Wheat | The Ewes going to Ram, removed the Rams from the Ewes to the Lambs on the Vetch's
7. Warm fine day, 7 Reapers own put on | Thwarting began the Old field | TURNIP green round seed, sow'd over the swede field where fail'd twice before in places, and rolled it in
8. Ploughing as yesterday & very warm ripening day | Reaping Mr Brown & Mr Wentworth began and generally in this Parish & finish'd Enford &c.
9. Very sultry warm, Plough'g as yesterday & 20 S- Wheat to Devizes & 20 € Coal returned | SWEDES, began hoeing, sow'd second time
10. Very sultry warm oppressive thin rain in the evening | Ploughing as yesterday | SALE=EWES, drew from the Stock and put to the RAMS with the Lambs on the Vetches, mornings & evenings & second Clover between,

all the Ewes, on the Down only, since waned the Lambs, and done well. [Pinniger then recorded his stock of sheep - see Appendix B.] SALE EWES, some Ram'd this day

11. Very fine day | Hurdles removed from Tan hill road to the young grass Nether Down Piece Ploughing as yesterday
12. Sunday, warm fine day | The young grass Sale Ewes, and all the Lambs began, & Vetches, morn'g & Eveg
13. Thwarting the Old field continued | Some Showers in the morning fine day
14. Ploughing as yesterday & Sultry fine day Wheat harvest finish'd Cowitch | Wheat load'd the Waggons to begin Ricking | Wheat, our last sowing green & the Straw BLIGHTED black, the Corn uninjured | The winter TURNIP'S began hoeing by Willm Strange, sow'd after the Swedes fail'd
15. Fine day, very little rain eve'g WHEAT=RICK made the first from the upper Piece, thick set 3A 3R 14P 2A 2R 21P in the Barn [Total] 6A 1R 35P = 837 Tythings = is 103 pr Acre & 4½ Loads
16. Fine day, very little rain afternoon | Ploughing as the 14th | NEW WHEAT, the first at Devizes market 12 S 18 li sold at 32s price of Wheat reduced, the Crop being larger & quality finer, than any remember'd | Sold this day my Old Wheat 12 S 2 li Nt @ 29s
17. Very fine harvest day | Wheat carried Whitelands 4A 1R 28P and part of the 3rd Piece tan hill road 6A 3R 26P together on lower small staddle, the remain'r of the piece for next Staddle 5A 2R 29P
18. RAIN at 4 O-Clock morn'g, top'd the Rick in the rain & thatch'd both, the best Wheat on these 2 small Staddles | Storms all day, to stop the Reapers & and finish reaping, Paid the Taskers (9) our own People to finish their lots Plough'g finish'd thwart'g the Old field
    This Rick slip'd aside, turn'd it found it much sprouted, being top'd in rain [Note in the margin probably added at a later date - see September 11, 1832]
19. Sunday, very fine day some rain morn'g
20. Plough'g began new field middle Down Piece for Rye Very drying fine day, rain eve'g | PEASE, began cuting, ful ripe, reaping wheat prevent'd before | Turnips, Nether Down Piece, large enough to feed, keep at present plenty. | having the remainer of the Vetches & the two Down pieces of grass (2nd Clover 18A)
21. Ploughing as yesterday | RAIN nearly all day at times heavy, the Horses return'd from Plough 1-O-Clock
22. Marlbro Fair, heavy hail storm & frequent storms drying between | Ewes=sale, exceed 20 ram'd | Draging & harrow'g the Plough'g yestery
23. Ploughing the Vetch land fed of | Partial heavy storms, many People carry'g Wheat, Yatesbury, Cleveancey &c | FOLDING, finish'd the Wheat land, all of it muckled
24. Rolling & harrow'g for Wheat & Cart g, muckle for Rye | Storms heavy

partially | FOLD'G began middle Down piece for Rye, and Turnips succeed for Wheat 1834

25. Heavy Storms wth hail during the day, dry evening, too wet to finish cuting the Pease, the Men diging holes in Garden for Clothes Posts | Ploughing, after the morn'g, Vetches fed off

[...] Sunday, dry till 1 O-Clock, when heavy storm, dry fine eve'g, Glass riseing, wind E

27. Dry night, began carry'g Wheat at 6 in the morning, stop'd wth rain at 7 O-Clock, begain at 10-O-Clock and stop'd again at 1-O-Clock, the Rick half up to the eves, steady rain remain'r of the day [191] | PEASE are hopeless of being saved having commenced sprouting & the kids open'g some few yet to cut

28. RAIN continued to last midnight, and storms in the morning Glass to M in much rain very warm & growing, the Pease in the greatest danger after 1-O-Clock, dry North winds & very cold, turnd the Pease & put up the Wheat, Plough'g as yester'y | Old watermead grass the Horses finish'd, with Vetches occasionally the last three weeks

29. Dry night, rain by day light, dry for 2 hours took a load of Wheat Sheafs from Rick, and covered it again with dry straw, RAIN all the the day after 8-O-Clock (2 Turnip hoers & 3 other men, no employ for them | Ploughing for Rye

30. RAIN heavy in the night & morning & frequent Storms in the day, Glass rising slowly & finer eve'g | Ploughing 3rd time the rough part of Wheat field by down Harvest Men work'g at Avebury by Stable

31. Fog'y morning drying after turn'd the Pease in hope of carrying tomorrow, turn'd the wet Sheaves outside to dry, and began carrying Wheat at 12 O-Clock & <u>FINISHED</u>, <u>WHEAT HARVEST</u> by 8- in the evening, dry except a few wet in the bond, finish'd the Rick, began Monday & 4 loads remaining, put in house & covered up. BARLEY began mowing VETCHES, the sale Sheep finish'd Ploughing as yesterday.

September

1. Ploughing as yesterday. | VETCH'S, the young drill'd, Horses began (in Kid | Showers of RAIN in succession all day, much Wheat not carried say 500 Acres, in Parish's of Winterbourne & neighbourhood, Rt Philpot no Wheat carried, my Wheat just sprouting yesterday

2. Sunday, Dry fine day glass riseing

3. Very fine drying day - , the Pease turn'd soft yet 20 S- Wheat to Devizes & removing Hurdles to Turnips

4. Draging & Harrowing before the Fold | Extraordinary fine day <u>PEASE</u>, carried 14 Loads, some soft yet <u>TURNIPS</u>, began feeding, Down Piece

5. Very drying fine day, 4 loads dung to Fold Turnips & Swedes finish'd hoeing

---

191 Begain - began again.

1st time BARLEY, began carry'g SOW farrowd 7

6. Exceeding fine day, Barley carry'g from 8 in the morn'g to 8-O-Clock evening
7. Barley carried 5 loads to top the 1st Rick little RAIN till 11-O-Clock, dry after & carried from 4 to 8-O-Clock
8. Beautiful fine day, Barley began carrying at 8-O-Clock & finish'd at 4, dress'd & top'd the 3 Ricks & FINISH'D HARVEST
9. Sunday, fine warm day
10. RAIN heavy early in morn'g, drying afternoon | Ploughing Clover lay, Down Piece for Rye
11. Draging & harrowing Down Piece 4 Horses to KENNT to help Barley Cart | The Wheat Rick slipped abroad turned and Harvest SUPPER | Fine drying day
12. Fine drying day WILTON FAIR, Prices 2s low'r, Britford Ninety Thousand Pend Ewes 24s to 38s Wethers 26s to 32s Lambs 16s to 22s Mr Boys of Kent, renowned Glynde breed of Rams exposed for Sale | Bot 2 Ram Lambs of Mr Saunders | Ploughing Vetch land mow'd for Horses
13. Ploughing Vetch land Rain in morn'g, dry'g after | 16 S- old Wheat the last to Devizes & 1T 2C 3Qr, Coal retd
14. Fine dry day, Plough'g as yesterday & Draging 4 Men began thrash'g Wheat by the Sack | CALF weigh'd 187 li @ 6d Fatting 11 Weeks
15. Fine day, Drag'g & harrow'g Wheat land
16. Sunday - Very fine day
17. VETCH'S, the Horses, finish'd, and began the GRASS, Silbury mead | Fine drying Day | Draging & harrowing & burning rowet
18. Draging & harrowing & burning rowet | Drying winds storm mid day
15. 6 best Ewes put to the 2 new RAM Lambs
19. Very sharp white FROST, beautiful fine day, cloudless Sky, like a second Summer | Horse raking the rowet & burning Wheat field Rolling & harrowing Wheat field
20. Rolling & harrowing Wheat field | Weather as yesterday Glass high to R in fair
21. Weather as yesterday Glass high to R in fair | 13 Qr Oats to Devizes, Rolling, harrow'g &c
22. Plough'g Vetch land 2nd time, Couchey for Wheat | Fine dry day, not so warm | SALE Ewes & Sale Lambs, put in Silbury mead lay back 3-four tooth & 1- two tooth RAMS, exchanged wth Mr Simkins of Stanton for 1-four & 1 full mouth
23. Sunday, very fine day, The Delias Kidney beans &c, cut up by the late Frosts
24. Fog'y morning, the WEATHER, very UNUSUAL exceeding warm fine day, much finer weather since the harvest days, only one shower since 10th the weather Glass fixed as see the 20th | The small Wheat rick, carried 31st Augst wet in bond, continue so, & sprouted, put the whole Rick, about the

Yard to dry ⋆ | Ploughing, couchey spots in Wheat field

25. Weather equally warm, Plough'g &c, Couch'y land Clover Seed began mow'g, Wheat Rick, put into Barn, see yesterday
26. Ploughing &c - as yesterday, Weather equally warm
27. 26 Sacks Wheat to Devizes 800 Bricks retd | Market very dul & low sold my Wheat thick set 12 S 1 li Nt at 24s & Wheat 12 S 16 li Nt sold at 27s 6d
27. Weather & Plough'g as yesterday
28. Weather equally warm, Plough'g athwart for Rye
29. Weather equally warm, Plough'g athwart for Rye | 16 Qr Oats to Devizes
30. Sunday, some thin fly'g Showers

October

1. 16 Qr Oats to Devizes & Plough'g as 29th Warm & Showery, our Clover seed begin very grass'y, no prospect of carrying it, the Clover seed in general being clean, are carried well & harvest finish'd
2. Plough'g as yesterday, turn'd the Clover seed very damp, but drying wind, rain in the night
3. RAMS, put to the Ewes 2 Old ones of Mr Simkins, to young Ewes 2 Lambs of Mr Saunders to Old Ewes Mr Brown & Mr Wentworth put Rams 20th Septr Fog'y morn'g drying winds, began carry'g Clover seed tolerably dry at mid=day, damp air at times after, finish'd carry'g 8-O-Clock | Ploughing finish'd as yesterday | Tiles at the INN, repairing 1st 2nd & this day
4. Draging & harrow'g, the Plough'g yester'y | RAIN, heavy before & after day light.
5. RAIN, heavy in night, & rain all day til eve'g | Plough'g began for Wheat sowing | Ram sent to Cowitch | Sale Sheep removed fm Silbury=mead
6. BRIDGE, over the Carriage Silbury=md finishd Dry day, Plough'g as yesterday
7. Sunday, Dry morn'g thin showers eve'g
8. RAIN heavy & very boisterous in night & day to uncover the Ricks, dry eve'g & stil | Plough'g after 9-O-Clock. | BAKE=HOUSE, began building for Carter
9. Very dry'g day, Plough'g as yesterd'y | COACH=HORSES, came to the Inn, to serve with Straw 15-Viz The White Hart, York House & Exeter Mail all Mr Halcombs | [Note in margin] See July 18th
10. Rain heavy many times in night, & slight showers in day | Plough'g for Wheat
11. Plough'g for Wheat Fine drying day GRASS, finish'd mow'g for Horses Silbury mead
12. HAY, the Horses began, the new Field Hay Down Piece. [Pinniger then recorded entries and prices at Weyhill Fair - see Appendix H] 17 Sacks Wheat 12S 3 li Nt to Corton & 12 Sacks returned 11S 18 li Nt for Seed | Very high dry'g winds in morn'g, mizly rain at times after, the Clover Seed

put in a Rick | Plough'g as yesterday

13. Plough'g as yesterday high winds in morn'g, fine till eve'g | HAY, Sale Sheep began with Turnips, the Lambs do ill & the Ewes not well on the Turnips Down Piece. the other new Ram put to the young Ewes

14. Sunday, Fine day some thin rain afternoon * | Awfully sudden DIED, Mr Robt CAMBRIDGE 54 *

15. POTATOE'S began diging in Field Very fine day, Plough'g for Rye (3rd)

16. 2 PORKER'S put to fatten Plough'g for Rye (3rd) | Very fine dry day | Isaac Gilbert & his 4 Men, began diging out the Water Carriage's Silbury=mead

17. PEASE=RICK, put into Barn, Very fine day Sow'd 2 S- of Rye, 1 S- of Barley & 1 S= of Vetches on 4½ A= mid:Down Pie | LAMB'S, dipd in a solution of Arsnick & soft Soap, to kill the Ticks, and prevent ther tucking, Viz 1 li of Arsnick & 2 li of Soft Soap to 30 Lambs [192]

18. WHEAT, sow'g BEGAN, Piece 20 A, adjoin'g Down | Mizly morn'g dry after

19. Wheat sow'g finish'd from Potatoes to Down | 2 Loads of Green Sand fm Cherhill for Hatches | Jno Chivers & Son, trim'd 2nd time & markd 100 Ewes & tail'd & markd 100 Lambs | Mizly rain till 3-O-Clock afternoon THUS END THIS BOOK.

[Volume 5 of the diary ends with a page of notes on Church property and tithes, followed by prices for blue slate and deal timber on the inside back cover: diary turned landscape for these entries]

Church Property, There are Ten Thousand, four hundred & twenty Church livings in the Kingdom, and nearly five Thousand of these are Vicarages not Rectories, that is to say, the Tithes belong to some lay Landlord, and not to the Parson of the Parish, it is therefore, evident, that nearly one half of the Tithes of the Kingdom are in lay hands. – add to this, the further consideration, that at least three parts out of four of the whole number of these Ten Thousand livings are in the patronage of Laymen, and that their right of Presentation comes into the market almost daily, like other Property. N B on the Tithe question March the 12th 1832, see Bells Paper

Prices of Blue Slate, J. Fowler & Co. on the Quay Bath, 5% Money | Queens 28 to 36 In deep at 5£ pr Ton | Duchess 12 by 24 In 12,,1,,6 pr Thousand & weigh 50 ₠wt | Countess 10 by 20 In 7,,12,,6 pr Thousand & weigh 35 ₠wt | Lady's 8 by 16 In 4,,1,,6 pr Thousand & weigh 20 ₠wt | [Pencil note in margin] Mr Shack 8-15-0 [193]

Mr Hartnell, Bristol | Duchess 12 by 24 In at 10£ 5s 0d pr m weigh 50 ₠, Cover 7 Square | Short 12 by 16 In at 6,,5,,0 pr m weigh 32, | Countess 10 by 20 In at 6,,10,,0 pr m weigh 34, Cover 5 Square | Plantation 12 by 14 In at 0,,0,,0 | Lady's 8 by 16 In at 3,,10,,0 pr m weigh 20, Cover 3 Square

---

192 Tucking probably refers to the action of the lambs, when irritated by the ticks.

193 Whilst Pinniger wrote *Queens 28 to 36 In deep*, he probably meant to write *Queens 28 by 36 In deep*.

Price of Deals, James Palmer Spring Gardens Bath | Red Deal Plank all lengths 11 In wide 3 In thick 6½d pr foot, forward | The Pieces – 12 In Square 2s 6d pr foot, forward | Lathes for Slates, 6d pr foot forward pr doz, in a Bundle, or ½d pr foot single | [Pencil note in margin] 17 June 1828 [Volume 6 opens with a guide on the number of coops required for sheep at Marlborough, Tanhill and Devizes fairs from 1830 to 1845. This is followed by a record of the weights of various family members on 14th December 1837: these have been transcribed under **Garden, family and health** in the BECKHAMPTON section above.]

20. Devizes Fair, Sharp frost, very fine day. Sold 80 Ewes to Thos Handy, Norton, Nr Bristol @ 26s Sold 70 Lambs @ 16s Sold 14 Lambs @ 13s } To Sister Smith N.B. 14 Coops, a proper No for the Sheep (12 not enough | Harrow'g off Wheat sowd 2 last days | The Grafton Men, cuting out Mr TANNERS Water course
21. Sunday, fine dry day
22. WILLM TANNER of KENNETT, forfeit'd his word, and careless of his Honour forbid the Grafton Men, proceeding with the Work in his Meadow as agreed with me on the 18th Instant, to clean out his Water carriage for a sum, not exceeding 3£, and in all probability would not cost 40s | Sharp frost & fine day | 93 Lambs, sent to Mr VINES, Catcomb to winter at 6s each | The Old & Young Ewes, put together, on Turnips | Raftering the Old field
23. Raftering finish'd the Old field for Winter Vetches Very fine day | The Grafton Men, finish'd the Carriages in my Silbury Mead
24. Beautiful fine day, Sheep tread'g Wheat | Mason, making new Hatch work in Silbury mead, to keep back the flooding caused by WILLM TANNER'S unlawfully useing Hatches, & fixed 18 Inches to high to let the water draw off Silbury mead. | The GRAFTON men (5) fill'g Cart with the trench'g & return'd home
25. Cold dry day. Plough'g for Wheat
26. Very fog'y morn'g & fine warm day. Plough'g for Wheat
27. Very fog'y morn'g & fine dry day. Plough'g for Wheat | POTATOES finish'd diging 130 Sacks Viz Red - 27 Black 35 Prolifficks - 27 White Apple 41 [Total] 130 [Pencil note alongside potato record, as follows] Pigs - 13½ Pigs - 17½ Pigs - 27 Pigs - 14 [Total] 72 Cook - 25 [Total] 97 @ 3s 6d
28. Sunday, Rain in night & all the morn'g
29. John Pinniger alias Willm Golding of London, Transported for life for Horse & Gig stealing [This entry written within a thick black border] Rain all the morn'g, Plough'g as the 27th
30. Fine drying day, Ploughing as yesterday | Rye, Barley, Vetches & Wheat, appearing | The Masons finish'd the Carters Bakehouse
31. Fine day, rain in even'g, Sow'g Wheat, home side | SOW & the 4 young Pigs put to fatten Ewe died of Stop<p>age

November

1. Boisterous stormy night, fine drying day | Ploughing Down Piece, Turnips fed off for Wheat
2. Ploughing Down Piece, Mizly rain nearly all day.
3. Wheat sow'd to Potatoe land, good work, being a fine day.
4. Sunday, rain in night, fine day.
5. Mr Wentworth, began giving HAY to all his Sheep | Bitter cold SNOW storm in morning, very cold drying winds remainer of day. | Wheat sowing began Nether Down piece. 16 cul Lambs @ 10s to Mr Jno Large
6. The 7 spring PIGS taken from stubbleing and put to fatten. The 4 young Pigs, wean'd for Pork | A Down Turnip 15½ li & 32 Inch round | Dry cold day, Plough'g & draging the Potatoe land
7. Dry cold day, Plough'g & draging the Potatoe land a 2nd time for Wheat
8. Dry cold day, Sowd the Potatoe land & the headlands & finish'd Wheat sow'g up to Turnip
9. Rain night & morn'g dry after Harrowing Rafters. | 28 Sacks Pease to Devizes & 25 Sacks Potatoes red Kidneys from Roundway @ 3s 6d | Fold removed from mid:down Piece to next
10. RAIN no intermission all day, 1T 2C 2Qr Coal 3 Horses in Stable & 3 to Devizes wth WOOL
11. Sunday, Dry day & mild'r
12. Rain in night, dry cold day. | WELLS both empty | Raftering mid:down Piece 2nd time athwart for Vetches | Rt Philpot finish'd his Turnips & began Swedes
13. Fine day, Ploughing Turnip land for Wheat
14. RAIN all day, Ploughing Turnip land, Horses in Stable, aft noon | 15 S- Pease winnowd, NO Corn in Barn | Pease=haulm, Ewes began with Turnips
15. Mr Brown diging GRAVEL for ChurchYd | RAIN all day & night & yesterday 36 hours Horses in Stable, 2 Men choose to stay at home no work in the dry
16. Fine dry day, Wheat sowd 2 Acres, up to Hurdles Down piece
17. 15 Sacks Pease & 7 of Brother Wills to Devizes & 1T 11C 2Qr Coal returned | Ploughing as the 12th, Storms | No Corn in Barn, too wet a time to put more in, the 2 Men dig'g Chalk for Garden Walk but little for them to do.
18. Sunday, dry day, <u>FORM PRAYER</u>, read in Church's for abundant Harvest.
    Novr 17th Died Mr John Nalder, of Berwick 41 [194]
    19 Died Miss Fras Maskeline of Bupton [...]

---

194 Berwick – Berwick Bassett. John Nalder of Barwick (*sic*) buried on 24 November 1832 at Berwick Bassett, aged 45, *Wiltshire, England, Church of England Deaths and Burials, 1813-1916 for John Nalder* [accessed via Ancestry, 24 July 2020].

20 Died Mr Chas Stagg of Netheravon 67
22 Died Mr John Hunt Godwin of Holt […] [This entry and the previous three entries written within a three-sided thick black border]

19. Dry day, Raftering athwart finish'd Down P The young SOW, brim'd by Mr Wentworth Boar | [Note in margin] TURN'D 21st Decr
20. Damp air, harrowing Rafters
21. Some heavy Storms, Ploughing the Rafters for Vetches
22. Very fine day, Plough'g & Pressing for Vetches | Men diging Gravel, no other work to do.
23. Ploughing & Press'g for Vetches, finish'd
24. VETCHES, sowed 2½ Sacks on 4½ Acres middle Down Piece | First Pig Kill'd Porker 5 S 18 li | Fine morn'g, damp afternoon
25. Sunday, Rain heavy in night, fine D
26. Rain, very much in night, fine till evening | Fallowing began, the Pease Stubbles [Pinniger then recorded wool weights for 1831 and 1832 - see Appendix E]
27. Rain nearly all day, Fallow'g & Cheese fin Cle:[195]
28. Rain heavy in night, dry day, Fallow'g
29. Rain heavy in night & heavy storms in day, Fallow'g | 1 Winterer died & another, cast a Lamb at Catcomb
30. Dry night, rain afternoon, Chalk, Carting for Garden walk, below the Gravel

December

1. Rain all day, - Horses in the Stable
2. Sunday, rain in night, dry morn'g, storm of Hail in evening
3. Cold dry winds, Fallow'g finish'd Pease Stubl | Wheat last sowing, appearing.
4. The Ewes, finish'd Pease haulm & began the new HAY, Down Piece 2 Truss's only at night with Turnips | Very dry'g day, Fallow'g Wheat stubble,
5. Frost morning, very fine dry calm day | WHEAT RICK the 1st of the four put in Barn on small Staddle 6A 3R 26P | Stubbles carried Whitelands & began fallow'g it
6. Fallowing as yesterday | Rain in night, damp air at times in day
7. Sharp frost morn'g, fine day. Fallowing | BARLEY, the first half a Rick put into Barn.
8. Damp air all day, Fallow'g Whitelands
9. Sunday. - Half pass 2-O-Clock, morn'g my Daughter Martha, born (died at Linden Grove, Calne 1907 or 8) [Note made by E.M.P. 1979] [196] | Mild day, damp air

195 Cle: - Clevancy.

196 On the inside cover of Volume 6 of the diaries, there is the following inscription: *Our great grandfathers diary. (E.M. Pinniger) 1979*. Martha Pinniger died at Linden Grove, Calne on 24 December 1906 and was buried in Calne Non-conformist cemetery, WSA 4381/2/1, *Pinniger genealogical memoranda book, c.1820-1943*, p. 61.

10. Mild day, damp air, TURNIPS, finish'd feeding for Wheat & Plough'g for Seed
11. Mild dry day - Plough'g for Seed | The Ewes feeding the Downs & 4 Truss s of Hay in Fold nights, further Down Piece
12. WHEAT sowing FINISHD, Nether down Piece after Turnips - all about 29A - at 2½ Bu pr A full thick | Fine mild Day, more like Summer
13. Fine mild Day, Harrow'g wheat off sowd yestery
14. Fine mild Day, Wheat Stubbles finishd Carry'g
15. Rain all the morn'g - dry after Fallow'g Wheat Stubbles Tan hill Piece
16. Sunday, Dry day, rain in evening
17. Rain in morning, dry after | Fallow'g as 15th (the Grey Horse ill) Duck
18. Fallow'g as 15th Rain in the morn'g dry after
19. Fallow'g as 15th Fine mild day | Vetches sow'd 24th Novr appearing
20. Ploughing as yesterday Sharp frost & freezing all day in the Shade | The Stock Sheep, counted, Viz 238 - Ewes 2 - 2 tooth Rams 2 - Rams of Mr Simkins 2 - Rams of Mr Saunders [Total] 244
21. Rain in the night, very mild dry day | Ploughing as yesterday
22. Ploughing as yesterday Very mild fine day. | [Note in margin] Sheep bit'g Dog
23. Sunday, Rain nearly all day
24. Fine dry day, rain in evening | Carting Dung from the Inn, to further Down Piece, old Field, for Spring feed & Vetches & Wheat after
21. Young SOW, turn'd & brim'd again
25. XTMAST DAY, Rain all morn'g, fine aftr
26. Fine dry day, Plough'g Wheat Stubbles uper end of Tan hill piece for Swedes | Barly began thrashing, to straw the Sheep
27. The first bacon Pig Kill'd 9 Score | White Frost in morning, very fine mild & clear warm day | Carted the remainer of Dung from the Inn to further Down Piece. | The Fold removed from Down Piece to Wheat stubbles for Swedes Tan hill Pie
28. Very fine dry day | Carting Dung, on home Meadow
29. Rain all the morning, Horses in Stable
30. Sunday, Dry day
31. Rain in night, Dry day | Fallowing Wheat Stubbles | STRAW Barley the Ewes began in the Yard, fold'd with Hay, for Swedes

1833

January

1. Fine day, damp evening | Fallow'g Wheat Stubbles Tan hill Pie [197] | TURNIP'S - by Tan hill road, began by the two tooth Ewes, fold'd wth hay
2. Plough'g as yesterday Dry morn'g, rain all afternoon
3. Damp air in morn'g, dry after | Ploughing as yesterday - wth 4 Horses 1

197 Pie – Piece.

Plough | Silbury, mead, in water last week first time

2. Revd W Willmot Chippenham Died aged 71 Years
4. Very cold dry day, Ploughing as yesterday 2nd PORKER, Kill'd 64 li
5. Ploughing as yesterday. Very sharp frost all day.
6. Sunday - Fine dry mild
7. Beautiful mild, Spring like day | Fallow'g as the 5th
8. Fallow'g as the 5th | Unusual fine mild weather for the Season
9. The weather as yesterday | Fallowing as yesterday & FINISH D except 2 Fold drifts left
10. The 2nd Bacon Pig Kill'd 8 Sco 18 li | For Jas Howell 2 Kill'd 8-15 & 9,,4 [Total] 17 Sco 19 li [Total] 26 Sco 17 li | Very hard frost, freezing all day | Carting Dung on the new Clover field
11. Carting Dung on the new Clover field | Frost in the morning, thaw in the evening | 3 Porkers Kill'd for London 46 li 54 60 [Total] 160
12. Thaw & cold winds | 13 Qr Barley to Devizes & 2T 10C 2Qr Coal retd
13. Sunday, much rain in night, Fog'y day
14. Dry day, Waggon & 4 Horses to Devizes for the first 60 Bushels Ashes
15. Dry cold day 60 Bushels Ashes
16. Dry cold day 60 Bushels Ashes | The 1st Lamb - from the 18 Sale Ewes | The last sowing Wheat, Down Piece appear'g
17. Ash's sow'd 3 loads & 2 loads dung fm Inn | 2 Fat Pigs 9 9,,1, Jas Howell [Total] Sco= 18,,1 | Very unusual mild dry weather for Jany the old ewes doing much better in the Straw Yard & young Ewes on Turnips Hay nights only 8 Truss's for 244 *
18. 60 Bushels Ash's from Devizes | Cold dry day, remaining half of the first Barley rick put into the Barn
19. Very cold dry day, 240 60 Bushs Ash's frm Devis [198]
20. Sunday - Frost & Cold dry day
21. 60 Bushls Ash's fm D | Very sharp hard Frost
22. Exceeding sharp frost, to stop the Horse Pump | Sow'd 3 Loads of Ash's
23. Frost not so sharp - 60 Bushls Ash's fm Devi | Grass SEEDS, began thrashing for Sheep & Horses
24. Very sharp hard Frost - 60 Bu= Ash s Dev
25. Very sharp hard Frost 60_Bu Ash s. Own 24 Bu: Edmonds 564 [199] | ASHES, finish'd sowing ful 40 Bushls pr Acre sow'd about 13 Ac
26. Frost morning, thin rain after | Carting Dung, from Inn, and finish'd dunging, further Down Piece
27. Sunday, Frost morn'g, fine mild day
28. Mild dry day, Carting earth from Tan hill road, to Garden Meadow
29. Thin rain nearly all day | Carting earth finish'd in morning Horses in Stable

---

198 The figure of 240 is a running total of the number of bushels of ashes collected from Devizes to date.

199 The figure of 564 is the total number of bushels of ashes.

after

30. Damp morning, very fine after Clover seed rick, put into Barn | Fallowing a Fold drift Tan hill Piece
31. Fallowing a Fold drift, Frost morn'g SNOW all day after lightly

February

1. Frost in morning, thaw & storms after | HORSES in STABLE, but little to do,
2. HORSES in STABLE, Carters, cuting Chaff | Very Stormy night & rain nearly all day
3. Sunday, damp morn'g dry'g afternoon
4. Some rain in morn'g dry'g after | Plough'g the 2nd Fold drift
5. Rain in night & morn'g dry after | Horses in STABLE, Carter & Boys beat'g dung on Clover
6. Fallow'g 2nd Fold drift finish'd | Rain in morn'g dry winds after
7. Very fine mild day Carting Dung on home Meadow
8. Carting Dung on Clover, finish'd | Rain in night, fine mild drying day
9. Rain heavy in night, fine mild drying day | Hurdles, Carting to Turnips TURNIPS, the old Ewes, began feeding mid=day, Straw mornings & thrash'd hay evenings, have done badly with Straw the Turnips inclining to green, must be fed, 19 LAMBS, from the cul sale Ewes
10. Sunday - RAIN heavy in night & nearly all day
11. Fine dry morn'g, heavy HAIL storm after | Horses in Stable, Carter & Boys, throw'g dung.
    Died Mr John Crook of Avebury at 12-O-Cl: Aged 84 Yrs Octr last
    Died Mr Stephn King of Overton at 6-O-Cl=eve Aged 82 Years [This and the previous entry written within a thick black border]
12. Rain in morn'g dry afternoon | Waggon to Wroughton for Poles
13. Boisterous storms | Carting long Muckle from the Inn, to bed up a standing Pen to lamb the Ewes | 3rd Fat Pig 13 Sco.,4 li
14. Boisterous storms of rain & hail morng dry & mild after | Carting Muckle & Hurdles to standing Pen
15. Carting Thatchd Hurdles &c. to standing Pen, and Ewes, pen'd in it, having a Lamb perish'd in the Fold last night, by the inclement weathr | Very boisterous night RAIN & HAIL, great FLOOD, Silbury mead covered, dry day
16. SNOW heavy in night frost morn'g cold drying winds, in the day | The small Barley Rick, put into Barn | Horses in Stable | Mr Crook. Buryed at W= Tytherton [This entry written beneath a thick black line] [Pinniger then recorded the planting of Flowering Trees and Pot Flowers in the Garden - see Appendix O]
17. Sunday, Drying Winds with storms between
18. Drying winds with storms between | 12 Qr Barley to White hart Calne

19. Fine drying day, sharp frost in morn'g very boisterous stormy evening. | Ploughing the last Fold drift, for Swedes
20. Exceeding boisterous winds & rain all night & day | the SPRINGS, bursting & rising rapidly, the spring water enough for the whole of my Silbury md | Horses in Stable, geting the Potatoes from the Pits
21. Boisterous night, fine drying day | Ploughing the Fold drift
22. Ploughing finish'd, and no Ploughing to do, till, thwarting for Barley | Fine drying day
23. HAY, the old stack of Clover, the Ewes began, expecting that with the Rick of Meadow will serve them till May, having had seed hay since 23rd Jany is become very poor, are now mending on Turnips | Sharp frost, very fine dry'g day. | Carting earth to fill the Gravel Pit
24. Sunday. Cold rain in morn'g, dry eve
25. Fine drying day. Waggon & 4 Hor to Pickwick for Border Stones & Paving Remaining piece of dry bank Silbury mead, made to water
26. Cold storms in morning dry after | Carting earth to Gavel Pit
27. Carting earth to Gavel Pit | Frequent thin storms in morn'g dry after
28. Exceeding boisterous winds & rain all day Horses in Stable, Carters cuting Chaff EWES time for Lambing, have now 50 Lambs, including those from 17 Sale Ewes [Pinniger then reproduced an article from the Devizes and Wiltshire Gazette relating to the depression in the corn market - see Appendix I]

March

1. Drying boisterous winds | Carting earth to Gravel Pit finish'd
2. Calm mild spring day, Carting Hay & Straw to Sheep, & turf from the Down to Silbury mead.
3. Sunday, fine mild dry day
4. fine warm grow'g day, the Turnips & Swedes going fast to green the Women half drawing them. Thwarting began the Pease fallows for Barley.
5. Rolling the new field grass | Very fine mild day | Fat PIG, 15 S.,5 li, the last for own use
6. Rain heavy in night, cold drying day storm in evening | Rolling the field grass finish'd
7. Horses in Stable, land being rather to thwart for Barley 135 Lambs 8 Ewes gone by 1 of which died 95 to lamb [Total] 238 their time out only a week | Frost in morning, very fine warm day
8. 20 Lambs last night & this day | Pierceing cold drying winds. Ploughing athwart for Barley, wth 2 Teams made wth Colt bot at Corsham yesterday
9. Extreme cold winds wth flying storms of SNOW | Ploughing as yesterday & 3 Horses to Quemfd for Seed Beans & Malt | 21 Lambs last night & this day 3 Lambs died last night & 2 Ewes this day one of them drawing a lamb doubled, and the other suddenly from no apparant cause

10. Sunday, very sharp frost & cuting winds
11. very sharp frost frequent light | storms of hail & Snow, cold drying winds | 3 Horses to Pickwick for Paving Stone &c | 3 Horses - thwarting for Barley
12. Ploughing as yesterday, and DRILLING Kidwell BEANS, Sack pr Acre @ 18s | Very cold drying winds
13. Very cold drying winds | Ploughing & drilling as yesterday | Clover, began seeding
14. Very cold harsh winds | Very hard frost to prevent Harrowing and Drilling Beans till 12 O-Clock | The EWES, time up a fortnight for lamb'g only 34 remaining to lamb from 236
15. Very hard frost till mid:day, when drilling Beans & Plough'g as yesterday | Very cold harsh winds
DIED 7th Mr Robt Smith of Broad-town A 54 | 10 Mrs Tayler Chippm Widow of WT- 62 [200] [This entry written within a thick black border]
16. Exceeding cold harsh winds | Harrowing for & drilling BEANS, the the lower side of Tan hill Piece, about 5½ Acres wth 6 Sacks of B P of TYN KIDWELL'S [201]
17. Midlent Sunday, rain in night. cold dry day
18. Damp air & much mild'r day | Plough'g athwart the Turnip land fed off | 3-Ewes died since began lambing 17-Ewes Cast their Lambs & lambs died 28-Ewes yet to lamb or barren 188-Ewes with lambs [Total] 236
19. Cold dry day, Ploughing as yesterday
20. Cold dry day, Ploughing as yesterday and one Horse to Barnbridge for 4 Sacks Beans
21. Very sharp Frost & heavy Snow & hail storm in day, Plough'g as yesterday, & 3 Sample Sacks to Devizes, 13¼ ₠ Coal returned | Drilling Mrs Crooks Beans 1 Sack pr Acre
22. Drilling BEANS=FINISH D with 10 Sacks of Kidwells | Ploughing as yesterday & 10 Qr Bents seeds to Mr Pavy Wroughton | Storms of Snow & Rain & more mild
23. SNOW, deep in the morning to cover the land mild thaw & gone in the evening | Horses in the Stable all day.
24. Sunday, SNOW, again to cover the land in the morning, thaw & gone in the eve g
25. SNOW, again to cover the land in the morn'g mild thaw & gone in the evening | Horses in the Stable
18. The Meadow HAY - Rick, began truss'g for the Sheep, having about 2 Tons of the Old Clover Hay left for them & about 2 Tons left of the Old Clover hay for Nag Stable
25. Bad time for feeding Turnips & Sheep require much more hay. & but little

---

200 WT – West Tytherton.

201 B P of TYN - Broome Pinniger of Tytherton, b. 21 September 1788, died 22 July 1866. WSA: 4381/2/1, *Pinniger genealogical memoranda book, c.1820-1943*, pp.7-8.

left.
26. Plough'g Turnip land fed off | Cold storms of Hail
27. Cold storms of Hail sunshine morning | Ploughing as yesterday & Seed hay geting home to be thrash'd
28. Stiring for Swedes, having Plough'd to the hurdles for Barley | Rain in the night fine drying day
29. The Turnips finish'd feeding 6 Acres began 1st Jany by about 100-two tooth Ewes, & the 9th Feby by about 140 Ewes & Rams | SWEDES, began feeding about 6 Acres | Very fine mild day rain in even'g | Garden seeds, Cabbage Plants & Potatoes &c. planted in the Garden | Ploughing as yesterday
30. Ploughing Whitelands, Harrow'g the Dung on Meadow & 19 Qr Bent Seed to Mr Pavy Wroughton Very fine drying day. DAHLIA'S Plant'd
31. Sunday - Very fine day

April
1. Rain with intervals of Sun shine all day | Horses return'd from Plough'g mid day & put into Stable | Swindon Fair, exchanged the brown Gelding ris'g 8 Yrs bought of Thos Hillier 10th Feby 1829 at 33 £ with Hillier Reeves for a Rone Gelding at 28 £ rising 4 Yrs & gave 11 £ sinking 16 £ from the cost of it
2. Rain morn'g & evening dry between | Remainer of Bent Seed brot from Down to be thrashed | Carting Dung from the Inn to make a second standing Pen, the other heating
3. Fine drying day, some little rain | Carting Dung as yesterday, finish'd
4. Rain heavy in night & morn'g & heavy Showers in the Day, Horses in Stable all day, Carters &c making a new stand'g Pen, the old one dirty & heating, the Sheep do bad & waste the Swedes | Ray grass finish'd thrash'g. 35 Qr Best, 7 Qr Tail [Pinniger then presented a single page analysis on the profit or loss of fattening pigs - see Appendix A]
4. COW=CALVED, took the Bull 18th June (41 Weeks 1st April)
5. GOOD FRIDAY, rain last even'g, very fine mild drying day | The land to wet for any Plough'g, the Horses in the Stable, except 2 - bushing the Meadow
6. Fine drying day, hasty Showers eve'g | 4 Horses Plough'g after Turnips, and 2 Horses to Wroughton wth 12½ Qr Bent Seeds
7. Sunday, fine mild drying day
8. Tegs retuned from wintering, Catcomb 89 4 - died & kill'd with Dogs [Total] 93 sent at Michms | Beautiful fine morn g, heavy storm of hail in the evening.
9. Frost & fine mild day, Plough'g Turnip land finish'd & 8 Qr Barley to Bremhill | HAY, discontinued giving the EWES mornings & give the TEGS, 2 Truss's even'gs & fold'd on muckled, Down Piece for Potatoes
10. Thin rain nearly all day, with some heavy cold storms | Ploughing began

Whitelands for sowing Barley, very wet & very unfit work for sowing Barley, no other work for the Horses

11. Ploughing as yesterday, the land very wet the Horses better in the Stable | Some very cold storms of Hail & Rain | SOW, farrowed 11 - 3 died, (first litter)
12. Cold drying winds in morn'g, very cold storms in evening | Hurdles removing to Kennett & 111 Tegs &c. began feeding Brother Ws Swedes | Hatches, drawn in all the Watermeads preparatory to feeding them, my grass home side of the Marsh very good & lodged the other part flood'd by Willm Tanner, spoil'd
13. Grass seeds sow'd on the new made Meadow Silbury, the Turf kill'd in Summer | Ploughing finish'd Whitelands for Seed | Very drying day, some cold showers
14. Cold drying morn'g, Storms of Hail & rain in the evening | Sunday
15. Frost every morning, heavy Storms of COLD rain & hail afternoon | 4 Horses draging for Barley & 3 Horses to Stockley COPSE, for Thorn Bushes | Clover Seed, finish'd thrashing Best 3 € 2 Qr 0 li Tail 0 € 2 Qr 14 li Ray Seed 1 Qr with some Hop & Clover Pips 120 li the Sack. (116 li Nt) [202]
16. SNOW early in morn'g heavy, rain after drying eve'g, Horses in Stable | Men & Boy's through'g up dung Sheep pen | Barn full of thrash'd seed hay, no fine weather to Rick it & put a Wheat Rick in in the Barn, the Straw nearly all gone
17. Frost & foggy morn, warm sun shine till mid=day, Storm of hail & rain after. Thrash'd Bents hay, removed from the Barn, & made a Rick of it, & salted. Mr Richd Hickley's Funeral, died 10th in his 87 Year [203]
The Revd Rowland Hill, of Blackfryers road Chapel, died the 11th aged 89 Years [204] [This entry and the previous entry written within a thick black border]
18. Fine drying day, with some Showers of hail & rain. | Harrow'g for Barley | Brother Willm at Kennett, bad luck with his Lambs, having lost 20, the last

---

202 Pip - The bud of a flower. Dartnell & Goddard, *Wiltshire Words,* p.118.

203 Richard Hickley of Avebury buried on 17 April 1833 at Avebury, aged 87, *Wiltshire, England, Church of England Deaths and Burials, 1813-1916 for Richard Hickley* [accessed via Ancestry, 24 July 2020].

204 Rev. Rowland Hill established the Surrey Chapel, sometimes known as Rowland Hill's Chapel, about 500 yards from Blackfriars Bridge in London. The building was octagonal-shaped, cost £5000 to build and opened in June 1783. For the next fifty years until his death, Hill was the non-conformist minister of a congregation of Calvinistic Methodists, Edward Walford, 'Blackfriars Road: The Surrey Theatre and Surrey Chapel', in *Old and New London: Volume 6* (London, 1878), pp. 368-383 *British History Online* http://www.british-history.ac.uk/old-new-london/vol6/pp368-383 [accessed 24 July 2020].

week on Swedes, suppose to eat some penricious weed, the cause, the Ewes & Lambs removed to the Mead, and having half his Swedes yet to feed, must give them to others to feed in time for Barley sowing [205]

19. Fine day, with exception of heavy hail Storm | Ploughing Turnip land for Barley | Ewes & Lambs discontinue in Standing Pen at night & lay back on Swedes wth 6 Truss's of Hay, evenings
20. Mild rain in morning, fine mild grow'g day | Plough'g as yesterday
21. Sunday, Beautiful fine growing day Beans in Field shewing out of ground
22. Wheat Rick the 2nd put in Barn, 2 now out Fine warm grow'g day Plough'g for Barl | Rolling RYE, winter Barley & Vetches for mow'g the rye grow'g to spire, fit to feed
23. CUCKOO, heard on Down Fine warm day, plough'g above Swedes, finid
24. BARLEY, sowing began, Whitelands, the land yet too wet & clung, fine morning, gave the Barley two tine with Harrows when RAIN'D heavy put the Horses in Stable.
25. Fine day, Ploughing the Pease land for Barley
26. Fine day, Barley sow'd 7 Sacks upper end of Tan hill Piece
27. Fine grow'g day some little rain | Rolling & harrowing the Barley land sow'd
26. Cow turned to grass
28. Sunday, Fine grow'g day, partial Showers
29. Fine grow'g day, partial Showers | ROUND TAIL, the Sheep & taild the (seered) Lambs 200, by John Chivers his Son, & 2 Shepherd boys, their victuals & Jane picking up the Wool | 4 Horses to Devizes wth 25 S- Wheat & to Westbrook for 10 Doz Sheep Gates | Grass seeds, sow'd Whiteland & top of Tan hill piece | Rolling & harrow'g the Barley sow'd 3 horses
30. Sharp white frost, partial heavy Showers, Ploughing for & harrow'g off the Barley sowd | SWEDES, finish'd feeding by the Ewes & Lambs

May

1. HAY, the Ewes, discontinued & began feeding the Old field, Chert bents, further down piece, very good, about 4 Tons of Meadow HAY & 3 Tons of Field hay left | Hurdles removed from Swedes & Ploughing | Cold driving rain all day from 10-O-Clock
2. RAIN all night & all this morning, the land very wet, & like middle of WINTER, water stand'g on the Down.
   April 19. Mr Kemm & Brown, began feed'g Watermeads
   April 22. Mr P_ Kennett began feed'g Watermeads
2. Robt Philpot, began feed'g Watermeads
4. Mr Geoe Brown finish'd feed'g Watermeads & began New field
5. Mr Kemm finish'd feed'g Watermeads & began New field

---

205 penricious - pernicious.

14. Mr Wentworth & Robt Philpot, finish'd feeding Silbury=mead. | Feeding, began my Silbury=mead, having had a good old field & Rye spoiling
2. Horses in the Stable, too wet to Plough
3. Ploughing last Swede ground fed off | Thin cold rain all the morn'g, drying after
4. Oppressively warm, Plough'g as yester'y Seeds 22 Bushls sown Down Piece 9 A in the Wheat
3. Robt Philpot, inclose'd the waste, bottom of my Tan hill piece
5. Sunday, Warm growing day, Barley appeg [206]
6. Warm growing day | Barley, sow'd 9½ Sacks Tan hill piece | BEANS, began hoeing GRASS, began mow'g Silbury mead for the Horses, very heavy crop
7. Turnips, Mr Canning of Rackly began sowing after Swedes [207] | BEAN Kidney, & 3rd Pease sow'g garden | Harrow'g & Rolling Barley sow'd yesty | Very fine day
8. Warm fine day & harrow'g &c, as yesterday
9. Warm fine day Plough'g & sowing Barley | Swedes, Kennett, all fed off, my Sheep returnd from Kent (1-Ewe died there)[208] | All my Sheep on the Old field, geting hard
8. Seeds harrow'g in Wheat Down Piece
9. WHEAT standing in DEVIZES=MARKET 8 Weeks had only one biding, sold this day 12 S 2 li Nt @ 25s | And BARLEY same standing 10 S 17 li Nt @ 28s Qr
10. 15 Sacks Barley to Devizes & 21 € Coal returned | Ploughing & sow'g Barley, Roll'g & harrowing Very fine growing day
11. RYE, the Sheep began feeding shew'g the Ear | Warm fine day, Barley sow'g | W P- Kennett, finish d Barley sow'g
12. Sunday - very warm fine day
13. The Ewes & Lambs & Tegs DO=ILL, on the the Rye, being too old 3 feet high, and the Ears bursted half out, One land of Winter Barley 1 foot high mixt with Vetches sow'd same time, Sheep feed to the ground and not touch the Rye, could not feed the Rye before, having the old field, Chert bents spoiling | BARLEY, finish'd sow'g, Plough'g & sowd the 2 headlands Tan hill Piece making dry churly work, will not grow till have rain [209] | Mr Wentworth sow'g Vetches & Rape for Sheep by my Down | One of the old Lambs died, the 1st since finish'd lambing
14. Very warm fine day, Carting Hurdles from Kennett and Down Rye, to Silbury mead | SILBURY=MEAD, all the Sheep began feeding the Marsh never so dry & good by opening all the Carriages & sinking to Wm Tanners

---

206 appeg - appearing.

207 Rackley – this is probably Rockley in the parish of Ogbourne St Andrew.

208 Kent - Kennet.

209 Churly - Dry, stiff, hard, as applied to the soil. Dartnell & Goddard, *Wiltshire Words,* p.SA241.

mead, drawn the Water off. too late to feed the mead, other People having finish'd could not feed <the Rye> it before, having the old field & Rye spoiling | Finish'd harrow'g off the Barley

15. Rolling the Barley & began Ploughing the old lay further Down Piece for Potatoes | Sultry warm day with hollow winds | [Diary turned landscape for the remainder of this entry] Seed Barley, sown on 22½ Acres 28 Sacks | Seeds sown in Whitelands 4½ A Chert bents 8 Bu and Clover Pips & hop 116 li | Seeds sown Tanhill Piece 12 A & in Wheat Down Piece 9 A } 21 A Hop & Ray W P- Kennett @ 28s Qr 40 Bu Clean Ray Mr Pavy Wroughton @ 30s 16 Bu Clover - own best 2 € 0 qr 0 li Tail 0 € 2 qr 14 li } say 5 Bu [Total] 61 Bush is near 3 Bushl of Seed pr Acre
16. Ploughing as yesterday | Sultry warm day, some small drops of rain | Bean hoeing 1st time finish d
17. Ploughing as yesterday Sultry warm, some small rain for for a minute
18. Ploughing the old field as yesterday Very warm, at times oppressive
19. Sunday, warm day, slight Shower eve'g
20. Warm dry day, Plough'g as 18th | 98 Lambs cut by Abednego Blackman 102 Ewe lambs [Total] 200
21. Warm fine day, Ploughing as yesterday
22. Warm fine day, Finish'd as yesterday
23. Warm fine day, Draging & harrow'g as yesterday for Potatoes
24. Very warm weather continue. Plough'g 2nd time for Potatoes, whole furrow, nearly
25. Draging, rolling & harrow'g Potatoe land | Very warm fine day.
26. Sunday, Cool air, fine pleasant day | Beans, only about 6 In high & bloomg
27. FROST, morn'g, cut the Kidney Beans & Dalias | Witmonday, Very fine pleasant day | Thwarting the Fallows, began for Swedes | The 2nd Wheat=rick finish'd thrash'g | Barley, the last sow'g, cannot appear for want of rain
28. Press'd 3 Lands further Down Piece, & sow'd 5 Bushls Vetches | FROST, morn'g, cut the Potatos, fine day | MOWING, began home meadow the grass drying for want of rain.
29. FROST again & fine day POTATOES, Plough'd in about 13 Sacks & about 1½ Acres further Down Pi - 3 Horses, 5 Women droping [210] | Hurdles removed 2 load from Silbury mead to Down Piece & fold'd on Vetches sowd yestery | SHEEP, removed from Silbury mead the grass too heavy, do not eat half have kept all the sheep a fortnight the Marsh & enough left for 4 days
30. 20 Sacks Wheat to Devizes & stiring for Swedes | Fine pleasant day | Mowing began Silbury Mead for Hay
31. Ploughing as yesterday & Cart'g Hurdles from Mead | Very warm fine day | WHEAT EARS, Mr Browns half out & Mr Wentworths Talevera quite out WINTER=BARLEY, in full ear

---

210 Pi - Piece.

June

1. Warm fine day, Plough'g as yestey | HAY, carried the home Meaday | Silbury=mead finish mowing | Painted out side of the House by Ward 7 days @ 4s

2. Sunday, Very warm fine day, till eveng when a heavy storm of RAIN to to fill the Water=Barrel, the first RAIN since the 3rd of May, the Barley not yet out of ground where land work'd rough, and all other Corn & Grass want rain

3. Fine day till 4=O-Clock, when a storm of large HAIL & succeed'd with a heavy storm of RAIN, Plough'g as the 1st | My Wheat shewing the EAR, half out May 13th, Horses began the Rye, the Sheep discontinued the Rye, doing ill with it & May 22nd, Horses do ill with the Rye, being in full Ear & very hard, returned to Silbury mead grass May 28th, Winter Barley geting in Ear, began cuting it for Horses & June 4th Horses finish'd Winter Barley, in full Ear & do ill with it, the Rye a bad food, unless eat very early, Winter Barley not so bad as the Rye, but both soon get too old for Sheep & Horses, forward Winter Vetches, very preferable, the Horses have done ill with the Rye & Barley & improved when returned to Mead=grass

4. Plough'g as yesterday, some showers to check hay making

5. Some little rain, the Hay put into Rick carried Saturday Ploughing as yesterday

6. Ploughing as yesterday & 25 S- Wheat wth 3 Horses to D | Fine day, thin rain in evening

4. Mr Kemm, sow'g Swedes

7. Ploughing, finish'd thwart'g for Swedes the ground covered with Charlock in full bloom & docks, no time to clean it before.

7 & 8. Very fine days, Draging & harrow'g for Swedes

8. Mowing <finish'd> finish'd Mead & field | WHEAT EARS, full out part of ours

9. Sunday, Very fine, Barley shewing, the 2nd coming, edge'd grow'd

10. Beautiful, fine calm day, HAY, carried Silbury=mead in excellent condition 8 Loads from 3A 1R 0P, the other part of the Mead kept 507 Sheep & Lambs 14 days, & enough left to keep the Horses a Month | Drag'g & harrow'g for Swedes | Willm Duck, from Calne came

10. W P- of Kennett, Swedes & 12th sow'g Turnips Vetches, Lambs began feeding Down Piece the Ewes, follow & feed the Rye, the 2nd, growing and in ear. | Rt Philpot sowing Vetches, home side

11. Rolling & harrow'g for Swedes.

11. Very Tempestuous all day, with some slight Showers, Ricks & Buildings in part uncovered, Hay blown out of the Fields, the Wheat just in Ear, crippled, by being bent in half the ear on the ground, in some places a tenth

of the Crop The Element darkened with the DUST blown from the roads, to appear almost awful Tremendous Hurricane in London and all parts of the Country [Diary turned landscape]

11. [Note in margin] Rt Philpot, distrain'd, rent 185 £ Tythe free [211]
12. Rolling & harrowing the Swede land | Tempestuous Night & day, not so bad as yesterday, all the dust being blown from the Roads. 3 Men drawing docks from Silbury mead, no Corn in Barn
8. SHEEP=WASH'D at Horton 327 @ 4d pr Score, 2 loaves, Cheese & Bacon, 5 Qts Ale & 5 Qts of Beer, James & Stephen Davis sent the Lambs sent, not wash'd
12. The COW=BULL'D, at the Hail Farm
13. Rain in morning, Thunder, Lightening and heavy storm 1-O-Clock, very acceptable for working the Swede Land & all, ex the hay, Rolling, harrowing &c.
14. Fog'y morning fine day, Roll'g &c as yester'y VETCHE'S, spring, sow'd the 2nd drift | BREMHILL=Farm advertised for Sale [212]
15. Fine day rain in even'g Rolling & harrowing as yesterday Elder = being in full bloom, harvest expected in six weeks, observed as a true rule.[213]
11. OBIT Miss Mary King of Calne [...] [This entry written within a thick black border]
15. SHEEP=SHEAR 16 Score & 7 @ 1s. by 6 Men began half pass'd 7 in morn'g finish'd half pay 7 - even'g, helpers, the Shepherd & Geoe Pope, Stephn take the Sheep to the Field, two Women wind & carry the Wool to stack it, all keep
16. Sunday, Rain heavy in night storms after
15. Mr Wentworth, began sowing SWEDES Mr Brown began Whitelands Mr Philpot began after a good piece of Vetches just fed off, good work & a Crop gained good farming
17. SWEDES, began Ploughing & Sowing, left hand of Tan hill road
17. Barley 1st sow'g, short & picked shewing the beard the last sow'g but just appearing, in part, having lain in the ground dry till the rain came | Frequent Storms, hay spoiling, all Corn improving, Ewe died suddenly & 1 died 15th - ill | VETCH'S, 1st drift, appearing after the Fold
18. Potatoes, in Field, appearing out of ground | Fine dry'g day, Plough'g & sow'g Swedes & sow'g Swedes, on the thwarting, harrowed being clean,

---

211 Distrain - to seize (someone's property) in order to obtain payment of rent or other money owed.

212 Bremhill House and Farm, ring-fenced land extending to 173 acres of pasture and arable land to be sold at the end of July or early August in London at the direction of the executors of the late Mr Richard Sadler Smith, unless sold privately beforehand. Devizes & Wiltshire Gazette, 13 June 1833.

213 That is, 27 July - Mr Wentworth commenced reaping on 31 July, whilst Pinniger started on 5 August.

good work

19. Very little work for 3 Men, no Corn in Barn & continual wet, | Rain to keep Horses in the Stable till 9-O-Clock, frequent storms after | Plough'g & sow'g Swedes | Thos Caney, boy, came from Tockenham as under Shepherd [214]
20. Ploughing & sow'g Swedes, the Hay nearly spoil'd, being a fine morning turn'g it - at 3-O-Clock, heavy storm of HAIL & RAIN, wth lighten'g & 3 DREADFUL peals of THUNDER
21. Fine day damp evening. Plough'g & sowing Swedes
22. HAY. field began carrying, nearly spoild sow'd 1½ Galln Salt to each load very dry'g day RAIN evening. People in general, began carry'g Field hay this afternoon
23. Sunday. RAIN, heavy in night & very heavy Storms in the day | Vetches sow'd only 8 days appearing
24. Fine day, except very little thin rain & much rain in night | Ploughing headlands & finish'd Swede sowing | TURNIPS, began sowing, Norfolk whites harrow'd in after the Fold, Down piece, being old field muckled in winter, fed & fallowd Rt Philpot began sowing Turnips
25. PEASE, Nimbo Taylers, Mr Wentworth sowing for experiment after Vetches fed off | Carting Dung, after Rye & winter Barley fed off to be fallow'd & fold'd & Turnips sow'n on the back | Fine drying day | Hay, carrying from 3-O-Clock till 9-O-Clock very dry, and brown as old thatch sow'd 3 half Pecks Salt to each load, the Rick much injured, by the Rain Sunday, put 3 loads of wet hay off the Rick
26. Drying morning, Carting dung till 11-O-Clock when began carrying Hay, and stop'd carry'g by the RAIN at 2-O-Clock, & very heavy Storm in the evening, about 7 loads remain to finish carry'g the Piece of 18 Acres
27. Rain heavy in night, dry morn'g. Violent storm 1-O-Clock, drove every Person from Devizes Market, Cart'g Dung as yesterday LAMBS=WANED, on Vetches, & Upper Down, very good, not stock'd before | Pigs, ring'd & waned
28. Glass sinking all day, yet fine, little rain eve'g | Cart'g Dung as yesterday & began Plough'g it
29. Plough'g as yesterday & FOLD follow Turnips, sow'd the remainer of Drift after the FOLD | HAY, FINISH'D carrying the Clover, began began mowing a month continued gales of wind all day, danger of being blown off the Rick | This Month, had many very Tempestuous day, the hay spoild, very quick, having rarely two dry days together

---

214 The only Thomas Ca[i]ney in Wiltshire in the 1841 Census was living in Lyneham, aged 20, labourer: that is he was aged between 20 and 24 which would mean he was aged between 12 and 16 when he came to Beckhampton as shepherd boy - however, see entry for 19 July 1833. TNA: HO107/1179/7, *Census returns. 1841 census. Wiltshire. Hundred: Kingsbridge including Parish: Lyneham or Lineham ED12.*

Sacred to the memory of the beloved Mary Pinniger of Cowitch, who changed this mortal for a blessed immortality on Monday morning 24th June 1833, at 1-O-Clock born 1st Novr 1806 | Bury'd Saturday 29th at Bremhill with her Brother William & Grandfather [215] [This entry written within a thick black border]
22nd June Obit Jas Wickham Devizes
25th June Obit Robt Rumming Hillmarton {Aged 58 years} [This entry and the previous entry written between two thick black lines]
30. Sunday, Dry day

July

1. Frequent Storms, 6 Horses & 2 Waggons to Bowood for 1 Hd Faggots
2. Dry day, storm in evening | 3rd Wheat Rick put into Barn. Plough'g the dung'd ground Down Piece
3. Ploughing as yesterday & rolling Vetches | Fine day with some little bligh'g rain
4. Fine day with some little bligh'g rain | 4 Horses Thrash'g & 3 to Devizes wth 20 S Wheat
5. 4 Horses Thrash'g & 3 Rolling & harrow'g | Warm fine day
6. Warm fine day 4 Horses Thrashing & 3 Horses to Tockenham, for Fancy Stones
7. Sunday, fine morn'g, rain evening
8. Fine morning, Storms afternoon | Thrashing, finish'd the Wheat Rick in 3 days | Turnips sow'd a drift below the Rye harrow'd in after the Fold
4. Calf 45 li Qr want 3 li, 13 Weeks old fm 4th Apl
8. Mr Kemm, hoe'g Swedes, & a Piece of Turnips hoed out | My Swedes in rough leaf, size of 6d
9. The last Wheat Rick put into Barn, and thrashing half a day with Machine | Rt Philpot began mowing Silbury Mead | Wheat much improved, got even at head, & a prospect of an average Crop | Barley, first sowing a fine Crop, the last sow'g not in ear, much improving | Beautiful fine day, the evening shewing the finest weather this Summer
10. GRASS, finish'd mowing Silbury mead for the Horses, began May 6th | VETCH'S the Horses began Down Piece, sowd 24th Novr much too late, after the Rye fed, 2nd crop | 13½ doz Hurdles from John Meadham, Chisbury | 4 Horses thrash'g, very fine day | CALF, a suckling, brot, WPxK, Turnip hoe g began
11. 4 Horses thrash'g, & 20 S- Wheat to Devizes 21 ₠ Coal returned, fine dry day
12. 4 Horses thrash'g 2 Ricks in 5¾ days [...] | Fine morn'g thin storms after, Hurdles remove'g from Vetches to 2nd Clover for Lambs to begin it with the

215 Fifth child of John & Catherine Pinniger of Cowage [Cowitch]. WSA: 4381/2/1, *Pinniger genealogical memoranda book, c.1820-1943*, pp.11-12.

little Down, Vetchs about 2 Acres reserved for the Horses, being very bear of flesh | Mr Wentworth, hoe'g Swedes, having hoe'd a Piece of Turnips. | PINK, pipes planted 230 | CAPTAIN, the horse resting for sale

13. Very fine day, top'd, Clover hay Rick | POTATOES, earthing, with 2 Horses Swedes rolling & harrowing | Lambs, fill & do well on 2nd Clover, Ewes follow
14. Sunday, fine morn'g - blight'g rain after
15. POTATOE'S, finish d earthing with the Plough, made much better work than the hoe, & save expence, part of them not hoe'd, no hands to do it | Very fine warm day, harrow'g Swedes | SWEDES, began hoeing, St Anns hill Piece by 2 Devizes Men & Boys = bad workmen & discharged them
16. Very warm fine day | Turnips, new green rounds, Mr Wentworths sow'd part of a Fold drift by the Rye | Carting Dung on Vetch land cut for Horses
17. Thos Clements, Jno Brewer, Jas & [...] Holly began hoe'g SWEDES, very warm day | ECLIPSE of SUN, very clearly seen | Ploughing a Vetch drift fed off for Turnips
18. Ploughing a Vetch drift and Pressing & harrow'g Turnips | Very warm fine day | SILBURY mead, began MOWING, finish'd feeding with the Sheep 29th May
19. Heavy shower at noon, dry before & after | Shepherd Boy - Tom Caney, desert'd. | Turnips green rounds, sow'd a Vetch drift fed off | Muckle, began carting from the Inn, to new field, fed by Lambs, to be fold'd, for Turnips & Whe [216]
20. Fine morning, showers afternoon. | Barley=Rick, put into Barn, no Corn out. Horses, in the Stable, but little work to do
21. Showers in the morning, dry & warm after | Sunday
22. SOW, brim'd, Mr Wentworths, Boar | RAIN all morn'g till 1-O-Clock, Horses in Stable, except removing FOLD, from Down Piece to new field muckled for Wheat
23. RAIN, very heavy Storms, till afternoon Turnips sowd the half of last Fold drift down | Muckle, Carting from the Inn to New field
24. Fence put up at Stable Avebury & Carting Stones from the Stable, home | RYE, began REAPING | Fine day
25. Very fine day, Carting FLINTS for high way & 20 Sack Wheat to Devizes
26. Horses in the Stable all day, no work for them Carter & Boys hay making | Broccoli, Savoys-&c Planted, | TURNIPS, further Down Piece BEGAN hoe'g my first too late, Warm day
27. Very warm fine day | DIPING, 200 LAMBS, to destroy the Tick, put into 5 Gallons of boiling water 2 li of soft Soap, when dissolved add 1 li of Arsenic boild together a short time, and put into a Tub with 25 Gallons of cold water 2 li of Soft Soap & 1 li of Arsenic is allowed for 30 Lambs, but 10 li of Soft Soap & 5 Arsenic was enough for my 200 Lambs [Pinniger then recorded

216 Whe – Wheat.

his stock of sheep as at 24th July 1833 - see Appendix B]

27. HAY, carried, the Marsh of Silbury mead 2½ days making, | Horses in Stable, Carter &c haymaking | W P K, Wool weigh'd 750 Sheep 60 Tod @ 40s about 12½ to the Tod
28. Sunday, very warm fine day
29. very warm ripening day | Carting Muckle from the Inn & removed the Rye, to the Down, to dung the ground
30. SWEDES, finish'd hoe'g second time Carting Dung from the lambing Pen to the Rye ground for Turnips & Wheat
31. Very fine ripening day, Reaping Mr Wh Talav [217] | Carting dung from the Lambing Pen to the Rye ground for Turnips

August

1. Shower in morn'g, warm fine day | Ploughing the Rye ground dung'd yesterday
2. TURNIPS sow'd Rye ground for Wheat Dry dull day | Carting dung from lamb'g Pen to new field
3. Carting finished & Plough'd the gd for Turnips RYE carried in Barn, Fine day | HORSES, fast improving with old Vetches & 1 Sack of mix'd corn. (Oats & grittle'd Beans mix'd) for 7 Horses, baited eve'gs on the young Clover (pr Week) [218]
4. Sunday, very fine day, Reaper flocking from the Bourne
5. REAPING, began in the evening by 5 of our own People to green to put on Taskers | Warm fine day, FALLOWING, new field began with 4 Horses, stiff plough'g resting 2 Horses Turnips sow'd Lambing Pen
6. Turnips hoe'd 1st time, sow'd 8th July, after the after the 1st feed'g Rye. | Very fine day fallow'g as yesterday
7. Very fine day fallow'g EXCEED'G sharp FROST to cut up the Potatoes & D[...]
8. Frost & fine day, fallowing as yesterday
9. Fine day, fallowing the Vetch grd cut for Horses
10. fallowing new field | The 98 Wether Lambs put in Silbury mead for Marlbro Fair, the grass hurdle'd in ties | The Wheat ripening fast, put on 5 Taskers - Fine day, very heavy Thunder STORM, in the evening, very acceptable for the Turnips & grass
11. Sunday, very fine drying day
8. New Wheat, Devizes=market 12 S 14 li Nt @ 30s S
12. Very fine day, fallow'g New field, the Fold following | Rt Philpot, began breaking up his DOWN next Mr Browns Whitelands by raftering it Wheat began carrying generally in this Parish
13. WHEAT, began carrying from the ½ land 8A 0R 5P = 18 Loads, the Straw

---

217 Wh – Wentworth; Talav - Talevera.

218 grittle'd – probably griddled.

dry and warm, the corn mixed hard & Soft | The prospect for Sheep keep bad, the last sowing Turnips not move unless where well dung'd and work fine

14. Very calm fine ripening day, fallowing
15. Reaping began the Down Piece, all others in the Parish finish'd reaping Very fine day fallow'g before the Fold
14. BREMHILL FARM, offered for Sale by Auction. The highest true biding 11550 Bought in at £12300
16. Very fine day, some thin blighting rain, and storm in evening | Fallow'g as yesterday, Mr Wenth mow'g Silbury Marh [219]
17. Exceeding sharp frost & fine day, fallow'g | VETCH'S, the Ewe Lambs began feed'g, last sow'gs | VETCHS, the old discontinued giving the Horses being well Kid'd & nearly ripe. | CLOVER, began cuting for Horses | Reapers Paid off & made 2nd Wheat Rick 5½ Sacks, of Mouse knaw'd tail Wheat sent to Mill, made 17½ Bushls & 8 li Flour worth 7s 6d pr Bushls the Wheat not worth 18s pr S
18. Sunday, thin blighting rain at times
19. Very fine day, Wheat carried the remainer of the 20 Acre Piece | Wheat finish'd reaping Down Piece sowd 12th Decr after Turnips, very green, good at head. | Turnips 1st sow'g, hoeing the 2nd time sowd 24th June

    Augst 6th Died Miss Mary Dorothea Buxton Aged [...] Years [220]

    Augst 19th Died Mr Henry Headly Gaby Aged [...] Years [221] [This entry and the previous entry written within a three-sided thick black border]
20. Very boisterous winds, threatening rain | Wheat carried part of Down Piece 6A 2R 11P and part of 20 Acre Piece next Barley 0A 3R 31P [Total] 7A 2R 2P
21. Exceed'g boisterous winds, nearly equal to winds on the 11th of June, to prevent people, carry'g hay, <u>RAIN</u> from 9 to 11-O-C eve | Fallow'g & 22nd Fallowing & fine day
22. <u>Marlbro Fair</u>, [...] | Being a very dry time & no prospect of rain & short of Sheep Keep, took all Sheep for sale sold 54 Ewes @ 30s, over full mouth 10 Ewes @ 26s broken & over full mouth 50 Lambs @ 22s 18 Lambs @ 13s 6d culls of all 30 lambs, the middle return'd bid 15s
23. Fallow'g & fine day, some little rain even'g A two tooth Ewe Kill'd 28 li giddy
24. Turnips began hoe'g sow'd 16 July

---

219 Marh - Marsh.

220 Mary Dorothea Buxton of Huntercombe, Bucks. buried on 12 August 1833 at Mildenhall, aged 32, *Wiltshire, England, Church of England Deaths and Burials, 1813-1916 for Mary Buxton* [accessed via Ancestry, 24 July 2020].

221 Henry Headly Gaby of Calne was buried on 25 August 1833 at Calne, aged 31, *Wiltshire, England, Church of England Deaths and Burials, 1813-1916 for Henry Gaby* [accessed via Ancestry, 24 July 2020].

25. Sunday, very fine day
26. GLASS, rising beyond E in change, very fine | VETCH'S, seed began hacking | Rye, winnowed 9½ Sacks from 1A 0R 3P weigh 11 S 12 li Nt (sold at 17s to 20s) | Fallowing up to the standing Clover
27. Thwarting the Fallows, began | WHEAT, carryed the remainer 2¼ A, very dry very green & soft when cut, good berry - for Seed in Barn. | BARLEY, began mowing Whitelands by [...] Thick fog and fine day | Turnips hoe'g sowd 23rd July
28. Very warm fine day, thwart'g as yesty
24. Fallow'g & very fine day RAIN, Show'r even'g *
29. Very warm fine day, Thwarting with 2 Ploughs, have use'd but 1 Plough the last Month, resting 2 Horses, the fallowg being being heavy *
[The lettings of the turnpike tolls recorded here]
1833 Beckhampton GATE, let 1st July last to Isaac Selman Low for [...]
1832 The Gate let last Year to Ezekiel Evans for 1325£
1834 for 1362 to Ezekiel Evans
30. Very fine pleasant morning, till 4-O-C when very driving heavy RAIN, till 10-O-Clock at night, very acceptable | Plough'g as yesterday | Barley, mow'd upper part of Tan hill piece
31. Ploughing as yesterday, the RAIN gone as deep as the Plough | a very gracious rain, Turnips &c renewed with growing life | Very boisterous Winds all day to uncover the Ricks

September

1. Sunday, very fine dry'g day
2. Fine day rain in evening | Barley, began mow'g, bottom Tan hill P | Thwarting finish'd the Fallows [Pinniger then recorded the weighing of this year's Wool and presented comparisons from 1829 to 1833 - see Appendix E]
3. Heavy Storms & a very heavy HAIL storm Ploughing began Wheat stubbles for Vetches, intended to be fed in May & SWEDES succeed in June, to gain a Crop see the 15th last June | Too wet for mowing Barley, and began BEAN cuting, & too wet also, put the Men thrash g Wheat & old B
4. Ploughing as yesterday Storm'y morning, dry evening
5. Very fine day, Ploughing as yesterday
6. Very fine day, very drying & wind'y | BARLEY, began carry'g, Whitelands Rick by Waggon house VETCH'S seed, carried VETCHS sow'd 3½ Sacks of old on the Wheat stubble, to be fed in May for Swedes to follow, to gain a Crop
7. Barley carrying all day from side of the hill, Tanhill Rt hand Piece. | Very drying Winds, to impede the loading
8. Sunday, little rain in night & frequent Showers in the day
9. Dull day, damp air at intervals | Harrow'g off the Vetches sow'd the 6th & Plough'g the Vetches ground seeded. | Turnips - Just appearing, sown 5

WEEKS (5th Augst after the Rye & standing Pen.

Septr 2nd Died Mr Willm Cambridge, of Chippenham Aged 88 Years [This entry written between two thick black lines]

5. My Sister Sarah, left Calne, to live with her Son John at Cirencester
10. Ploughing as yesterday, Fine day Barley not dry enough to carry
11. Ploughing after the young Vetches fed off | Rain in the Night, storms in the day
10. The last sowing Turnips, finish'd hoeing
11. The RIDGLING, changed for a ROAN Colt [222]
12. Ploughing for Vetches & harrow'g Wheat land, Very fine drying day | Very dull MARKET, 27s only offered for Wheat 12 S 14 li & 13 Score sold for 28s Sack
13. Drying windy day, some few drops of rain | Barley Carting all day
14. Thin rain all the morning | Harrowing the Wheat land
15. Sunday, fine drying day
16. Very heavy RAIN in the night & frequent heavy Storms in the day, the Barley in danger of growing in the field and the Ricks not thatched Plough'g for Vetches & harrow'g the Wheat land The young Turnips, dieing before & after the hoe, generally, altho, the land moist the old Turnips & Swedes do well
17. Drying day, some little rain. | Harrowing the Wheat land
18. Some light Showers in the morning to prevent carrying Barley till 1-O-Clock | Rolling & harrow'g as yesterday
19. Barley, began carrying at 9-O-Clock, heavy Storm 2-O-Clock, to prevent finish'g Barley harvest, 10 loads to carry.
20. VETCH'S, finish'd feeding, about 4½ Acres began them 17th Augst and a tie of Clover every day. | CLOVER, the Ewe lambs finish'd feeding, the 12th July began it wth Sale Sheep | Clover enough remaining to keep the Horses a month, Ploughing the Vetch land for Wheat | SILBURY=MEAD, the LAMBS, began - the only feed for them, till the Turnips are large enough, the Ewes live hard only the Down, for them | RAM, turn'd the best old of Mr Simkins to the Stock ewes | Dry dull day, the Barley not fit to carry
19. BEANS, finish'd cuting
21. Very thick Fog, all morn'g, fine drying afternoon | BARLEY, harvest FINISH'D, in making the round Rick, the 2 Ricks by Waggon house bright, some damp loads in the other 2 Ricks | Plough'g & Rolling as yesterday
22. Sunday, Very fine drying day
23. Very fine drying day | Finish'd Ploughing the Vetch land fed off
24. Rain nearly all day, & Horses in in the Stable
25. Very drying winds, & heavy storms alternate | Ploughing began for Wheat

---

222 Ridgling or Rig - A horse which has not been 'clean cut', that is, is only half gelded, owing to one of its stones never having come down. Dartnell & Goddard, *Wiltshire Words*, p.133.

sowing | VETCHES, sow'd the 2nd drift, 6 Bushls own new ones Vetches the 1st sow'g up very fine

26. Storm in morning, drying after | Plough'g as yesterday & 10 Qr Barley to Devizes Coal retd
27. Beautiful fine Summers day, (Mr Brown some Barley & 2 Pieces Oats, not carrid) | Ploughing for Wheat, Rt Philpot, break'g up his little Down. (Raftering)
28. Ploughing as yesterday | BEANS, began carry'g 6 O-Clock morn'g after a dry night, RAIN at 9 O-Clock to stop'd carrying, after 5 - Loads | very heavy rain afternoon | T P, sworn a Commissioner of the Kennet & Avon CANAL Navigation with Guy Warwick of Milton John Washbourne Yatesbury John Brown Burdrupt John Gale Savenake Park
29. Sunday, very fine day
30. Warm & fine day, Plough'g for Whe

October

1. Very warm fine day, Plough'g & harrow'g for Whe
2. Unusually fine, a second Summer BEANS, finish'd carry'g, very dry On Staddle 5A 3R 4P, on ground 3A 2R 0P
3. Foggy morn'g, very fine day Plough'g, Fallows for Wheat | RAMS, turn'd, other two, to the Ewes
4. Ploughing as yesterday, very fine
5. Press'g the 18 Acre Piece, Plough'd for seed the land too dry for Wheat sow'g | Very fine day & an unusual fine Week | Painting front Gates, Iron fences &c - Dung from the Inn, finish'd Cart'g & about 2 Acres to finish Muckle & Fold for Wheat
6. Beautiful fine day Sunday
7. Beautiful fine day Press'g finish'd as 5th. Press about 9 Acres pr day
8. Ploughing for Seed Down Piece | Very foggy morn'g, & very fine day

7\. TURNIPS, began feeding with the Lambs, Down Piece & the Ewes, put in Silbury Mead, still living very hard & too poor | Bot a 6 Tooth Ram, Mr Saunders Stock and a 2 Tooth Ram, Mr Saunders Stock at the late Mr Kings Sale, at Overton and put to the Stock Ewes, removed the other 4 Rams

9. Very fine day, Ploughing as yesterday
10. Very fine day, Ploughing as yesterday
11. Very fine day, Ploughing as yesterday | CLOVER, finish'd mowing for Horses | PEASE, Nimbo Taylers, Mr Wentworth carried, sow'd 25th June Obit 4th Octr Mr John Slade, Glazier &c Chippenham Aged 58 Years [This entry written between a double black line and a short thick black line]
12. Fine morning, little rain in evening Harrowing & Rolling Down Piece GRASS, began mow'g Silbury mead for the Horses | DUNG'D, finish'd the Clover field for Wheat, the Fold follow 3 Weeks to finish, NO dung left in the Yard, or at the Inn

13. Sunday, - fine drying day
14. Rain in Night & morning, dry after | Plough'g the Bean Stubble, too dry to Plough for Wheat or Sow | RAMS, turn'd 3 resting ones to Ewes & took the 2 Overton ones away
15. Fine morn'g, rain heavy in night & heavy Storms afternoon | Plough'g as yesterday
16. Ploughing the Clover lay, muckled & fold'd, one earth for Wheat | Wheat sow'g, generally, began in in this Parish, the land now wet enough & work to a powder | Rain heavy in night & afternoon
17. Fine drying day, Ploughing as yester'y | CABBAGE, Plants, plant'g for Spring
    9th Octr died suddenly John Bailey Wool stapler Calne
    12th Octr died after a lingering illness Mr Benjn Edwards, Calne [This entry and the previous entry written between two thick black lines]
18. WHEAT, began drilling, the Down Piece 6 Acres & 2 Acres of the 18 A: piece 2 Bushls pr A, at 1s 6d pr Acre | Fine morn'g rain in evening [Pinniger then recorded details of Weyhill Fair held on 10th October - see Appendix H]
19. Heavy Storm afternoon Treading the 6 Acres of Wheat drill'd on Down Piece with 250 Ewes in 3 hours, firm work | DRILL'D, about 8 Acres, as yesterday
20. Sunday, - Very fine drying day
21. Rain in Night & rain at times all day | Ploughing the whole ground after Fold
22. Ploughing Wheat x Stubbles - dry day x Bean
23. Rain heavy in night & morning - Horses in the Stable, Brewing of Grains from Kennett | TURNIP'S, the Ewes, began feeding the young on Down no heads where fail'd, to sow it to Wheat.
24. Very fine drying day. finish'd treadg the Wheat sown, harrow'd after & the furrows struck
25. Fine drying day, 3 Horses Plough'g & 4 Horses to Tytherton for 6 Hogsheads of CIDER [Pinniger then recorded numbers and prices from the great Lamb Fair held at Overton on 18th July - see Appendix H]
26. Fine day, heavy rain in evening Fallow'g Bean Stubbles | HAY, the Sheep began, the Seed thrash'd in Spring & Rickd salt'd
27. Sunday. dry morn'g rain even'g
28. Rain morn'g, dry even'g | 4 Horses Plough'g Bean Stubbles the others resting
29. Plough'g as yesterday, Fogg'y morn'g very warm & fine day.
30. Plough'g as yesterday | POTATOES, began Ploughing out out of the ground, a bad crop | SULTRY WARM & fine day
31. SULTRY WARM very fogg'y more, like June weather | 14 Qr Barley to Devizes & finish'd Plough'g up Potatoes

November

1. Rain in morn'g, very windy & dry after | Plough'g a drift after the fold, New field | Pressing the one earth Wheat land
2. Very fine mild day, rain & windy even'g | 4 Ewes & a 2 tooth Ram, Blown on Turnips, drop'd dead going to Fold - one after another
3. Sunday, - fine dry day
4. FOLD, finish'd the one earth land for Wheat & removed to the Potatoe land | Ploughing the Fold drift & harrow'g after the Presser | Lambs at Kennett, dip'd for the Tick | Fine drying cold day
5. Fine morning rain, evening | Ploughing, finish'd, the one earth land
6. Ploughing Turnip Land Down, morn'g & finish'd Press'g the one earth, afternoon | Rain in morn'g, drying after | 7 Store PIG'S. put to FATTEN, 4 Score ea farrow'd 11th April
7. Harrow'd the one earth land for Drill'g Dry morn'g, HEAVY RAIN, from 12-O=Clo to night, to prevent Drilling | WELLS, empty, see 12 Novr last Year

6. Fat CALF, Kill'd 60¼ li pr Qr 18 Weeks old = see 4th July, the other Calf

8. Ploughing Down Piece for Wheat | Cold drying day
9. Fine calm drying day | DRILL'D, the remainer of the 18 Acre Piece about 2½ Bushls of Wheat pr Acre
10. Sunday, fine morning, dampe eve
11. Rain heavy in night & morn'g the | Horses in Stable & Plough'g Down Piece after | FIRES, frequent, one at Urchfont last night & one at Wroughton in the night Monday the 4th Incenderies [Pinniger then included a copy of a table regarding the 1833 corn yields - see Appendix I]
12. Ploughing Down Piece, young Turnips fail'd fed off | Fine dry day
13. Fine dry day finish'd Ploughing as, yester'y | SOW, farrow'd 7, time up 11th being 16W= from 22nd July | BARLEY, finish'd thrashing the old
14. Dry cold day | DRILL'D 2½ Sacks of Wheat about 5A next the other Piece on Down About 30 Acres now drill'd with 15 Sacks of Mr Cannings Seed 2 Bu. pr Acre, equal to 2½ Bu: broad cast sowing TURNIP'S, finish'd feeding further Piece on Down 2 Acres began 7th Octr
15. Ploughing Turnip drift, Cold dry day
16. Ploughing Turnip drift & harrow'g it | Dry morning, thin rain after | TURNIPS, finish'd a Piece about 1 Acre mid: furlong Down, the Sheep have done ILL on the Down Turnips with 6 Truss's of Seed HAY nights | SWEDES, began feeding Tan hill Piece with the Lambs, Ewes to follow | BARLEY new, the first put part of Rick into Barn, when the rain stop'd

    17. Sunday, dry day

    8th Suddenly, the Wife of John Flower, Langley

    10 Mr Geoe Pinckney, Wolf hall

    10 Mr Chas Hitchcock Monckton Agd 79

14 Miss Sarah Locke, Rowde Ford Aged 20 much admired Beauty [This entry and the previous three entries written within a three-sided thick black border under the heading "Died"]

17. GRASS, Silbury mead, the Horses finishd & HAY, began, the old thrash'd seed | FOLD, finish'd the Potatoe land for Whet & all the Wheat land dung'd

18. FOLD, began the Barley stubble, Tan hill piece for forward Turnips & Wheat | Very mild day, damp morn'g Hurdles removing, from Down Pieces, & Ploughing Potatoe land | Ewes, feeding, Silbury mead, having finish'd mowing it, for Horses

19. Ploughing - as yesterday | Foggy & very mild dry day, the Grass growing, more like May, then Novr | 6 li Ginger @ 1s 4d, & 19 li Treacle @ 3d with Flour enough to make 28 li of Gingerbread @ 8d - is 18s 8d, put into 4 Hogsheads Cider & 2 Hogsheads of Cider=Kind [223]

20. Ploughing finish'd for Wheat the Turnip land fed off & harrow'd | Fine dry day

21. Fine dry day | The 2nd SUCKLING CALF, brot | WHEAT sowing FINISH'D, by DRILL'G 5 Sacks of Talevera (fm Hazleberry) after Turnips Potatoes & Rye seed'd Down Pieces on 7 Acres

22. High winds & dry in morning, thin driving rain after | Barley remainer of Rick began 16th put into the Barn | Sheep treading Wheat sow'n yesterday ONE Team Plough'g Bean Stubbles, & the other Horses RESTING, from this day One Team will now finish the fallow'g

23. WELL, in Rick yard, deepening, not having any water | Treading last sowing Wheat, Ploughing Bean Stubbles | Very fine day, Marlbro Fair 2s head Ewes lower than Devizes Fair

24. Sunday, dry morn'g rain after

25. SNOW morn'g, Fine dry cold day. | Ploughing Wheat Stubbles
Novr 25th Died Mr Jno Hopkins of Lyneham Aged 69 Years [This entry written within a three-sided thick black border]

26. Very sharp Frost & fine day, Plough'g Bean Stubbles | PORKER 5 S 16 li

27. Plough'g as yesterday | Fine day, cold drying winds | WELL, in Rick yard lowerd to Water [...] feet in 3½ days at 1s 6d & keep

28. Finish'd Plough'g Bean Stubbles | Dry morn'g, very DRIVING & heavy rain after

29. VERY boisterous winds, some rain | Plough'g Wheat Stubbles & Carting Trenching Silbury mead | Ewes, do no longer well in Silbury Mead

30. EWE'S, began SWEDES, the Straw in Yard & 4 Truss of Seed hay for 389 Sheep in Fold | Fine dry day, Plough'g as yesterday - Ewe Lamb or Teg Killd giddy | Thrashers, lowering, Water carriage mead

223 Ciderkin.

December

1. Sunday - Dry morn'g, rain after Hazlebury
2. Rain morn'g Fine day - Bristol | Plough'g Wheat Stubbles
3. Plough'g Wheat Stubbles Rain morn'g Troubridg
4. Dry morn'g, driving rain after | Carting Muckle before the Fold
5. Fallow'g Wheat Stubbles, Fine day | HAY, the Meadow Rick, Cow & Nag begn | SWEDE'S, the Horses & Cow began having a prospect of plenty SHEEP=COUNTED [see Appendix B]
6. Rain in Night dry day Carting earth from Watercarriage Silbury mead to the flat | Drew 40 Ewes very poor & put with the Lambs on Swedes
7. Fallowing Wheat Stubbles | Boisterous Winds & heavy rain
8. Sunday, Fine dry'g day, damp eve'g
9. Very Tempestuous, winds & rain all day Fallow'g afternoon | Stephn Jacob & Geoe Davis, ill Scarlet=fever
10. Calm drying day | Fallowing as yesterday | Talevera Wheat sow'd 21st Novr appear'g
11. Cold drying morn'g, large blossoms of SNOW & Sleet afternoon | <Fallow'g> making Water carriage Silbury=mead | One Plough, assist'g W P- Kennett Wheat sow'g near home (new Farm) | Married at Compton Bassett William Stiles of Up: Lambourn Ann Pinniger of Cowitch [224]
12. Sharp Frost morn'g, cold dry day | Carting Muckle before the Fold, 5 Loads from the Inn
13. Fine dry day, Fallow'g
14. Fine dry day, Fallow'g
15. Sunday - damp mild day
16. Thin driving rain all day | The 3 Boys (see the 9th) recover'd & at work The Horses Fallowing

12\. 1st Bacon Pig - 7 Sco 9 li

17. Fallowing, Very boisterous cold dry winds
18. Fallowing, FINISH'D, except 6 lands to to receive the Fold (for Swedes) | Dry & cold boisterous winds
19. Boisterous Winds & rain Carting earth from Tan hill road to stone heap Decr 15th died at Warminster Richd Noyes, Aged 63 Yrs [This entry written within a three-sided thick black border]
20. Wheat stubble fallow'g being finish'd the Horses in Stable, Carter Winnow'g | Calmer morning, continue Boisturous winds & heavy storms after Horse keepers
21. SNOW & heavy rain in the night, to cause a FLOOD, Silbury mead under water, dry till now & the Wells continue empty | Horses in Stable

---

224 Ann Pinniger, fourth child (second daughter) of John and Catherine (née Crook) Pinniger of Cowage (Cowitch), born 14 June 1805, lived at Lambourne, no issue, died 29 December 1858, aged 53, buried at Hilmarton. WSA: 4381/2/1, *Pinniger genealogical memoranda book, c.1820-1943*, pp.11-12, 28.

22. Sunday, thin rain nearly all day
23. Incessant RAIN all day, the land too dirty for the Ewes to go on Swedes from the Straw yard to the Down & fold [Pinniger then recorded the killing of a fat sheep and the meat produced - see Appendix R]
24. Horses in the Stable | RAIN, heavy storms in succession all day the Ewes on the Down, the Swede land being exceeding wet, the Lambs on the Swes | Men & Boys making Chalk Floors to Mow in the Barn. & Thrashing.
25. Cold dry day, Ewes on the Down
26. Frost & very cold dry day, rain in even'g | 12 Qr Barley to Devizes & 22 € Coal for J-Andrew Ewes on Swedes again, land dryer | Xtmas day the WATER returned to the two WELLS, having been since the 7th of Novr empty
27. Carting 4 Loads of Muckle from the Inn | Dry morn'g, heavy rain after
28. Dry day, Frost morn'g, Ewes on the Down, too wet on Swedes | Horses in the Stable
29. Sunday, Boisterous morning Storms after
30. Driving rain all morn'g, dry eve'g Horses in the Stable
31. Horses in the Stable except Carting 3 loads of Chalk to Barn, for the Mow | HURRICANES of Wind & Rain all day Very Wind'y great part of this Month, The Ewes do very bad too wet to go on the Swedes & starved on the Down, wth Straw mornings & Seed hay nights. Men & Boys making Chalk floors to the Mows, to find work in the dry.

1834

January

1. Carting Chalk, for Floors of Mows | Cold dry'g winds, all day. | Ewes, follow'g the Lambs, on Swedes, again
2. Frost Sun shine & cold drying winds. | Cart'g Muckle finish'd the 6 Acres for forward Turnips & began Muckleing Wheat Stubbles for SWEDES
3. 2 Plough Teams, to KENNETT to finish my Brothers Wheat sowing | Thin driving Rain
4. Very warm morn'g, not a breath of Air, stiring, Storms after | ASHES, the first two loads from Devizes 100 Bushls for Whitelands
1. Potatoes, the Pigs finish'd, now feed'g on Barley meal only, 5 Stores and Sow with 6 young
5. Sunday, dry mild day, damp even'g
6. ASHE'S, began sow'g - Whitelands. 100 Bushls Ashes from Devizes 2 Waggons & 6 Horses | Barley finish'd thrash'g 1st Rick, 7½ weeks | Wind & Rain all day
7. Wind & storms at intervals | Ashes finish'd sowing Whitelands 150 Bu & began sow'g Tan hill Piece Rt hd
8. 100 Bushls Ashes, from Devizes | Rain very heavy night & morn'g, dry & very mild after | 2nd Bacon Pig 10 S 1 li | And a Pig - for John Chivers 10 S

5 li | The Ewes, 3 truss's of Seed hay this morn'g, the Barn empty of Straw, and the weather to wet to put a Rick in The Ewes DO ILL, being continally wet the last month, very dirty at Swedes would DO WELL in dry weather, at Straw morning, Swedes noon & Seed hay 4 Truss's night. The Lambs DO BETTER, having a new tie of Swedes in the clean every day, & 2 Truss's of Seed hay nights only. | Hatch put to new Water carriage, and Meadow first WATERED, the bank

9. Frost morning, fine mild day | 50 Bushls Ashes from Devizes | Ashes began sow'g Nether Down Piece 100 B | FOLD, finish'd 6 Acres, & muckled for forward Turnips & Wheat
10. FOLD, began Wheat stubbles for Swedes & muckled The Fold will be 28 Days dunging 3 Acres The EWES time up for Lambing the 14th Feby being 21 Weeks from 20th Sepr 100 Bushls Ashes from Devizes Fine mild day, little rain evening
11. Fine mild morn till 11 - rain remain'r of day Ashes sow'd 3 loads 150 Bushls in 3 hours by 3 Men, myself filling Seedlips Down Piece & 3 loads of Muckle fm Inn before the Fold.

[12]. Sunday, Thin driving rain at times

13. RAIN all day Bath | 100 Bushls Ashes fm Devizes The Men cuting Chaff
14. RAIN very heavy all morn'g, dry after Ashes sow'd 50 Bush finishd Do: piece and 50 Bush Tan hill P 50 Bushls Ashes fm Devizes[225]
15. 100 Bushls Ashes fm Devizes | RAIN heavy night & morn'g, dry after
16. Rain & TEMPESTUOUS, night rain morn'g - dry at times after Ashes sow'd 3 loads, Tan hill Piece, & 3 loads Muckle before Fold
17. 100 Bushls Ashes fm Devizes | Boisterous night with rain, and a Tempestuous storm in the morn'g with storms after
18. Ashes sow'd the last 2 Loads Tan hill P | Muckle, 2 loads before the Fold. | Thunder & lightening & boisterous night storms in the Day, rather finer day
19. Sunday, fine day, damp evening | The SPRING'S rising rapidly, the Meadows flood'd more than any time before and the Wells, over full
20. The first LAMB, from the cul sale Ewes (not sold being too poor) | 15 Qr Barley to Devizes wth 4 Horses & 2 <C> 4C 1Qr Coal returned | Some thin Rain at times
21. ASHES, finish'd sowing on Clover with W-Edmonds 21 Bushls & 750 Bu: from Devizes on 25 Acres, about 31 Bu: pr A | Some thin Rain, finer weather | Seed Hay, finish'd cuting the Rick for Horses | Carting Muckle, before the Fold
22. Carting Muckle & finish'd the Piece 3 Acres for Swedes | CLOVER=HAY, the Sheep began | Rain all the morn'g, dry even'g.
23. The 3rd Bacon PIG - 12 S 18 li and a Pig to Mr Jas Large 12 S 18 li (say being same size & alive | Driving Rain all day. | Carting Dung behind the Fold, for Turnips

---

225 Do: - Down.

24. Carting Dung behind the Fold & finish'd | Driving rain all day, with little exception Tuesday night the 21st the FAT WETHER Sheep, (or Ram Stag) stol'n from the Fold (with the Tegs) The Swede Greens at Kennett, some shewing the yellow blossom, in general ready burst into blossom, at Beckhampn the Purple short tops, not runing to seed

21. Tuesday night Incendiary FIRE, at Lower Bupton, Barn & Buildings, distroyd
Very distressing time for the Sheep, having had very boisterous, and Rain'y weather the last Two months with few exceptions, the Sheep nearly whelmed in the Mud feeding Swedes, the Tegs having a tie before the Ewes do better. their Wool never dry, and treading the Swedes in the Mud, with bad thrash'd hay till now, and dirty Fold on Wheat stubbles muckled are become very poor, and very unfortunately the Sale Ewes kept, are now lambing on the cold wet ground, the Lambs perish as they fall, some of the Ewes no milk, no Straw to make a standing Pen, & no dry weather to put a Rick into the Barn since the 8th Inst which make altogether a melancholy prospect, having no means of doing better for them. Floods every where, and the Springs seldom higher. January 24th 1834.[226] [Diary turned landscape for this entry]

25. Dry mild calm day, after a very boisterous night, the 1st fine day last fortnight | Barley rick, on Staddle put into Barn | Horses in Stable, land too wet to Plough | Mr Crook, Cn Plant'd Potatoes & B=Beans [227]

26. Sunday, Rain night & morn'g dry after

27. RAIN, nearly all day, Load Thorns from Yatesbury

28. RAIN, heavy Storms, cold dry eve'g | Carting, muckle & Hurdles, for a Starn'd Pen [228]

29. FROST, exceeding hard, Glass high a cheering change in the Weather | The SHEEP, very much INJURED, by walking to their belly in dirt, and being on each other like Pigs in the dirty Fold this morning some of them lamed with the dirt & unable to walk out of the Fold. put them in a STANDING PEN by the Swedes, hay morn'g & eve'g & Swedes with Barley Straw, a hope of doing better, if dry wear [229] * | Fallow'g Wheat stubbles behind the Fold, the first since the 18th Decr having had almost a continued rain.

30. Fallow'g as yesterday, 2 other Lambs | Mild Frost & very mild, pleasant morn'g some thin rain after | The 3rd CALF, 11 Weeks old 32¼ li pr Qr (cost 24s 6d @ 8d li £4 6s 0d The 2nd Calf (cost 21s 6d) 18W: 60¼ li pr Qr 6d li £6 0s 6d The 1st Calf fin 4th Apl 13W 44¼ li pr Qr 6d li £4 8s 6d The

---

226 Whelm - To engulf, submerge or bury. Oxford English Dictionary, *Whelm* https://www.oed.com/search?searchType=dictionary&q=whelm&_searchBtn=Search [accessed 1 May 2020].

227 Cn - Corton.

228 Starn'd - Standing.

229 wear - weather.

keep of a Calf 1 Week 6s od [Total] £15 1s od The Butter & Milk from 4th Apl equal to the cost of the 2 Calves 46s od

31. Thin rain at times, fine day, Plough'g as yesty

February

1. SOW, brim'd, time out 23rd May | Fine drying day, the DUST flying on the Turnpike Road, great change since 28th Jany Carting earth from Tan hill road, on the Turnip Land, fold'd &c - 9 Lambs
2. Sunday, drying day, some little rain
3. Very cold drying day Carting earth from Tan hill road to home Meadow
4. Carting earth from Tan hill road to home Meadow, Cold dry'g day, rain eve'g | 12 Lambs, from the cul sale Ewes
5. Cart'g Earth till 11-O-Clock, finish'd | Fallow'g for forward Turnips | Dry morn'g rain after
6. Dry morn'g Ploughing as yesterday | Giddy Teg, 28 li
7. Rain in night, very fine calm day | Fallow'g as yesterday
8. Rain all day. Horses ret'd fm Plough
9. Cold and thick Fog all day
10. Frosty morn'g, Fine dry day Fallow'g for Turnips
11. Fallow'g for Turnips much rain in night dry day | The Tegs, finish'd Folding the muckeled grd for Swedes, their Wool cloted & mired in dirt no place to fold them in the clean
12. STANDING FOLD made for the Tegs with the dung from the Inn | 4 Loads of Muckle from the Inn for standing Pen, removing Fold, Hay &c | Boisterous rain'g night, very fine day
13. Frost morn'g, fine drying day | Ploughing as yesterday
14. Ploughing as yesterday | EWE'S, time up for LAMBING. 30 - Lambs includeing those from the cul'd sale Ewes, not sold | Rain in night, dry day | The SPRING'S sinking, the water disappear on Meadow, front of the House, and lower'd from Silbury mead to the tops of the grass, and prevent'd drawing quite off by Willm Tanners, hatches only, being 2 feet too high.
15. Fogg'y morning, some little rain occasionally | Fallowing for Swedes after the Fold
16. Sunday, Frost in morning, fine day 10 Lambs last night & 7 this day
17. 10 Lambs last night & 7 this day | Fallow'g for Swedes | Frost in morn'g, very fine day
18. BARLEY RICK - the 3rd put into Barn | Fine mild dry day | Fallow'g as yesterday
19. Basingstoke. Fallow'g as yesterday Fine drying day
20. Very fine drying day, the roads dusty Fallow'g after Swedes fed of began
21. Fallow'g after Swedes fed of began | 100 LAMB'S, One Ram only turn'd to the Stock Ewes, see the 20th Septr and two others 13 days after see 3rd Octr | Rain heavy in night, very drying day

22. Beautiful fine mild day | Earth to young hedge top of Whitelands & Plough g as yesterday
23. Sunday. - Damp day
24. Ploughing as 22nd damp afternoon Freegrove, 1 Piece of Wheat fed with Sheep to the ground, another feeding | [Note in margin] PORKER 45 li
Feby 22nd 1834, Died Jane Pinniger at Cirencester. Born 5th April 1812
Feby 25th Died Mrs Walter Crook of Tytherton Aged 82
(1835) Feby 20th Died suddenly Mrs Fell of Tytherton Aged 65 [This entry and the previous two entries written within adjacent thick black borders]
24. SWEDES, growing fast to green, draw'g them halfway to break the tap root & check their growing too fast
25. Very fine day - Thwart'g Bean Stubbles
26. Beautiful fine day | Bremhill, Sale of 19 COWS, average £12 16s 6d Ploughing as yesterday
27. Damp morn'g fine day Ploughing as yesterday
28. Rain, morn'g fine after Ploughing as yesterday | Died 26th Dame Everett & Isaac Gale aged 27

March

1. 150 Lambs, about 70 Ewes not lamb'd 20 Ewes no Lambs & 3 Ewes DIED | SWEDE'S about 6 Acres remaining | Sheep eat less hay, a prospect of enough food for Sheep, Swedes being good | Thwarting Bean Stubbles | Fir trees - Planted | Rain thin in morn'g, mild & fine after
2. Sunday - Very fine day
3. Plough'g as the 1st - Very fine day | POTATOES, planting in Garden
4. Boisterous winds & dry | Ploughing as yesterday
5. Ploughing as yesterday Boisterous with rain
6. Ploughing as yesterday exceeding Boisterous night dry morning some rain after.
7. Calm mild fine day, Plough'g as yester'y
8. Cold boisterous winds Plough'g after Swedes fed off The 3rd giddy Teg Kill'd | Picking Stones & Carting them from Clover to the bottom of Tan hill road | Water discontinue runing through Meadow
9. Sunday, mid Lent, very fine dry day
10. Beautiful fine mild day | Ploughing as the 8th & Bushing Meadow
8. 3 Porkers 13 Stn 2 li at 4s 4d 1 Porkers 4 Stn 7 li at 4s £3 16s 11d Com & Porte 4s Carr: [...] The 4 Heads 16 li £3 [...]
11. Beautiful fine mild day, Plough'g [...]d 2 Teams, 8 Load of Rubble to Tan hill road
12. Ploughing as the 10th | Frost morn'g, warm fine day like June | Asparagus, Planted
13. Ploughing as yesterday, warm dull day | 12 Qr Barley to Devizes & 1T 4C 1Qr Coal retd

10. DAHLIA'S planted
Died this morning March 14th Mrs Broome, aged 78 at East Kennett [This entry written within a thick black border]
14. Ploughing Swede ground fed off | Rolling Meadow & Field grass began | Cold dry day | The Ewes, all nearly lambed, about 30 being late in Lamb or barren put with the Tegs, and about [...] gone without Lambs | HAY, spent as Truss d The Seed Rick - 14 Tons Part of Clover Rick 9 Tons Part of Meadow Rick 6 Tons [Total] 29 Tons | HAY, remaining to spend Part of Clover Rick say 11 Tons Part of Meadow Rick say 6 Tons The whole Meadow Rick 4 Tons [Total] 21
15. Fine cold dry day Ploughing the Fallows athwart for Swedes 20 Sacks Wheat to Devizes & 10 doz Fleaks from Bromham [230] | Pease & Beans, - plant'd in Garden Parsnips Carrets &c
16. Sunday, fine drying day
17. fine drying day Ploughing as the 15th | COW milk'd twice a day for the Lambs discontinued milking her this day expecting her to calve 31st bull'd 12th June
18. Sharp frost, very cold cuting winds. Ploughing as yesterday.
19. Ploughing as yesterday. | Exceeding sharp Icey frost
20. Sharp frost, harsh cold winds | Plough'g as yesterday & 11 Qr Bar: to Devs | Barley Rick the last of 4 put into the Barn | Mrs Broome of Kennett = bury'd at Cliffe [231]
21. Frost morning, fine mild day. Ploughing athwart for Swedes. ROUND TAIL'G, the Tegs & cuting the Knobs of dirt from their belly's | Porker 44 li the last
22. Ploughing as yesterday | Boisterous cold winds
23. Sunday - Exceeding boisterous
Died this evening 23rd March Mrs Thomas (late E=Chivers) Aged 58 Years[232] [This entry written within a three-sided thick black border]
24. Mild morn'g & even'g, windy day | Plough'g athwart for Swedes | Wheat began harrowing | Ewes & Lambs, lay out of Standing Pen back on the Swedes | Potatoes Purple, planting Cottage Garden | HAY, the Ewes & Lambs mornings only | [Pencil note in margin] Philpot Drillg Barley Down
25. Cold windy day, Plough'g & Harrow'g as yesty | ASH'S - with Chamber lye, Soot & Hen dung sow'g over the thin Wheat further Down Piece,

230 Fleaks – Flakes.

231 Susanna Broome of East Kennet buried on 20 March 1834 at Clyffe Pypard, aged 78, *Wiltshire, England, Church of England Deaths and Burials, 1813-1916 for Susannah Broome* [accessed via Ancestry, 27 July 2020].

232 Elizabeth Thomas of Bremhill buried on 29 March 1834 at Bremhill, aged 57, *Wiltshire, England, Church of England Deaths and Burials, 1813-1916 for Elizabeth Thomas* [accessed via Ancestry, 27 July 2020].

harrow'd after & hoe'd

26. 2 Men began hoe'g the Down Wheat | Very sharp frost, cold drying winds | Ploughing & harrowing as yesterday
27. Ploughing & harrowing as yesterday | Mild morn'g, some thin rain
28. Good Friday - driving rain | 1 Ewe on its back died in lamb | 1 Ewe with bad mouth died, | Ploughing as yesterday
29. Ploughing finish'd thwarting for Swedes | Rain heavy in night, Showers in the day
30. Sunday, rain heavy in night, frost morn'g
31. Rain heavy in night drying day. | Plough'g Bean Stubbles 3rd time for Barley

April

1. Frost in morn'g fine mild day | Plough'g Bean Stubbles 3rd time for Barley
2. Plough'g Bean Stubbles 3rd time for Barley | COW CALVED, bull'd 12 June is 42W | Rain in night, fine day
3. Very fine day, Plough'g as yesterday 10 Qr Barley to Devizes
4. Frost in morn'g, very fine day Plough'g as yesterday & harrow'g 4½ A_ of Wheat hoe d, very tuft'y | PIG, the last 18 Score, weigh'd when put to fatten 6th Novr about 4 Sco
5. Ploughing as yesterday & harrow'g & roll'g the Couch'y land | Frost & very fine day
6. Sunday - Calm mild day
7. Frost & Sultry warm, exceeding fine weather for the land, burn'g Couch | Plough'g Bean Stubble 3rd time for Barley [Pinniger then wrote about the loss made by grazing pigs - see Appendix A]
7. The Wheat & field grass, turn'd yellow with the frost, the Swedes, also check'd, not yet in flower, having been half drawn 24th Feby affording now plenty of green for the Lambs, and enough to keep all the Sheep 18 days to come, those who drew their Swedes out of ground or mowed them are witherd & the green dried, & Sheep keep will be scarce, if we have not growing rain soon. The short top Purple Swedes, the best to keep | Mr Browns Swedes Whitelands [...] Acres finish'd, began feeding lambing
8. Very cold drying Winds | Plough'g as yesterday
9. Plough'g as yesterday & cold dry winds
10. Plough'g finish'd the Bean stubbles 3rd time for Barley | Sharp Ice'y Frost, cold winds & hail storm eve'g
11. Sharp frost, exceeding cold winds wth SNOW storms | Rolling & harrow'g for Barley | Round TAILING the EWES, & cuting the knobs of dirt from their belly's | The LAMB'S tail'd seer'd by J Chivers & Son the Shephers, their food & a Boy 100 - Ewe Lambs 89 Ram Lambs} [Total] 189 from 250 Ewes, (see 5 Decr) 4 Ewes died, at lambing 1 Ewes died on her back 1 Ewes died decline About 35, Cast their Lambs or Lambs died About 20, Barren

or late in Lamb 189 with Lambs [Total] 250 | The Ewe Lambs punch'd the near ear

12. Pierceing cold winds & hail storms | Roll'g & harrow'g Bean stubbles for Barley the Barley land now prepared to begin plough'g for Seed, some large Farmers have finish'd sow'g (better time after rain)
13. Sunday - Cold dry day
14. Sharp frost dry day | Plough'g began for Barley, and wait for rain, the ground being very dry
15. Ploughing as yesterday, Frost & milder
16. Ploughing as yesterday, Frost & milder
17. Ploughing as yesterday, Frost & milder | Barley finish'd thrashing
18. Ploughing, one earth for Barley Swedes fed off | Warm fine day
19. Very warm fine day Plough'g as yesty
20. Sunday - Very warm fine day | Ewes & Lambs removed from the Swedes to SILBURY=MEAD, 3 days keep left of Swedes for the Tegs

Mar 21st. Mr Rumbolls - Sheep began feed'g Meadows

7. Mr Geoe Brown - Sheep began feed'g Meadows
11. Mr Kemm - Sheep began feed'g Meadows
18. Mr Wentworth - Sheep began feed'g Meadows
19. Mr W_ Pinniger - Sheep began feed'g Meadows
20. T Pinniger - Sheep began feed'g Meadows
21. Rt Philpot - Sheep began feed'g Meadows
20. HAY - discontinued giving the Sheep having half the Clover Rick, and the Small meadow Rick except 1 Ton, left
21. Sharp frost & warm day, harrow'g for Swedes | Carting dung from Standing Pen
22. Carting dung and harrow'g as yesterday | Frost & warm day | WHEAT=RICK, the 1st put into Barn | CUCKOO, first heard
23. BEAN=RICK, the small put into Barn | Fine day, harrow'g for Turnips and Carting Hurdles from Swede ground | The young Sheep Tegs, finish'd the Swedes and Fold'd on the Clover hither Down Piece to be kept with that, the Down & Water the Down dried up, no grass | The 6 Yr old Black Horse exchange'd for a 3 Yr old Blue Roan Colt
24. Fine warm day | Ploughing the Swede land for Barley and 13 Qr Barley to Devizes the last Coal returned 1T 18C 2Qr
25. Ploughing as yesterday, sharp frost, fine day
26. Plough'g finish'd through the stand'g Pen | Very dry day
27. Sunday, Glass sinking rapidly, dry day rain a little in the evening | Vetches [...] Rt Philpot began feeding by young Sheep in Fold SEE the 31st March, but little RAIN since
28. RAIN in the night & showers in the day | BARLEY sowing BEGAN, bottom of Tan hill Piece, considered better doing it, than that done before in the dry

29. Frequent heavy Showers | LAMB'S 86, cut by Abednego Blackman | Barley sowing as yesterday | [Note in margin] W-P 473 Lambs fm 574 Ewes
30. 28 Bushels Hop & Ray 87 li Nt the Sack & 112 li Marl grass Clover sow'd on about 10½ Ac lower part of Tan hill piece. | Fine growing day, some light showers.

May

1. Some thin rain morn'g Beautiful fine day | Barley sow'd 7 Sacks | VERY=GROWING, the Wheat & field grass mending very fast | Kidney Beans & 2nd Crop of Pease Planted
2. Fine dry day. Barley & Seeds sow'g | Grass Silbury mead good R Philpot mow'g for Horses
3. Very fine dry day, Sow'g Seeds, roll'g & harrow'g Barley | At a meeting at Mr Merrymans Marlbro consisting of Mr Baverstock Merryman Mr W R Brown. Mr Geoe Brown, myself (T Pinniger) & Rt Philpot, to decide the right to a Piece of Land at the bottom of my Tan hill Field, when Mr Bavk Merryman for & in behalf of Mr Halford, relinquished all claim to the right of a Road, through my Tan hill Field to Mr Wentworth's Meadow, which was 35 Years since possessed by Mr Jefferies to a Barn in the said Meadow. & not since that period [233] | Mr W R Brown being appointed a Referee, by myself, and Messrs Merrymans in behalf of Mr Halford, at this Meeting, agreed to abide by his decision of the above Piece of Land. | The follow g is a Copy of Mr Browns Letter | Chilton Foliat 17th May 1834 Sir In the matter referred to me between yourself and [...] Halford Esqr as to a small corner of Land at Beckhampton I shall direct Mr Halfords Tenant to give you Possession of the said Corner of Land, and that you do pay me the expences of the reference. Your Obt Servt W R Brown To Mr Pinniger Beckhampton. | [Note in margin] See June 11th | Hillier Reeves returned the Black Horse as lame & a Roarer.[234]
4. Sunday Fine day showers in morning
5. Warm day shower evening | Ploughing & sowing Barley
6. Ploughing and FINISH'D sowing BARLEY 30 Sacks & 1 Bushl Barley 50 Bushls of Hop & Ray 2½ Cwt of Red Clover} on 23½ Acres The Seeds mix d is 2½ Bushls pr Acre The hole in Machine nearly full | Fine warm day
7. Fine warm day Rolling & harrow'g off Barley
8. Fine warm day Rolling Barley & a load of Stones from Calne

---

233 Baverstock Merriman – Thomas Baverstock Merriman was one of the partners of Messrs Merriman, attorneys in Marlborough, J. Pigot, 'Commercial Directory 1842' in WRS47, p.124.

234 Roarer - A horse which produces the sound of turbulent air within its larynx when worked hard. Horse & Hound, *Q&A: What is a roarer?* https://www.horseandhound.co.uk/horse-care/vet-advice/qa-what-is-a-roarer-35076 [accessed 1 May 2020].

9. Fine warm day Rolling Barley finish'd | Ploughing for Turnips | WATER to Silbury mead for EWES & LAMBS do very ill for want of it | Wheat growing rapidly (above my Knees) | Ploughing &c. for Turnips
10. SEED'S sowing in Wheat. home side of the middle Down Piece 4¾ Acres with 8 Bushs of Hop & Ray & 56 li Red Clover mix'd, half hole in Maching, the Wheat too high & ground too dry to harrow
11. Sunday warm fine day VERY sharp FROST
12. warm fine day Plough'g for Turnips
13. Rain in morn'g, soon dried up | Plough'g for Turnips
14. Plough'g for Turnips & Rolling & harrow'g Thin Showers, do but little good
15. Thin Showers partially, harrow'g & roll'g for Turnips | x Markets very low [235] | Wheat 12 S 3 li Nt offered only 22s Sack Wheat 12 S 13 Nt sold at 23s Sack
16. Warm dry day | Plough'g began for seed, Turnip sow'g wait for Rain, the land now work fine

5th. Died suddenly, fall'g from his Horse in a Fit Christr Pinniger of Pottern aged [...][236]

6th. Died at her Daughters at Tytherton Mrs Pinniger of Wandboro=leigh Aged 78 Years [This entry and the previous entry written within adjacent thick black borders]

[Diary turned landscape for the following eight entries]

April 27th Mr Kemms Sheep left Meadow

1 Mr Browns Sheep left Meadow
5. Mr Wentworths Sheep left Meadow
12. Rt Philpot Sheep left Meadow began Vetches
19. W P_ Kennett Sheep left Meadow
20. T P_ Beckhn Sheep left Meadow began Newfield
19. Mr Kemm left newfield for Down
7. Began building Keepers House by the the Fox Covert on the Down

[Pinniger then wrote notes on the supply and demand for Wheat - see Appendix S]

May 8th Obit, Farmer Thos Cole Kington Langley. [This entry underlined with a thick black line]

17. Showers at intervals all day | Ploughing for Turnips
18. Sunday, Storms of Rain, with Thunder & hail storm, afternoon.
19. Ploughing for Turnips, lowering day Swindon Fair
20. TURNIPS, begn sowing Norfolk Whites 2 li pr A, 3 holes every 3rd place 3 Acres at bottom of Tan hill piece Rt hand | Fine warm day | Ewes &

---

235 x – The following entry made at the foot of the page: May 20. WHEAT 12 Sco: Nt allowd at 21s, at Calne no buyer.

236 Christopher Pinneger of Potterne buried on 9 May 1834 at Potterne, aged 43, *Wiltshire, England, Church of England Deaths and Burials, 1813-1916 for Christopher Pinniger* [accessed via Ancestry, 27 July 2020].

Lambs, removed from SILBURY MD having been 29 days feeding it, & not so good as other years till now, exceed 400 Ewes, Lambs &c | E & Lambs, put on Field grass - Down Piece with the Tegs & lay back night & day, for Vetches & Whet Confined to the Field grass & Water, the Tegs worsted on the Field grass & Down, the Down sweet, would not eat the field grass [237]

21. Warm fine day | Ploughing finish'd the 6 Acres for Turnips
22. TURNIPS Green rounds, sowd 8 li on 3½ Acres 2 holes every 3rd place to come in drill, next the Norfolk Whites, and to succeed them. rain wantd for Seed to grow | Dry windy day | Rolling & harrow'g for Potatoes
22. Ploughing for Potatoes 3rd time | Warm windy day | FIELD GRASS, began mow'g for HORSE'S Nether Down Piece, being tire'd of hay and very Poor | BECKHAMPTON Tolls let to Ezel Evans for 1387£ Let last year at 1362£ to Isaac Selman Low [Increase] 25£
23. The SOW'S time up this day being 16 Weeks & farrow'd 15 fine live Pigs | Rolling & Plough'g for Swedes
24. Plough'g Rolling & harrow'g for Potatoes | Cold blowing harsh winds
25. Sunday - Cold blowing a large Limb split from the Ash Tree, in the Meadow eveg
26. <T> POTATOE'S, began planting in Field Cold blowing day Rolling for Swedes
27. Potatoes, finish'd Plough'g in, with 3 horses by 10-O-Clock & yesterday 1A 1R 20P with 13 Sacks fill'd (cut) 5 Women droping | Ploughing for Swedes 3rd time, warm day
28. Ploughing for Swedes 3rd time, warm day | WHEAT EAR'S, quite out in Albourn Parish | The Wheat in general very short & picked many places no higher than Knees, the Spring Corn much not up want rain, the Field grass thin & Mowing almost universally
29. WHEAT=EAR'S, burst'd nearly, Talevera on my Down Piece, sowd 21st Novr and the red Wheat sowd 18th Octr adjoin'g Down | Warm fine day, Plough'g for Swedes Married {Mr John Brown of Lower Upham, & Miss Kate Brown of Berwick Basset} Also Mr James Hale of Calne & Miss [...] Harris
30. Warm fine day, Plough'g for Swedes
31. Warm fine day, Plough'g for Swedes | Mow'g in general this week in this Parish

    Extraordinary dry, Spring & Summer. Dry half February Dry all March Dry all April Dry all May} see with few exceptions [Diary turned landscape for this entry]

June

1. Sunday, warm fine day
2. warm fine day - 2 Waggons & 6 Horses to Bowood for 1¼ Hd Faggots

237 E - Ewes.

3. COW bull'd at Mr Browns [Note in pencil] again 21 Sepr | Rain in morn'g, dry'd up immediy | Plough'g for Swedes
4. Finish'd Plough'g for Swedes | VETCH'S, all the Sheep began feeding the Ray grass Down Piece, now too hard have made good work till now, began it the 20th May | MOWING, began home Meadow, the Grass nearly dry'd up, no prospect of being better, RAIN at times all the even'g
5. RAIN even'g dry day | SWEDE'S, began & sow'd 12 Acres with 20 li 2 holes every third place to come in drills, the short Purple tops
6. 3 Horses Roll'd the 12A of Swede's sown & | 3 Horses Harrow'd the 12A nearly 2 tine | Fogg'y morn'g, warm day, shower even'g | The Vetches just flowering very strong above the Sheep, the Lambs a tie forward
5. MOWING, the Field grass began
7. Warm dry day | Roll'd the 12A Swedes the second time | Harrow'd the 12A Swedes the made fine good work to retain the moisture, expect too dry to grow
8. Sunday. Warm fine day
9. Plough'g began new field fed, Nether Down Piece for Vetches, Warm fine day | HAY, began carry'g, 2 ladds, left by Sheep being too old, Nether Down Piece | SHEEP=COUNTED 379 Sheep 189 Lambs [Total] 568 to Summer | [Note in pencil] Mr Connington, wish Sheep Shorn end of June | Talavera Wheat, nearly even at head The Red Wheat, about about half in full ear
10. Ploughing as yesterday, Storm mid=day Soon dry'd up | [Note in margin] See 3rd May | RT PHILPOT. have persist'd in keeping possession of my Piece of Land, awarded to me by Mr W R Brown till this Evening, altho: directd by his Steward Mr Merryman to Surrender it, and after repeatedly complaining to Messrs Merryman & Brown of his obstinacy.
11. POTATOES, Planted the Piece of Land which Rt Philpot seized from me & since award'd to me | [Note in margin] see 3 May 1833 [238] | VETCHS, Summer, sow'd 3 Bushls: old and 4 Bushls of New with 2 li of RAPE on full 3 Acres Nether Down Piece after Grass fed off (for Wheat) Thin Showers soon lost
12. Showers in the Day, acceptable 3 hours rain in the evening | 27 Sacks Wheat to Devizes & Plough'g for Vetch's
13. Harrow'g off Vetches, & 12 doz Hurdles @ 7s from Sevenacke Mr Mills some light Showers | SHEEP=WASH'D at Clatford 18 Score & 18 @ 6d pr Score by 4 Men 2½ hours, usage 1 Loaf Bacon & Cheese a 5 Qt Bottle (3 part Beer & 1 part Ale) gave satisfaction
14. Swedes appearing, Turnips fit to hoe | Plough'g for Vetches - thin Showers | Mow'g Whitelands, finish'd Tan hill Piece
15. Sunday, fine dry day (hay dry)

238 There is no reference to this matter in the diary for 3 May 1833: however, it is covered at length in the entry for 3 May 1834 – see above.

16. SHEEP=SHEAR. 18 Sco & 18, by 9 Men, being 42 each, professing to shear them better @ 2s 6d each Man pr day, instead of shearing them by 6 or 7 Men at 1s pr Score began 8 O-Clock morn'g, finish'd at 9 O=Clock in evening Helpers 3 Men & 2 Women & stack the Wool The Sheep better out of their Wool, than any precedeing year | Showers in morning, heavy rain & Boisterous winds in the evening | Faggots ¾ Hd from Bo=wood, heavy load
17. Storms & dry'g between, Plough'g & Press'g for Vetches. | Mowing finishd Field grass & began | TURNIP=HOEING, right of Tan hill road
18. Dry morn'g, prepared | for Hay Cart'g. Rain all the afternoon, thin | Vetches, sow'd a 2nd drift about 4 Acres with 2 Sacks and Rape about 2 li
19. Glass rising, fine day | Carting Muckle from the INN to a store heap by Tan hill road collecting since the 21st Jany
20. HAY, Rt of Tan hill began carrying at mid day, stain'd a little | Very fine day, Cart'g Muckle, morng
21. The most delightful Summers morn'g this year RAIN in evening | Hay carryed the remainer of Tan hill Piece in warm dry condition | VETCH'S sow'd the 11th June appearing | The Wheat at the present time in full bloom & at its grandest appearance, being level at head, upright and high, and the Winter Vetches also look beautiful being as high as the Hurdles & heavy Crop * | The Barley, a good colour & thick, promising a good Crop, the first sowing, showing the Beard
22. Sunday, very fine warm day
23. Beautiful fine warm day. | Cart'g Muckle from the Inn, finishd | HAY, the Field finish'd carry'g - (Whitelands
24. Dry meadow mow'g, very little Grass and a prospect of less, very fine day | Plough'g Down Piece, next to Vetch's, grass fed.
25. Plough'g Down Piece, next to Vetch's, grass fed. | 4 Turnip hoers, put on, grow'g rapidly & 2 Men hoe'g Potatoes | Wheat Rick - put a second in Barn (2 out) | Ploughing as yesterday
26. MARKET'S, bad sale for Wheat, sold 15 S_ 12 sco: Nt @ 23s the last of 1832 Crop. and 20S 12,,9 li Nt @ 24s 3d the first of 1833 Crop | Damp blight'g day | 20S_ Wheat to Devizes & Plough'g & Harrow'g
27. Cart'g Dung for Turnips Down Piece | LAMBS WANED on Vetches & the little Down not stock'd before, the Ewes on the great Down & fold'd on the Dung for Turnips Down Piece
28. Cart'g Dung finish'd about 3A: Down P Fine day, little Rain in even'g | HAY=MAKING, finish'd home mead | Turnips, finish'd hoe'g 6 Ac | Swedes, in rough leaf,
    [Note in margin] 28th CUCKOO, disappeared
29. Sunday, warm fine day
30. Plough'g Vetch land fed off, for Winter Turnips Very fine warm day

July

1. Plough'g as yesterday 1 Team - Very windy day | HORSE'S resting 1 Team, having very little work, for them
2. Plough'g as yesterday, dry winds | 1st CALF 13 Weeks old to Butcher 34¼ li pr Qr | 2nd CALF 4 weeks old fm Tytherton
3. Warm windy day | 2 Hd Faggots from Overton for Labourers
4. 1 Team Plough'g as the 2nd warm dry day
5. Harrow'g as the 2nd warm dry day | The black Horse sold at Newbury Fair for 20 £ bought Augst 1830 for £31 10s 3 Yrs old | little rain in evening
6. Sunday, RAIN heavy in night, and Thunder & lasting Storms in the day to soak the land, a very good & providentil rain for every thing, the heaviest Crop of Wheat till now upright, (now lodged) - on all dry soils the Crops partly dryed up
7. SWEDES, began Ploughing up, about 2 Acre covered with Charlock | SWEDES, began hoeing, strong Plants Fine morn'g rain heavy afternoon
8. Plough'g as yesterday, finish'd Rain heavy in night, storms afternn
9. SWEDE'S sowing second time where Plough'g yesy TURNIP'S sowing new field, dung'd & fold'd drag'd only after the Fold | Turnips hoe'g 2nd time, growing rapidly Fine drying day
10. Harrow'g & Rolling the Vetch gd for Turnips Carters & Women hoe'g Potatoes - Fine day
11. Carters & Women hoe'g finish'd - Fine day
12. Plough'g Vetch Gd & burn'g the Haulm &c. Fine day
13. Sunday, heavy Storms afternoon
14. Very fine, Ploughing Vetch land fed off | CLOVER, the Horses finish'd Whitelands & began SILBURY MEAD
15. Turnips finish'd seconding | Turnips appearing Down Piece | Swedes appearing the second sowing | Swedes growing rapidly 6 hoe'rs on | Plough'g as yesterday, fine day
16. Plough'g as yesterday, fine day
17. POTATOES, began earthing with the Plough | The warmest & most ripening day
18. TURNIP'S began Drilling after Vetchs fed Thunder lightening & HEAVY=RAIN commenced at 9 O-Clock, continued all day Chevalier Barley 11th July in Southampton Market Talevara Wheat 14th July Mr Pinniger Coombe Bissett began Reaping Wheat Reaping 19th July, began at Up=avon (stopd)
19. Rain in Night with Thunder & lightening | RAIN ALL DAY, water on the Land, as the the middle of Winter | Hurdles removing from Vetches to 2nd Clover | All the Sheep on Down, too wet to feed Vetches

    Sacred to the memory of Mr Henry Crook of Corton who departed this life the 7th July 1834 Born the 8th March 1775 Bury'd at West Tytherton 11th July Funeral Sermon at Hillmarton 20th July by the Revd Mr Guthrie from 37th Chaptr of Ezekiel and 3 Verse can these Bones live &c. Sermon in the morning by the Revd Mr Barry from the 2nd of Samuel 12th Chapr Ver: 22 & 23 while the Child was yet alive &c. Anthem from the 19th Chapr Job Verse 25th for I know that my Redeemer liveth &c [This entry written within a thick black border]

18. Rt Philpot hoe'g Swedes, after Vetch's fed (= a Crop gain'd)
19. Fold finish'd for Turnips Down Piece & removed to Vetch ground for Turnips
20. Sunday, Dry morn'g, storms after, heavy RAIN in the evening | 3 Ties of Vetches remaining, too wet to feed them & too old, the Lambs do ill with them | CLOVER, the second the Lambs began Rt of Tan hill Road. Horses finish ye Vetchs
21. Rain nearly all day, Horses in the Stable The 10 young Pigs waned
22. TURNIP'S sow'd the 2nd Fold drift Down Piece 1½ A with 2 li Seed 1 hole every other place | Very fine warm day - Glass riseing
19. Mr AMOR, began serving the Mail Horses with Straw at the Inn having 15 of Mr Halcombs, cannot serve more
21. 10 Pigs waned, Sow go to Boar in a week, and every 3 Weeks after, if not stand
23. 9 PIGS cut, Kept one wth 12 Tets, Farrowd 23rd May proper time for cuting - & not before 8 weeks old | TURNIP'S sow'd about 4 Acres after Vetches fed off, one hole every two places | Just finish sow'g ½ pass'd 1-oClock when commenced most awful Thunder & Lighten'g with exceeding heavy rain for two hours such as have not had before, Water stand in Pools on the Down & strand the Land after a dry & sultry morning - every thing grow'g rapidly | Turnips 1st sow'g as large as Cricket Ball
24. Reaping many Places, sultry day 20 Sacks Wheat to Devizes 12 S 8 Li Nt 24
25. Reaping Mr Wentworth began Talavera blightd not ripe | Harrow'g Turnips dash'd 23rd wth heavy rain | Very fine day
26. 6 Men finish'd hoe'g Swedes 1st time except 2 Acres full of Charlock as high as knees dispair of & refuse doing | Muckle, Carting from the Inn to New field Fine day rain heavy in the even'g
27. Sunday - Rain in morn'g dry after [Pinniger then recorded his stock of sheep as at 25th July - see Appendix B]
28 FOLD removed from Vetches fed off for Turnips to New field muckled for Wheat Rt of Tan hill | CHARLOCK 6 hoe'rs 3 Women & Carters draw'g from 2 Acres of Swedes | Dry morn'g rain evening, Horses in Stable | Reaping, began every Farmer except myself
29. SWEDE'S, finish'd hoe'g 1st time and the oldest 2nd time | Plough'g the last Vetch drift for Turnips | Drying & warm Sultry day, THUNDER=LIGHTENING all the evening
26. BROCCOLI, Savoys & Buxcoli planted [239]

239 Buxcoli written in pencil.

30. Exceeding heavy Thunder & RAIN in the Night, storms in the day, dry even'g | Horses in the Stable too wet to Plough
31. Very mild calm moning | REAPING began, the Talavera, Down | RAIN very heavy 8-oClock | heavy storms in the day, the Wheat cut this morn'g in grip, the Corn near grow'g standing | MARKET, this day at Devizes advd 2s Qr Plough'd remainer of Vetch ground for Turnips = too wet | [Note in margin] The first NEW Wheat at Devizes @ 23s very damp

August

1. Very heavy Thunder Storms partially | Reaping discontinued the Talavera being sproutd since yesterday in grip, 12 Reaping red Wheat | Carting Muckle before the Fold - from the Inn
2. RAIN early in morn'g, to stop the Reapers, very fine day the Wheat RESCUE'D from destruction, being sprouted in shok | TURNIP'S finish'd sow'g, (the last of Vetch grod)
3. Sunday - very warm fine day
4. Fine day mizly rain evening | FALLOW'G, began after Clover fed off for Wheat | All hands (12) cut'g the Talavera Wheat all the Wheat want cuting
5. Some mizly rain early in morn'g & noon, dry after | Draging the Swedes that are hoe'd only once & & the Charlock set, do well & refresh the swedes | Draging the Turnips Down Piece sow'd 9th July | Turnips sow'd 23rd July eat wth Flea, except where the ground was so hard after the Fold, that could not heal the Seed, there thay are healthy
   Augst 2nd Died Miss Edith Potter of Little Bedwin [This entry written within a thick black border]
5. WOOL at Thetford Fair 42s Ewe & 58s Teg. 52s & 60s Mr Coke. 43s & 58s Duke of Norfolk at Lewes Fair Duke of Richmond 55s & Mr Ellman 58s all round
6. St Anns Hill Fair, heavy storms through the day Fallow'g as yesterday | Lambs from [...]
7. RAIN driving storms all day, my Talavera Wheat in great danger of sprouting Wheat advanced this day at Devizes 6s pr Qr New Wheat 12 S 5 li @ 27s Wheat harvest finish'd in Enford Parish & Neighbourhood & that badly only 4 Ricks made in this Parish & myself only half done reaping | 20 Sacks Wheat to Devizes
8. RAIN heavy storms, frequent, put the Reapers to leaseing 15 Acres to cut The Talavera sprouting in Shok, must soon be lost unless the Weather become dry.
   Trefolum=incarnatum a New French Grass to be Draged into Wheat & Barley Stubble 10 li pr Acre @ 1s pr li ready to feed 14 days before Clover, in the Spring [240] [undated and written at the top of the page finished with

240 Trifolium incarnatum - known as Crimson Clover, Carnation Clover, Italian Clover or Scarlet Clover. IUCN Red List of Threatened Species, 2020-1, *Crimson Clover,*

a double line]

8. Harrow'g a Fold drift & Cart'g Muckle fm Inn
9. A merciful change in the weather The Talavera Wheat very much sproutd in the Sheaf this morning, going fast to destruction, but hope with the blessing of God almighty, by removing every Shok and seperateing the wet heads of the Sheaves and exposing them to the Sun & wind this fine day to become quite dry, this even'g they appear comparetively to have sustain'd but little injury - The Red Wheat began sprout'g removed in the same way by 6 Reapers in half a day, and began Reaping again - Muckle Carting from the Inn before Fod Turnips after Vetches destroyed by the Flea harrowing more Seed in the ground [241]
10. Sunday - very fine drying day
11. very fine drying day WHEAT began Carrying, in good condition The 7 A: of Talavera & 8 A: of Red from 5 in morn'g to 9 O-Clock even'g
12. Finish'd the 2nd Rick & put 7A in Barn | Beautiful fine dry'g day
13. Fallow'g Clover fed off
14. REAPING finish'd & Cart'g Wheat finish d Very fine day | CLOVER, began mowing for Hay, too old for the Sheep right of Tan hill road | TURNIP'S began feeding, full ready some measure 21 Inch's round & 25 In | Turnips, began hoe'g Down Piece sow'd 9th July, full large (reap'g hindered) | Swedes began seconding the small, the first hoeing too large to hoe 2nd time | MARKET Devizes decline 6s Qr Wheat refused 27s pr Sack last Thursday, sold 24s this day, no prospect of Wheat being dearer an immense quantity being secured in excellent condition these last 4 beautifully fine days SEE 1821 - Septr 14th wet harvest Wheat advanced to 50s pr Sack & See Novr 1st new grow'd Wheat sell'g at 12s pr Sack, the probability was last week, that Old Wheat would be dearer and new cheaper.[242]
15. Very fine harvest day | Swedes, Draging, those sow'd the 2nd time | Cart'g Muckle before the Fold | Clover Whitelands, began mowing for Horses Died this morning 16th Augst Mr Joseph Edwards Aged 84 Yrs late of Beaversbrook [This entry written within a thick black border]
16. Very fine day, fallow'g Clover fed Swedes began hoe'g those sow'd a second time
17. Sunday, very fine warm day | Up: Lambourne, Barley but little mow'n in neighbourhood & but few late Turnips & Swedes hoe'd, hinder'd by the rain

---

Trifolium incarnatum https://www.iucnredlist.org/species/176390/7231548 [accessed 12 May 2020].

241 Fod - Fold.

242 Diary entry for 14 September 1821 included the following: *The Wheat now in Tythings, in great danger of being all spoild all being in a state of VEGETATION ... MARKETS, rose yesterday to 5£ pr Qr.*
Diary entry for 1 November 1821 included the following: *Sold first old Barley @ 35s old wheat 11 S 16 li Nt @ 37s New grow'd Wheat selling @ 12s pr Sack.*

& harvest

18. SOW, hog'd, the 16 Weeks expire 8th Decr Mr Wentworths Boar | Fine morn'g some thin rain afternoon, Fallow'g
19. Very fine day, Fallow'g as yesterday SILBURY=MEAD, began Mow'g for HAY Young SWEDE'S finish'd hoe'g 1st time, see, sow'd 7th July Turnips the 2nd drift on Down, hoe'g
20. VETCH S, sow'd 11th June, Sale Sheep and Lambs began feed'g, wth Turnips | MACHINE, haymaking Silbury Mead - Fine day, Cart'g Dung before Fold
    Died this morn'g, suddenly at Avebury Farmer Richd Crook Aged 59 Yrs [This entry written within a thick black border]
21. TURNIP'S destroy'd twice after Vetches began Plough'g & sow'g the 3rd time the six weeks, Stubble Turnip | Very fine day, boisterous afternoon | HAY, the 2nd Clover Tan hill Piece carried four Loads evening
22. HAY, at 2-o-Clock very dry - prevent'd carry'g by heavy Storm, dry boisterous morn | The Hay Whitelands fit to carry, and Half Silbury mead hay, work'd twice wth the Machine & hatched up once only[243] | Ploughing & Sow'g Turnips as yesterday
23. Ploughing & Sow'g Turnips & finish'd | The young Swedes, that is hoe'd once, drag'd | Storms to prevent carry'g hay | BARLEY, began Mowing, left of Tan hill road
24. Sunday, heavy Storms
25. Young Swedes, hoeing the 2nd time Glass low, partial Storms & very dry'g | Carting Dung before Fold
26. CLOVERHAY, carried remainer from Tan hill & 9 Loads from 2A 3R 17P A:Whitelands Beautiful fine harvest day
27. All Silbury mead hay fit to carry, began at 11-o-Clock, stop'd at noon by Thunder Storm carry'g again evening, little Frost & fine fog'y morn'g
28. HAY, carry'g till 2-O-Clock, when stop'd wth rain Draging young Turnip'd (see) sowd 2nd Augst | Turnips sowd 21st appearing
29. A-two tooth Ewe died, bowels ruptured | Storms all day, Fallow'g
30. Storms all day, Horses in Stable, Carters winnow'g
31. Sunday, - Storms in morn'g

September

1. BARLEY, sprouting in the Swath RAIN a storm, pouring in the morning, dry after, fallow'g before Fold
   Septr 2nd Obit Mrs Philpot Aged 88 [244] [This entry written within a thick

243 Hatch – In this context: Noun: A wallow or line of raked-up hay; Verb: 'To hatch up' - to rake hay into hatches. Dartnell & Goddard, *Wiltshire Words*, p.76.

244 Betty Philpot of Avebury buried on 6 September at Avebury, aged 88, *Wiltshire, England, Church of England Deaths and Burials, 1813-1916 for Betty Philpot* [accessed via Ancestry, 27 July 2020].

black border]

2. HAY, finish'd carry'g Silbury Mead Some light fly'g Storms, fallow'g
3. BARLEY, began carry'g Tan hill Piece very unpleasant boisterous Day some thin rain | Turnip hoe'g finish'd all, except those just up
4. Barley carry'g from 6-o Clock morn'g to 7 in even'g, 8 loads in Barn | Fine warm day
5. Barley, Carry'g fm 6-in morn'g to 7 eve'g Exceeding boisterous all day.
6. BARLEY, carry'g fm 6 in morn'g to 7 eve'g and finish'd, the 23½ Acres | HARVEST=FINISH'D, Boisterous winds The 1st Rick from the home end of the Piece. The 2nd Rick is the best by Waggon house The 3rd Rick is thinest by Granary, uper end of the Piece.

[7.] Sunday, Beautiful fine day, Glass sinking

8. Dry morn'g topd one Barley Rick, after it rain'd very heavy. Muckle Cart'g before the Fold
9. CHALK, began Cart'g from the Pit, to further further Down Piece Storms in the Day (little for Horses to do)
10. Top'd a 2nd Barley Rick, when RAIN came & continued all day. 5 Horses harrow'g some Wheat Stubbles (too short to mow) to top the Hay Rick, not having enough Straw to Thatch 2 - Hay Ricks & 3 Barley Ricks now receiving injury | CALF - the 2nd sold 14 Weeks old 42¾ li pr Qr @ 6d Cost 24s at 4 Weeks old £4 5s 6d

    On Sunday last the 7th Inst by an att[...]of the Cholera whilst at Church, terminated in 12 hours the life of my esteemed Friend Mrs Hale the Widow of my late valued Friend Mr Saml Hale of the Poultry London [245] [This entry written within a thick black border]

    Septr 10. Died suddenly after eating hearty at Tea this evening Mr Willm Lawrence of Idstone Farm Aged 67 Yrs [This entry written within a thick black border]

11. Tremendous RAIN to prevent the Market taking place at Devizes, and driving rain nearly all day. Horses in Stable & Sheep on the Down only
12. FLOOD on all the Meadows at Avebury from the rain fallen yesterday | Fine warm drying day. Hay removed a load fm Mead rick to dry it & topd the rick | Plough'g athwart Down Piece, Vetches fed off
13. Plough'g athwart beautiful fine day like a second Summer, removed 2 load of Barley from Rick to Dry in Meadow (Jno Chis Junr insolt[246]
14. Sunday, beautiful fine warm day, Glass rose from much rain to A, in fair
15. Beautiful fine warm day, Plough'g as 13th
16. Beautiful fine VETCH's finish'd feed'g Nether | Down Piece 6 Acres began see 20th Augst with 300 Sheep & Lambs & Turnips.
17. Mr Brown, draging in, TREFOLUM=incarnatum little Whitelands barley

245 att[...] - attack.

246 John Chivers Junior insolent.

stubble 15 li pr A [247] | R. P_ Plough'g in Vetches Down Piece & the 20th Rye | Very warm day, Plough'g after Vetches | Kellaways, Estate, bot in by Auction 7600, bo [...] [248]

18. Very hot day 20 S_ old Wheat to Devizes | Finish'd Plough'g Vetches fed off [Note in margin] T1,,4,,2 Coal
19. 20 S_ Old Wheat to Calne. Drag'g & harrow'g -D- piece | Very foggy morn'g & very warm day
20. Very foggy morn'g & very fine day | RAM'S, turn'd Mr Kings two, to the Stock Ewes & mark'd the Ewes & countd them | Drag'd the young Turnips, Roll'g & harrow'g & couching after Vetches fed off, for Wheat
21. Sunday - Beautiful fine day COW=bull'd 2nd time see 3rd June
22. Very fine day, not so warm as last week | Fallowing before the Fold & Cart'g Muckle
23. Plough'g 2nd time Down Piece, fine day | TURNIP'S began hoeing, see sow'd 21st Augst
24. TURNIP'S finish'd, Plough'g as yesterday, very fine
25. Harrowing, Couchey Down Piece, Fine day
26. Rain in Night - Storms in the Day | Ploughing a Drift before the Fold Sheep finish'd feeding the Clover, the Wheat & Clover in Barley Stubble only left for Sheep to feed with Turnips, (feed not plenty)
27. SOW, hogd the 3rd time, last time 1st Septr | Rain heavy in morn'g fine day | Ploughing Wheat Stubbles for Vetches began
28. Sunday, Very fine day
29. Ploughing for Vetches Very fine day | The Two Tooth Ewes put in Silbury mead to have fresh Rams
30. Put 2 of my Brother Ws Rams to the young Ewes & another to the other Ewes (Mr Pumphry | VETCHES, sowd 2 Sacks of own Old @ 10s Bu on about 3¾ Acres | Very fine warm day

October

1. Very fine Plough'g before Fold | HAY the 1st load from a Meadow Rick bot at the late Mr Kings Sale Overton 1T 17C 2Qr the yellow Wagon 1T 3C 0Qr [249]
2. HAY the Overton began giving the Sale Ewes & all the Lambs, laying back on Turnips & feeding Stubbles & new field | EWES, triming 1st time for

---

247 The number 15 is written in pencil.

248 Kellaways House & Estate was offered for sale by auction on 17 September 1834, the farmland extending to 125 acres: the tenant was Mr Broome Pinniger. Devizes & Wiltshire Gazette, 28 August 1834.

249 The remaining farming stock of Stephen King Esq deceased was auctioned on 15 September 1834 and included twelve ricks and stumps of hay, amounting to upwards of 250 Tons, on Overton and Lockeridge farms. Devizes and Wiltshire Gazette, 11 September 1834.

Sale (Devizes 20th) Plough'g as yesterday, beautifully fine

3. HAY, the second Clover the HORSE'S began the Clover grass finish'd in Whitelands | Fog'y morn, oppressively warm afternoon | Ploughing as yesterday
4. FALLOWING finish'd the Clover lay for Wheat | Fog'y morning & very warm
5. Sunday Fog'y morning & very warm
6. Fog'y morning & very warm Ploughing as 4th
7. Fog'y morning & very warm Ploughing as 4th | WHEAT RICK the 3rd of last years put into Barn
8. Cartd the heap of earth on the Bank thin soil Silbury=mead | CONTINUATION of foggy mornings & fine days
9. CONTINUATION of foggy mornings & fine days Cart'g the heap | of Dung for Wheat Tan hill Pie [250]
10. Foggy mornings & fine days and finish'd Tan hill Pie | 12 Sacks of Seed Wheat exchangd wth Sister Crook on Berwick Common | PLANTS, Cabbage, planted
11. 20 S_ Wheat to Berwick Comn Mr J_ Large | Dung all, Cartd fm the Yard & the Inn, the 12 A: Piece cover'd Rt of Tan hill road Very fine day (Rt Philpot drill'd Wheat Downs)
12. Sunday - Very fine day VETCH's appear'g sowd the 30th
13. Plough'g after the Fold for Wheat. | POTATOES, diging in Cottage Garden | Very fine warm day
14. TURNIPS, finish d feed'g the first sow'g (20th May began 14th Augst wth 300 Sheep & Lambs 6A 1R 25P | Turnips - began feeding Down Piece, hoe'd 14th Augst Fine morn'g very heavy rain mid=day | NO RAIN, till now since see 11th Septr Plough'g as yesterday | Potatoes began diging in Field, black ones only ripe, the red & white not ripe The Sheep countd 243 Stock Ewes 90 Ewe Lambs 5 Rams 2 Wethers fat [Total] 340 to winter at home 100 Sale Ewes 95 Wether Lambs 5 Cul Ewe Lambs [Total] 200 for sale | The Ewe Lambs, drew from the Sale Sheep and put in Silbury mead, tied out & fold'd, the 2 tooth Ewes put on Down wth others | The Two HOUSES OF PARLIAMENT[251] [This entry written below a triple line] | This evening an evil spirit abroad [This entry written in margin at foot of opposite page]
15. Some slight showers of Rain, fine day | Plough'g as yesterday, Potatoes dug 17 Sacks filld by 2 Men & 2 Women

---

250 Pie - Piece.

251 On 16 October 1834, a fire destroyed both Houses of Parliament along with most of the other buildings on site. The House of Lords first sat in their new chamber in 1847 and the House of Commons in 1852, https://www.parliament.uk/about/living-heritage/building/palace/architecture/palacestructure/ [accessed 27 July 2020].

16. Ploughing Turnip land for Wheat Some thin Showers, drying between
17. Ploughing stubbles for Vetches Very boisterous winds & dry'g
18. VETCH'S, sowd a second drift 1 Sack | Exceeding cold boisterous winds
19. Sunday - some slight Storms
20. Ploughing Turnip land | Devizes Fair, dry morn'g, heavy Storm mid:day exceeding warm & dry after [Pinniger then made an entry about Devizes Fair - see Appendix H]
21. Heavy storm mid day, drying after | Plough'g as yesterday
22. Plough'g and finish'd (the Turnip Piece) | Boisterous drying winds | SHEEP do well, the Ewe Lambs do better in Silbury mead & folded then laying back on Turnips & hay & field grass
21. The Sale Ewes - do well on Down Turnips & field grass WHEAT sow'g Mr K & W P= of K= began [252]
23. Dry'g winds Storm mid day | Plough'g Couchey part of Down Piece
24. Plough'g Couchey part and harrowing | SNOW, a little in morn'g very dry cold day [Note in margin] 24th FIRE at Bushton a Skilling
25. POTATOES, finish'd diging 85 Sacks of White 45 Sacks of Black 10 Sacks of Prolifficks 8 Sacks of Red 15 Sacks from Meadow Garn [Total] 163 [253] | Sharp frost & fine day | 2 PORKER'S put to fatten | CHALKING, began Nether Down Piece, having Plough'd all that is ready for Wheat and waiting for Rain to sow,
26. Sunday, Very fine day DAHLIAS, cut up by Frost | FIRE at Knighton, Mr Spearing Wheat Rick (11-0 Clock Sunday morn'g) [Pinniger then recorded notes on Weyhill Fair, held on 13th October - see Appendix H]
26th Died this evening 6 O-Clock Mrs Watts of Beckhampton Inn. Aged [...] [254] [This entry written within a thick black border]
27. Sharp frost & fine day HORSES, baiting in Silbury mead evenings, the grass too short to mow & too cold for Stable | Chalking as 25th
28. Chalking as 25th Beautiful mild calm day
29. Chalking as 25th Beautiful mild calm day
30. Chalking as 25th Beautiful mild calm day
31. Chalking as 25th Beautiful mild calm day

November

1. Chalking as 25th Beautiful mild calm day
[2.] Sunday, Beautiful fine mild Day
3. Foggy morn'g like mid:summer, mild Day | Chalking continued. | Mr Brown diging GRAVEL my Meadow

---

252 Mr K & W P= of K=: Mr Kemm and William Pinniger of Kennet.

253 Garn - Garden.

254 Mary Watts of Beckhampton buried on 31 October 1834 at Avebury, aged 46, *Wiltshire, England, Church of England Deaths and Burials, 1813-1916 for Mary Watts* [accessed via Ancestry, 27 July 2020].

4. Chalking & very fine day
5. My Sister Ann Crook, left Corton, to live at West Tytherton | Very boisterous Winds all night and all day rain all afternoon | Chalking continued now 10 days cover'd 5 Acres 6 Men filling - 3 Carts & 5 Horses allmost double the quantity put, more than when done by the Acre at 42s by Men only the 9 Men & Boys employ'd with the Horses &c. cost 8s 4d pr day | FOLD, finish'd Tan hill Piece & put on the Chalk Nether Down Piece for Wheat
6. Fine day, too slippery for Chalking (in Pit) | Ploughing last Fold drift
7. Ploughing as yesterday, very driving rain in the night & morn'g, dry after | FIRE at Chute, Mr Bethels, Hay Rick Sunday eve 2nd Novr | DIPING the SHEEP in Arsnic & soft Soap to Kill the Tick & prevent tuckg 243 Ewes 91 Lambs 5 Rams 2 Fat Sheep [Total] 341 3 Rams WP_ K} @ 11s pr Hd began 9-oClock morg finish'd 5 even'g, 3 Men helping all their Keep [Pinniger then recorded prices at Appleshaw Fair, held 4th November - see Appendix H]
8. WHEAT=SOWING, began, about 6 Acres upper side of Rt hand of Tan hill road wth 4 Sacks, good firm work, much better prospect than what has been done before by others in the dry state, the Wheat being thin having perish'd in the ground, a very gracious rain the 5th & 7th, and a beautiful Summers day this for Sowing | 5 Men spreading Chalk | As a Tribute of gratitude to the giver of every blessing I here record his mercies in the benefits we now enjoy in cultivating the little Farm, and beholding the Land covered with green verdure at this late season All the Sheep healthy and doing well The Stock Sheep on the Down & Wheat Stubbles. The Lambs a small tie in Silbury mead only. The Sale Ewes on Down Turnips & Grass in the Wheat stubbles adjoining & a small quantity of hay T P
9. Sunday, RAIN in the night & all day
10. Rain in the night & day too wet for Wheat sow'g FALLOW'G began Wheat stubbles
11. Very sharp frost & fine drying day | Wheat sow'd the Turnip Piece, bottom of Tan hill piece 6A 1R 25P with 4¼ Sacks, good work | Hay W P of K & Mr K[...]m, began giving to Lambs | Miss Kitty Buxton Mr Willm Dean} Married [255]
12. Ploughing for Wheat, Cold dry day 1 Man diging Gravel & 1 Thrash'g Wheat.
13. Plough'g finish'd Tan hill Piece for Wheat | Cold dry day
14. Wheat Sowing finish'd Tan hill Piece 18A 1R 2P with 50 Bu: of Seed fm Corton | TURNIP'S the 100 Sale Ewes finish'd feeding the Down Piece began 14th Octr 2A 3R 10P and kept the 100 Sale Lambs 6 days | Porke[...]

---

255 William Dean, of the parish of St Michael, Wood Street, London and Catherine Elizabeth Buxton of the parish of Mildenhall married on 11 November 1834 at Mildenhall, *Wiltshire, Church of England, Marriages & Banns, 1754-1916 for William Dean* [accessed via Ancestry, 27 July 2020].

the 1st Pig 3 S 14 li | Porke[...] the 2nd Pig 5 S 3 li - Kill'd by the Sow Sharp frost, cold dry Day

15. A Sale Ewe Killd being ill | CHALK, Cart'g - 6 young Men out of employ fill'g @ 7s pr W: & if employ'd by the Parish 3s only | WHITE=HART COACH began serving with [Note in pencil: 6 Horses] Straw Mr Amor giving it up | Very sharp frost, fine day | TURNIP S the Sale Ewes began the 20 A_ Piece | HAY, began giving the Lambs evenings
16. Sunday, - Beautiful fine mild day
17. Chalk_Cart'g as 15th Beautiful fine mild day | 8 STORE PIG'S put to fatten
18. Chalking as yesterday & fine mild day
Died this morning Novr 1st Mrs Gray of Newbury Aged 91, beloved by her Family and esteemed by her Friends and Neighbours for her amiable and benevolent disposition [This entry written within a thick black border]
19. Very sharp Frost & Cold day Chalking
20. Very hard Frost & Cold day Chalking | STRAW-Barley the Ewes began in Yard
21. Very cold morn'g, Storm of large blossoms of SNOW, Chalk'g as yesterday The 3 borrow'd Rams retd to WP Kennett
22. CHALKING FINISH'D Nether Down Piece A8,,3,,6 in 17 days, by 6 young Men out of employ @ 7s pr Week fill'g - 4 Carts & 7 Horses - for lower end of the Piece | Thin rain nearly all day | 100 Ewes trim'd the third time for Marlbro Fair
23. Sunday - Very fine day
24. Plough'g the Chalk'd Land for Wheat | Very mild fine day Marlbro Fair [Pinniger then recorded a note on the Price of Malt - see Appendix T; the trees planted by the front Gates - see Appendix O; prices achieved for sheep - see Appendix R]
24. Lambs finish'd feed'g Silbury Mead began feed'g it 29th Septr with the Young Ewes of great value, having Kept 100 - Sheep or Lambs all all the time
25. Very fine day - Plough'g the Chalkd Land LAMBS, began feed'g TURNIPS, succeeding the Sale Ewes LAMBS on New Field mornings & Hay in Fold Nights
26. Very sharp Icy Frost - Plough'g as yesterday
27. Very sharp White Frost - Plough'g as yesterday
28. TALAVERA Wheat began sow'g the Chalkd Land | Fine day RAIN all even'g
29. RAIN in night too wet for sow'g, Carting Dung from the Inn to Whitelands made since 11th Octr covered about 1A 2R 14P | Gates removed fm Silbury mead & hay to Sheep HAY, the Ewes began (old Clover) night Straw mornings & the Down &c
30. Sunday Very fine dry day

December

1. Very rain'g boisterous night & storms in day | Finish d fold'g Down Piece for Wheat & Chalk'd | FOLD removed to WHITELAND'S, muckled for Please & Wheat follow'g | Plough'g Down Piece for Talevera Wheat
2. Ploughing finish'd Down Piece for Wheat | Rain in night, dry windy day.
3. Beautiful calm mild day, like mid=sumr | WHEAT SOWING FINISH'D. on the Down Piece 8a 3q 6r wth 7 Sacks Talavera extraordinary fine season, all weathers have combined to make a late good Wheat sowing, and healthy for the Sheep [256]
4. Most beautiful fine mild day | Cart'g the BANK by Silbury hill to the Mead
5. Finish'd harrow'g the Wheat sowd the 3rd | Another equally fine warm day.
6. CHALKING began Wheat stubbles further Down Piece, the remaining of the Piece from where the Man was kill'd W Paradise see 29th Jany 1830 [257] | Damp thick air all day | WHEAT just appearing sow'd 4 Weeks
7. Sunday, some rain in night & day
8. Chalking - as the 6th, very fine time for it, very dry'g day and Glass geting high
9. Sharp white Frost, clear fine day | Chalking as yesterday
10. Chalking as yesterday, many People finish'd fallowing, whilst oweing to Chalking I have fallow'd but one day, stubbles | Rain in night, dry day | FAT WETHER Sheep 57 li @ 7d li & Skin is £1 14s od in house for own use very good
11. Very sharp frost, clear day, sharp frost even'g | Chalking - & 4 Loads dung fm Inn, before Fold Lambs hay morn'g (grass frosty)
12. Chalking, sharp frost & dullfoggy day | Commissioned Mr B P_ of Chipp[e]n[ham] to proceed against W T of Kennet for injury sustained in Silbury md [258]
13. CHALKING FINISHED, the remaining part of the further Down Piece 4A oR 18P by 6 Men fill'g @ 1s 2d pr day - & 8 day and spread'g 10s is £3 6s od - in addition to the Carter 2 Boys & 6 Horse The other part of the Piece Chalk'd by hand in 1830 - 4A 3R 1P - cost £10 12s od, & about half as thick 8A 3R 6P 4A oR 18P } Chalk'd in 25 days, unusual thick dress'g [259] | Thaw and dry'g day
14. Sunday - Cold drying winds 11th Decr Married at Cliffe Josh Smith of

---

256 q - quarter (of an acre, same as a rood); r - square rod ($^1/_{160}$ of an acre, same as a perch): see entry for December 13 1834.

257 Whilst the diary entry for 29 January 1830 records the efforts that were made to get William Paradise out of the chalk pit, there was no indication that he had died.

258 BP - Broome Pinniger, WT - William Tanner.

259 The diary entry for 1 February 1830 records that "The Chalking, being dangerous work discontinued it, 4¾ Acres done out of 9 A", but there is no reference to the costs.

Chilvester=hill Eliza Smith of Thornhill

15. FALLOW'G Wheat Stubbles began again having Plough'd but one day 10th Nov | Damp cold air
16. Dry fine day Plough'g as yesterday | The Ewes having fed the Wheat Stubbles to the ground, have only the Down wth Straw morn'gs & 5 Truss hay evenings
    DIED 16th Dr Hill at Chippenham Aged 92 18th Mr Edwd Michell Chippenham Monkton House [This entry written within a three-sided thick black border]
17. Fallowing, fine dry day
18. Fallowing, fine dry day
19. Fallowing, fine dry day | Laurels, and Tress planted
20. Fallowing, fine mild day
    Saturday eve, the 20th Decr - Died by accidentally fall'g into the Quarry in Widcomb road & drown'd aged 71 Mr Stephn Jeffrey of Spilmans Farm [260] [This entry written within a thick black border]
20. Wheat all up well Tan hill Piece finish'd the 14th Novr
21. Sunday - Fine dry day
22. Frost in morn'g - Fine dry day - Fallow'g | The Tegs scour, either by feed'g the grass Tan hill Piece or the young Turnips wth Charlock discontinued both & began the SWEDES - with hay mornings & evenings in Fold 1 - Ewe Carriage ran over it & died
23. Very sharp Frost & very fine day Part of Barley Rick put into Barn Fallow'g
24. Very sharp Frost, & fine day | Fallow'g & 6 Loads Muckle to Whitelands
25. Xtmas - Thick air morn'g cold dry The EWE'S, want better keep, having had the Down only till now with Straw mornings & 5 Truss's of Hay evening, now began TURNIP'S | FAT SHEEP Killd for Labourers (Ram stag cut two Yrs) very good 89 li, Carcass Caul 11 li [Total] 100 li | 100 li @ 7d is £2 18s 4d Head & Hie 8d Skin [...] and loose fat for kill'g [261] | Talavera Wheat sowd 28th Novr appearing
26. Damp morn'g dry day | Plough'g finish'd, fallow'g the 18 Acre Piece & began fallow'g further Down Piece
27. Fallow'g as yesterday and fine dry day
28. Sunday - fine dry day
29. Frost morn'g, fine day, fallow'g as 27th
30. Rain in night & very damp day, fallow'g as 27th
31. Rain night & storms in day, fallow'g as 27th | The Ewes finish'd the drift of TURNIPS to make a Road to the Swedes - to begin feeding SWEDES tomorrow (only half enough) wth Straw morn's & 6 Truss hay evenings The

---

260 Stephen Jeffreys (yeoman) of Spillman's Farm buried on 26 December 1834 at Hilmarton, aged 71, *Wiltshire, England, Church of England Deaths and Burials, 1813-1916 for Stephen Jeffreys* [accessed via Ancestry, 27 July 2020].

261 Hie – hide.

young Turnips reserved for the E & Lambs [262]

30. KELLAWAY'S Estate sold

[Pinniger opened his diary for the year with a note describing the impact of the introduction of Imperial Measure on his farming business - see Appendix U]

1835

January

1. Exceeding fine mild day | 10 Qr Barley (my first) to Devizes, 2T 9C 2Qr Coal | Fallow'g further Down Piece | FOLD finished Whitelands, Muckled began 1st Decr 4A 1R 28P
2. FOLD removed to the middle Down Piece muckled for spring Vetches | Very fine mild dry day, Fallow'g as yesty
3. Very fine mild dry day, FINISHED fallowing the further Down Piece all Chalk'd for Turnips. | NB. The half of this Piece was Chalk'd in January 1830 with Barrows is now clean from Couch, Plough freely the other upper half chalk'd last month is full of Couch and Ploughs a Horse heavier on this half of the Piece was a bad Piece of Wheat & the lower half a very good Piece last year. such encouragement for Chalking ★
4. Sunday, Sharp frost & fine day
5. Sharp frost & fine day | Fallow'g Whitelands Clover stubble, Muckled & Fold'd either for Pease or Turnips & Wheat after
6. Very sharp frost & fine day | Fallow'g as yesterday
7. Mr Sandell, applid | Fallow'g as yesterday | to rent the INN | Very sharp HOARFROST all day, stop'd force Pump
8. Very sharp HOARFROST the Road covered with ICE my Horse unable to walk on it, gone to Devizes Market on Foot, and walked home from the Bell Devizes Green 6¾ Miles in 1 Hour & 35 Minu | Fallow'g finish'd Whitelands & finishd fallow'g | Mr Watts, propose leaveing Beckhampton Inn

   8th Died Miss Ann Smith, Calne aged 64 [This entry written within a thick black border]
9. Thaw & mild day, Cart'g muckle before Fold
10. Thaw & mild day, Cart'g Trenching fm Silbury mead
9. The 1st Bacon PIG 7 S 6 li.
11. Sunday - Mild damp day
12. Mild damp day | The remaining part of the 1st Barley Rick put into the Barn, to supply the Ewes wth Straw mornings | Mr B= Pinniger of Chippenham with Mr Jno Wentworth & Geoe Maslen both of Beckhampton & myself (TP) surveyed the Hatches &c. in Willm Tanners Meadow.
13. Very mild fine day, the Horses in Stable except Cart'g earth fm Gravel Pit to

262 E - Ewes.

Garden having but little to do with them

8. The SPRINGS, are become too low to use the Horse Pump, water the Horses & Sheep from the Rickyard Well The House Pump with difficulty draw Water

14. Mr Viveash of Berwick Basset 3 Waggon loads of Gravel fm my Meadow | Earth, Cart'g from the Down to fill the Gravel Pit, thin rain at times, very mild

15. Earth, Cart'g and finish'd fill g the Pit | Cartd the Potatoe Haulm, very warm day

16. Very boisterous winds & rain all night & till mid:day, drying after, Horses in the morning, Carting muckle before the Fold afternoon

17. Very sharp Frost & very fine day | 6 Waggon load of Dung, before the Fold, wth one Team, middle Down Piece, the Colt & Sale Horse resting | The 2 lots of Straw to the Inn pr week, make 4 lots of Dung, on Waggons pr week.

18. Sunday, very sharp, freezing all day Thaw & damp evening

19. Rain in night & SNOW from 9 to noon, dry after, Horses began Plough'g after the Fold return'd to Stable, when snow'd, Ewes in | Mr Treene) | the Straw yard all day | (inspected the INN

20. the Straw yard all day | SNOW, drifed in the night above the Hurdles in the Fold, cold freezing winds all day | ASHES, the 1st load fm Devizes 60 Bushls | SOW, farrow'd in the night (4) time up 17th

21. ASHES 2nd Load 60 Bu: Freezing all day Ewes to Swedes again

22. ASHES 3rd Load 60 Bu: Snow & Frost in night clear warm Sun all day

23. Very mild fine day, the Snow disappear except in the hollows &c. | ASHES, began sowing bottom of Tan hill Piece the Marl grass 180 Bushels in 2¾ hours by 4 Sowers - 40 - Bu: pr Acre | Beans, put part of the Rick in Barn for the Horses | 2nd Bacon PIG 9 S 9 li - the one Killd the 9th 7 S 6 li larger at the time

24. 4th load of Ashes, 60 Bu: beautiful mild day | MR WATTS, served Notice to quit the Inn Michaelmas next

25. Sunday - fine mild day

26. 5th load of Ashes fine mild day

27. 6th load of Ashes, Beautiful mild day the Glass very high & no Frost

28. ASHES sow'd the 3 loads 180 Bushels in 3 hours by 3 sowers N.B. the 6 Loads 360 Bu: sowd 9 Acres 40 Bu pr A | Fine mild day

29. Beautiful spring like mild day | 7th load of Ashes 60 Bu: & 15 Qr Barley to Dev | The 1st New WHEAT RICK put into Barn

28. The level taken from the Sill of the Hatch in Mr Wentworths Silbury Mead to the Sill of the Hatches in Mr Tanners meadow which is 600 feet in length, and the fall only 2¾ Inch's | The fall from the Sill in Mr Wentworths Meadow to Mr Tanners Ditch is 12 Inch's, from the Ditch Mr Tanners Carriage rise 7½ Inch's in the first 10 Feet | It is 116 feet from the Ditch between Mr W and Mr T_ and Mr T_ Sill and the rise is 9¼ Inches from the

Carriage in Mr Wentworths Mead to the Sill in Mr Tanners Mead | 484 feet from the Sill in Mr Ws mead to Ditch 116 feet from Ditch to Mr Tanners Sill. [Total] 600 | It is nearly a dead level round Silbury Hill to Hatch in Mr Wentworths Mead

30. Fine mild day | Ashes the 8th load of 60 Bushls

31. <Ashes> 9th load of <60> Bushls Error | Dense fog all day Mr TRIN, looking over the Downs, lost on Tan hill | 2 Loads of BOG EARTH from Pen

February

1. Sunday. - Dull dry mild day

2. Ashes the 9th load of 60 Bushls | Cart'g earth Silbury mead, fm gutters | Thin driving rain morn'g, fine eve'g

3. Ashes the 10th load 60 Bushls, fine day Barley part of 2nd Rick put in Barn

4. Ashes the 11th load 60 Bushls Bushing the Bank Silbury mead | Valuation at Corton for Jno & Executors Beautiful fine mild day [263]

5. Ashes the 12th load 60 Bushls Damp driving rain morn'g - dry aftr

6. Ashes the 13th load 60 Bushels | Boisterous Winds

7. Ashes the 14th load 60 Bushels Very boisterous damp day | Gooseberry &c Trees - Removed, to New Gard

[...] An additional Horse brot to the Inn of Mr [...] Mr [...]

6 & 7. 20 Sacks Wheat to Devizes each day

8. Sunday - Very cold hail storms

9. Frost & very fine drying day | HAY, the old Clover finish'd by the Ewes began the 29th Novr is 10 Weeks. & 11 Tons | HAY, the Ewes began the new Clover, wanted better hay before, The young Sheep Nag & Cow having it all the Winter | Ashes sow'd 3 loads 180 Bushls in 3 hours 2 loads to finish the Piece 23½ Acres | Ploughing, middle Down Piece for Vetches muckld and fold'd

10. Ploughing as yesty very sharp frost all day

11. Ploughing as yesty very sharp frost & thaw after | LAMBS the first 3 this morn'g, the Ewes time expire the 14th and no Ewe yet cast a Lamb from 240 Ewes

12. The Springs, very low, no Water in Pump | Ploughing as yesterday rain all the morn'g dry after | Standing Pen made for Lambing the Ewes but continue fold'd on Down Piece whilst the weather so favorable

13. Ploughing as yesterday, Fine day, damp eve | 3rd FAT PIG 10 S 6 li 4th FAT

---

263 The valuation relates to the estate of the late Henry Crook of Corton (died 7 July 1834, aged 59), husband of Ann Pinniger, Thomas Pinniger's sister, WSA: 4381/2/1, *Pinniger genealogical memoranda book, c.1820-1943,* pp. 8, 21-22. Henry Crook appointed Christopher Pinniger, Thomas Pinniger and his son, John Crook as joint executors, TNA: PROB 11/1835/77, *Will of Henry Crook, Gentleman of Hillmarton, Wiltshire, 22 August 1834.*

PIG 9 S 11 li Mr Kemm 5th FAT PIG 10 S 19 li Jno Chivers | Mr Treen sign'd an Agreement for Renting the Inn of me

14. Very fine day, Falllow'g Potatoe Land for Barley | EWE'S time expire for Lambing, now 5 La [264]
15. Sunday - fine mild day
16. 20 Lambs, and no Ewe without a Lamb good luck, no Ewe cast Lamb | Plough'g as the 14th some thin rain
17. Beautiful fine mild day | BOG=EARTH, Waggon & 4 Horse from the Sicily Clump Bedwin Common for American Plants
18. Hurdles 12 doz from Marlbro | Rain morn'g very fine after | Mr Saml Wentworth, took Mr Amors Farm of Mr Hopkins, of Cheltenham
19. 50 Lambs, no Ewe gone without a Lamb, and no Ewe died | 13 Qu: of Barley to Devizes & 60 Bu: Ash's retd | Dry cold winds in morn'g, rain after | HURRICANE almost alarming at 5-0-Clock eve to strip the Porch of Barn. with heavy rain hail & sleet in the night
20. FLOOD this morning, to water the whole of Silbury mead, and a prospect of enough in future to supply us with a good crop of Grass in due season, till now a dread of a very short supply to take the Ewes & Lambs after the Swedes, being a very unusual dry winter, none so dry in the 38 Years pass'd This another proof of the goodness & power of the Almighty [265] | Horses in the Stable, except sow'g a Load of Ashs | Good Wheat 12 S 8 li Nt 20s pr Sack | Vetches not good - 15s pr Bushl | Barley the best - 42s pr Qr - see no proportion in price
21. SNOW, a storm of very large blossoms in the morn'g, cold dry'g winds after | Horses in Stable, but little to do, too wet to Plough | BREW'D, 12 Bu: Malt, strong Beer
22. Sunday, dry morn'g, rain remainer of day
23. The 1st two loads of Stones fm Calne for building the Raceing Stables at the Inn for Mr Treen | LAMB'S 102 - no Ewe died & no Ewe gone without a Lamb | N B - The Ewes in lamb fold'd to this eveng having finish'd fold'g middle Down Piece muckled & HAY only evenings, and walked up to the 17th Inst a mile to home Yard to Straw | The Ewes in a Standing Pen & the Tegs began fold'g Turnip ground for Barley | This morning usher'd in with Tremendous HURRICANES, to uncover the Ricks &c. and the greatest difficulty in remov'g the Fold
24. Very fine clear day, - the 2nd two loads of Stone
25. the 3rd two loads of Stones | Very boisterous weather to injure the Buildings and rain all afternoon | SWEDES, Keep well, just geting fresh to gather greens but not to run to seed began half drawing <half drawing> them to check
26. The 4th two loads of Stone from Calne to Inn | Boisterous Winds and rain

264 La - Lambs.

265 The number 38 written in pencil.

27. The 5th two loads of Stone from Calne to Inn | The Weather continue very Tempestuous
Obit the 15th Feby H- Hunt aged 62 (late of Littlecote) [266]
Obit the 26th Feby Jno Maslen aged 48 drowned at Avebury (lame) [This entry and the previous entry written with a short thick black line above the first and a long thick black line below the second]
28. The 6th two loads of Stone from Calne to Inn | Fine clear day.

March

1. SNOW in the night to cover the gound in the morn'g (Sunday) and storms of Snow in the day, dry even'g
2. LAMBS this morn'g 186 in 15 days from 240 Ews, and 63 Ewes yet to lamb no Ewe gone without a Lamb no Ewe died This almost a general good lamb'g season | 26 Posts fm Devizes for Sister Crook & Coal 1T 4C 1Qr | Frost morn'g fine day
3. 26 Posts to West Tytherton & 20 Bush's & 30 Bdls of Rods Retd fm Bremhill grove | Very Boisterous Winds in the Night & all day | CHALK began diging for walling up the Waggon=house, & build'g Stables (Inn)
4. 32 Bushels Lime & 1 load of Stone fm Calne | Hail storms & drying between
5. Fine drying day | 2m Bricks & 4 Posts fm Devizes wth 5 Horses
6. Chalk _ Carting to Waggon House | Very Boisterous Winds in the Night and all day [Pinniger then presented more information in the economics of fattening pigs - see Appendix A]
7. Very boisterous Night & day wth Storms | 5 Horses wth 3 Waggons brot 6 Loads of Rubble fm Mr Wentworths Down to Tan hill road, - 5 Loads, a days work | EWE'S 205 wth 212 Lambs in 3 Week no Ewe without a Lamb & no Ewe died, 35 to lamb | Teg giddy Kill'd 30 li | 2 Young Pigs wan'd 6 Weeks old to fatten the Sow
8. Sunday. exceeding Sharp frost fine day
9. Boisterous Winds & rain | MASON'S, commenced enclose'g Waggon=house | 5 Horses wth 3 Waggons brot 6 Loads of Rubble as see above the 7th
10. Barley, remainer of 2nd Rick put into Barn a 3rd Rick remaining | Very sharp frost morn'g, cold storms of Snow & rain afternoon | Dung. Cart'd from Inn to Tan hill Piece in Store heap for Wheat after Grass
11. Very boisterous & heavy rain in morn'g, dry eve | Cart'd 7 loads of Rubble fm my Down to T hill rod
12. 10 Qr Barley to Devizes & 500 Slates returned | Rain heavy & driving all day to prevent People doing their business in Devizes Markt
13. FLOOD this morning, to water Silbury Mead Frost & beautiful mild spring

266 Henry Hunt 1773-1835 (also known as Orator Hunt). Born at Widdington Farm, Upavon, he was a radical Member of Parliament, Oxford Dictionary of National Biography, *Hunt, Henry* https://doi.org/10.1093/ref:odnb/14193 [accessed 27 July 2020].

like day | Cart'g earth to Garden & Straw to the Sheep
Sacred to the memory of my lamented Mother-in-law Mrs Mary Large of Tockenham Court Farm, who depart'd this life at 5-o-Clock Saturday morn 14th March 1835 Aged 74 Years The Funeral Wednesday the 18th on Sunday 22nd The Funeral Sermon, by the Rev Mr Colton The Text 14th Chaptr Revelation 13 Verse And I heard a Voice from Heaven saying unto me blessed are the Dead which die in the Lord The 16th Psalm 8-9-10 & 11th Verses I strive each action to approve, &c The 39th Psalm 4-5-6-7 & 13th Verses Lord let me know my term of days &c. [This entry written within a thick black border]

14. RAIN heavy in the night & all morning drying after, Horses in Stable all morning, Cart'g Chalk, to Inn after
15. Sunday - Rain heavy in morn'g, dry after
16. Cart'g Straw to Sheep &c in morn'g, Horses in Stable, the Land to wet to Plough across
17. Rain all morn'g dry after | Carting Earth to garden for Flower Bed American Plants &c
16. The 3 Shepherds discontinued their Suppers having it a Month in lamb'g | 29 Ewes to Lamb, a part barren put with the Ewes & Lambs
18. Drying day, except some cold light Storms | Funeral of Mrs Large | Died Willm Brewer Aged [...] Beckhampn | Thwartg commenced, Whitelands for Turnips | Rolling commenced, the Clover
19. Ploughing & Rolling as yesterday | Sharp Frost & Beautiful fine day | A Ewe with a stopage & mortification died the first lost all the Winter
20. Beautiful fine mild day | Ploughing and Rolling as yesterday
21. Ploughing as yesterday & finish'd Rolling Fine mild growing day | TURNIP'S began feeding the last sowing, Kept good till now, now runing to seed, | SWEDE'S, enough remaining to Keep the Sheep Aprl | POTATOE'S - the Ash leafe planted in both Gardens
22. Thin rain morn'g, cold dry afternoon | Funeral Sermon, for the late Mrs Large Lyneham
23. Thwart'g finish'd Whitelands for early Turnips & began thwart'g further Down Piece for Turnips. Cold dry'g Winds | A 2nd Ewe died (suddenly) blood Vessel ruptured
24. Ploughing as yesterday, Cold dry'g Winds | The Masons, finish'd enclosing Waggon House
25. Ploughing as yesty - Frost & very fine day
26. Ploughing as yesty Very fine mild day | 15 Qr Barley to Devizes & 2T 12C 0Qr Coal retd | Barley part of last Rick put into Barn
27. Ploughing as yesterday, Sharp frost, fine day Turnips, the Sheep finish'd, about 4 Acres of Swedes remaining to Keep the Sheep 5 Weeks till Silbury Mead is fit to take them
28. AMERICAN PLANT'S, Planted in Bog earth [Pinniger then left the

remainder of the diary page blank] | Plough'g as yesty Frost & fine day | WATER, taken to the Sheep in Standing Pen drank the trough full the 1st night returned to Swedes & Hay in Standing Pen

29. Sunday, DIED this morning Timothy Caswell Smith of Avebury aged 50 Yrs | Sharp frost, and fine day
30. Swindon Fair, Sharp frot & fine day | Ploughing athwart began Fallows for Swedes
31. Ploughing athwart Thin RAIN all afternoon

April

1. Ploughing athwart Foggy morn, beautiful fine day
2. Ploughing and 20 Sacks to Devizes & 800 Slates retd | Very WARM fine day, the Ewes & Lambs, discontinued in Standing Pen - and lie back on Swedes, Hay & Water nights, all the Sheep doing well HAY evenings only 1st day | The Wheat & Grass improving very much cut up with the late cold winds & Frost The Sow Kill'd 20 Sco 10 li her young waned 26 dys
3. Partial heavy showers, warm & growing | Thwarting as yesterday for Swedes
4. Thwarting as yesterday some partial Showers
5. Sunday, very growing some partial Showers
6. Very mild, fine growing day | 12 Qr Barley to Devizes & 800 Slates retd | Thwarting for Swedes | Pease & Garden Seeds Planted
7. Very warm fine growing day | Barley, put a 2nd part of 3 Rick in Barn | Ploughing as yesterday. Mr Watts [Pencil note] decline Paying his Rent
8. Ploughing finish'd thwart'g for Swedes | Very fine warm day
9. 20 S: of old Wheat to Devizes & 5 Qr Seed Barley retd | Cold drying Winds N-B See the 26th Feby part of the Swedes half drawn out of the ground, are become mild & not injured, the Lambs feed on them freely & not the others side by side
10. Ploughing began for Barley sowing | Cold harsh Winds
11. Cold harsh Winds Ploughing as yesterday | Barley the remainer of last Rick put in Barn
12. Sunday, Very sharp frost, fine day
13. Very sharp frost Ploughing as 11th | 8 Qr Barley to Mr Giddings all Cannings | Eves Spouting put up at the Inn.
    Sacred to the Memory of my Niece Anne Smith who departed this life at Bremhill this morn'g du-obl: the 2nd of April Born the 19th January 1811 [This entry written within a thick black border]
14. Sharp Frost, warm fine day | SHEEP the Tegs, discontinued Swedes & Hay turned upon the Down with a Fold of Grass on Down Piece, muckled & fold'd before Lambing. Sheep Keep, scarce, Swedes a short allowance for 8 days only for Ewes & Lambs, have done well to the present time, not the least green on the Swedes for the Lambs, expect now, will not do well The Watermead backward | Ploughing for Barley too dry to sow

15. Ploughing for Barley and harrowing Down Wheat Talevera | Dry morning, Cold storm after [Pencil note in margin] Apl 15 - Watts & Treen not agree
16. Ploughing as yesterday, EXCEEDING hard Frost & pierceing cold winds wth heavy HAIL storm.
17. GOOD FRIDAY, very unusual hard FROST | The ICE in the Sheep trough in the Field this morn'g 1 Inch & ⅛ of an Inch thick, one nights freezing, exceeding Cold morn'g & evening, wth SNOW Storm in the day | Ploughing as yesterday [Note in margin] ICE in some places 1½ In thick, & the Pumps stop'd
18. The ICE, this morn'g ¾ In thick in Sheep Trouh mild'r evening | 20 Sacks Old Wheat to Devizes & Plough'g as yesty
19. Sunday, some little rain in night mild day
20. Rain in night, fine mild day Devizes Fair | Ploughing for Barley & harrow'g Down Wheat | The EWE'S do very ill, appear starved, Knit-up with the cold frosty weather and want of sufficient Swedes, not having half enough the remaining Swedes will be gone in two days eat 12 Truss's of worst hay evenings with Water ALARMING prospect for Sheep Keep, but little feed in Silbury mead, and the Clover field. prepare to keep sufficient Spring feed, even to spare
21. Fine mild dry day. Ploughing as yesterday
22. Mr Wentworth's Ewes & Lambs finish'd Swedes and began Silbury Mead | Ewes & Lambs finish'd SWEDES, and put on the grass Down Piece (Muckled &c.) with Hay & Water, to keep them a few days from the Mead, it being very backward | The Tegs on the Down very grassless, and so little Hay, and fold'd on Down Piece for Potatoes | Heartless time seeing the Sheep do ill, having very little Keep, the grass frosted not grow drying harsh Winds | Ploughing as yesterday & removing Hurdles
23. Ploughing as yesterday Cold harsh winds
24. Ploughing as yesterday Cold harsh winds | The 5 Year Old Black Gelding exchanged with Mr Hillier Reaves for a Grey Gelding rising 3 Years payd 6£, (the Colt 26£) the Horse £23 10s at 3 Years old
25. Cold harsh Winds, Ploughing for Barley | Barley finish'd Thrashing
26. Sunday - Cold storms HAIL & SNOW
27. Very Cold storms of HAIL & SNOW , and exceeding sharp FROST morning | 31 of the Calne MASONS, pass up the road for work their pay 15s pr Week & strike for more | Jno Chivers & Son, Jas Andrews, Jacob Davis & Geoe Lad Round Tail the SHEEP, and Seer'd the Lambs Tails, Punch'd the Right Ear [Pencil note in margin] 221 | CAUTION Not to Coop the Ewes & Lambs too close together, having Suffocated A Ewe & two Lambs this morning, and several others all but dead before we could be aware of it | Ploughing for Barley
28. EWE'S & LAMB'S removed from Grass Down Piece to SILBURY MEAD, the Tegs on the Down & Fold'd (Potatoes) | HAY, all the Sheep discontinued

| Draging for Vetches & removing Hurdles to Mead Sharp frost Cold harsh Winds | HAY=CONSUMED sinced 8th Novr 13 Tons Meadow 11, Tons old Field 16 „ Tons New Field 13 „ Tons 2nd N-Field [Total] 53 HAY, not spent 6 Tons Rick New Field 4 Tons Piece New Field [Total]10 8th Novr Sale Sheep began hay 15 Novr Stock Lambs began Hay evenings 29th Novr Stock Ewes began Hay evenings | [Pinniger then describes an unnamed building [267]] The Building possess's a large Hall in the Centre 33 feet by 28 feet and 27 Feet high The Criminal & Nisi Prius 57 by 32 feet & 26 feet high The Grand Jury Room 27 by 25 feet

29. Exceeding, boisterous, Stormy day, to prevent cuting the Lambs | Ploughing for Barley

30. Ploughing for Barley RAIN in night & Storms all day, no rain since the 6th & very little this month | LAMBS=CUT 105 - in 1½ hour, including Cooping three times by Ebednego Blackman, of Charlton @ 6s pr Score & 6d - old Rams 105 Wether & 116 Ewes Lambs 221 & (2 stifled) | CUCKOO, - first heard at Beckhampton 30th at Tockenham 15th at Tytherton faintly 7th [Pinniger then recorded the weighing of his wool - see Appendix E]

May

1. Frequent warm growing Showers | Ploughing finish'd for Barley

2. BARLEY SOWING, began, with 10 Sack Shewclear @ 38s pr Qr 10 Sco 4 li, the land till now too dry for Sow'g (very good work) | Fine dry warm growing day

3. Sunday - warm growing day

4. warm growing day | VETCH'S drill'd 2¾ Ac: with 5½ Bushls Foreign small & dry weigh 13 S 4 li Nt | BARLEY sowing FINISH'D, 20 Acres with 10 Sacks of Shavelear 13 Sacks of own 168 li of Broad Clover 60 li of Mill'd Hop 5 S 6 li Nt 12 Bushls Black Hop 4 S 6 li Nt 24 Bushls Ray & Hop 3 S 13 li Nt 24 Bushls Ray only } together 64 Bushls 54 Bushls of mix'd Seed the hole full size sow'd the 20 Acres not quite 3 Bushls pr Acre, the other 10 Bushls sow'd over the light parts again wth 1/3 of hole in the Machine N.B: made mellow kind Barley work

7. The Barley sow'd the 2nd is rooted & that sow'd the 4th is striking & the Vetches sow'd the 4th are struck

5. Kidney Beans, Black Beans, Potatoes and Cabbage Plants, - in Cottage Garden | Flower Seeds, Drying harsh Winds | Harrowing the Barley & sowing the Seeds

6. Harrowing the Barley & sowing the Seeds | Some rain in the Night, fine growing day | Mr Wentworth finish'd feed'g his Meadow and put it in Water at the same time flooded mine under my Ewes & Lambs to the injury

267 The building is the new Assize Courts at Devizes, under construction at this time: it is likely that Pinniger obtained these details from a report in the Devizes & Wiltshire Gazette, 23 April 1835 (page 3, column 1).

of the Sheep & loss of my Grass, unless the Water taken off again, which was done as soon as could by Waterman (G. Giddings)

7. Fine grow'g day. Rolling & harrow'g the Barley BEANS, put remainer of Rick in Barn (2 Yr old)
8. Rolling & harrow'g finish'd the Barley land | Damp blighting air
9. Dry harsh winds. damp evening | Harrow'g the Fallows further Down Piece
6. HAY the Horses finishd the second Clover began the Rick Octr 3rd | The Horses began Hay in Field (best Clover)
10. Sunday - Dry harsh winds
11. Dry harsh winds | 6 Horses & 2 Waggons to BOWOOD for 1¼ hd Faggots heavy loads
12. Harrowing & Rolling Whitelands for Turnips | Dry windy day (2 Lambs
13. Harrowing for Swedes, Dry windy morning RAIN all the evening, very acceptable
14. Copious Showers till evening | Dung, Carting from the Inn to Tan hill Piece | Mr Watts refuse paying his rent, & or allow me to build for Mr Treene
15. Very heavy Showers, dry evening | Ploughing & harrow'g the Potatoe Land, Down muckled & Fold'd & now fold'd again Fold removed to adjoin'g Piece for Turips (Tegs)
16. Heavy Shower in morn'g dry after | TURNIPS, Norfolk whites, sow'd Whitelands 4A 1R 28P with 12 li Seed 1 hole every place, sow'd on the stir, harrow'd first, very mild fine work | POTATOES, began Planting middle Down Piece by 4 Men diging a channel to cover the last row, 2 Women placeing the Sets, about half an Acre with 5 Sacks fill'd [Pencil note] 1-0-34 Labourers 0-1-20 [Total] 1-2-14 [268]
17. Sunday Damp air, growing day
18. Potatoes, finish'd Planting Down Piece 5 Sacks [Pencil note: 11S] for [Pencil note: 1-0-34] | The EWE'S & LAMB'S, removed from Silbury mead, having been there 20 day 220 Couples the meadow never so light in grass, want of Water all the Spring | FIELD GRASS, the left, Tan hill Piece, the E & Lambs began [269] | Hurdles removed from Silbury mead | Fine mild growing day
19. Fine mild growing day | Rolling & Harrow'g Couchey part of Down Piece for Turnips & Wheat | WHEAT=RICK, the Talevera, put into Barn (1 New & 1 Old Wheat now in Yard)
20. Cloudy growing day | Rolling Barley & Cart'g Dung fm Stand'g Pen
21. Finish'd Rolling Barley & Cart'g Dung (to Tan hill Piece for Wheat) | Very warm growing Day

---

268 On the basis of this calculation, these figures are acreages recorded in acres, roods and perches.

269 E - Ewes.

20. HORSE'S, began having a small tie of GRASS to bait them before the Ewes evening | The Wheat, Vetches & Grass growing rapidly | The Vetches fit to feed, see sow'd 30th Septr N B OBSERVE, Vetches fed off from this time may be succeed'd by Swedes in good season at middle of June, which will be a Crop gaind the Sheep improved the land Kept clean & in heart
May 19th Died Aged 89 Yrs Willm Savory of Calne Died Aged 57 Yrs Willm Pinnegar of Calne Millwright [This entry written with thick black lines to the left and bottom]
22. Very fine day little rain in evening | Granary, the Thatch taken off, to be Slated | Ploughing &c the Standing Pen for Potatoes
23. Ploughing &c second time &c for Potatoes | Ploughing, Harrow'g &c Down Piece for Turnips & Vetches | Very fine warm day
24. Sunday - Very fine warm day
25. Very fine warm day high winds | Rolling & harrow'g Down Piece for Vetches & Turnips | Swindon Fair, with Mr Hillier Reeves Wroughton { Sold may Gray Mare for [...] Bot my Bay Horse at [Pencil note] 31£ | Sold the 6 Yr Rone Cart Horse [...] Bot A 3 yr Grey Cart Gelding [Pencil note] 28 Gui [270] | Potatoes planting in the Standing Pen
26. Some slight showers in morn'g, Thunder & light'g eveg | Ploughing for Vetches Down Piece
27. Ploughing for Vetches Down Piece Rolling & harrowing | Fine morning, Thunder, lightening & steady rain after
28. Fine morning, Thunder & Rain evening | Plough'g 5th time couchey part of Down Piece for Turnip | Vetches, Drill'd 5 Bushls further side of Down Pieces 2 Bushls pr Acre (@ 1s 6d pr A | Exceeding, heavy storm of HAIL & RAIN at Bremhill, Calne &c.
29. COWITCH sold to Mr Poynder | Brother John removing his Goods to live at Calne, his Son John remain at Cowitch | Plough'g Down Piece for Turnips Fine day
30. Plough'g for Swedes 3rd time Fine day
31. Sunday. Very sharp Frost to destroy the Kidney Beans & cut the Potatoes | The Lambs on a Tie of Clover many of them SWOLN & 4 DIED immediately at 10-0 Clock morning, none swoln before | Rain mild all the evening, fine dry morn'g

June

1. Rain fine mild all night & morn'g dry after | Plough'g for Swedes
2. Plough'g for Swedes drying day, except HAIL shower | Ewe lambd this morning Twins | GRASS, Clover began mowing for HORSE'S
1. The Clover began mowing & Caging for the Ewes & Lambs, to prevent their BLOWING, and began FOLDING them ˣ, to prevent their breaking out, having lain back since the 16th Feby ˣ on the Clover Tan hill Piece

270 may - my. Gui - Guineas (£1 1s 0d).

3. Ploughing as yesterday, Very fine day
4. Ploughing as yesterday, Very fine day
5. Ploughing as yesterday, Very fine day
6. Ploughing as yesterday, Fine morn'g Thunder lightening & Rain all the evening Cows & Horse Killd
7. Whit=Sunday, Fine mild & very growing morn'g Thunder & lightening & exceeding heavy Storm at noon to lodge my Wheat not in Ear & the Grass
8. Sheep counted 238 Ewes 2 Rams 3 Stags 86 Tegs [Total] 329 223 Lambs [Total] 552 | Whit=munday - TURNIP=HOEING began Whitelands sow'd 23 day (16th May) | Ploughing finish'd for Swedes, Fine morn'g DREADFUL=THUNDER & lightening wth deluge of RAIN during divine service & after, fine even'g
9. Sheep removed from Field grass Tan hill Piece having fed 7¾ Acres in 22 day, to the Field grass Down Piece fed before (the 28 April) Grass mornings, Down mid day & Vetches evenings | Hurdles &c removing all day | Sultry warm, Thunder & lightening | WHEAT=EAR'S The Talavera just bursting sow'd 28th Novr - last year burst'd 29th May sow'd 21st Novr 1833
10. SWEDES began sow'g lower Furlong next Farmer Philpots 3 holes every 3rd Place | VETCH'S began feeding, in Cages see yesterday Ewes & Tegs put together & Folded on Dung further Down Piece for Turnips | MOWING began Clover Tan hill Piece, heavy but not in bloom, (last year June 5th) | TURNIPS, broad as a Saucer 2 Men hoeing Sultry warm Thunder & rain in the evening
11. Rolling & harrowing the Swedes sow'd yesterday | Sultry warm with Thunder evening
12. Sultry warm with Thunder evening | Sow'd a second 6 Acres of SWEDES, 4 holes every 3rd Place in the Machine, sow'd 35 li on 12 Ac
13. Rolling & harrow'g the Swede ground | Very fine day not so warm
14. Sunday, - Warm fine day | The Red Wheat as well as the Talevera in part the Ear out.

7, 8, 9. See heavy Thunder, lightening, hail & Rain Cattle & Sheep Kill'd, & great injury sustained by breaking Glass, 2'000 Panes broke in Mr Heales Green house Calne [271]

15. SWEDES appear sow'd the 10th | Potatoes in the Field, the Women hoe'g | Rolling and harrow'g the Swedes sow'd the 12th Beautiful warm fine day
16. Beautiful warm fine day | Ploughing a Fold drift further Down Piece & Carting dung before the Fold for Turnips
17. Beautiful warm fine day | Ploughing 4th time Couchey part of Down Piece | Vetches winter shewing the blossom | Mowing finish'd Field grass & began dry meadow.
18. Very fine pleasant day | TURNIP, green round, the first, Plough'd & sowd

271 William Heale & Son were nurserymen of Calne and Devizes: the main nursery was at [Potterne] Wick near Devizes, Devizes and Wiltshire Gazette, 27 October 1836.

2 lands further Down Piece, Dung'd & fold'd | HAY, began carrying Tan hill Piece, very dry and good, not a drop of rain upon it

19. Hay carrying, fine drying day
20. Turnips sow'd 5 WEEK'S grow rapidly the greens spread 18 Inch's | Ploughing a second Fold drift Down Piece | Very fine day Hay carrying Tan hill Piece
21. Sunday - Very fine day | Swedes, sow'd the 10th shewing the 3rd or rough leaf
22. BOISTEROUS winds all day, RAIN evening | HAY, Tan hill Piece finish'd carrying, a heavy crop in excellent condition without a drop of rain, or heating, - Ploughing as yesterday | Chippm Fair, W P & T P, reconc[...]'d by Mr B P [272] | BREMHILL Estate sold for [...] the Timber [...] s administer'd & sign'd the Agreement | Field grass Down Piece 2nd too old for the Sheep mow'd for Hay
23. SHEEPWASH'D at Clatford 16 Score & 7 usage as last year | LAMBS=WANE'D, having plenty of Vetches & the up'er little Down very good cover'd with Yellow=craise, Ploughing for Turnips | Some slight Showers during the Day | Potatoes in the Field, earthing | Turnips sowd 18th appearing | The Wheat, in general just in full ear, the high Winds & heavy rain in night LODGED the Wheat injuriously, many pieces - one half down.
24. RAIN, very heavy mid:day, to drive the Teams home, boisterous winds, dry evening | The Wheat LODGED much more this day, the whole of the 12 Acre Piece blown down | TURNIP'S Plough'd & sow'd a 2nd drift Down P
25. Ploughing the grass Fed of Down Piece Very heavy storms
    18th Died Willm Spackman Chippenham Agd 72

    18th Died Willm Cobbet [...] Agd 73 [273][This entry and the previous entry written between a triple line above and a triple line below]
26. Dry night RAIN steady from 10-0 Clock till eveg | Plough'g as yesterday till the rain came | Sheep suffer very severely after being shorn from the exceeding cold rain
27. Carting Dung from the Inn, some Storms | The Wheat by the winds & Rain yesterday beat about in every direction, apparently injured 2 Sacks pr Acre, the Markets advance in consequence 3 to 4s pr Qr
28. Sunday, - Fine day, Glass rising
29. Plough'g grass fed off Down Piece for Turnips | Fine day
30. Very fine day - Ploughing as yesterday Hay dry meadow carried & Wheat finish'd Thrash'g | 1 Wheat Rick of each of 1833 & 1834 in Yard to be Kept

---

272 reconc[...]'d – reconciled.

273 This is William Cobbett, political writer and farmer (1763-1835), Oxford Dictionary of National Biography, *Cobbett, William* https://doi.org/10.1093/ref:odnb/5734 [accessed 27 July 2020].

till after Harvest 40 Sacks of Wheat year 1833 at Devizes 84 Saks of Wheat year 1834 over Waggon House 44 Sacks of Wheat year 1834 in Granary 32 Sacks of Wheat year 1834 in Barn [Total] 200 [Pencil note in margin] 2 Mill

July

1. Very fine day | TURNIP'S sow'd uper half of 4 lands Down Piece two lands ash'd & fold'd & two dungd & fold'd
2. TURNIP'S Ploug'd & sold the 3rd Fold drift | Very fine day, - Topd the Field hay Ricks none made so well since at Beckhampton took no rain & no heat | COW=CALVED 40 weeks fm 21st Septr to 30th June | CUCKOO, cry on the Down, last time heard
3. Died Mrs Edwards Bell Devizes Green aged 47 | SHEEP=SHEARD 327 by 8 Men short cuts Beautifully fine day | Mowing Silbury Mead the bank, begin coarse | Plough'g as yesterday & draging
4. Harrow'g as yesterday very fine warm day | The Sheep cull'd for Sale and Stock & mark'd Stock 67 - two tooth - Punch'd near Ear 91 - four tooth - No mark 64 - six tooth - Punch'd the off Ear [Total] 222 2 - Rams 3 - Stags 223 - Lambs [Total] 450 Sale - 100 - Viz 73 of 8 tooth 5 of 6T 4 of 4T & 18 of 2T [274]
5. Sunday, rain in night dry day
6. Swedes, harrowing, for hoeing, Plough'g athwart down Piece, very fine day
7. Plough'g athwart down Piece, very fine day
8. Plough'g athwart down Piece, very fine day heavy rain night | Potatoes finish'd earthing | This evening ¼ before 5 my Child & Son born Abbot Large
9. SWEDES, 3 Men began hoe'g, sow'd 10th June NB, the Swedes sown on the thwart being a fine rain, considered better, than losing the chance & defering the season by Plough'g again have a good plant and clean | Some rain in night, fine day Plough'g as yest'y | Hay carr'd from Bank Silbury mead.
10. Plough'g for Swedes, Vetches fed off, Storm of rain in night fine day
11. SWEDES, sow'd 3 Lands Vetch's fed, adjoin'g Nether Down Piece | TURNIP'S Plough'd & sow'd last Fold drift further Down Piece | Warm fine day
12. Sunday, fine day, rain evening
13. Storms all morning fine after | SWEDES, Plough'd after Vetches & sowd 2 Lands | 5 Swede hoers, Plants grow & do well
14. Draging & harrow'g, grass'y Down Piece for Turnips Fine dry day [Pencil note in margin] Brocoli
15. Draging & harrowing as yesterday | Fine dry day
16. Fine dry day 20 Sa Talavera Wheat to Dev & harrg | SWEDE'S finish'd hoe'g the 12 Acres first time never a finer hoe'g - and a good plant Saturday

274 6T – six tooth; 4T – four tooth; 2T – two tooth.

11th Married at Calne Mr W B Darley, Burton upon Trent To Miss Ann Noyes Chippenham

17. TURNIP hoe'g began sow'd 18th June further Down P Fine warm day Draging harrowing and burning the sward of grass Down Piece | MOWING began remainer of Silbury mead, the grass not grow, the land dry & chopd | VETCH'S finish'd feeding the Winter for Swedes
18. VETCH'S began feed'g sow'd the 4th of May in Bloom | VETCH'S began feed'g wth Ewes sow'd 28th of May not Bloom full of Charlock lower end, the top of the Piece destroy'd by the RABBITS, from the Preserve | Complaind to Mr Estcourt by Letter the 16th - Rain thin Showers frequent soon disappear | Swedes sow'd 2 other Lands after Vetches
19. Sunday warm fine day
20. Swedes sow'd 1 other land & removed the Hurdles from Vetch land to Vetches Down Piece | Very fine day & warm, WOMEN striking for higher pay & Jas Andrews dareing, grass all burnt Down Piece
21. Very dry scorch'g day, carried the first 3 Loads of Hay from Silbury Mead cost 6d pr load only mak'g turn'd only once | Plough'd & Sow'd an other 3 Lands wth Swedes
22. Plough'd & Sow'd the remainer of Vetch land to Swedes | Turnips, hoe'g the second drift Down Piece | HAY, carried 3 Loads from Silbury Mead | Very boisterous troublesome Wind load'g hay and very sultry warm.
23. Thunder & lightening wth Rain in morn'g, hot & dry after | 20 Sacks Wheat to Devizes & Plough'g burnt ground | 2nd CLOVER, MOW'G, fed with Sheep removed 9th June
24. Ploughing as yesterday, exceeding warm
25. Ploughing as yesterday finish'd exceeding warm | 100 SALE EWES drew & put wth Lambs and the RAM'S, on Clover Tan hill Piece mornings (began) and lie back on Vetches Down Piece evening's 23rd Mr Geoe Brown 25th Mr Estcourt } Inspected the Injury done to my Vetches & Turnips further Down Piece, by Rabbits Mr Estcourt promised to have the Rabbits shot & prevent their injury'g me in future
26. Sunday, warm ripening day | Reaping many places, this week
27. Very warm day Draging & harrow'g Down Piece | Wheat Rick the last of 1834 put into Barn, one of 1833 still in the Yard
28. Draging & harrowing as yesterday, scorching warm, Thunder in evening | Clover hay carried.
29. Rolling & harrow'g as yesterday, sultry warm
30. harrow'g & 20 Sacks Wheat to Devizes 1T 14C 2Qr COAL | The MARKET'S depress'd, the Weather unusually dry & ripening, Reaping began & finish'd Enford Parish this week | 4 Samples of new Wheat at Market to day sold 23 & 25s & very good old Wheat much better 12 Sco 12 li Nt sold @ 20s Sa | Sold my Talevera 12 S Nt @ 20s & the 16th June sold @ 22s Speculations in Keeping Wheat this summer even at very low prices

unfortunate | Swedes, finish'd hoe'g 2nd time, the young Turnips Down Piece grow rapidly without rain being well dung'd, the Swedes after Vetches want rain for the seed to strike, the Lambs get poor, or Vetches Down Piece no pro [...] in the Downland - & have a tie of Clover Tan hill Piece

31. Very fine ripening day, Hurdles removed to Tan hill Piece & Raking the rugget Down Piece 3rd time [275] | 3 HORSE'S RESTING 1st time having but little work

August

1. Very warm ripening day | TURNIP'S all hoe'd and second'd that is ready | Muckle Cart'd to Tan hill Piece, Clover fed & Fold'd | Fold removed from Down Piece Vetches fed
2. Sunday at Chippenham 4th Chapr Hebrews 9th Verse by the Revd Mr Morgan | Fine warm morning, Thunder lighten'g & heavy RAIN from 5 to 8 O-Clock evening
3. Fine warm day, Cart'g the tuffets of Down Piece | Reaping began generally, in this Parish
4. Vetches finish'd feed'g, middle Down Piece | TURNIP'S sow'd mid: Down Piece just burn'd & cleand | TURNIP'S, hoe'd further Down Piece sowd 11th July | Very fine harvest Day
5. TURNIP'S, began Plough'g & sow'g Vetch gd fed mid Down Piece | Some thin mizley rain afternoon | Jno Chivers trim'd 107 Wether Lambs 116 Ewe Lambs [Total] 223 | Vetch's began feeding next Horton Down
6. REAPING began, very lodge'd & green - Very fine day, Sow'g &c as yesterday | 80 Lambs at Tan hill sold @ 20s 6d, Mr Buckland – Crl [276]
7. Finish'd Plough'g & Sow'g the Vetch land to Turnips Some blight'g rain in morn'g, dry & fine after
8. Sharp Frost morn'g, very ripening day. | Muckle, Carting before the Fold. NO WORK FOR THE HORSES, till have finish'd a Fold drift on the Clover lay, Monday morn. | My reapers 5 Men & 3 Women only, no good Reapers call, yesterday 18 impudent ones, call'd, demand'd 13 to 16s pr Acre, fortunately my Wheat not ripe.
9. Sunday, Very warm ripening day
10. Watermeadow grass the Horses finish'd | Clover, the second began cuting for the Horses | Clover, the second began Mowing for Hay, being too old for the Sale Sheep, by Willm Davis the Carter & James his Son (only 1

275 Rugget - Ridged. Oxford English Dictionary, *Ridged,* https://www.oed.com/view/Entry/165679?rskey=TNVsss&result=1&isAdvanced=false#eid [accessed 1 May 2020].

276 Crl – this is possibly Crudwell. In 1841, John Buckland (age 35) was a farmer at Crudwell, TNA: HO107/1181/8. *Census returns. 1841 census. Wiltshire Hundred: Malmesbury, including Parish: Crudwell, ED20.*

Team at Plough | Fallowing began for Wheat Tan hill piece Muckle & Fold second Clover fed off | Wheat ripening rapidly, put on 4 Taskers

11. Wheat ripening rapidly, put on 4 others & not enough | Plough'g as yesterday
12. Plough'g as yesterday Fine harvest day
13. Plough'g as yesterday Fine harvest day | TURNIP'S began FEEDING Whitelands, wth sale sale Sheep & Chilver Lambs wth a tie of Clover & Vetches | CORN=MARKET'S depress'd, Old Wheat 4s Sack lowr New Wheat 12 S 19 li Nt 22s 6d Sack | Clover, finish'd mowing for hay
14. Ploughing as yesterday, very fine, Carter reaping
15. WHEAT began Carting, the 6½ Ac: Tan hill Piece (31 loads) Rick by Barn | Very fine harvest day, the 10 Reapers Taskers Paid reaping nearly finish'd
16. Sunday, - fine day
17. The Clover hay, carry'd, very good without a drop of rain - Very fine day
18. Wheat the Talavera carried 8 Acres by Well | Very fine - 3R 7P in Barn
19. Very fine Reaping finish'd | Wheat 1¼ Acre above the Path in Barn | Wheat Rick by Hen house 4A 2R 9P
20. Wheat Rick the middle one 3A 3R 0P | Wheat Rick the small one 2A 0R 27P | WHEAT=HARVEST. FINISH D | No rain since the 2nd Instant, a most Providential harvest, the Wheat bright and very thin, - being entirely lodged if it had been a wet harvest, it would have been quite spoild. | Warm fine day, Vetches finish'd Down Piece | Sale Sheep Fold'd, Keep short, Turnips & the Clover stubble only, the Stock Sheep, the Down only, & done well, hetherto
21. Cart'g Muckle & hurdles, Fine day
22. Cart'g Muckle to further Down Piece after Vetches | Thunder, Lightening & Rain in the evening | The young Turnips hoe'd out Down Piece was the 16th very gross, are now almost quite destroy'd except the rib of the leaf, by a black Maggot an Inch long, call'd the Negro Maggot, bre'd by a Yellow Fly depositing Egg's, like the yellow Dung Fly, many methods of destroying them suggest'd, but the most likely to succeed is that of removing them from the leaf by two Men drawing a Waggon line over a land at a time 50 Maggots being on one Turnip (the leaves) Some People say 20 others 40 Years, since we were visited by this destructive blight Insect | The other Turnips that are large enough to feed, are fast decaying for want of Rain, an alarming prospect for Sheep food. Sheep this day at Marlbro 2s pr head lower than Tan hill
23. Sunday - Dry day, storm in evening
24. Rain in night, and at times all day | Never so many Wheat Ricks made in the Fields and so few Thatch'd - Cart'g Muckle
25. Dryer day some rain - Cart'g Muckle
26. Fine dry cool day damp eve, Plough'g Vetch land muckled Down Piece for Turnips

27. Muckled Down Piece lowering stormy evening | BARLEY began MOWING, many finish'd harvest CORN=MARKET'S still more exceeding depress'd, Two year old Wheat S 12„9 li Nt before Harvest I refuse 22s pr Sack, now sold @ 18s Good new Wheat sold @ 15s & beautiful new 12 S 16 Li Nt @ 18s 6d
28. Press'g the Vetch ground muckled & sowd to Turnips | Turnip seed harrow'd in, where Negro Maggot destroy'd the Turnips | Very fine day
29. Very fine day BARLEY, began, load'd 4 Waggons | Cart'g Muckle before the Fold Tan hill Piece
30. Sunday Very fine day
31. Beautiful fine day, borrow'd 3 Men of Mr Pumphrey, to mow my Barley, my men at Publick house half the 2nd day, - Carting Barley

September

1,2 & 3. Beautiful fine days, Carried all the Barley in 4 Day's from 7 o'Clock morning to 7 at even'g & dress'd & top'd the 3 Ricks with Mr Pumphrey's Men helping, my 3 Men thrash'g for Straw to thatch the Barley Ricks The Wheat also carried in 4 days, all the Corn cut & carried without the least rain, for which I acknowledge and adore a graciouse Providence
4. Ploughing Clover lay Muckled & Fold'd one earth for Wheat Sowing | Fine day, heavy storm in evening
5. Fine day, Ploughing as yesterday
6. Sunday - Very fine day
7. Very fine morning heavy Storm after | 20 S: Wheat to Devizes & Ploughing as the 5th
8. Ploughing as Yesterday & Storms at times all day | SWEDES, sow'd see the 18th July after Vetches finish'd hoe'g - The Swedes sow'd finish'd 22nd July being a dry time are not large enough to hoe, if had been a wet grow'g time would have been early enough after Vetches (see a guide for next year)
9. HORSES, began baiting in Silbury=meade to save the Clover - Ploughing as yesterday SOW=hog'd, Very heavy storm of hail & rain mid day
10. Heavy Storms all the morn'g, enough rain now for the young Turnips to grow | A=Sale 2 tooth Ewe Kill'd 35 li, - taken ill suddenly Ploughing as yesterday | NEW=WHEAT, began thrashing, for Straw to finish thatching the Barley Ricks & for the Stables
11. Ploughing as yesterday, Very cold boisterous Winds
12. Rain in night, - and very heavy hail & rain storm afternoon with heavy Thunder | CHALK, Carting, for Building at the Inn. The black Maggot appear again destroying the Turnips hoe'd on the burnt ground
13. Sunday - Fine day
14. Carting Chalk & Stubbles - Storms
15. Plough'g Wheat Stubbles began for Vetches - Rain in night & afternoon
16. Plough'g as yesterday - Fine dry day

17. Plough'g for Wheat - Storms in morn'g
18. 1st SALE, of Jno Watts, Beckhampton Inn - Dry morn'g rain all afternoon - Plough'g as yesterday
19. Plough'g Wheat Stubbles - fine day
20. Sunday - fine day | TURNIP'S, finish'd feed'g Whitelands see - began 13th Augst
21. Plough'g a Fold drift for Wheat - Rain all the afternoon | TURNIP'S began Down Piece wth Sale Ewes & Chilv:L [277] | RAM'S to Stock Ewes, 4 of Mr Pumphreys Lambs
22. COW, bull'd (to calve 1st July) | Plough'g began Whitelands for Wheat | Rain all the afternoon
23. Plough'g as yesterday - Storms
24. Plough'g as yesterday exceeding heavy Storm
25. BREMHILL - last SALE, - fine dry day - Plough'g finish'd Whitelands
26. CLOVER=grass, - the Horses finish'd - Rain in night and evening
27. HAY the second Clover, Horses began & continued baiting evening's in Silbury=mead | Sunday - Fine dry day
26. Cart'g muckle before Fold, below Hay Rick [Pencil note] My Wheat 12 S & 8 li Nt = Mr Large 11 S 15 li Nt
28. BANKING began my further Down Piece, by the Hunters on Mr Browns Down | Cart'g Muckle from the heap before Fold, Rain in morning dry after
29. Carted the heap of Earth Silbury=mead, on the bank | Fold removed to lower part of Tan hill Piece, fine day
30. Exceeding, boisterous winds, heavy rain morn'g & eveg | Cart'g Muckle from the heap

October

1. 2nd SALE of John Watts, Beckhampton Inn by his Crediters | Cart'g Muckle as yesterday, Storms
2. SALE at the Inn close'd & Watts persist in Keep'g Possession, altho: have Paid his Rent (after a threat of Distress) to New Michaelmas, when due | Ploughing Tan hill Piece for Wheat, Storm even'g
3. Ploughing Tan hill Piece for Wheat, Storm even'g | MR TREEN, entered Beckhampton Inn, by John Watts giving him Possession at 8-oClock in the evening, after extorting from him five Pounds to give possession, New Michls quiting
4. Sunday, much rain in night, fine day
5. Fine day, Plough'g for Vetches | Waggon & 3 Horses to Marlbro, for Mr Treens, goods
6. Ploughing for Vetches, partial Storms
7. Ploughing for Vetches 15 Sacks Wheat to Mr Jas Large | Fine dry Day - 2nd

277 Chilv:L - Chilver Lambs.

Day Devizes Races
Sacred to the memory of Miss Martha Large of Tockenham Court, who departed this life 4th Octr 1835 Funeral Sermon, Text 26th Psalm Bible 8th & 9th Verses The Psalms sung, the same as the late Mrs Large
21st Octr Died Wadham Locke Esqr Rowde Ford House Aged 55 Yrs 6th Micha 9th Verse [This entry and the previous entry written within adjacent thick black borders]
This day Sunday Octr 4th have been Solemnized being the third (1535) Centenary of the Reformation when the first entire English version and publication of the Bible were accomplished by Miles Coverdale Bishop of Exeter, as a day of special Thanksgiving to Almighty God for the blessings of the Reformation [Pencil note] See Saturdays Magazine for September

7. Began preparing for the foundation of the Racehorse Stables at the Inn. for Mr Treen
8. Very fine mild day | WHEAT=SOWING = began Whitelands 4¼ A- wth 11 Bu of Mr Jas Large's seed
9. FUNERAL of the late Miss Martha Large of Tockenham | Rain heavy from 11 to 2-oClock, dry till 6-oClock rain all the evening | Carting Chalk & Stone from the Wall in road to Inn Garden for New Stables, Plough'g for Vetches | HAY, began giving Sale Ewes & Chilver Lambs, the Turnips & Down not enough to do them well | Stock Sheep have Down only, bad Keep do ill
10. Dryer Day, Storms, Cart'g Sarsens & Road Earth for Stables | Ploughing for Wheat
11. Sunday - Cold dry'g day, Funeral Sermon of late Miss M L
12. Cold drying day | BUILDING began the Race Horse STABLES at the Inn for Mr Treen | VETCHES, began Sow'g, bottom of Rt of Tan hill road 4 Sacks on [...] Acres
13. Finish'd harrowing bottom of Rt of Tan hill road fine dry day | POTATOE'S, began diging in Field
14. Very fine dry day, Glass high | PRESSING, began Tan hill Piece for Wheat
15. Very mild fine day | 2 Waggon load of Lime from Calne
16. Cart'd the Calne Stone from the Road to the Garden & Cart'g Marter earth | Beautifully calm mild day Beautifully calm mild day
17. WHEAT, began sowing Tan hill road Piece, above Rick
18. Sunday - very fine mild day
19. very fine mild day | Waggon & 4 Horses to Cleveancey for 11 S_ Wheat &c
20. Pressing uper side of Tan hill Piece - Sharp Frost, fine day, little rain even'g | 60-head Sale Ewes sold at 21s 6d, last Yr best @ 30s - W P=K, sold 100 head Ewes 20s seconds 13s 200 Lambs @ 15s
21. WHEAT RICK, the small new put into Barn | Sharp frost very fine day | Wheat sow'd uper part of Tan hill Piece
22. Harrowing uper part of Tan hill Piece, Rain heavy afternoon

23. Ploughing for Wheat, dry day rain in evening
20. The sharp Frost cut up the Dhalias & Potatoes [Pinniger then recorded prices at Weyhill Fair - see Appendix H]
21. HORSES 3 of the STAR Coach taken from the Inn and 6 of the WHITE=HART Night Coach brot to the Inn [Note written in pencil] 6-Mail Mr Parsons 17 - Mr Halcomb
24. Wheat harrow'g off, Tan hill Piece 1st sow'g, Dry day
25. Sunday, Rain from 10- oClock morn'g, incessantly
26. Boisterous rain'y night, dry day, partial storm | Plough'g for Wheat Cart'g Chalk
27. Wheat harrow'd off, top of Tan hill piece | Dry morn'g storms evening | Wheat up Whitelands sow'd the 8th
28. Press'd & sow'd Wheat to the Fold - Fine dry day | HAY, the Ewes began evenings 3 Truss's, do very ill have the Down only - & that very bare | VAXINATED - the 2nd time for safety Mary, Ann & Tom from their Brother Abbots Arm
29. Rain heavy in night, fine day, Muckle before Fold
30. Wheat harrow'd off sow'd the 28th - Frost & dry morn'g Rain all afternoon
31. Rain in night & morn'g dry day
29. COWITCH, Convey'd to Mr Poynder
30. BREMHILL, Convey'd to Marquess of Lansdown
31. 6 Horses to Honey St. for Carriage load'd with the Roof of New Stables at the Inn.

November

1. Sunday, fine day
2. Waggon & 6 Horses to Honey St Wharf for Roof of Stable, Dry morn'g rain after
3. Chalk for Stable, Cart'g & 48 Bu: Grains fm K | RAIN all day
4. RAIN all day | 8 Waggon load of Rubble for Mr Treens road
5. CALF 48 li Qr @ 6½d - 18 Weeks old | CALF Young from Tockenham - Rain in morn'g dry after, Carting Hurdles to Silbury Mead &c. | TURNIP'S the Lambs finish'd Down Piece
6. SILBURY=MEAD, the Lambs began feeding Dry day, rain in evening | VETCH'S, sow'd a Sack, Rt of Tan hill Road
7. Fine day heavy rain in evening | SHEEP DIP'D, in Arsnic & Soft Soap (from 9- oClock morn'g to 5 o-Clock evening) 261 - Ewes 143 Lambs 4 Ram Lambs 2 Rams 3 Stags [Pencil note] [Total] 413 | 2 Waggons & 6 Horses 1 to Box & 1 to Pickwick for the Inn Stables | Dry morn'g very heavy rain evening
8. Sunday - dry day
9. Cold dry day, Cart'g Muckle &c - 3 Load of Rubble to Inn
10. DUNGING finish'd Wheat Field with Muckle from the Inn, & began

dunging the Down Piece Stubbles for Turnips from the Inn for Turnips next May | Exceeding cold drying winds

11. The 1st very hard black FROST | 20 Sacks old Wheat to Devizes & Timber returned for the Roof & Stalls of new, Inn Stables | Carting Dung, Stubbles, Hurdles, Chalk &c. | The Walls of the Inn Stables up, & the Roof framed
12. The Roof, began puting on the Walls of the Stable | Ploughing a Fold drift for Wheat | Very Cold Winds
13. Very Cold Winds Ploughing as yesterday
14. Very Cold Winds | 2 Waggons & 6 Horses to Pickwick for Pitchen Stones to cover 20 Yards @ 1s 6d pr Yd (from 11-0-Clock last evening to 8-0-Clock this evening).
15. Sunday - Fine day damp morn'g | TURNIPS, see sow'd 28th Augst are very small not worth keeping, put the Stock Ewes on them, to sow the land to Wheat
16. Press'g & sowing 2 Fold Drifts to Wheat Tan hill Piece | Cold dry day
17. Plough'd & Press'd another Fold drift bottom of Tan hill Piece - Fine dry day
18. Wheat sow'd & finish'd to bottom of Tan hill Piece the 3 short lands to finish folding | The Roof of New Stables at the Inn finish'd and the ends of the Wall ends up | Slating began New Stables | Dry morn'g, very boisterous winds after
19. WHEAT=RICK 3¾ Acres put into Barn - Very fine day | 20 S- old Wheat to Devizes &c. 12½ € Coal retd & Manger &c for New Stables | Ploughing began further Down Piece for Wheat Turnips fed off
20. WELL began sinking at the INN, by Lawrence Chivers | 20 Waggon load of Flints, Carted fm Meadow to the INN=road | Cart'g Dung to further Down Piece for Wheat = Damp morn'g, fine mild day
21. Cart'g Dung as yesterday, fine mild day | Cart'g Chalk to Wall the WELL at the Inn
22. Sunday -, Fine day, rain evening - FOLD, finish'd Tan hill Piece
23. FOLD, removed to fur: Down Piece, muckled the 21st | WHEAT SOWING finish'd the usual quantity of 28 Acres, by Plough'g, Press'g & sow'g the short lands bottom of Tan hill Piece, the Down Piece Extra to sow, rain all night & morn till 8-oClock dry after
24. 20 Sacks new Wheat to Devizes the first & 1m Slates retd - The Wheat sowd yesterday harrowd off | Fine drying day
25. Fine drying day Plough'g Down Piece for Wheat
17th Novr Obit Miss Elizah Brown Avebury Aged 66 Yrs 24th Novr Funeral. 29th Novr Sermon 25th Mattw 13v [This entry written between two thick black lines]
25. T.P Elected Guardian of the Marlbro Union for the Parish of Avebury
26. Plough'g as yesterday, Fine day rain in even'g | New Stables, cover'd in Garden side and began other side | POTATOES, finish'd diging in Field

27. Rain in night, and Storms all day | Plough'g as yesterday
28. Plough'g as yesterday wth 2 horses & removed FOLD from Down Piece for Wheat to Nether Down Piece muckled for Turnips to sow next May | 1m Slates & 20 Bundles of Lathes fm Devizes wth 4 Horses Rain in morn'g dry after | WELL, finish'd 22 feet found Water, began, Wall'd, & finish in 6 days by 1 Man at 2s pr Dy (The Man his Keep) & 1 Young Man 3½ days @ 10d pr day
29. Sunday - RAIN all day
30. RAIN, nearly all day, very heavy storm at noon - Cart'g Muckle from Inn to Down Piece for Turnips

December

1. FLOOD, the first this season, the flat in Silbury mead fed wth Sheep, now under Water, still feeding the other part wth the Lambs, the Ewes yet on the Down & 5 Truss's of Hay evenings, want to get them in Straw Yard, have not hands or time, the Building requiring the hands, 2 Men thrash'g Wheat for Straw. | Fine drying day | Ploughing Down Piece for Wheat
2. Ploughing Down Piece finish'd | Fine drying day
3. Rain in night storms in day | SLATING, finish'd the New Stables WHEAT SOW'D fur: Down Piece with 5½ Sacks of Corton seed 12 S 17 li Nt |Whitelands 4½ Tan hill Piece 23½ Down Piece - fur 9 [Total] 37A} With Mr Jas Large seed 18 S[acks] - 11 S[core] 15 li Nt | Mr Jno Crook seed 6 S[acks] 12 S[core] 17 li Nt | T P Seed 0½B[ushel] 12 S[core] 8 li Nt | [Total] 24 S[acks] & ½ B[ushel] 2½ Bushls pr Acre & 1 Sack over | The 8 Years croping Down Pieces 1829, 30, 1-2-3-4-5-6 Home Piece 1830 Wheat 8A 3R 6P 1831 Oats 8A 3R 6P 1833 Wheat 8A 3R 6P 1835 Wheat 8A 3R 7P | Midd: Piece 1831 Wheat 8A 2R 39P 1832 Ray Seed 1834 Wheat 8A 3R 19P 1836 Oats [Pencil note] Barley | Furr Piece 1829 Wheat 9A 0R 32P 1831 Oats 8A 3R 0P 1834 Wheat 8A 2R 21P 1836 Wheat Measured by one Person Jno Gale Turnips, Vetches, Grass & Hay between
   Obit Miss Elizh Rumboll of Wotten Basset 1st Decr Aged [...]
   Obit Miss Ann Vincent of Calne 27th Novr Aged 50 Years [This entry and the previous entry written within separate thick black borders]
4. Ceiling Lathing began the New Stables Harrow'g off the Wheat sow'd yesterday - Storms after morning
5. HAY, finish'd the Stack of old in the Field and began the Old Rick in Meadow for Ewes & Horses, reserving the 2nd Clover for the Horses in the Spring | 3 Horses & Waggon to Calne for 8 Qr Lime | Hurdles removed from Silbury meadow to Swedes - White Frost morn'g fine day
6. SWEDES, the Tegs began, having finish'd SILBURY MD | Sunday, mild fine day
7. Very thick fog all day, 5 Horses harrow'g Wheat Stubbs | HORSE, rested, the Yellow Rone to make up for Sale

8. Dry day, 4 Horses to Devizes Board & Coal | LATHING finish'd New Stable for Plastering
9. Window put in Edmonds's Cottage | Stubbles finish'd harrow'g Down Piece, Dry day
10. Horses in Stable, Very sharp FROST, & fine day | HAY, gave the Ewes this morn'g first time no Straw - BARLEY, put the first in Barn to Straw the Ewes, no good weather for it before. | Plastering began the Ceiling & Walls of New Stables | Glazing Windows of New Stable
11. Exceeding sharp FROST, to bear up Sliders on Ponds | 6 Horses & 2 Waggons - 4 Loads of Sarson Pitch'g Stones for New Stable
12. Stubbles Cart'g from Down Piece to Lamb g Pen | Very sharp Frost fine day | STRAW the Ewes, began in the Yard, and 7 Truss of Hay in the eve'g & walk on Down
13. Sunday, - Very mild Thaw
14. Muckle Cart'd before the Fold a days work from the Inn, 14 days making it | PITCHING began new Stables at the Inn - Very mild fine day
15. Plastering the Ceiling 1st time finish'd, and began Plastering the Walls of Stable | Harrow'g Stubbles Tan hill Piece - Very mild fine day
16. Platered the Wall by Garden & began Plastering the Ceiling - the 2nd time | HATCH, began puting in Carriage of Mead | Stubbles finish'd harrow'g | Beautiful fine mild day
17. Beautiful fine mild day Stubbles carried (6 Loads) | ½ a day Carting Chalk &c from Stable to Meadow | PITCHING - finish'd the 8 Stalls in 3½ days
18. Fine mild day, 4 Load of Pitching brot fm Monkton Down for Mr Treen | Glazing finish'd the Inn, the Old & New Stables
19. Rubble, Cart'g fm New Stables finish d. | Road earth Cart'g to Meadow to fill Flint holes | Exceeding cold drying Winds
20. Sunday, Exceeding sharp Frost
21. Exceeding sharp Freezing Winds | Beckhampton Gates let to Mrs [Pencil note: Watson] at an advance of [...] for 1400£ | Cart'g earth to fill the Flint holes for Mr Treen
22. 6 Horses & 2 Waggons to Overton Copse for 20 Pole of Wood for Mr Treen | Very sharp Frost & fine day
23. Very sharp Frost Cart'g Muckle & hay to Fold
24. Xtmas=eve. New STABLE at the INN, finish'd by the Plasterers & 4 Stalls only to board to make a finish of the Stables | 6 Horses & 2 Waggons to Overton Copse for 20 Perch of Wood for Mr Treen | Sharp Frost and dense Fog all day.
25. Sharp Frost and fine, the Roads as dusty [...] Summer | HAY, the Sheep eat much 14 Truss's pr Day evening's only
26. NEW=STABLES, occupied this evening by 6 Horses (before finish'd) the Racehorses brot from Marlbro | Mr Saml Wentworth began serving the Race Horses with Straw | Muckle Cart'g before the Fold | Calm mild

Frost, incline for Thaw

27. Sunday mild Frost, incline for Thaw
28. mild Frost, incline for Thaw 20 S: New Wheat to Devizes & 50 Bu: ASH'S, returned - the 1st load
29. 2nd 50 Bu: Ashes Fine mild dry thaw
30. Fine mild dry thaw | BARLEY, the 2nd part of large Rick put into Barn | Horses in Stable, Frost prevent Plough'g
31. 3rd load of Ashes 60 Bu 4 Horses - Mild damp day | 4A 0R 2P Muckle'd & fold'd since 28th Novr for Turnips Nether Down Piece

1836

January

1. SNOW, all day, drifing | Horses in the Stable
2. Very sharp Frost, Muckle Cart'g before Fold
3. Sunday, - cold rapid thaw | HAY, the old finish'd (Round Rick) Kept the Sheep & horses since 5th Decr
4. NEW STABLE FINISHD, (at the INN, by the Carpenters finishg the Stalls (Mr Treen have two of his Boxes to finish) | 13 Qr Barley to Devizes, 4th load of 60 Bu Ash's retd | Boisterous drying winds | HAY, the Sheep & Horses all began the Tan hill road Rick, all hay remain'g new & good
5. 5th load 60 Bushls Ash's - Fine mild day
6. ASH'S began Sowing, 5 Waggon load 280 Bushls (1 load an hour 4 sowers) | OLD Xtmas, like MAY, beautiful mild calm day
7. Very cold chilly morn'g damp eve'g | FALLOW'G began Wheat Stubble, Build'gs & Frost prevent'd before. | Teg, Kill'd - taken ill in head (ewe) [Pinniger then made some observations on the price of wheat, entitled Depression in Wheat - see Appendix H]
8. Mild day, damp at times | WHEAT sow'd 23 Novr but partly up, bottom of Tan hill Piece | Muckle'ing before the Fold
9. Very sharp Frost morn'g - damp evening | 4 Horses at Plough & 2 horses to Honey Street for Timber for 2 Pr of Gates & Posts for Inn
10. Sunday, dry morn'g SNOW all afternoon
11. Frost morn'g, Snow & Sleet in evening | Horses in Stable, the land being covered with Snow
9. Mr TREEN, finish'd his part of the New Stables, (the Boxes) at the Inn
12. 6th Load of Ashes 60 Bu: from Devizes | Frost morn'g thaw after, dry day
13. 7th load of Ashes 60 Bu: from Devizes Frost morn'g thaw all day, little rain 2 PIG, began to fatten with Meal & Potatoes
7. The POOR, began relieving by the New LAW, & relieving Officer Mr Budd
14. Rapid thaw, snow all gone, high winds 8th load of Ashes 50 Bushls 3 horses The Colt ill with the Fret gave it ½ Pint of Hicra &c mixture & took 2 Gallons of Blood is now doing well, alarming, 10 Horses have died in the

Parish & adjoining within a month [278]

15. Ash's sow'd 3 loads & Cart'd Muckle to Fold | Cold drying Winds
16. Very sharp Frost, Fine day | Fallowing Wheat Stubbles
17. Sunday - Frost & Fine day
18. Sharp Frost & Rapid thaw, Fallowing as 16th | Ram & a stag, lost half their Teeth do ill at Swedes brot home to have cut Swedes.
19. Fallowing as yesterday, fine drying day | BARLEY, the 3rd & remaining part of the Rick put into the Barn
20. BARLEY, Winnow'd 14 Qr & 2 Bu, thrash'd by a Man aged 67 and a under Carter aged 19 at 5s pr Week ea: to keep them off the Parish Books 15 days each, prices given pr Qr vary 1s 1s 2d to 1s 4d N B. Detected the one dictating to the other writing a Letter instead of working | Fallowing as yesterday, damp mizly day
21. Good Ram Stag 92 li gave labourers | 2 Giddy (ewe) Tegs Kill'd | Ploughing as yesty damp morn'g dry after
22. 12 Qr Barley to Devizes - 42 € Coal returned Muckle Cart'd 6 Load before the Fold - Drying morn'g Storms after
23. Boisterous winds and storms | Ploughing as yesterday
24. Sunday - Dry morn'g damp evening | WHEAT, last sow'g 3rd Decr appearing
25. Ploughing as 23rd | Beautiful - clear & calm Day, like May. | Lambs - coming at Grove (Littlecote &c)
26. Plough'g as before, fine dry day | Messrs Halcombs inform'd me, they have Notice from Mr Treen to quit the Inn Stables
27. Ploughing, finish'd fallowing Tan hill Piece | Brewing 12 Bushls - Fine drying day
28. 12 doz HURDLES, from Grove - Dry morn'g little rain after
29. Cart g earth to fill Gravel Pit | Frost morn'g, heavy rain & Snow, afternoon
30. Snow, in night, Frost morn'g, Thaw after | 60 Bu: the 9th load of Ashes | Brewed 12 Bushls
31. Sunday, - Rain all morn'g dry after

February

1. 60 Bu: the 10th load of Ashes | Very cold winds, with storms
2. 60 Bu: the 11th load of Ashes | The Weather this morn'g commenced with a slight frost, - & a little rain till 9 o-Clock when succeeded by the heaviest fall of SNOW, (blossoms large as half Crowns) remembered continuing all

---

278 Fret – another name for colic. Delabere Blaine, *A Domestic Treatise on the Diseases of Horses and Dogs, 4th edition* (London: T Boosey, 1810), p.46.
Hicra – this is probably hicra-picra or hiera picra made from powdered aloes, Virginian snakeroot and ginger, used as a purgative for the stomach. The prescribed dose was 1½ ounces made into a ball and given with syrup of blackthorn or a pint of warm ale. Francis Clater, *Every man his own Farrier* (Newark: J. Tomlinson, 1783).

day | [Note in margin] Deep SNOW | The Coaches from London in the evening nearly buryed on both roads below Beckhampton, and no Coach or Mail up in the evening, both Inns at Beckhampton full of Travelers and Two Lady's remain in the York house Coach all night on the Down, no possibility to geting to the Coach with with any Conveyance to bring them away.

3. great number of hands employ'd this day the 3rd in diging the road through the Snow from Overton to Cooks Cottag the old Company s Coach, Waggon and Carts not released (by Mr Browns Whitelands) till 3-o-Clock this afternoon, - the last night Coaches and Mails not up till mid:day. the road from my new Stables at the Inn, all the way of Mr Browns great Whitelands full of Snow level with the banks The Calne Coach's this afternoon turn out of the road down Farmers Philpots new broke ground to the Devizes road - and again turn out into Kennet Meads and again turn into Mr Tanners Plough'd Field by the Cottage The Ewes confined in the Yard on Straw only as well the Horses, having no Hay at home the SNOW to deep to get to the hay Rick The Young Sheep in the Fold Nether Down Piece partly fill'd with Snow high as the Hurdles

4. The Ewes time up for Lambing the 14th Inst must now remain in the Yard till the Weather will allow to take them to a standing Pen on Swedes | 2 LAMBS, twins this morn'g from a Cul sale Ewe Rain a little & thaw, the Horses in Stable except get'g a little hay home for Sheep | 30 Men opening a road through the Snow 3 feet deep top of Folly hill

5. Dry day - gradual Thaw, the Snow remain remain deep in every hollow | 2 other Lambs

6. Fine mild day, gentle thaw | Carters winnowing, the Snow prevent Horses workg
28th January Died Mr Thos Henly of Cheltenham, Wine Mercht Aged 41 Years. [This entry written within a thick black border]

7. Sunday - Fine morn'g damp even'g

8. Dry morn'g, some rain even'g, the [Written in pencil: Snow] remain deep in all the Valeys | Ashes sow'd by 3 Men (self fill'g) 180 Bushls 3 Lds [279]

9. Dry morn'g, damp eve'g, the Snow remain in the road Harts hill 6 feet deep | Plough'd the headland Tan hill Piece Rt finish | FIRE, evening - Bean Rick Mr Jno Brown of Chissleton (a Guardian of the Poor) | The Ewes, have improved in the Yard the last week, Straw morn's, Hay evenings and drink much water, (Sheep do best when drink)

10. Barley, put part of 2nd Rick in Barn damp at times rain evening | Fallowing, began Nether Down Piece, muckled &c for Turnips

11. 13 Qr Barley to Devizes and 60 Bu: Ashes 12th load returned | Cold winds, very drying day. N B. the Roads exceeding good nearly dry, and this day week was almost impassable by deep Snow, the Snow yet remaining about the Hills

279 Loads.

12. Fine mild day, frost morn'g, Plough'g, as 10th
13. Fine mild day, frost morn'g, Plough'g, as 10th.
1809 SNOW, 21st April 12 Inch's deep on level 5-o-Clock morn [280]
1814 SNOW, 20th Januy Roads impassable, Sheep bury'd [281]
1823 SNOW, 31st Octr lodged on the Trees, destroyed great numbers & high Floods [282]
1830 SNOW, 20th Januy Roads impassable, see Diary [283]
1836 SNOW, 2nd Febuy Roads impassable see Diary
12. Chippenham Agricultural Society, form'd | The Calf weigh'd 16 weeks old 197 li @ 7d is £5 15s 0d cost at Tockm 14 days old £1 8s 0d for keep £4 7s 0d
13. A Calf from Tockenham 6 days old £1
14. Sunday, fine dry mild day
15. Frost morn'g, beautiful clear sun shine day | Fallow'g muckled ground for Turnips, Down Piec Ashes sow'd 2 loads (48 Bu: own & 20 Edmds one load of 60 Bu: will finish sow'g [284] | Wheat Rick, by Barn, fill'd end Mow with half | EWE'S time up for Lambing, HAY, only evening | Hay from this time twice a day & Straw in the Yard, too cold to lamb on the Turnips, and likely to want Hay
16. Dry Cold day - Plough'g as yesterday
17. FROST very hard and pierceing cold boisterous Winds, Load of Flints to West Tytherton 3 €wt Cheese, half a Pig, 25 Bdle Rods & 10 Bush's retd [285]
18. Very cold harsh Winds, Roads very dusty | Fallowing as the 16th | The MARKET at Devizes this day more animated, the first good Market since the 27th Augst see Wheat sold then at 18s worth now 22s pr Sack, - Sold my Talavera 12 S 8 li for 25s for Seed, next highest price 23s pr Sack
19. Very cold drying Winds | Barley, remainer of the American put into Barn 25 Sacks Wheat to Devizes & 60 Bushls of Ashes returned the 13th load | Lambs come slowly, only 20
18. Complain'd yesterday to Mr W Brown of Horton on the bank being made to prevent my Waggon passing on his Down, when would not order order it to be leveled. N B. on my return from Market, found it LEVELED. wrote to Mr Estcourt 23rd Novr last, on the impropriety of it. ★ | Nether Down Piece Muckled & finish'd Folding
19. The young Sheep lay back on the Swedes
20. Very sharp frost & harsh March winds | ASHE'S sow'd the last load. 820

---

280 This date is before the start of Pinniger's diaries.

281 Diary entry for January 20, 1814: *SNOW drifed, the roads impassable - SHEEP DUGOUT of the Snow, Ewes fold in Gate Close Horses in the Stable.*

282 See diary entry for October 30, 1823 above.

283 See diary entry for January 20, 1830 above.

284 Edmds - Edmonds.

285 Bdle - Bundles.

Bushls on 20 A is 41 Bushls pr Acre @ 3d pr Bu | Fallowing as before

21. Sunday, fine dry cold day
22. fine dry cold day Carted Muckle fm Inn to make a standing Pen Nether Down Piece, when feeding Turnips adjoining wth E & La [286] | WALTER CROOK to Albany Street London, to be Apc [287]
23. Ploughing as yesterday | Sharp Frost, cold sleet & rain af afternoon | LAMBS, the first 40 Couples, taken to Down Turnips and standing Pen
24. Frost morn'g rain even'g, Plough'g as yesterday
25. Frost morn'g Fine dry day Plough'g as yesterday
26. SNOW, commenced 7-o-Clock morn'g & continued ALL DAY, fully employ'd in making a Shed for the Lambs Down Piece, & houseing those at home, - nearly as much Snow as the 2nd Inst
27. SNOW, small all night, & nearly all day. Men employ'd diging a road through the Snow Mr Tanners Road Waggon stop'd wth 6 Horses - 6 Ton - 7 <Cwt> assist'd with my 4 Horses to top of Cherhill-hill, the fore wheels druging by the Snow nearly all the way.[288]
28. SNOW, heavy in the morning and evening, Sunday - The Ewes time up a fortnight, 100 Lambs
29. Very mild thaw, the face of the earth this morning continue entire White | Attending the Sheep with Hay and beding, Muckle from Inn, to bed the Ewes in Yard, having no Straw

March

1. Rain in night and very boisterous winds in the morn'g commencing March, very cold storms of SNOW, sleet & rain in the day & boisterous evening Carters cuting Chaff and attending the Sheep | The Ewes & Lambs confined in the Waggon house, the rough nights, the Lambs in danger of perishing
2. KILL'D Instantaneously coming down hill Cannings by the Waggon, (Mr Josh Larges) Carter, Whyr [289] [This entry underlined with a thick black line] | The rain yesterday dissolved the Snow, except in the hollow places, and being a very drying morn'g took 70 Couples to Down Turnips, together 110 Coupls | Ewes, counted 126 with Lambs 126 not Lambed 7 - lost Lambs, turn'd out 2 - Died [Total] 261 | The Sheep fill themselves at Turnips, would do well if weather fine, Swedes very backwards have not the least

286 E & La - Ewes & Lambs.

287 Apc – apprenticed.

288 To drug a wheel is to put on some kind of drag or chain (see Glossary) - in this case, it appears that the snow as the cause of the druging.

289 From the burial register for Winterbourne Bassett: Esaw Waite. Killed by a team of horses running over his body. Coroner's verdict: Accidental death. Esaw Waite of Winterbourne Bassett buried on 6 March 1836 at Winterbourne Bassett, aged 55, *Wiltshire, England, Church of England Deaths and Burials, 1813-1916 for Waite* [accessed via Ancestry, 27 July 2020].

green to them

3. BARLEY, Shaveleer, put into Barn, the first [290] | Dry morn'g, heavy rain evening | Ploughing began athwart. Swedes fed off | Tegs began feeding the young Swedes
4. Ploughing as yesterday, Rain in night dry day
5. Ploughing as yesterday, heavy storms of hail & rain | CHIMNEY, ON FIRE, - The Ricks & Premises in great DANGER, fire falling on them, but being damp escaped (7-0-Clock evening
6. Sunday - cold damp day - some storms
7. Ploughing as the 5th - some storms | 190 Lambs (160 at Turnips Down Piece
8. Sharp frost, fine dry day | Ploughing as yesterday
9. Ploughing as yesterday Cold rain all afternoon
10. Ploughing as yesterday Dry morn'g, rain evening
11. Ploughing as yesterday Cold storms of rain & hail | Barley, remainer of Rick put into Barn
12. Heavy storms of Hail & rain | 10 Qr Barley to Devizes & 1T 16C 3Qr Coal retured
13. Sunday - Dry day
14. Rain heavy in night & all morn'g | 2 Load of Stone from Calne to repair Inn Stables | TREE'S Lyme, Planted 6 in Meadow @ 1s ea | 218 Ewes with Lambs 17 Ewes not Lamb'd - put wth Ewes Lambs 24 Ewes no Lambs put wth Tegs 2 Ewes Died [Total] 261
15. Boisterous winds & rain, thwart'g after Swedes
16. Cold storms of Snow & hail, thwart'g after Swedes | The 2 Shepherd Boys their suppers a month
17. Very cold boisterous, uncomfortable day some rain | Ploughing the Potatoe land for Oats
18. Ploughing the Potatoe land for Oats | The Weather changed, become very mild suddenly the Glass very high
19. Beautiful mild day - all Vegitation but just move'g cannot gather a boiling of Swede greens, the Turnip just ready (greens) to gather (alter almost hourly) | Carting Dung from Standing Pen, Rick Yard
20. Sunday - Very mild spring day x
21. Very mild spring day x | Rolling the Field grass, & Ploughing for Oats
22. Rolling finish'd Field grass 20A in a day & half wth 2 horses Very mild spring day x Plough'g as yesterday, Down Piece after Turnips | RABBIT'S, Mr Wentworth Hurdle'g in the COVERT with 20 doz, to Keep them from my Wheat
23. Carting Muckle from Inn to make a Standing Pen on Swedes, & Cart'g Gravel to Inn | Dry morn'g very cold rain after, Hunters cross'd the whole Farm 150

---

290 Shaveleer - Chevalier (see Glossary).

24. Wheat improve in price, sold 40 S- of old 12 S 5 li & 12 S 12 li and 20 S of new 11 S 13 li } @ 23s 6d N B - worth only 17s since Xtmas | Very cold wind & hail storm | 12 Qr Shavelear Barley to Devizes & Plough'g as before | TURNIP'S - the Ewes & Lambs finish'd Down Piece began feeding them the 24th Feby
25. SWEDES - the Ewes & Lambs began feeding about 7 Acres to feed in the Standing Pen all the morn'g, the RAIN being very heavy, dry after noon. The Horses removing Hurdles and Hay for the Sheep
26. Carting Earth to fill Gravel Pit for Mr Treen | Exceeding boisterous winds & piercing Cold with very heavy storms of hail & Snow
27. Sunday, Frost, & cold storms of rain & Snow
28. Most winterly day, heavy Snow morn'g, cold storms after, Glass sunk to E in very dry | The Springs, rising in Meadow rapidly | The Sheep do very ill, - the Teg Knit up with the wet & cold, - The Horses Cart'g Muckle to Stand'g Pen
29. Frost morn'g, Piercing cold storms after | 15 Qr Seed Barley from Marlbro with 3 Horses | 1st FAT PIG - 14 S 10 li | The Tegs finish'd the young Swedes began feeding them the 3rd March
30. Rain heavy all the morning & very cold storms after | Horses moving Hurdles morn'g - in Stable after
31. Very cold hail storms morn'g, dry after | 10 Qr Barley to Devizes, & Ploughing

Sacred to the memory of my Sister Mrs Smith late of Bremhill who departed this life at West Tytherton at my Sister Ann Crooks Residence 24th March 1836 Born 20th March 1776 } Aged 60 Yrs Funeral Sermon (31st March Funeral) Text 2nd Ch: of Acts 26th & 27 Verses [This entry written between two thick black lines]

5th March died Mrs Susah Mary Crook of Barn Bridge Aged 71 Years [291] [This entry written beneath a thick black line]

April

1. GOOD FRIDAY. SNOW, very heavy from 8-0 Clock to 4 o-Clock afternoon, extreme cold winterly weather, the Sheep do very ill 2 Ewes & 2 Lambs died no apperant cause, except chill'd by the weather, distressing times with many Farmers in buying Hay
2. SNOW heavy cold storms, have no Straw to bed the the Ewes & Lambs, used the Dung from the Inn in lieu | 3 Horses & Waggon started at 8-0-Clock last evening to BATH, for a Thrashing Machine, returned home at half pass 10-0-Clock evening | No Corn in Barn, or weather to put it in

---

291 Mary Susanna Crook of Barn Bridge was buried on 13 April at Tytherton Lucas, aged 71, *Wiltshire, England, Church of England Deaths and Burials, 1813-1916 for Susannah Crook* [accessed via Ancestry, 27 July 2020]. It would appear that Pinniger wrote *5th March* when he intended to write *5th April.*

3. Easter Sunday, Very cold windy morn'g mild eve'g
4. Frosty morn'g, fine calm day | Wheat Rick, the remaining part put into Barn see 15th Feby began it, have taken all the rain since, a Waggon load of sheaves spoild for want of Barn room, have a Machine in consequence
5. Dry morn'g rain all afternoon, Plough'g as before | Thrash'g Machine began puting up
6. Thrash'g Machine put to Work | Waggon & 4 Horses to Grove, for 12 Doz Hurdles for Mr Jno Large | TEGS removed from Swedes & put on Down, as bare as a Floor wth Hay, & Fold'd on Down, the Fallows being too wet to fold them, Sheep Keep exceed'g scarce, The Swedes reserved for the Ewes & Lambs in hope with short allowance to Keep them April a bad prospect in Silsbury mead, the Springs being very high will not be dry 1st May for the Sheep
7. 29 Sacks Wheat to Devizes, Very rain'y morn'g dry after
8. Chalk Cart'g round the Machine, some thin showers of rain
9. SNOW, a Storm of very large blossoms, cold storms of HAIL_ & rain after | Cart'g Rubble for Mr Treen from Mr Browns Down | Cart'g Rubble from my Down to Thrashing Machine
10. Sunday, some Rain in morn'g - drying day after
11. Frost morn'g - very fine mild day, an apparant CHANGE in WEATHER | Ploughing finish'd middle Down for Barley, one earth Turnips fed off [Pinniger then recorded the weighing of his wool - see Appendix E]
12. Rain thin all morn'g drying after | THRASHING with Machine 1st Day, with 4 Horses about 20 Sacks, very damp wheat, Rick open to the weather 6 Weeks | FURZE, began cuting - joining my Down Piece
13. Thrashing as yesterday, Boisterous drying Winds [Note in margin] 13th W Pearce quited Cottage
14. Fine mild day, Market depressed in consequence | WOOL to Devizes, Seeds retd Plough'g Swedes fed off
15. 3 Mr Wentworths Horses assist'g harrow'g mid: Down Piece for Barley | Fine growing Day
16. 2 Mr W P of K Horses assist'd Drill g Barley | 3 of my Horses Plough'g for Mr Wentworth | Very fine growing Day | Ewes & Lambs, discontinue in Standing Pen on Swedes, short allowance and Hay & Water, 7 Truss's night, 2 Morn'g
17. Sunday - Fine dry day
18. 3 Horses Rolling & harrow'g Barley Down | 2 Horses Ploughing - for Mr W P of K | Beautiful fine grow'g day, to Littlecote
19. Beautiful fine grow'g day, Plough'g began the Swede ground for sow'g Barley | Thrashing finish d the damp & grow'd Wheat
20. Ploughing as yesterday | Boisterous winds morn'g. Thin rain evening | The 5 Yr old Rone Horse cost 28 Guineas 11th Septr 1833 sold by by Hillier Reeves for 27£ at Devizes

21. 20 S_ old Wheat to Devizes 1 Ton Coal Retd | Plough'g for Barley, Cold harsh Winds
22. Plough'g for Barley and sow'g Hop & Ray 3 Bushls pr A in Barley mid-Down Piece 11 Score with the Bag the Qr | Rain in night, cold harsh winds all day. | CUCKOO, first hear'd here, - 10 days since at at Bremhill | LAMBS Tail'd 222, Ewes 261 put to the Ram 4 Ewes died in the lambing 1 giddy & some barren (no EAR MARK)
23. Hop & Ray sow'd 3 Bushls pr A in the Wheat further Down Piece, fed to the ground by the Rabbits
8. Mr G Brown finish'd Swedes & began Meads & New Field
12. Mr Kemm finish'd Swedes & began Mead
22. Mr Wentworth began Mead
23. Mr P_ Kennett began Mead
26. Farmer Philpot (over Shoes in Water) began Mead
May 14. T P began Mead
23. New Mangers put up Inn Stables
24. Sunday Rain in night, and cold heavy rain from 11-0 Clock morn'g to night
25. Ice'y Frost, and very fine mild day | 136, Tegs 24, Ewes [Total] 160 } Sent to Grove and forward'd to Littlecote tomorrow, to Keep on Watermead @ 60s | 100 - Couples (Ewes & Lambs) sent to New Mill, to Keep on Watermeads @ 60s pr A | Swedes enough left to Keep the Sheep at home till May when must go on New Field, Silbury Mead being under Water, by the SPRINGS, being very high on my Down viz- from Fromer Philpots to the end of the Roman road by the Road side, DISTRESSING time for Sheep Keep, many Farmers buying HAY, all this month and great numbers of Lambs - DIE with Scour, the weather having been very unkind for Sheep & Lambs. * | Ploughing for Barley | JNO DAVIS, entered the Cottage (W Pearce left) Potatoes in New Garden & Small Seeds
26. BARLEY began sow'g the 18 Acre Piece | Cold harsh winds, violent storm of HAIL at 10-oClock night
27. SNOW & HAIL very heavy Storm wth heavy Thunder & Lightening at 1-oClock, very cold harsh winds in the morning | Harrow'g & rolling Barley sow'd yesterday
28. COLT the 3 Year old black gray rone received of Mr Hillier Reeves | Frost in morn'g, fine day, cold rain evening | Ploughing for Barley | The SPRINGS, on my Down sinking, the Mead continue under water
29. Ploughing for Barley, DAHLIAS planted | SNOW, large blossoms, very cold storms evening very cold winds all day | New Mill, mead, the 100 Couples began, on Turnips since Monday Mr Chs Pinnigers
30. SNOW & hail storm, evening, cold harsh winds Barley sowing the 2nd Piece - in 18 Acres

May

1. SNOW & hail afternoon, Cold winds, Sunday
2. 3-two horse Ploughs for Barley | HURRICANES of Winds to uncover the Buildings and Ricks - and very cold | SWEDES, finish'd feeding, with 143 Sheep & 122 Lambs | 143 260, from home 1, Stag Kill'd Xtmas 2, Stag sold 7, 4 Ewes died 3 tegs Killd since 7th Novr [Total] 413 10 [Total] 403 222 Lambs [Total] 625 Present Stock
3. The Springs continue runing through Silbury Mead over Shoes in water to prevent feeding it, good grass and lodged
4. Field grass began feeding with the 143 Sheep & 122 Lambs | HAY, NIGHT & MORNING, continue 6 Truss's night & 2 mornings, the exceed'g cold harsh boisterous Winds, check the Grass & Wheat from growing, the Turnip Kept for Seed are but shewing blossom the Swedes, have not appeared to blossom, see 1834 Jany 24th the Swedes in blossom

1. Mr G Brown finish'd Mead & began Field grass
1. Mr W Kemm finish'd Mead & began Field grass
3. Three, 2 horse Ploughs for Barley | Continuation of very boisterous weather & very cold
4. Barley sowing, Rain morning & evening | New Mill mead, added 50 Couples to the 100 there
5. Many cold storms bad barley sowing | Ploughing & sowing Barley
6. Some cold storms, rather more growing | BARLEY sowing FINISHD about 25½ Acres - with 30 Sacks of Mr Church, Dudmore Lodge 2 Sacks of Own - is 5 Bushls pr Acre 96 Bushls of good hop & Ray 11 Score Nt the Qr and 4 Bushls of Cow grass (2¼ Cwt) sow'd with the Barley & [Total] 100 the Wheat further Down Piece about 8½ Acres - together 34 Acres about 3 Bushls pr Acre, the Machine full hole
7. Finish'd harrow'g the Barley & Cart'g Muckle from Stand g Pen Dry & more growing Day, harsh winds
8. Sunday, - Fine dry day
9. Very fine mild day, Cart'g Muckle from Standing Pen, & Rolling Barley
10. Harrowing Wheat, Rolling Down Barley & Down Wheat | Very fine day Died this morn'g [Written in pencil: April 30] John Chivers of Calne Died this even'g April 30th Mr Orial Viveash of Calne J C_ aged [...] OV aged [...] [This entry written within a three-sided thick black border]
11. Rolling Barley & Wheat & harrow'g Wheat Tan hill Piece | Frost & fine day
12. Fine day, harrow'g Wheat & Plough'g | GRASS, began mow'g for Horses Silbury mead
13. Cart'g Dung finish'd from Stand'g Pens | Frost & very fine day Plough'g Down Piece | WILLM TANNER, himself put in his Hatches, and covered my Grass with water round Silbury hill, full two Tons pr Acre
Died this [...] May the 13th at Nuthills Farm Mr Richd Pinniger Broome

Aged 59 Years Buryed in Cliffe Church the 20th of May [This entry written within a three-sided thick black border]

14. Very fine mild day Ploughing as yesterday | 70 Couples & 23 odd Sheep began feeding SILBURY Mead, all other Sheep from home | All the Mead except the Bank, under Water by Willm Tanners hatches & Robt Philpot watering his Mead
15. Sunday, Fine mild day
16. Died this morning May 16th at Cirencester Mr Willm Jenner late of Charlcot Aged 56 January last Buryed May 20th at Maisey hampton [This entry written within a three-sided thick black border]
16. Plough'g as before, Diging Chalk for Edmonds House | Very fine warm day
17. Very warm day 6 Horses & 2 Waggons to BOWOOD for for 1½ Hd Faggots
18. Plough'g Down Piece for Turnips | Carting Dung from Down Piece standing Pen | Very warm fine day
19. Cool morn'g fine warm day, Plough'g &c as yesterday
20. Plough'g &c as yesterday | Funeral of Mr Broom, Very warm fine day | 150_Couples Ewes & Lambs returned from New Mill Mead 4 Acres at 60s & 1 Ton of hay 4£ is 16£ 29th April 100 Couples began the Mead & 4th of May 50 Cou
21. POTATOES, began planting middle Down Piece 4 Diggers (trenching) 2 sitters | Rolling & harrow'g Potatoe Land, Dry harsh Winds | Mr Eastcourt, inspect'd with me, the injury my Wheat sustaind by the Rabbits which is at this time nearly as bare as a piece of fallows
22. Sunday, cold harsh east winds
23. very cold harsh east winds | 159 (1 died) Dry Sheep, Returned from Mr Rumboll of Littlecote on 3 Acres of Mead Grass 9£, Shepherd 2s 6d, - 28 Days feeding it Fold'd on middle Down Piece, Muckle'd & fold'd before for Turnips, no other land to fold, feed to Down only All the Ewes & Lambs in Silbury mead (see the 14th) Very FORTUNATE in puting the Sheep out to Keep the last month, would eat all the Field grass if had remained at home, very distressing time to provide Sheep Keep | Witmonday, Cuting the Grass in my Water carriage (lowered the Water 10 In) to W Tanners hatches & took the Water off Rt Philpots Mead, which rendered the left of my Meadow feedable, round the Hill remain under water, the Springs still rising | Preparing Potatoe land Down Piece
24 WHEAT=RICK, by Door 4A 2R 9P put into Barn Cold harsh winds, harrowing Down Piece for Turnips
Died May 24th James Butler aged 80 Yrs Parish Clerk of Lyneham [This entry written between two thick black lines]
25. FROST, to cut & destroy, Kidney Beans, Potatoes, Shrubs, Trees &c. Cold harsh winds | CALF, 49½ li pr Qr @ 5½d from Tock[enha]m 15½ Weeks old Roll'g Down Piece for Turnips & Plough'g for Swedes | LAMBS, cut 107 in 1½ hour by Ebednego Blackman & 2 Rams bot at Mr Kings Sale

Overton 1833

26. Plough'g athwart for Swedes Rt of Tan hill road Cold harsh East Winds
27. Cold harsh East Winds & Ploughing as yesterday
28. Cold harsh East Winds & Ploughing as yesterday | Waggon & 2 Horses to Tockenham for Furniture
29. Sunday, Cold harsh winds
30. Cold harsh winds | Ploughing athwart for Swedes
31. Ploughing athwart for Swedes finish'd, NO Plough'g to employ the Horses, intending to sow the Turnips and Swedes on the thwart first harrow'd & Roll'd waiting for Rain, - others sow'g the Seeds in the dry earth, thinking it may grow, or be ready when the rain come | This Day closes the Month of May which has been remarkable for its cold harsh drying Winds, Keeping Vegitation almost at a stand, the ground is seen through the Wheat, generally, which is not half way up my legs, consequently no appearance of Wheat Ears | No rain since the 5th of May and no warm Shower at any time, the Spring we feel to have lost, have gone from Winter to Summer, the CUCKOO feeling it has become mute, very rarely hearing it chearing Note | The prospect of Sheep Keep is become alarming the Field grass being fed (but little left for Hay) the Downs also which have not moved, consequently there are nothing left on many Farms, but to walk over the fed ground Day after day, Keeping the Lambs and Ewes poor, which will give us very little Wool | Lambs good luck, not one died since Lambing when 222, since 1 Kill'd at Calne, by Waggon going over it some breeders have lost a third, by scour &c.

June

1. Wheat Ears, shew'g at Upavon, the stalks not a foot high | Less wind & mild | Thrashing began Wheat R: harrowing for Swedes [292]
2. Thrashing began Wheat R: harrowing for Swedes | Mild morning, glass sinking, some little blight'g rain
3. Mild morning, glass sinking, some little blight'g rain | Rolling & Harrow'g for Swedes
4. Some heavy growing Showers of RAIN to do good see the 5th of May, no rain since
3. Freegrove, Christened Abbot
5. Sunday, heavy showers of rain in the day
6. Silbury mead, finish'd feeding all except the wet part of the Marsh, still too wet to feed or mow the SPRINGS, continue to rise, see 14th May began feed'g | LAMB'S WANED 220, on Vetches, Rt of Tan hill road and a tie of grass home mead | EWE'S put with Tegs on the Down, both very poor, & & the DOWN so grassless, will only Keep them alive no other Keep for them, the Lambs want all the Vetches (will cut very little Wool) VETCHS,

292 R: - Rick.

began feeding (too short to Cage) Hurdles removing from Mead, dry day | SWEDES, began sowing Rt of Tan hill road 9 Acre 21 li (Mr Wentworth Slateing his House) Sheep Counted 401 (1 died at Littlecote & 1 giddy Ewe Kill'd)

7. Rolling & harrow'g Swedes, sow'd yesterday | Some thin blight'g rain
8. Rain heavy in night to wet the land sufficiently for Turnips, warm dry grow'g day after | Ploughing Nether Down Piece for Turnips | Thrashing finish'd the Wheat Rick (the horses)
9. TURNIPS began sow'g, Nether Down Piece Norfolk W Fine growing day [Note in margin] SPRINGS discontinue running from Beckhampton through Silbury mead
10. Rain morn'g, Horses in Stable, Cart'g Muckle afternoon from Inn to the Yard, being dry
11. Plough'g & finish'd sowing about 4 Acres Turnips Nether Down Piece with 10 li Norfolk White Seed - Some thin misly showers
12. Sunday, Cold dry day
13. Very warm mild day, finish'd roll'g Swedes, and 12 Doz Hurdles from Grove The Lambs now waned and the VETCH's CAGED do better, - the Ewes do exceeding bad as well the Tegs, to appearance will not cut 2 li Wool ea have only the Down & bait on little Down which is fresh'r N B, have done well some Years with same Keep.
14. 2 Load of Stones from Calne for Mr Edmonds house | WHEAT=EAR'S, burst'g half out Whitelands Fine Day
15. Wheat Rick the last new (Talavera) put into Barn being full of Mice, Kill'd, half a Bushel 8 Acres from 6-o-Clock morn'g to 8 at Night with 2 Waggons puting into the Barn | Very Warm fine day - 4 HORSES in Stable, nothing to do in the Field
16. Very Warm fine day 20 Sacks old Wheat to Devizes and T 2.2.1 Coal Retd
17. Thunder & heavy RAIN, afternoon to soak the Land very acceptable for Turnips & Swedes, (Fly very strong) | Machine began Thrash'g Talavera Wheat | WHEAT=EAR'S, some few Whitelands quite out
18. Horses Thrash'g - two resting
    June 10th Died Jonathan Pole Tytherton Aged [...] [This entry written in pencil between two thick black lines]
19. Sunday - Storm at noon dry after
20. Dry day, Horses Thrashing | Wheat uper end of Tan hill Piece, but few ears bursting
21. SHEEPWASH'D 19½ Score @ 6d, - 3½ hours going from little Down to Clatford Mill, washd in 2½ hours, usage as last year satisfied | Thin Showers & thrashing, POTATOES appear'g in field
22. THRASH'G, finish'd the Talavera Rick in 5 Days Showers at times all day, Rain more or less every day, some Farmers have all their Field grass cut and spoiling

23. Ploughing a Fold drift for Turnips, fine dry day | VETCH'S - the EWE'S, began feeding in the Fleaks geting large & shewing blossom having enough for all the Sheep & Lambs
24. 5 PIG'S bot and put to fatten on old boil'd Potatoes and offal meal Midsummer day, not a Wheat ear to be seen between Beckhampton & Devizes, my Whitelands Wheat out fully in Ear, the forwardest Piece in the Neighbourhood | SILBURY=MEAD, the Marsh began Mowing for hay by Wm Hampshire, very heavy too wet to feed or mow before | Rt Philpot, mow'g his Mead (21st) fed till [...] | Fine dry day. Plough'g as yesterday
25. TURNIPS, sow'g - the 2nd Fold drift, Nether Down Piece RAIN heavy to drve home the Teams | Very fine day, Cart'g the Grass frome Silbury mead to home mead to make it hay, the Crop very heavy & wet
26. Sunday - fine day
27. Fine day, Cart'g Grass from Silbury mead
28. Fine day, Cart'g Grass from Silbury mead
29. Fine day, Cart'g Grass from Silbury mead
    Died June 21st Mr Willm Standish, Bowood Aged 60 Years
    Died June 22nd Mr Willm Seager late of Highway Aged 48 Years [This entry and the previous entry written within a three-sided thick black border]
28. Barley Down Piece very forward shewing the Beard nearly a Yard high, higher than the Wheat adjoining - injured by the Rabbets
    Died June [...] Joseph Pepler Hail Farm Aged [...] Years [This entry written within a three-sided thick black border]
30. Finish'd mow'g Meadow & began FIELD grass too ripe the ray, no time before | Cart'g Grass finish'd from Silbury Marsh to the Down 20 Loads about 3 Acres | Fine day

July

1. Fine day SWEDE'S fail Rt hand Tan hill Piece Plough'g & Sowing again
2. Fine day SWEDE'S Plough'g & Sowing again
1. SHEEP SHEAR [...]
2. Wheat rapidly burst'g the Ear & grown much high'r than expectation, all other Corn improving very fast HAY, the first began carry (the Marsh) from home Mead
3. Sunday, Very warm growing day
4. Exceeding warm, Hay carry'd home mead and part of (Marsh) from Down | Ploughing & sow'g the Swedes fail'd | COW, CALVE'D, the time expire 1st July
5. Exceeding warm, alarming Thunder & lightening all even'g and all night | Field grass finish'd mowing | Swedes sow'g as yesterday & Cart'g hay
6. Swedes sow'g as yesterday & Cart'g hay | 4 Men hoe'g Potatoes - fine day
7. Plough'g and sow'g Swedes after Vetch's fed off, Rt of Tan hill road | Clover hay Cart'g, - Very warm fine day | Down Wheat not half in Ear, kept back

by the Rabbits | Barley in general geting in Ear | Turnips Down Piece - fit to hoe & the Swedes choaked with Charlock, all hands with hay & Potatoes

8. Plough'g & sow'g Swedes as yesterday, very fine day | Harrow'g Swedes & Turnips for hoeing
9. HAY, finish'd carry'g, - very fine day Rolling & harrow'g Swedes sow'd yesterday | TURNIP=SEED, cut
10. Sunday very warm fine day
11. SWEDE'S began hoeing very warm fine day Swede sow'g Vetches fed off
12. Swede sow'g finish'd to the Path | TURNIP'S - began hoeing Nether Down Piece, fine day
13. Turnips sow'd the picked Nether Down Piece with own new seed rubbed out in the Field, very fine day | Turnip seed carried two good loads
14. 20 S: of Talavera Wheat to Devizes | Rolling & harrow'g Turnips sow'd yesterday, warm fine dy [293]
15. 20 S= of Talavera Wheat to Moss s Mill for Mr B Bailey & 2 loads of Stone from Calne | RAIN heavy storm afternoon to lodge my Barley & W [294] | Potatoes finish'd earthing
16. SWEDE'S sow'g below the Path Vetches fed | Very fine day
17. Sunday - Very fine day
18. Overton Fair 17 Miles (Very fine day) beyond Forsbury | TURNIP'S, began sowing below Path Vetches fed off
19. Rain a storm afternoon | Plough'g & harrow'g, the Turnip seed ground 18th Died Mr Saml Viveash, of Calne Aged 76 Years [This entry written within a three-sided thick black border]
20. Rain all night & all day, very heavy hail & rain eveg | Horses in Stable
21. Chalk, Cart'g heavy Storms
22. Chalk, Cart'g - Storms
23. Barley rick put into Barn, Very fine day
24. Sunday, Rain nearly all morn, dry after
25. Swedes, seconding, Fine day damp morn'g | Swedes & Turnips appearing after the Rain | 19 Sacks of Wheat to Mr Bailey s 7 Mills, and 2 Loads of Stone return'd from Calne | EDMOND'S COTTAGE, began repairing and rebuilding
26. Barley began Thrashing with Machine, fine day
27. Barley began Thrashing with Machine, fine day
28. Barley began Thrashing Windy dry day | 2 Cart loads of Lime from Calne
29. Rain heavy night & morn'g, thin Showers after Horses Thrashing
30. Turnips, Sow'd a drift Tan hill Piece
31. Sunday Very fine day [Pinniger then recorded the weighing of his wool - see Appendix E]

---

293 dy - day.

294 W - Wheat.

August

1. Very fine day, Plough'g & Sow'g Vetches fed off to Turnips Tan hill Piece | Turnips the 2nd sow'g began hoe'g Down Piece
2. Wether Lambs drew & put on 2nd Clover with the old Vetches 104 - in the evening 1 died blown on the Vetches Caged | 4 Horses to Honey St[reet] for Roof of Edmonds Cottage | Cart'g Marter Earth | Very fine ripening day
3. Very fine ripening day Horses Thrash'g Barley | GRASS, Silbury Mead Horses began, the Clover finish'd
4. Barley finish'd Thrashing with the Horses | Swedes finish'd hoe'g the 2nd time by 1 Man | Turnips finish'd hoe'g the 2nd Sow'g Down Piece | Sultry warm morn'g Thunder Storms afternoon | 2nd EWE died murrain | 1st Sack of NEW WHEAT in Devizes Market 30s
5. 20 S_ Wheat to Devizes & 1m Slates returned Fine day
6. St Anns hill Fair, no Wheat to be seen from the hill ripe, my White=lands only ready for Reaping in the Parish | Chalk Cart'g for Edmonds Cottage | LAMB'S sold 80 at 16s and 20s at 12s to Messrs Baker & Lanfear of Sutton [295] | Fine ripening day | ROOF, put on the back part of Edmonds's house
7. Sunday, beautiful fine ripening day
8. beautiful fine ripening day | REAPING, began Whitelands, quite ripe 8 hands | Chalk, Carting for Cowhouse
9. Hurdles removing from Vetches to Clover & Carting Marter earth to Cottage, Fine ripening Day | VETCHE'S finish'd feed'g 9 Acres, began 6th June 600 & odd Sheep feed'g it from 23rd June
10. CLOVER all the Sheep began, fine ripening day Hurdles the remainer removed from Vetches MOWING Silbury Mead that part fed by Sheep over the Carriage (see left 6th June) | Rubble Cart'g from Cottage | Whitelands Wheat finish'd Cuting, no other ripe, the Down Wheat in blossom as green as grass, Kept Kept back by the RABBITS | NEW COTTAGE by Inn began Slating

11,12,13. Warm ripening Day's & finish'd Plough'g & sow'g the Vetch land fed to Turnips

12. EDMOND'S COTTAGE, unrooff'd
14. Sunday warm ripening day Thunder lightening & storms partially in evening
15. Swedes, began hoe'g the 2nd sow'g where the 1st fail'd | WHEAT=RICK from Whitlands made the 1st in the PARISH 342 Tythings 4,,0,,38 Pole [296] | Muckle Cart'g from Inn the FIRST before the Fold on the 20 Acre Piece for Wheat Fine morn'g some rain evening
16. HAY, Carried 3 loads from Silbury Mead & Cart'g muckle, very fine ripening Day
17. HAY, finish'd <finish'd> carry'g Silbury mead 3 Loads Very warm fine day

295 Although Pinniger wrote 20/- (20s), it is likely that he meant 20.

296 4,,0,,38 - 4 acres, 0 roods and 38 perches.

Waggon & 3 horses to Box for Chimney Top &c for Edmonds Cottage | REAPING began again middle of Tan hill Piece ripen very uneven, very green in furrows | by 5 Taskers from Bedwins & 5 own People | Roofing began Edmonds's Cottage

18. Fine day, rain little in evening | HORSES, no Plough'g & very little work in the Field, now in the STABLE, Carters & Boys load'g Straw from the Rick yard &c.
19. Men & horses employ'd as yesterday, fine day
20. The Walling Edmonds Cottage finish'd, the Horses Carting away the refuse | RAIN after the morn'g nearly all day very acceptae [297]
21. Sunday, Fine dry day
22. Dull cloudy day Glass sinking, Cart'g Earth to Cowhouse | Wether Teg, Died Murrain
23. RAIN all day, Carters Winnow'g Barley
24. RAIN all night, very dry'g after morn'g | Wether Teg died Murrain
25. 400 Sheep, gave a Table spoon full of TAR to each to prevent their die'ing by the MURRAIN consume'd and fix'd Tar'd Stakes every night in the Fold Carter helping, Horses in Stable all morning Very drying all day
26. RAIN in night & morn'g to stop Reapers, fine after | EDMOND'S COTTAGE, complately cover'd and finish'd with Slate
27. Wheat Rick began making above Potatoes Tan hill 15 Lands [...] Acres 1081 Tythings | Fine dry morn'g, thin driving Rain after the Rick was half made to the finish of it (by Barn)
    27th Died Mrs Jno Brown of Avebury Aged 80 A Lady truly good and very much esteemed [This entry written within a three-sided thick black border]
28. 25th the Horse Punch rec'd the injury and died the 28th | Rain in morn'g, very drying after, (Sunday)
29. Very fine day, Carting Muckle | 4 Men began hoe'g young SWEDE'S after the 1st fail'g
30. Reaping finish'd Tan hill Piece - 10 Men hoe'g young Swedes & Turnips left of Tan hill Road | 3 Horses to Honey Street, fine day
31. Very fine harvesting day | Wheat Rick from uper side of Tan hill Piece on the middle Staddle by Wall in good condition contain 7A 2R 10P 342 Tythings Whitelands 1791 Tythings Tan hill Piece [Total] 2133 is 77 Ty: pr Ac: on 27A 3R 12P | Edmonds, began occupying Sleeping Room after Plaster'g Plastering the other part of the House
30. SHEEP, sorted for Sale & put Ewes to the Rams with the Chilver Lambs, and mark'd the Stock Ewes [Pinniger then recorded his stock of sheep - see Appendix B]

September

1. Carting Rubble for Mr Treen | Damp morning and windy, very drying

297 acceptae - acceptable.

after

2. Driving heavy rain till mid:day, fine drying after | FUNERAL of Mrs Brown [298] | Carting Muckle
3. BARLEY, began Mowing, Down Piece | REAPING began the Wheat Down Piece, very green Kept back by the Rabbets and injured to the amount of 50£ | Carters Winnowing Barley, Horses in Stable | COTTAGE, finish'd - fine drying Day
4. Sunday, some slight showers Funeral Sermon for Mrs B: 30th Job 27v
5. Rain in night, Showers in day | 3 Qr Lime fm Calne, Horses in Stable, Point'g Cottage | COW=BULL'D
6. Dry morn'g, Wheat carried 4 Loads bottom of Tan hill Piece, very heavy rain after RACES
7. BARLEY, began mowing the 18 Acres | REAPING finish'd Down Piece | Fine Day, Carting Dung, Lamb - blown
8. Rain morn'g, fine day, Cart'g Muckle
9. Rain heavy all morng, dry after Cart'g Muckle
10. Storms night & day, heavy wth hail afternoon Cart'g Muckle | EDMONDS House, finish, BUSH'S accident
11. Sunday, fine dry'g day, very little rain
12. Wheat, carried the remainer of Tan hill Piece 8 Loads, see the 6th Inst 4 Loads damp, by Well | Very fine drying day, Plough'g Whitelands for Vetches
13. 800 Bricks from Totterdown, Plough'g Whitelands for Vetches | BARLEY began carrying, 6 loads mid:down Piece Very drying Winds, Sale Ewe Kill'd ill
14. Barley carryed 18 loads down Piece & 4 load left Dry morn'g, Storm mid:day, very dry'g after
15. BARLEY, carrd 4 loads Down Piece & 15 load from long Piece - Barley finish'd mow'g, fine dry'g day
16. Barley, 4 loads in Barn & roofing 2nd Rick | Fine morn'g RAIN after 3 O Clock till evening
17. RAIN at times all day, - Poor Ewe died - Ploughing Whitelands for Vetches | CLOVER, began mowing for hay, too old for Sheep
18. Sunday, dry day, storms at distance
19. Barley carrd morn'g, Dry day, damp eve'g Hurdles removed & draging Turnips | COW=HOUSE, finish'd & began Shed over Inn Well
20. Ploughing Whitelands & drag'g Swedes | Rain morn'g dry after (Philpot sow'g Vetches on Down
21. RAM'S, turn'd 2 Sour two tooth of Mr Geoe Brown to 230 Stock Ewes: NB, Flock too Kind. | Sheep discontinued Clover, too old, about 4A

---

298 Catharine Brown of Avebury buried on 2 September 1836 at Avebury, aged 80, *Wiltshire, England, Church of England Deaths and Burials, 1813-1916 for Catharine Brown* [accessed via Ancestry, 27 July 2020].

mowed Very sharp Frost and drying day | BARLEY, FINISH'D Carrying, put into Barn | Well shed finish'd at the Inn [Pencil note in margin] Turnips began feed'g about this time

22. WHEAT, Carried Down Piece 8A 3R 20P, 515 Tythings 13 Loads on Staddle by Wall Door | HARVEST=FINISHD, Fine day rain evening Two tooth giddy Sheep Kill'd | MASONS=FINISH'D, by Walling up the Door Ways in New Cottage at Inn
23. RAIN heavy in night & driving rain all day Hurdles removed to Silbury Mead, and removed the refuse Chalk &c from Meadow after Build'g
24. Drying day, damp air at times | Dung Carted the heap from Down to Wheat land
25. Sunday fine day
26. Silbury=mead, Sale Ewes, & Lambs began feeding Very fine warm drying day | Dung, Cart'g from heap in Field to Wheat land
27. Thin Rain at intervals & heavy in evening
 27th Septr Died Susannah Pinniger at Calne 21st Octr 1816 born at Cowitch [This entry written within a three-sided thick black border]
27. Barley Ricks Thatch'd, much not carryed | Ploughing began one earth, Clover lay Dung'd & Fold'd for Wheat | Draging Young Swedes, hoe'd after reaping
28. Ploughing as yesterday, - Warm Showers
29. Ploughing very mild dry morn, rain all even'g
30. Cart'g Dung in Field, on Wheat land | Rain in night, very cold morn'g, rain after
 Septr 17th Mr Thos Calley of Burdrupt Park Aged 56
 Septr 22nd Mr Benjn Glass of Worton Aged 56
 TP_ Aged 56, mercifully spared a little time [299] [This entry and the previous two entries written between a double line above and a double line below, under the heading "Died"]

October

1. Steady rain all the morn'g dry after, & rain steady in evening | Horses in Stable
2. Sunday, Dry day (wth Partial Showers
3. Rain heavy in night, Dry day (wth Partial Showers | Ploughing for Wheat
4. Ploughing Very sharp Frost, fine day
5. Cart'g Dung fin Inn sharp Frost, fine day | COBWEBB'S cover the land in even'g sign of Rain [Pencil note in margin] 5th The 1st Pig kill'd Porker 6 Sco 5 li
6. Fog & dry morn'g RAIN all afternoon | 25 Sacks Wheat to Devizes and 1T 11C 2Q Coal retd | 2 Horses Plough'g for Wheat
7. 5 Horses 2 Ploughs for Wheat Rain night, Showers Day | Robt Philpot &

299 TP – this is Pinniger referring to himself.

Jno Wentworth Wheat Sowing began

8. Plough'g as yesterday, Rain night & morn'g, dry mild day
9. Sunday. Dry day
10. Rain very heavy till 8 o-Clock, dry till eve'g, when rain very heavy | 16½ Qr Old Barley to Marlbro
11. ALARMING, boisterous WINDS in night & morning to uncover Ricks & Building & break limbs from Trees, Dry day mild eve | Carting Dung for Wheat | POTATOES began diging in Field for Pigs
12. Dry morning, rain heavy evening | 17 Qr Old Barley to Marlbro 4 Horses
13. Rain heavy morn'g to Keep horses in Stable heavy Storms afternoon | Horses Plough'g afternoon
14. VETCH'S, Sowd Whitelands wth 11 Bushls old @ 4s 6d no dry day before (4A 0R 38P) | Very fine drying day
15. RAIN in night, damp morn'g - drying after and damp eve'g, the second CLOVER see began mowing A=MONTH 17th Sepr and no chance of carrying it since nearly spoil'd | Muckle, Cart'g before Fold for Wheat
16. Sunday - Dry day
17. Glass - very high, damp fog'y morn'g not dry'g the HAY, 2nd Clover carried in my absence very damp, unfit to put together, but the land required to be fed & dung for Wheat | Plough'g Nether side of the Wheat field
18. PRESSING, began the one earth land for Wheat, Fogg'y mild day | Triming Sheep for fair, one Man only, apply Month before wanted
19. WHEAT=SOWING, began the 20A: Piece next next the Little Down | Beautifull fine mild day | 1 Man finish'd triming 156 Sheep & 15 Lambs.
20. Sharp Frost morn'g, fine day | Harrow'g the Wheat sow'd yesterday | SEE, the hay carried the 17th remain on the Waggons, heat to endanger to fire unload'd this morn'g on little Down to dry
21. The HAY, carried dry, being a fine dry'g day | Wheat sow'g, fine time for it, good work [Pinniger then recorded prices at Weyhill Fair - see Appendix H]
22. Unusual clear mild day, like a second Summer | Ploughing for Wheat
23. Sunday, - Very fine mild calm day
24. Thick Fog - fine mild calm day | Ploughing for Wheat
25. Carting Dung before Fold for Wheat | <Very> fog'y mild day
26. Ploughing for Wheat | Fine dry day
27. Rain in morn'g - dry boisterous Winds after | Carting Dung before Fold for Wheat
20. Sheep sold at Devizes Fair 100 Ewes @ 20s - 8 Coops 30 Ewes @ 15s - 2 Coops 24 Wethers @ 19s - 2 Coops 15 Lambs @ 9s - 1 Coop [Total] 169 } all proper
28. Cold Storms of Rain Snow & hail | Ploughing for Wheat & Pressing
29. Wheat sowing, Nether side of the Piece | Exceeding Cold SNOW Storm at day light & bitter cold winds all day. Deep Snow in London yesterday

30. Sunday - Very fine day
31. Exceeding severe FROST, to spoil the Potatoes & very cold dry day

November

1. Dung, Carting on Wheat land Frost morn'g & rain till mid:day
2. FINISH'D Dung Carting the 20 Acre Piece, very mild day. POTATOES, finish'd Diging 53 Sacks Red Kidneys 46 Red Kidneys 22 Prolifficks 11 Black 9 White 7 Red 3 Ash leaf Kidneys [Total] 151 Sacks fill'd only
3. Ploughing for Wheat, Cold dry day
4. Ploughing for Wheat Thin rain at times all day | The 100 LAMB'S DIPPED (at Kenntt
5. Ploughing as yesterday Rain in night, hail storms in day
6. Sunday, heavy hail storm afternoon Wheat the first sowing appearing
7. Carting Dung after Potatoes for Wheat | FOLD removed from 20 Acre piece to Down Piece after Potatoes dung for Wheat | Frost sharp in morn'g, Dry day
8. Frost sharp in morn'g, Dry day | Ploughing for Wheat
9. Ploughing for Wheat FINISH'D the 20 Acre Piece | Damp windy day rain in night
10. Rain heavy in night, Thunder & heavy Storm after noon | Plough'g Down Piece Turnips fed off, too wet to sow Wheat Plough'd yesterday
9. CALF 50 li pr Qr @ 6½d li 18 Weeks old Mr Clark Devs [300]
10. Very heavy THUNDER, lightening & HAIL Storm at Little=cote Kill a COW. of Mr Jaspr Maskelyne
11. Rain all night & thin Rain all day Horses in Stable & geting Hurdles from Silby=mead[301]
12. Ploughing Down Piece as the 10th - Fog'y dull day | GRASS, Silbury mead, the Horses FINISH'D | HAY, the Horses BEGAN, the second Clover with 2 Bushls grittled Beans the 5 Horses the same Corn with the Grass ALARMING prospect through the Winter, in providind food for Cattle Sheep, all now want Hay, the Ewes on the Down grassless, and in Silbury mead afternoon, the Lambs on young Clover & Turnips, fortunate in selling all the Sale Sheep at Devizes even at a low price, APPLESHAW and later Fairs being still lower for want of Keep [Note in margin] See Nov 8th 1834 Cheering prospect [302]
13. Sunday, rain heavy all night & rain nealy all day
14. The land too wet to sow Wheat, Carting Muckle from Inn to home yard for Ewes to stand on at Straw and removed all Hurdles from Silby mead | Dry night and day COMMISSIONER, of Beckhampton Turnpike Trust, Sworn and attend'd the first meeting, the TOLL'S let from [...] Evans this

300 Devs - Devizes.
301 Silby=mead - Silbury mead.
302 grittled – probably griddled.

year at 1395£ to James Coller of Bath at 1400£ to commence 1st January 1837 at 12-0-Clock

15. WHEAT, finish'd sow'g the 20 Acre Piece, with 12 Sacks of Mr W. Stiles's Seed 12 S 8 li Nt @ 27s pr Sack and 2 Sacks of Mr Geoe Browns Seed 12 S 8 li Nt @ [...] pr Sack seed Price - Frost and fine day | FAT PIG 8-14li FAT PORKER 6-5 Kill 5th Octr - Fold finish'd Potatoe Land Down Piece for Wheat and removed to Wheat Stubbles Tan hill Piece for Swedes
16. Harrowing off the Wheat sow'd yesterday some thin showers | Mr Kemm began HAY to Lambs & young Ewes and Swedes Thonhill a week since
17. Plough'd the Potatoe Land Down Piece for Wheat Very fine clear Day, rain even'g (LITTLECOTE) 1 Lamb, dead in Fold
18. WHEAT, FINISH'D sow'g, for present, sow'd 3 Sacks Mr Browns Seed on about 4 Acres of Down Potatoe & Turnip Land, about 4½ A of Turnips to feed for Wheat Nether Down Piece | Fine day, heavy hail storm evening
19. BARLEY, began Thrash'g with Machine for Ewes Frost morn'g, thin rain & wind'y Day
20. EWES, began BARLEY=STRAW in Yard mornings in Silbury mead afternoon, no Grass on the Farm, the Ewes do ill, too early to begin hay disheartening prospect for Sheep Keep. | Sunday, <Exceeding Sha> Fine mild day
21. Exceeding Sharp Frost, thin rain evening | Barley Thrash'g with Machine
22. Rain in Night, dry day | FALLOWING, began Wheat Stubbles top of Tan hill piece
23. FALLOWING, Very cold boisterous Winds
24. FALLOWING, Frost & cold dry day | Coal 13¼ €, Men & Boys cleaning Carrs: of Mead [303] | RAPID ADVANCE in Wheat of 10s pr Qr about the 10th Inst, and since an equal fall in price, yet this day the Market firm, sold my Wheat 12 S 10 li Nt @ 30s SEE, sold no Wheat at this price since July 1832 | Barley also advanced to 44s Qr
25. Very sharp FROST & very fine day | SNOW, fell in evening over Shoes | Ploughing as yesterday | Old WHEAT=RICK of 1833 put in Barn | 5 New wheat ricks in the Yard
26. RAIN, thin all night & nearly all day The Horses at Plough about an hour
27. Sunday, - Dry morn'g, rain evening
28. Very boisterous rain'y night, and boisterous day, the Ewes do very ill with Straw morn'g & Down after, dread giving them hay expecting shall buy hay in the spring very dear | Plough'g stubbles St hills Piece
29. Carting Muckle from Inn, to Yard to bed the Ewes [Pinniger then recorded details of an accident which occurred on 26 November - see Appendix V]
29. Rain & boisterous winds all the morning till near 12 oClock when followed the desolating sudden Gust, succeed'd by a calm dry evening | Great FLOOD in evening, all Silbury mead covered with Water except my level'd

303 Carrs: of Mead - Carriages of water meadows.

bank

30. Thin rain all morn'g | Wheat Thrash'g with Machine

December

1. Wheat Thrash'g with Machine Thin rain morn'g fine after
2. Wheat Thrash'g with Machine Damp windy day
3. Barley Thrash'g for Sheep in morn'g with Machine Boisterous storms after. Horses in Stable
4. Sunday Boisterous Winds & thin Rain
5. Fallowing Boisterous Winds & thin Rain
6. Mild pleasant morn'g, boisterous again in even'g | Carting Rubble for Mr Treen & Thrash'g Barley with Machine for Sheep at Straw
7. Fallow'g and Boisterous Stormy day | HAY. The EWES, began, doing very ill with the Barley Straw not good, (grass'y or sweet) and the Down as naked as a floor, their Belly's tucked up, a bad prospect of breeding Lambs, and equal bad prospect of the Hay Keeping them til May and the Swedes, unfortunately, partially failing allowing a short quantity 4 Truss's evenings only HAY, The Lambs began 25th Novr with half a Truss evenings for 100 with Turnips and walk on new field in Down Wheat Stubbles for 2 or 3 hours and do well, by selling the Sheep at Devizes, have Turnips left to Keep the Lambs till Lady day. One Land have Kept them since 20th Octr about ¾ Acre
8. WHEAT last sowing 18th Novr appearing | Turnips Norfolk Whites sow'd 8th & 25th June finish'd this day Kept good, all have had for Sale Sheep & Stock Lambs (270) 3A 3R 0P Nether Down Piece | [Note in margin] began feeding Turnips 21st Septr | Heavy Storms of Hail and rain | Fallowing
9. Fallowing and Thrash'g Barley for sheep | Storms of Hail snow and rain | Turnips, green rounds began, adjoining Norfolk W [304]
10. Very sharp Frost & very cold storm afternoon | Fallow'g & Cart'g Trenching Silbury Mead
11. Sunday - Frost morn'g cold dry day
12. Rain heavy in night & cold driving rain all day, Horses in Stable
13. Mild'r weather with partial Storms, Fallow'g
14. Frost & dry day, the 1st dry day for some time Fallow'g
15. Frost & dry day, Fallow'g
16. Frost & dry day Thrash'g Barley for Sheep with horses
17. Dry day, thin rain evening | 20 S Wheat to Devizes 28 Bu Grains [...] C Salt and Ton 1-17-1 Coal returned
18. Sunday, Very mild day, damp air. The Revd Mr Shorts, farewell SERMON at Tyth[erto]n 3rd Ch: Hebrews 5 & 6 Verses [305]
19. Dry mild day, Fallow'g Tan hill Piece

---

304 Norfolk W - Norfolk Whites.

305 Revd Mr Short – William Short, vicar of Chippenham.

20. Very thick Fog in the morn'g, Sun shining through exceeding mild pleasant morn'g, very like a fine Spring morn'g, & fine mild day, Fallow'g as yest=day
21. Fine mild day, Fallow'g as yest=day
22. Fine mild day, Fallow'g as yest=day
23. SNOW, several very cold Storm FROST sharp in even'g | Thrash'g Barley wth Horses for the Sheep
24. Horses Fallow'g continue Tan hill Piece | FROST very sharp & Cold, some little Snow | The Ewes do very ill with 3 Truss of Hay at night for 230. Straw morn'gs and walk on Down allow them this eve'g 4 Truss being severe Frost and the 100 Lambs 1 Truss instead of half a Truss, night only and Turnips short allowance
25. Xtmas Day, exceeding severe Frost to stop the Pumps with cuting winds & some drifting SNOW. the Ewes in the Yard all day, Straw twice & hay in Fold
26. A most inclement night & day Not a Passenger through the Turnpike Gate from 1-0 Clock in Night to 3 o-Clock this afternoon, when a Horse paid the first Penny | SNOW with Tempestuous Winds freezing, driving the Snow to fill all the low roads, have stopd all Carriages, part of the Coaches from London reached Overton and the morn'g Coaches to London came to Calne, or Cherhill, none could come to Beckhampton, nothing moving here all day | EWE'S brot from Tan hill Piece Fold to the Yard one bury'd in the Snow, & dug out LAMB'S brot from Turnips Down Piece to Rick Yd one left behind bury'd in the Snow.
27. SNOW, not so heavy but very driving all night and day, Men unable to bear the Weather and diging Roads through the Snow, A Horse only occasionally passing The Lamb dug out of the Snow bury'd since Xtmas Night | Tuesday 27th Decr. The first Carriage since Sunday night (Xtmas day) pass'g our road was the DUKE OF WELLINGTON'S, on his way to Badmington, who could not pass the Turnpike Gate, the Road being x of SNOW (x full) turn'd into the Fields by Silbury hill and through Mr Wentworths Rick Yard and the back of our Stable Yard at 4 oClock this afternoon. The Road opened soon after and the Coaches pass'd [306]
28. Coaches & Mails pass'g all hours of the day out of course, cold freezing winds, some little Snow | Horses in Stable
29. Horses in Stable Cold freezing Winds
30. Horses in Stable Calm freezing day, some little Snow in the morn'g, Horses in Stable | BARLEY, the first put part of Rick in Barn
31. Sharp Frost continue, some little Snow | 20 Sacks Wheat to Devizes [...]

---

306 Henry Somerset, seventh duke of Beaufort of Badminton House, was aide de camp to the duke of Wellington during the Peninsular War from 1812 to 1814: he was also the husband of one of Wellington's nieces, Georgiana Frederica Fitzroy, Oxford Dictionary of National Biography, *Somerset, Lord Granville Charles Henry* https://doi.org/10.1093/ref:odnb/26008 [accessed 27 July 2020].

Bushls Grains retd | SHEEP do better, by being whole'y in the Yard with Straw twice a day & 5 Truss's of hay evenings for 230, drink much water | Lambs do well in Rick Yard with 1 Truss hay ea; morn'g & eve and Water | HAY consumed, to the end of the Year for Sheep Cow & Nags about 4½ Tons HAY consumed, by the 5 Cart Horses about 2½ Tons | The Horses baited with 14 Bushls GRAINS pr Week & Chaff require much less hay 14 Bushls Grains 7s 2 Bushls Grittled Beans 9s } Cost pr Week for 5 Horses | 1 Ram Stag 96 li New Xtmas, & 1 Ram Stag 66 li Old Xtmas | HAY Consumed between 9th Octr 1835 and 13th May 1836 Paid for Truss'g Decr 11th 16 Tons Paid for Truss'g Feby 16th - 22 Tons Paid for Truss'g May 5th - 23½ Tons [Total] 61½ | Hay between 1st Octr 1834 and 28th April 1835 53 Tons Not spent 10 Tons | Hay between 17th Novr 1833 and 22nd April 1834 38¾ Tons | Hay between 4 Decr 1832 and 1st May 1833 44 Tons (Lambs wintered at Catcomb) | Hay between 16th Novr 1831 and 7th May 1832 28 Tons

1837

January

1. Sunday, Sharp freezing, some little Snow
2. Frost morn'g, mild & Thaw after | Horses Thrash'g Barley for Sheep at Straw | Tegs, removed from Rick Yard, to Meadow fold'd being mild and dry
3. Frost morn'g - mild Thraw after | Thrash'g as yesterday, Land too hard to Plough
4. Thrash'g as yesterday Frost, mild & Thaw after
5. Horses in Stable, - Cold dry day
6. Horses in Stable, Rain in night Cold dry day [Pencil note in margin] Philpot began Swedes
7. PLOUGH'G, stop'd since 24 Decr do well now | Frost little morn'g, cold drying day | The land being dry'r, the Tegs return'd to Turnips and the Ewes to Fold Tan hill Piece
8. Sunday, Frost and fine day
9. Rain in night, and dry day } Plough'g Tan
10. Thin rain nearly all day } Plough'g Tan
11. Very sharp Frost & fine day Plough'g Tan
12. Very sharp Frost till 10 oClock when SNOW'D till 3 oClock to compleatly cover the ground | The Horses Thrashing
13. Rain heavy in night & morn'g, Horses Ploughg afternoon
14. Sharp Frost & freezing all day in Shade Ploughing | Barley put in Barn, the 2nd part of Rick
15. Sunday, Sharp frost & fine day
16. Frost in Night SNOW to cover the Ground in the morn'g, damp mizly rain all day | Horses Thrashing

17. Ploughing and damp day
18. Ploughing and damp day
19. Carting Muckle from Inn | Dull damp day
20. SNOW, cover the ground in the morn'g thaw after Plough'g after the morn'g, the Plough'g all done up to the Fold for Swedes | [Pencil note] The Infuenza - universal - see Papers | NB My Labourers ill, none to work the Thrash'g Machine no Straw for the Ewes, Keep them with Hay.
    Obit John Harris, Calne Aged 52, Jany 17th [This entry written within a thick black border]
21. ASHE'S, the first load 60 Bu: 4 Horses, no Ploughg to do | Cold damp day
22. Sunday, Rain in night, damp day
23. Rain in night, driving rain nearly all day | Horses Thrashing Barley
    Obit John Goddard last night aged 40 Obit Willm Edmonds, after noon aged 79} Beckhampton NB Father, and Son=in=law [This entry written within a thick black border]
24. Dry day, ASHE'S, the 2nd Load 60 Bushls
    Obit Mr Steph Stiles late of Barbery Farm Aged 60 (14th Feby next) [This entry written within a thick black border]
25. Rain in night, damp day The Horses Thrashing Barley
26. RAIN heavy nearly all the Night & Day 4 Horses to Devizes on purpose for Coals 50£ for the Poor & Cart for 32 Bushls Grains | The Lambs brot Turnips yesterday, being very dirty, Pen'd on dry mead with hay The Ewes exceeding Poor, the Winter unfavorable continue at Straw twice a day & hay once
    25th Obit Mr James Young Hayward of Tufton aged 32 Yrs, of the Influenza, now prevailing through the Kingdom, 3 Men died in this Parish the last 3 days, my Men ill have not Men to do my work (in great difficultys) [This entry written between two thick black lines]
27. Cold windy day, Ploughing a Fold drift
28. Cold windy day, Ploughing a Fold drift | The Lambs returned to the Turnips Down Piece | The 2nd Bacon PIG 12 S 16 li
29. SNOW, in morn'g to cover the ground, Sunday
30. SNOW, deeper in morn'g, thaw after, the Tegs brot from Turnips to home mead | The 3rd load of Ashes 60 Bushls
31. 20 S_ Wheat to Devizes & 40 Bu: Grains returned | Rain heavy in night, damp day

February

1. Fog'y mild spring like day & dry | The 4th load of Ashes 60 Bushls
2. The 5th load of Ashes 60 Bushls | Mild dry day
    January 26th Died Mr Willm Rugg of Chippenham Aged [...] [This entry written between two thick black lines]
3. Rain heavy in night, damp day. | The 6th load of Ashes 60 Bushls

4. The 7th load of Ashes 60 Bushls | Frost morn'g Cold dry day Mr Kemm finish'd Thornhill Swedes by young Ewes & Tegs, the Ewes gone to hay = twice a day | My Ewes HAY, three times a day 16 Truss's, no Barley in Barn, no Weather to put Barley in Barn, and the Men ILL, influenza, the last 3 Weeks
5. Sunday, very cold drying day
6. BARLEY, put the remainer of 1st Rick into Barn (2 borrowed Men | Very fine drying Wind | The 8th load of of Ashes 60 Bushls | HAY the Horses, finish'd the 2nd Clover, Kept them 12 Weeks, must now have Barley Straw & Kerving, not having hay enough for the Sheep [307]
Died on the 27th Jany last, Eleanor relict of J Sutton Esqr late of New Park Devizes aged 88 Years & Sister to Viscount Sidmouth [This entry written between two thick black lines]
7. Cold dry day, Horses Thrash'g Barley for Sheep to have Straw twice a day & 7 Truss's hay evenings
8. Damp cold winds morn'g, Rain afternoon | The 9th load of Ashes 60 Bu
9. The 10th load of Ashes 60 Bu:
Feby 4th Died of the Influenza, Calne Mr Jas Fry, Aged [...] [This entry written between two thick black lines]
9. Dry day
10. Wind & rain in night & thin driving rain all day heavy rain evening, exceeding boisterous winds during the day, very DISHEARTENING, prospect for the Sheep, almost starved more by the weather than Keep, yet but little Keep & hay 6£ pr Ton | 12 Qr Barley to Devizes & 2T 7C 2Q Coal retd
11. Rain heavy night, DREADFUL WINDS, blowing a HURRICANE and very heavy rain before day to 12 o-Clock, impossible to serve the Sheep wth Straw from the Barn, had HAY, 3 times in Fold Tegs brot from Turnips, hay 3 times | CALAMITUS, time, my Ewes do so ill, expect many will die, can have no hope they will keep a Lamb, they appear hunger baned | Horses in Stable, FLOOD | CALF [Pencil note] 14 Weeks 9th Novr
12. Sunday, Storms of Snow, hail & rain
13. Standing Pen made for Ewes, being unable to walk from bottom of Tan hill Piece to Straw in Yard, through weakness, their time up the 14th, the 1st Lamb this morn'g - and dead | Cart'g Hurdles for Stand'g Pen | Exceeding boisterous Winds in night & morning & heavy rain morn'g, dry after
14. Great change in the weather, being very calm and mild | Horses Thrashing Barley
15. Dung from the Inn to Tan hill Piece | LAMB, the 1st to live | sharp white Frost fine day
14. Fras Pinniger, to London, on trial to be Apprene [308]

307 Kerving – Cavings – the chaff of wheat and oats; the broken bits of straw arising from threshing. Dartnell & Goddard, *Wiltshire Words*, p.25.

308 Apprene - Apprentice.

16. Fine dry Day | Ploughing the last Fold drift for Swedes
17. Ploughing the last Fold drift and fine mild day | HAY, began the Field Rick, The Ewes continue to have Straw twice a day & hay 7 Truss's evenings only in Stand'g Pen & drink 2 Hogsheads of Water daily and are much improved the last four dry days, regret Keeping them so badly but hay & Swedes being very short supply, are necessitated to make it serve them till have grass. | 4 Lambs only, Brewed 1st since July
18. Horses Thrash'g Barley, very Boisterous Winds and driving storms | STRAW to the INN, the last for the present, being unable for the Lambing season to serve the Coach Horses, the Sheep requiring all the Straw, Mr Wentworth commence serving them Wednesday next the 22nd for 3 or 6 Months [Pencil note in margin] Philpots Hay gone
19. Sunday, driving rain all day, very heavy in evening
20. FLOOD, very great, dry day | Ploughing finish'd Tan hill Piece for Swedes
21. Rain heavy night & morn'g, dry remainer of day. | Dung, Cart'd from Yard, to bottom of Tan hill Piece for Potatoes.
22. BARLEY, put the firt part of last Rick into the Barn to continue supply'g the Ewes twice a day with Straw, dreading the want of Hay. the continued inclement weather & want of sufficientcy of Food, begin to be seen in many Flocks, 45 Ewes having in the last (died fortnight at Whyre Farm (Jos L_e) [309] | Cold dry'g harsh Winds
23. Rain in night and boisterous winds & Rain all day | Thrash'g Barley & Cart'g Straw & hay to Sheep nearly whole time employ'd in providing for Sheep Brewed 12 Bu: Malt Hd Beer
22. Plough'g Turnip land Nether Down Piece for Talaa Wheat [310]
24. Very cold boisterous Winds & hail storm, Frost, morn'g | Horses Thrashing
25. SNOW, to cover the ground, fine dry calm day | Ploughing Down Piece
26. Frost & cold dry day, - Sunday
27. SNOW, to cover the ground, thin rain after - TURNIP'S finish'd Down Piece & hurdes removed to Swedes, too wet for Tegs to begin them fold'd in Meadow all day Thrash'g Barley the Horses [311]
28. Thrash'g Barley the Horses for Sheep Straw | SWEDE'S, the Tegs began, cold dry day 50 Lambs only in fortnight

March

1. Frost exceeding sharp, nearly to stop the Plough dry cold day, Plough'g | 4th PIG 11 Sco 4 li
2. Sharp Frost & cold dry day | 12 Qr Barley to Devizes 30 Bu Grains & 2 € Salt retd
3. Dry morn'g Damp eve'g, Plough'g for Wheat

309 firt – first. Jos L_e - Joseph Large.

310 Talaa - Talavera.

311 hurdes - hurdles.

4. Very cold harsh boisterous winds | Ploughing as yesterday for Talavera Wheat | DISTRESSING time with my two tooth Ewes, they bring a Lamb & have no milk, being reduced to Skin & bone only, consequently nearly all their Lambs die 15 of the Ewes have died and 30 not not worth 5s each turned out with the Tegs, for the cause see 7th Decr when starved & now supposed infected with the INFLUENZA, from the human species
5. Sunday, cold dry day
6. Plough'g for Wheat, cold dry day
7. Fine dry day | TALAVERA WHEAT, sow'd 5 Sacks on 4¾ Acres home Down Piece, Turnips fed
8. ASHES, began sowing, sowd about 500 Bushs 3 Men very sharp white frost | 40 Ewes to lamb only, continue at HAY, only evenings & Straw twice, - the old Ewes look as well before and and after lambing as years pass'd, but the two tooth Ewes are spoiled - Cold dry day
9. ASHES, finish sowing 600 Bushls from Devizes and about 60 Bushls of own on 18 Acres is 37 Bu pr A the long Piece adjoing Down furlongs | Ashes & Soot sow'd on Picked land of Wheat adjoining the Down above 20 Acre Piece | Cold storms
10. Cold boisterous winds and storms | SWEDES began feeding with Ewes & Lambs Dung Carting from Yard to store heap on Down for Turnips after Grass fed
11. Carting Dung as yesterday Very sharp Frost & fine day | SPRINGS, very high in Feby. now sinking still on my Down, nearly to the end
12. Sunday, SNOW, very cold & heavy storms of large blossoms to cover the ground
13. Tremendous hard Frost & exceeding cold wind. | Carting Dung as the 11th
14. 20 Sacks of Wheat of 1833 the last to Devizes & 30 Bu: Grains retd | Frost & very cold boisterous Winds
15. Frost & very cold boisterous Winds | BARLEY, remaing'g part of last Rick put into Barn | Mr Wentworth Drill'd his dwarf, marrow fat Pease
16. Damp cold winds all day | Carting Dung to heap on Down
17. Horses Thrashing Barley, for Sheep Straw | Very cold harsh winds
18. Very cold harsh winds Cart'g Dung finish'd
19. Sunday Very cold harsh winds
20. Very cold harsh winds very severe Frost to prevent harrow'g for Pease till noon, Pierce'g cold storms of hail & Snow
21. FROST, so very severe, with difficulty Plough plough for Swedes athwart (began) continued freezing till mid=day & steady heavy SNOW remainer of the day, the most winterly weather since Xtmas
22. SNOW this morn'g drifter above the Hurdles at Swedes, disheartening time for the Sheep, never done so bad, the two tooth Ewes can hardly live, many have died, and nearly all their Lambs, hay and Swedes not enough ⋆ | Freezing sharp all the morning & steady SNOW all day, to prevent the Team

going to Chippenham | Horses Thrash'g Barley

23. SNOW, nearly all night, and in morning almost ALARMING, being so heavy & driving on account of the Sheep, the Tegs removed from Swedes to Standing Pen, continued SNOWING without intermission all day, expect to lose some Lambs to night, the Ewes having not milk enough, the Snow deep and all the Coach's, have 6 Horses *
24. GOOD FRIDAY, FROST sharp in extreme the Sheep deprived of Water the Tap of Barrell being frozen, clear dry day freezing all day in shade, & in evening intense | The Snow continue deep over the whole face of the land, the Swedes & Turnips with the greens quite cover'd, all the Sheep in standing Pen, hay 3 times day
25. 15 Qr Barley to Chippm very sharp Frost & clear day
26. Sunday, Dry clear morn'g, frequent storms of hail & SNOW afternoon
27. Snow continue to cover the ground, Horses in Stable Very severe Frost, Warm water in Tea Kettle, taken to Standing Pen, to thaw the tap of Water Barrel to supply the Sheep with Water, Dry day.
28. SNOW piercing cold storms in morning wth sharp frost dry after - damp even'g, Horses Thrash'g | 12 Ewes remain to lamb or are barren, now 6 weeks from the commencement, came in very well but scarcely a two tooth Ewe have Kept a Lamb, and many of the Ewes have died, see observation the 11th Feby too true. (86 two tooths)
29. BARLEY, finish'd Thrash'g (all with Machine) | Frost slight, & thawing all day, damp eve | SWEDES the Sheep returned to, Kept from them 6 dy's
30. Clear Sunshine day, cold harsh drying | Horses in Stable, Snow yet on Fallows, too wet to Plough [Note in margin] 30th Mary to school
31. Plough'g athwart for Swedes, cold harsh dry day

April

1. 20 Sacks Potatoes to Devizes & Coal returned 1T 16C 2Qr | Frost & very Keen & cold harsh winds | RABBIT'S, distroy'g in the Covert adjoining me this week
2. Sunday, sharp Frost, cold dry day
3. Rain in night, very cold boisterous winds & continual storms of hail & Snow | 2 Ploughs the first time (5 horses) athwart for Swedes
4. 2 Ploughs athwart for Swedes | Pierceing cold dry'g winds, Frost morn'g
5. Extreme cold dry'g winds, Frost morn'g | POTATOE'S first Planted in Garden | 16 Qr Barley to Mr Belcher Chippenham Rt Philpots Swedes, finish'd [312] | 5th Fat PIG and the last 15 Sco 13 li 12,,16 11,,4 8,,14 6,,5 [Total] 54,,12 NB: Cost 20s each the 23rd June & put to fatten
6. Thwarting for Swedes | Sharp Frost cold dry day [Pencil note in margin] Mr Brown & Philpots Swedes gone

---

312 Mr Belcher – possibly Edward Belcher, baker & maltster, Timber Street, Chippenham, J. Pigot, 'Commercial Directory 1842' in WRS47, pp. 108-109.

The weather to this day extremely cold with all appearance of a continuance, more than half the Farmers have spent all their Hay, and buying at 6£ per Ton, many their Swedes spent and Sheep wintered from home returning, all combined make it a distressing time for many Farmers My small Meadow hay Rick left about [in pencil: 3] Tons and half my Field Rick about [in pencil: 8] Tons & the Stack of old Clover for Horses about [in pencil: 2] Tons and enough Swedes and small Turnip greens to Keep the Sheep all April on short allowance by sending the Tegs to Littlecote Watermead, the Meadows & Field Grass much worse than a month since & the small Swedes not the least Green on them, not the least move in vegitation [Diary turned landscape for this entry]

7. 12 Qr Barley to Devizes & 7 Sacks of Marrow fat Pease Returned Plough'g as yesterday | Cold cuting Winds, frost in morning, in the evening | SNOW storm blossoms nearly as broad as Crown Pieces with hail to cover the hills white
8. SNOW, frequent light storms wth pierceing cold winds all day & very hard frost Ploughing as yesterday | PEASE, began drilling top of Tan hill Piece | PEASE, planted in Garden
9. Sunday, Sharp Frost, and frequent flying SNOW Storms, exceeding cold
10. SNOW, intense cold storms, quick in succession Pease drilling & Plough'g across for Swedes
11. Pease drilling finished Plough'g across for Swedes with the single Drill put in 7 Sacks in [...] in 3 Days 2 Sacks 28s of long & 5 Sacks of Dwarf Marrowfats @ 36s S [313] | SNOW, frequent flying storms & piercing cold
12. Plough'g as yesterday, sharp frost, and | SNOW, Storms frequent & very cold
13. Plough'g as yesterday, very cold winds | SNOW, slight flying particles
14. Frost, clear warm Sun, the most comfortable day this year. Plough'g for Swedes across | Seeds, Onions &c sown in Garden
15. Frost & mild morn'g, cold damp eve, Plough'g as yesty
16. SNOW storm & Keen winds - Sunday
17. Very sharp frost, Tap of Water Barrel Frozen & Keen blow'g winds, finish'd cross Plough'g for Swedes | Plough'g began after Swedes for Barley
18. Plough'g Frost & cold morn'g warm after, mild eve | Barren Ewe (Fat) 13 li Qr Mr Clarke Devizes[314]
19. Frost & the first warm mild, the Sheep discontinue in Standing Pen, Plough'g as yesterday
20. SNOW, see, fall'n the 23rd March, remain on the Hills, and very cold since | Dew & mild morn'g, cold storm after | BARLEY began sow'g Right of Tan hill road 6 Sa

---

313 [...] – in this case, an undecipherable pencil note.

314 Mr Clarke Devizes – possibly Jacob Clark, butcher, Maryport Street & The Brittox, Devizes, J. Pigot, 'Commercial Directory 1842' in WRS47, p.114.

21. Cold storms of hail & rain to stop Barley work, harrowing for Swedes
22. 2m of Bricks from Totterdown for Garden wall | Cold dry day 10 Sacks Potatoes stol'<er>n
23. Sunday, very cold winds, some partial storms of hail & rain | Young Swedes above the Road finish'd feeding
24. Young Turnips 6½ Acres below the Road began feeding | Fine mild dry day, heavy storm eve'g | SPRINGS, sink, Water discontinue runing in the Carriage in Silbury mead | Harrow'g Seeds in Barley sow'd the 20th 6 Sacks | HAY the Meadow Rick the Sheep began
25. Fine morn'g heavy Rain in evening and night to bring up the Corn sown | Plough'g above the Road for Barley
26. Plough'g above the Road Rain heavy in night, fine day
27. Barley began sow'g above the Road, Rain stopd sow'g at 10-oClock, Cart'g Flints for Garden Wall
28. Barley took to Field again, when commenced raining & continued till even'g, harrow'g for Swedes | ROUND=TAILING the Sheep 310, by Henry Chivers, James & Jacob Davis, Boy & Woman assg [315]
    Died 19th April at his Sons [...] Stedhampton Oxon Mr Simion Viveash aged 69 Years, following to the Grave since the 17th July last, <his> his much respect'd two Brothers Samuel & Orial [316] [This entry written within a three-sided thick black border]
29. Dry night, RAIN all the morning to Keep the Sheep from Turnips, gave them HAY three times | Horses in the Stable
30. Sunday, Cold heavy driving Storms

May

1. Cold winds & Storms | Rolling began the Field Grass, & Plough'g Turnips fed | WALL, GARDEN, began building (the Mud Wall Blown down) | The Pease & Barley appearing & the Talavera Wheat appearing green | 13th March Mr Wentworth drill'd PEASE, dwarfs 8th April Thos Pinniger drill'd PEASE, dwarfs 2nd May Mr Pinniger Kent drill'd PEASE, dwarfs ★ [317]
2. Roll'g & Ploughing as yesterday | Mild May morn'g & fine day | 92 Tegs 19 Ewes } Sent to Watermead LITTLECOTE with 100 Tegs of Jno Crooks Corton
3. LAMB'S 5 died, on young Turnips & hay | Barley sow'd above road, fine mild day
4. Turnips finish'd feed'g, no blossom, one Acre of Swedes left to feed, shew'g

---

315 assg - assisting.

316 Simeon Viveash of Calne buried on 27 April 1837 at Calne, aged 69, *Wiltshire, England, Church of England Deaths and Burials, 1813-1916 for Simeon Viveash* [accessed via Ancestry, 28 July 2020]. WSA: 1171/138/31, *Will of Simeon Viveash, gent., Calne, 1837.*

317 Kent - Kennet.

for blossom Fine day rain heavy even'g, Plough'g below road | LAMB'S 7 died in 6 days

5. Barley sow'g below road, fine mild day
6. LAMB'S 3 died in morn'g 3 others <die>Kil'd would have died 2 others Kill'd would have died (afternoon) on Swedes & Hay, | Frost morn'g fine day Plough'g for Barley | COW, to grass
7. Sunday, fine day | LAMB'S 2 Kill'd, the Butcher cut up 8 - gave a Qr to as many Family's
This day May 6th at [in pencil: 2] oClock afternoon departed this life my Niece Sarah Pinniger at Kennett born 15th Septr 1807 [This entry written within a three-sided thick black border]
6th April Mr G Brown began feed'g Watermead
8th April Mr Chas Rumboll began feed'g Watermead
24 April Mr Jno Wentworth began feed'g Watermead
1. Farmer Philpot began feed'g Watermead
8. Mr Willm Pinniger began feed'g Watermead | T_ Pinniger began feed'g Watermead | CUCKOO, first heard here | 110 Ewes & Lambs began feed'g Mead | Fine day rain all the even'g, Plough'g as 6th | LAMB'S 2 died
9. LAMB'S 1 died Barley sow g bottom of Piece | Very cold day hail storms & exceed'g cold night
10. SNOW & hail frequent storms, FROST & exceeding cold harsh boisterous winds | 20 S_ Wheat to Devizes & 1m Bricks retd Plough'g for Pota [318] | LAMB 1 died, The frequent deaths, now become distressing, not Knowing a preventative, have died on Turnips & hay on Swedes & hay and now in Silbury Mead & return'd them to Fold on Fallows with HAY *
11. The Ewes & Lambs return'd to the old Swedes, the Dry Sheep first eat the green off the Swedes 20 of the best Lambs died or Kill'd the last 10 days weigh near 5 li pr Qr & sold the Skins @ 8d ea * | Frost white cut'g, dry & mild'r day | Plough'g, Rolling & harrow'g for Potatoes
12. Rain all the morning, Horses in Stable, dry after Rolling Home Meadow
13. LAMB 1, Kill'd LAMB 3, Kill'd & 2 died | TAR gave a Pap spoon full to eack the 1 <...> Lambs about 4li in ¾ hour John & Jas Davis the Boy & myself, made them exceeding SICK & ill much alarm'd though half of them would die
14. WitSunday, exceeding heavy HAIL Storm LAMB 1 died
15. LAMB 1 died, - Fine mild day | SWEDE'S quite finish'd feeding by dry Sheep | HAY, discontinued, the Ewes & Lambs, returned to Silbury mead again, no change prevent their dieing, altho not so frequent as before applied the Tar <...>, Rolling & harrow'g for Swedes
16. Tegs & dry Ewes put on the Down & have a Fold of Grass further Down Piece follow'd with Pot dung & Fold for my first Turnips | TEGS returned from Littlecote, being there 12 clear days 211 fed 3½ Acres at 3 guineas

318 Pota - Potatoes.

pr Acre is 1s & a half penny eack for the 12½ days | POTATOE'S, began plant'g bottom of Tan hill Piece Plough'g the last Swede drift for Barley, fine day | HAY, the remainer of the old 2nd Clover truss'd for the Horses 2 Tons | LAMB 1 died, Mr Wentworth finish'd his Mead & put water on it.

17. Barley sowing the last drift, warmest day | STRAW, commenced again serving the Horses for 5 COACH'S, discontinued from 18 Feby last
18. BARLEY, sow'g FINISH'D, storm morn'g cold eve'g
19. Rolling & harrow'g for Swedes | Exceeding cold winds, some rain & heavy hail storm
20. Cold harsh winds, harrow'g Swede ground & Carting dung on Down Piece for Fold & Turnips, grass fed off | Potatoes plant'g by Hay Rick
21. Sunday, harsh dry morn'g, some thin rain after
22. Exceeding cold harsh blow'g winds all day, colder than the debth of WINTER, Vegitation at a stand, the Trees not opened in bud | Ploughing a Fold drift for Swedes | WALL, finish'd in Garden, where Mud wall fell | Lambs doing better at present none die
23. Fine mild day, Rolling Barley | WHEAT=RICK, the 1st of last year put into the Barn. (full of Mice)
24. Fine mild day, Plough'g the rough spots, Swede Grod | LAMB 1 died, Potatoes finish'd plant'g in field
25. Very cold harsh winds | SEED'S sowing in Wheat Nether Down Piece & harrow'g it in, Rolling Barley | PEASE, began hoe'g 3 Men @ 4s pr Acre long Marrow Fats 4A 3R 16P Dwarf 2A 0R 28P with 5 Sacks @ 36s & 2 Sacks @ 28s of seed.
    Aged 74 Years Died 19th May Mr Jno Spencer of Barton Farm Bradford
    Aged 74 Years Married 23rd May Mr Algernon Brown of Broadhinton to Mrs. Smith of Broad Town about 50 Yrs Matthew the 24th Chaptr Verses 37th & 38th But as the days of Noe were, so shall also the coming of the Son of man be. For as in the days that were before the Flood they were eating & drinking, Marrying and giving in marriage until the day that Noe entered into the Ark, and Knew not until the Flood came [319] [This entry and the previous entry written within a three-sided thick black border]
26. LAMB'S - 2 died in morning Seeds sow'g in Talavera Wheat & Rolling it 9 Acres bottom of Tan hill Piece 6 li Clover 4 li Hop 2 li Dutch & 2 Bushls of Hop & Ray 9 Acres above the road 2 Bushls of Hop & Ray & 4 li Hop and 4 li Dutch 9 Acres in Wheat Nether Down Piece 6 li Hop & 2 Bu Ray Fog'y morn'g fine mild day | LAMB'S CUT & TAIL'D 102 - (50 Rams) 230 Ewes put to R [320]
27. Dry mild day | Chalk Cart'g to Garden walk by the new Wall & Carting Dung before the Fold
28. Sunday, <...> fine day, cold evening

---

319 Noe - Noah (Noe is the spelling used in v.38 of the King James' version of the bible).

320 R - Ram.

29. Fine warm day, Ploughing began for for Swedes | OAKTREE'S very backward no leaves, the ASH also only in bud & the ELM only appearing a little green
30. Very boisterous harsh winds, Plough'g, Pressing & harrow'g the drift of Grass fed further Down Piece Dung'd & Fold'd for Turnips Pease - finish'd hoeing.
31. Rain in morn'g, cold harsh winds, rain soon lost | Plough'g for Swedes & 1 horse Cart'g Dung. Stand'g Pen [Pencil note] 20 May Philpot finish'd Mead

June

1. Cold boisterous winds, Plough'g for Swedes | W Davis, my Carter, found 2 Hurdles & 6 Stakes conceal'd in his Bed Chamber
2. Ploughing for Swedes & Cart'g Dung, mild day
3. Ploughing for Swedes & Cart'g Dung, mild day | The Sheep in Silymead including Rams 134[321] | The Sheep on the Down 178 [Total] 312 The Stock see Augst 30th 342, 2 Stags Kill'd & 1 fat Ewe sold 3 [Total] 339 312 27 died since 30th Augst
1. LAMB 1 finest died
4. Sunday, warm mild day
2. GRASS Silbury Mead began cuting for HORSE'S heavy crop on the Bank
5. Thrash'g began 1st Rick & Plough'g for Swedes | SILBURYMEAD, the Ewes & Lambs removed from (having fed since the 8th of May about 3½ Acres) to the Down with the Dry Sheep and a tie of Field grass further Down Piece the Lambs runing forward & fold'd after for Turnips | Fine Warm day
6. THRASHING, began the first Wheat Rick & Plough'g for Swedes Warm dry day
7. THRASHING & Plough'g for Swedes bitter COLD evening
8. THRASHING & Plough'g for Swedes Pierceing COLD morn'g boisterous Winds all day
9. Mild day, some gentle RAIN afternoon Plough'g as yesterday & finish'd Thrash'g 1st Wheat R[322] | WHEAT=RICK, from Whitelands put into Barn
10. Exceeding BOISTEROUS day with some light rain Thrashing Whitelands Rick & Plough'g for Swedes Thrash'd all day for 8¾ Sacks instead of 25 or 30 S being half DESTROYED by SPARROWS in Field & MICE in the Rick.
11. Sunday, Dry morn'g, RAIN heavy storm after noon, the 1st Rain to do any good since now very acceptable
12. SWEDE'S began sowing Tan hill Piece left hand 16 Acres lower part very good work, sow'd only 2 Acr when rain'd all the morn'g to stop sow'g,

---

321 Silymead - Silbury Mead.

322 R - Rick.

Plough'g for Swedes [323]

13. Dry night & morning Swedes sow'd about 6 Acres when rain'd heavy stop'd sow'g, Horses to Thrash'g | LAMB 1 died, now only 100
14. Rain in night, dry day, the most mild growing weather yet this Summer | TURNIPS began sow'g, fur: Down Piece, the land one earth, til now too dry | Mr Wentworth, harrow'g his Turnips | Thrash'g finish'd Whitelands Rick
15. SWEDE'S finish'd sow'g, except headlands Very warm growing day
16. WHEAT=RICK, the 3rd put into Barn, (by large Doors Sultry warm day, close RAIN even'g | Rolling & harrow'g Swedes sow'd yesterday
17. Dry night, frequent heavy warm showers in the day | Carting Dung further Down Piece & Thrashing | LAMB, another of the best died, reduced to 99 from 230 Ewes put to the Ram
18. Sunday, rain in night dry day
19. Very fine growing day | WHEAT=EAR, the first have seen about half an Ace adjoining Stanley Bridge [324] | No prospect of a Wheat Ear in this neighbourhood before July, my Talavera Wheat sow'd 7th March no higher than the Barley | Ploughing the 2nd drift of Grass fed off, further Piece
20. Pressing the 2nd drift and sow'd to Greenrounds | Dry morning & frequent growing showers after THE COPIOUS SHOWERS on the 17th & 18th Inst and since have cause'd a great change on the face of the Earth, now a prospect of an abound't crop of Hay, Barley &c. before that period, an expectation of a very deficient Crop
    Expired this morning 12 minutes past 2 oClock the 20th of June Our beloved KING WILLIAM the 4th Born 21st Augst 1765 Marrd 11th July 1818 Ascended 26th June 1830 the Throne [This entry written below a thick black line]
21. Fine growing day, Swedes sowd the headland &c Pease finish'd hoe'g the 2nd time Thrash'g the 3rd Wheat Rick
22. Thrash'g Sultry warm day | MOWING began the Hop & Ray further Down Piece become too old for the Sheep to feed, gave the last tie | Potatoes began hoe'g Tan hill Piece, up fine Fogg'y morn'g, sultry warm day, Thrash'g
23. SHEEPWASH'D 312 began at 4 oClock, finished at half past 6 o-Clock, and left home at 1-o Clock morn'g, returned by 11_o-Clock | Thrashing Sultry warm day
24. MIDSUMMER=DAY, Sultry warm, the WHEAT in general just bursting the EAR, having made rapid progress since the RAIN the 11th -12 & 13th Inst, the universal opinion till lately was, that harvest would be a month later than usual, yet now forwarder than last year see 24th June 1836, the Crop of Field grass doubled since the rain with the heat = let us acknowledge and

323 Acr – Acre.

324 Ace - Acre.

admire & bless the wonderful Power and goodness of the Almighty towards us, in giving us plenty when we expected a scarceity. this extraordinary effort in the Vegitable nature, is the Theme of every one, see the Newspapers May we ever again put our Trust in God | Thrashing finish'd the 3rd Rick & put the 4th Wheat Rick into the Barn, this also full of Mice, the Down Rick only remaining 20 Sacks Wheat to Devizes the last of 1835 | LAMB'S, WANED, on Vetches Whitelands and the little Down.

25. Sunday, Sultry warm day Mr Brander Hillmarton } 4th Chaptr Jonas 7th & 8th Verse's
26. Very fine day, Thrash'g wth Horses
27. Very fine day, 22 Sacks Wheat to Devizes 12 S 17 li Nt the first of this Yrs Crop
Died at Lambourn Emma the Infant Child of Christr & Emma Pinniger of Hazleberry [This entry written within a three-sided thick black border] [Note in right margin] TPs accidt
28. Very fine day 18 S_ Wheat to Devizes 12 S 17 li Nt Coal returned 1T 17C 0 | Potatoes earthing Tan hill Piece
29. OBIT James Low, late of Beckhampton in 70th Yr Very warm HAY Field began carry'g, the Dry meadow grass dry'g up, land chop'd
30. Carry'g field hay, some cool'g winds | SHEEP=SHEAR, 310 by 6 Men, (30 half nakid) easy days work

July

1. Hay carry'g all day, warm windy day | Dry Meadow began mowing | Swedes do well in rough leaf half the Piece | Talavera Wheat backward, the Ear, not shewing
2. Sunday fine warm day
Died Mr John Young of Marden aged 64 [This entry written between two thick black lines]
3. Hay finish'd carry'g the 18 Acre Piece Warm fine day | Talavera Wheat grow'g rapidly, burst'g the Ear
4. Silbury Mead began mowing, the Grass in the Marsh 5 feet long | Swedes, grow rapidly, fit to hoe, warm fine day | 6 Horses to BOWOOD for 1½ hd (180) Faggots & 40 Poles[325]
5. Warm fine day, Hurdles remov'g & Plough'g Turnip seed Ground
6. Barley first sow'g shew'g the Beard | Plough'g &c as yesterday, some little blight'g rain | HAY, carr'd home mead, Mow'g finish'd Silbury md | Mr Wentworth & Rt Philpot, mow'g their spring fed Silbury mead, a good crop
7. Plough'g as yesty finish'd, fine day | SWEDES, 5 hoe'rs began, Tan hill Piece
8. Fine day, Hay began carry'g Silby med | Turnips, sow'd the 1st drift Whitelands, Vetches fed

325 In this case, Pinniger is using the long hundred of 120, rather than 100.

9. Sunday, fine warm day
10. Died Mr Josiah Wooldridge of Gt Bedwin Aged <...> [This entry written between two thick black lines] | HAY, finish'd carrying, Warm fine day
11. Horses Thrash'g, no Plough'g, Sultry warm day
12. Horses Thrash'g, Sultry warm day
13. Thin rain morn'g, Sultry warm day | Cart'g Dung from heap Down Piece | Swedes, finishd hoe'g first time The black Maggot distroy'g the Turnip.
14. SEED'S Ray, began mow'g, half middle Down Piece Blight'g rain morn'g, very dry'g after | Cart'g Dung as yesterday & finish'd the heap
15. Cartd 14 Waggon Load of Dung from the Inn to further Down Piece, and finish'd dunging the Piece, 5 Loads pr Acre the Fold, follow. some Showers of Rain in morn'g, soon lost by the warm boisterous Winds
16. Sunday, fine dry day
17. Fine morn'g, heavy Showers evening, soon dry'd | Plough'g further Down Piece for Turnips
18. Plough'g further Down Piece Rain heavy storm morning fine drying day | SWEDE'S began hoe'g second time, many Pieces <many Pieces> in the Parish, not begun 1st time
19. Ploughing as yesterday, some thin Showers
20. Pressing, Rolling & harrow'g further Down Piece for Turnips, fine dry day
21. TURNIP'S sowd the 2nd Drift Down Piece, the land too dry for the Seed to vegitate | RAY SEED, carried 4 Acres 12 loads Warm dry day
22. Warm dry day | Hurdles, 12 Doz from Chisbury Wm Meadham
23. Sunday, some Showers afternoon soon dry
24. NO PLOUGH'G &c for the Horses, in the Stable all day Very warm fine day
25. Carting Hurdles to fence Tan hill Road before FAIR | Fine morn'g some blight'g rain evening | The Talavera Wheat not yet whole'y in Ear now in bloom, very backward | The Ewes Kept very hard, have only the Down do not very bad, have no after grass Down & all dry'd up | The Lambs continue at Vetches Whitelands become old & go on Little Down do not expect to do well All things do better than could expect, after this long continuance of dry weather, the Swedes do well & the PEASE, Kid & fill well | The Barley uneven & check'd not so good as as the last 8 precedeing Years, yet better than many Countrys | The Wheat an average Bulk, but infected wth Red gum
26. SWEDE'S, 2nd hoe'g finish'd, no RAIN to wet the land since began hoe'g see 7th July | The Horses, Thrash'g Wheat, Warm fine day
27. The Horses finish'd Thrash'g Wheat, Warm fine day | TURNIP'S, began hoe'g Down Piece, sowd 14th June
28. Dry warm day | Horses in Stable, Carters Ricking Straw
29. 25 Sacks Wheat to Devizes | RAIN, none to do good since 14th June | Rain nearly all night to 9 o-Clock this morn'g when the wind rose to a

hurricane, blowing the Boughs & leaves from the Trees, and the Horses fright'd by blow'g the Gravel in their faces going to Devizes, RAIN again afternoon and night, the Wheat & Barley, laid flat & much scrall'd very much wanted

30. Rain in night, some showers in day, dry eve'g | Sunday
31. 25 Sacks Wheat to Devizes, some partial Showers

August

1. Turnips sow'd the 2nd Vetch drift Whitelands Dry morning heavy rain evening
2. RAIN, all night & nearly all day, heavy at times The Horses in Stable and Men at home | The London Market 3s Qr cheap'r, the Devizes Market this day 3s Qr dearer
3. NOMINATION at Devizes of Candidates for the County heavy storms in morn'g dry after | Horses in Stable, Carters cleaning Rick Yard
4. Some showers of Rain dry evening | VETCH'S Whitelands being in hard Kid, the Lambs discontinued, but little Keep for them, the Clover whole'y failing | 25 Sacks Wheat to Devizes 12S 17li Nt
5. Fold finish'd further Down Piece & began folding lower long Piece for Wheat fine day | VETCHE'S, Whitelands the Horses began, having no Clover, & Silbury mead too young | Plough'g further Down Piece for Turnips
6. Sunday fine day
7. Exceeding COLD windy morn'g fine day | Plough'g as the 5th
8. Plough'g finish'd the Piece, Windy dry day | ELECTION for North Wilts, closed Burdett 2349 Long 2183 Methuen 1868
9. Rolling & Pressing further Down Piece for Turnips Warm windy day | SHEPHERD, entered Geoe Smith
10. Warm shower mid day, dry after | Turnips 2nd drift, finish'd hoe'g | Turnips 3rd drift, just appearing | Turnips 4th drift & last sow'd Down Piece
11. Thunder, lightening & heavy storm early in morning, dry after | Hurdles removing from Tan hill Road
12. Muckle began Carting before the Fold on the lower long Piece from the Inn | Very warm fine day
13. Sunday Very warm fine day
11. PEASE, the dwarfs began drawing by the roots, and others hoe'g them up, being too near the ground to cut with hooks
14. Foggy morning, very warm fine day | Horses in the Stable, no work for them all hands drawing Pease | Reaping, my Neighbours began <began>, mine yet too unripe | VETCHES the Horses discontinued being too old
15. all hands drawing Pease Horses in Stable | GRASS, Silbury Mead, began mowing for Horses again & baiting them on Field Grass, having very little grass in Mead & no Clover, never so short of Horse Keep, as well Sheep,

sultry warm day

16. Sultry warm day, Horses in Stable | PEASE, finish'd, Cuting and drawing | VETCH'S, began Cuting for Seed Whitelands
17. Sultry warm day, Thunder & Storms partially, Cart'g Muckle before Fold from Inn | The fine harvest weather, cause a depression in the MARKET, Wheat 2 to 3s pr Sack Sold the last lot of old Wheat 20 Sacks 12 Score Nett @ 25s and 40 Sacks of first of last Yrs Crop 12 Sco: 14li Nett @ 27s 9d, all the Wheat in the forward Districts secured in good condition, causing an imprs on the Publick mind, that Wheat will continue to fall in price. my Wheat not ripe enough to reap [326]

    July 28th Obit Mrs Bartlett of Great Bedwin [This entry written above a single thick black line]
18. SALE EWES, drew & put with Rams & the Lambs on the second field Grass | 81 - Full mouth no mark for Sale | 59 - 6 Tooth near Ear mark for Stock of 1834 | 68 - 4 Tooth Off Ear mark for Stock of 1835 | 92 - 2 Tooth no mark for Stock of 1836 [Total] 300 | 2 - other Ewes with Cuckoo Lambs | 2 - Giddy Tegs | 4 - Rams | 2 - Ram Stags | 55 - Chilver Lambs | 44 - Wether Lambs [Total] 409 [Note in pencil] 40 [Total] 449 [Further note in pencil] 300 | 2 no lambs | 2 Giddy-Kill'd [Total] 304 The blind Kill'd | 80 - sold [Total] 224 [327] | Sultry dry day, Straw removing from Rick Yd | VETCH'S, finish'd cuting
19. Sultry fine day, Straw removing, the Horses continue in the Stable | REAPING, began, with 6 of own People, having very little ripe, all my Neighbours nearly finish'd none carried
20. Sunday - Dry windy day
21. Damp blight'g air, red morn'g & windy | PEASE, Carried one load when RAIN came | Muckle Carting before the Fold
22. Dry warm morning, Showers after & heavy Rain in the evening, to prevent tieing up the Wheat
23. RAIN in night & storms all the morn'g, dry afternoon & heavy rain evening | Discharge 7 bad Taskers, 4 of them three days reaping an Acre, with Beer, rain hindering | 10 Acres only of my Wheat reaped | Plough'd a drift Whitelands for Turnips
24. Sow'd a drift Whitelands for Turnips | Rain in Night & morning, drying after
25. Carting Muckle before Fold from the Yard | Very drying after the morning, put the Dwarf Pease in weak to dry in hope of carrying tomorrow & turned the long Pease & Vetches. Straw dry in even'g
26. Rain in night, drying morning, attempt'd to carry Pease, & prevent'd by Thunder, lighten'g and very heavy RAIN afternoon, the Pease haulm spoild

---

326 imprs - impression.

327 The further note in pencil appears to be accounting for the outcome of the ewes and tegs at some future date (undetermined).

| Carting Dung before the Fold

27. Sunday, Fine drying day
28. PEASE, turned, and carryed, Straw dry, being the third attempt, 5 Loads of Dwarf on 4A 3R 16P and 6 Loads of the Long on 2A 0R 28P | Turnips White lands, hoe'g, sowd 8th July - GLASS, sinking, very drying day, rain even'g | GRASS, Silbury Mead, the 80 Sale Ewes & the 99 Lambs began feed'g wth a Tie of Field grass
29. GLASS, fall'n to N, in much rain, heavy RAIN in night and RAIN all the morn'g, dry for a short time & rain again Reapers at home, Cart'g Dung
30. Reaping finish'd the Red Wheat, & began the Talavera, very green & blighted | Turnips sow'd further Down Piece, where fly, eat | Cart'g Dung, RAIN heavy frequent
31. RAIN, in night & morn'g, very drying after Cart'g Dung in morning | VETCH'S Whitelands carried in Barn

September

1. RAIN in night & morn, very dry'g after | WHEAT began carry'g, the Rick by the Wall, dry to the eves when rain came & topd | JACOB DAVIS, absent'd himself from work | JAMES DAVIS absent from his reaping, puting me to great inconvenience in carry'g both very bad characters | BARLEY began mow'g, the right of Tan hill road WINTER GREEN'S planted
2. Ploughing Whitelands, rain slight showers morn'g
3. Sunday, some Showers morn'g damp air after
4. RAM'S, Mr Wentworth put to the Stock Ews too early for me not having plenty Keep in the Spring | Rain a little in Night, dry day, GLASS rising from R in much rain to C in change, a prospect of fine weather
5. Very fine harvest Day | WHEAT carried the remainer of the Red 14 Acres in good condition in 2 Ricks by 8-oClo | JAMES DAVIS, finish'd reaping his land of Talavera Wheat and absented himself with Jacob see the 1st Inst leaving me only one Man the Mower to assist in making the Wheat Ricks
6. REAPING, finish'd the Talavera Wheat | Foggy morn'g fine clear harvest Day | Turnips, sow'd the remainer of Whitelands
7. Damp morning, drying afternoon | Finished harrow'g off Turnips Whitelands | All the Wheat in this part of the Country finishd carrying yesterday EXCEPT my TALAVERA
8. Rain in Night, very fine drying day | Ploughing the Pease stubble for Turnips, if not too late | TURNIP'S began hoe'g the 3rd Drift Down Piece sow'd August the 1st
9. Dry night, foggy morn'g RAIN from 6 oClock in morn'g with short intervals to the evening | Carting Dung at times before the Fold
10. Sunday, Dry night & very drying Day | The LAMB'S suffer much by the Fly & Maggot, seven very bad
11. WHEAT HARVEST finnish'd by carrying the Talavera 4¾ Acres on 4

Waggons large loads before 11 oClock very drying till then, RAIN the remainer of the day

12. Windy drying day Barometer sinking some little rain, and appearance of more rain | Ploughing the Pease Stubble
13. Ploughing the Pease Stubble RAIN, heavy in night & storms through the Day, the BAROMETER lower'd to O in Stormy, The Barley sprouted & none carried
14. RAIN in Night dry day rain in evening | Plough'g as yesterday
15. Plough'g as yesterday & finish'd, RAIN heavy night & rain in morn'g, exceeding drying winds remainer of the day, the BARLEY mowed the the 1st Inst not turn'd & the Grass grow'n through the Swathes, yet perfectly dry, and forked it this even'g for carry'g it tomorrow | WHEAT=RICK, made of the Talavera Wheat on Waggons since the 11th the weather too wet for it till now
16. Dry night, thin rain all the morn'g, dry eve'g Ploughing for Vetches
17. Tockenham Mr Young 24th Luke 46 & 47 Verses, rain in night, dry day
18. BARLEY, began carry'g, mow'd 1st Septr RAIN at 2 o-Clock stopd carrying
19. CONSERVATIVE, Banquet 1200 dined at the Pavilion Devizes [Pencil note] <...> feet by feet | Fine after the morn'g, some little dizly rain, Plough'g
20. Press'g for Vetches morn'g, Dry'g winds, Carting Barley afternoon
21. Plough'g Down Piece for Vetches morn'g, thick fogg'y morn'g, very drying after, Barley the 6½ Acres carried below the Road on the Staddle | CONSERVATIVE Ball at the Pavilion 1600 attend d [Note in margin] RAMS TO EWES
22. Thick fog till noon, dry after | VETCH'S began sow'g middle Down Piece 9 Bu: on 3 Acres one earth press'd good work | Barley, the late sow'g see the 18th May, began mowing light crop, dry seed time
23. TURNIPS began feeding, see, sow'd 14th June by sale Ewes, on young grass mornings on Turnips mid:day & l<...> Tie of 2nd Clover evenings | Lambs_Silbury Mead, tie'd out | Stock Ewes till now do well on Down, become grassless follow Sale Sheep on Turnips & Stubbles | Fog'y morn'g, exceeding drying, UNEXPECTED fine weather, enables all to finish Barley harvest well | Dung Cart'g from Inn
24. Sunday, drying wind'y day
25. FROST the 1st sharp white, dry windy day | HARVEST finish'd by carry'g the BARLEY, the finest Weather since the begining
26. PLOUGHING began for WHEAT sow'g, one earth grass lay, lower long Piece, next the Down Piece, fine morn'g, storm of rain after
27. Plough'g as yesterday, cold dry air
28. Plough'g as yesterday, cold dry air | COLT, 2 Yr off-Grey, & white face, of Mr H Reeves 32 £

29. Plough'g as yesterday & Press'g & sow'g Vetches finish d half the middle Down Piece | 40 Lambs from Corton @ 12s 3d each, very fine day
30. Vetches thrash'd 5½ Sacks wth Machine morn'g | Pressing Wheat land. - very fine day | Finish'd hoeing Turnips, Down Piece & Whitelands

October

1. Tockenham Sunday Rain morn'g fine day
2. Plough'g for Wheat, warm fine day
3. Plough'g for Wheat, warm fine day rain even'g
4. Carting Dung, before Fold warm fine day
5. 25 sacks old Wheat to Devizes & 12 Sacks Seed Wheat returned 12Sc 13li Nt @ 32s Golden drop from Upavon | Fine day cold Winds
6. 13 Qrs Barley to Marlbro. Lightening in the Night, heavy storm in morn'g, very dry'g after
7. 25 Sacks Wheat to Devizes 1T 17C 2Q Coal returned | Very fine Day
8. Sunday. - some thin rain afternoon
9. Plough'g for Wheat, exceeding fine day | Potatoes began diging, for fattening Pigs | Lambs removed from Silbury Mead to Turnips Down Piece to reserve the Grass for the Horses with the intention of serving them till Martinstide & Keep from Hay | 449 Sheep at Turnips, all doing well, and & much better than last Year
10. Ploughing for Wheat to the Fold | Beautiful fine mild, calm day
11. Beautiful fine mild, calm day | Cart'g Dung before Fold from heap by Tan hill road
12. Pressing for Wheat about 6 Acres | Unusual fine mild weather
13. Unusual fine mild weather the Barometer, beyond R: in Fair | Plough'g for Wheat, to the Fold turned, now waiting for Rain, being to dry
14. No Plough'g to do at present for Wheat & too dry to begin sowing, - harrowing the Pease land for Barley | CONTINUATION of Beautiful calm weather the Barometer now at Setfair, so dry a time can not remember, having been waiting the last three weeks to begin Wheat sowing
15. Sunday, very sharp Frost, to cut the Potatoes &c
16. very sharp frost | Press'g harrowing &c very sharp Frost and mild day
17. Days duty on the Road Stone Cart | H C Pinniger, opened his Drapery Shop at Calne
18. POTATOE'S began diging for Piting, the haum being frosted, - fine dry day | 25 Qr Barley to Marlbro H_ Chivers trim'd 80 Ewes & 20 Lambs for Fair
19. Frost very white, and warm fine day | Stubbles Cart'd & Plough'g Pease land 2nd time
20. Carting Dung from heap, before Fold | Very warm fine day, Devizes Fair, 6 Coops for 80 Ewes, rather too small, 1 for 20 Lambs right
21. Plough'g a Fold Drift | Very warm, calm & dusty, scraping the Roads a new Season, such not remembered

22. Sunday - Fine dry day
23. Barometer sink'g, very fine dry day Ploughing & Pressing
24. RAIN heavy in night, and damp day began Plough'g for Wheat sowing a month and no rain to wet the Land till now | Plough'g Pease land second time for Barley
25. RAIN, heavy in night, dry day WHEAT=SOWING=BEGAN. the land none too wet, geting green Plough'd a month one earth, the lower Piece adjoining Down Pieces the Golden Drop seed, from Mr Waters Upavon 2½ Bu: pr Acre
26. Fine dry day, sowing Wheat
27. Dry night & morn'g, Wheat sowing, RAIN at at 10 o Clock & Storms after, stop'd sowing | Thrash'g Vetches PORKER'S put two to fatten
28. Dry morn'g - rain afternoon | Wheat sow'd up to the Fold about 12 Acres
29. Sunday, - Dry morn'g, rain all afternoon
30. Rain in night & driving rain all day | Carting Dung from heap before the Fold | SWEDE'S to the COW to fatten began giving
31. Rain heavy in night & very cold hail storms in the day, too wet for Plough'g | Horses thrashing Vetches finish'd

November

1. HORSE'S in the Stable all day RAIN heavy in the night, and exceeding boisterous driving rain all day storms in torrents
2. Horses in Stable all day, Rain all night Lighten'g Thunder and continued heavy Storms of Hail and Rain some Snow, all the Day
3. Rain in Night, dry day | Carting Dung, from the Inn before the Fold
1. WATER at the lowest, none to be got from the PUMP
4. Plough'g for Wheat - Very fine day
5. Sunday - Very fine day
6. Ploughing for Wheat Very fine day
7. Pressing for Wheat Very fine day
8. Pressing & Ploughing for Wheat Very fine day
9. Wheat sow'd up to the Fold - about 3 Acres to sow the Grass stil tied out to the Lambs and the Horses baited evenings on it, Muckled and Fold'd 1 earth & Press'd, too late in the Season the system considered by good Farmers, yet made good work and may prove a good Crop (as before) Thin Rain all the morn'g, and damp day | 120 Lambs DIPED at Kennett | POTATOE'S finish'd diging 140 Sacks fill'd N B 14 Sacks of Purples the produce of 1 Sack
10. Rain heavy in night, & morn'g damp day 3 Loads only of Muckle before the Fold
11. Very fine drying day, Harrow'g off the Wheat sowd 9th | PEASE began thrashing, to serve the Sheep wth haulm
12. Sunday cold dry day
13. Dry mild morn'g, thin mizly rain all the day after | Wheat Stubbles began

fallow'g

14. Wheat Stubbles began fallow'g | Dry mild morn'g heavy rain after dry eve.
15. Fallowing Wheat Stubbles | Cold drying Day [Pinniger then gave observations about various sheep fairs - see Appendix H]
15. Wheat sow'd 3 Weeks just appearing
16. STRAW Pease the Ewes began in Stand'g Fold middle Down Piece | Blind Ewe Kill'd 44 li @ 7d - Skin 1s 6d, the worst of the sale Ewes by 7s | Plough'g & harrow'g after Potatoes
17. GRASS, Watermead, the Horses finish'd and finish'd baiting in Field | FROST, to bear up an horse (Wine fin Cleveancey) Plough'g Fold drift muckled
16. Dry cold frosty air
Died at Calne this morn'g Mrs Edwards Widow of the late Danl Aged [in pencil: 78] Also Sophia Wife of Mr John Broxholm Aged [in pencil: 32] Yrs[328][This entry written within a three-sided thick black border]
18. Plough'g & Press'g & Cart'g Dung before Fold | Another very sharp Frost, damp afternoon
19. Sunday - Dry morn'g Damp evening
20. Rain heavy, in Night & morn'g dry after | Field Grass the Lambs, finish'd Turnips the Lambs, finish'd Down Piece The Turnips 3A 0R 23P began feeding 23rd Septr Kept the Sale Ewes with Grass & the Lambs and the Stock Ewes follow'd on the Shells | 4 Horses to Tytherton for Cider 6 Barrels
21. SWEDES and HAY, the Lambs began Tan hill Piece (dry day) Fold removed from Wheat land to middle Down Piece for Vetches & Wheat 1838 | Hurdles removed & Dung Cart'd to finish Wheat field | DUNG of Standing Pen remain & considerable at the Inn, more than want'd for Wheat field
22. Plough'g for Wheat, Damp mild day
23. Plough'g for Wheat FINISED, too wet to Press & Sow | Boisterous rain'g day, heavy rain afternoon | COW, began give g, Barley Meal & Chaff (& Swedes) Potatoes stoln from home Skilling
24. Fallow'g Wheat stubbles, - dry mild day
25. Frost morn'g clear mild day | SWEDE'S the EWE'S began Tan hill Piece fold'd mid: Down Piece, and Pease haulm mornings SWEDE S the 5HORSE'S 1 Bushel when sliced each pr day and <...> Truss SWEDE'S the COW 1½ Bushel when cut pr day and 1½ Bushel of hay Chaff with <...> Swedes, the 15 Acres Tan hill Piece being very large and thick on the

---

328 Entry probably taken from Salisbury & Winchester Journal, 20 November 1837: *Nov. 16 at Calne, after a few days' illness, Mrs Edwards, relict of Mr. Daniel Edwards – Nov. 16, after a lingering illness, in the prime of life, Sophia, the beloved wife of Mr. John Broxholm, of Calne.* Mary Sophia Broxholme of Calne buried on 23 November 1837 at Calne, aged 33, *Wiltshire, England, Church of England Deaths and Burials, 1813-1916 for Sophia Broxholm* [accessed via Ancestry, 28 July 2020].

ground, expect will keep all the stock till the middle of May | HAY, the Horses finish d the two Year old second Clover & began the New Field hay | Hurdles removed from Down Piece to Swedes Tan hill Piece | Press'g finish'd for Wheat & fallow'g Stubbles

26. Sunday - Dry morn'g, cold damp even'g
27. Fine dry day, Wheat sowing
28. Thin rain part of the Day | WHEAT SOW'G FINISHED, the lower long Piece with 12 Sacks of Mr Walters of Upavon Golden drop thick set Wheat and 6 Bushels of thick set Worcester Wheat on 18 Acres being 3 Bushels pr Acre, SEE 14th NOVR 1833, 2 Bushs pr Acre DRILL'D, on the same Piece, was a light Crop Carting Hurdles & Stubbles 76, Chilver Lambs earmark'd 43 Wether Lambs [Total] 119 } Tails squared and marked }
29. Very sharp Frost, & clear fine day, very cold freezing evening, fallow'g Stubbles | Water carriages in Silbury Mead finish'd clean'g Man out of employ reak'g Wheat Stubbles @ 5s pr Week, (the Stubbles worth the Pay)
30. Thaw & damp morn'g dry evening Fallow'g

December

1. Thrash'g Seed hay for Horses, - fine mild day | Fallowing Stubbles
2. Fallowing Stubbles Sharp frost, dry day
3. Sunday Sharp frost, dry day Chalfield
4. Fallow'g Sharp frost, dry day
5. Fallow'g Sharp frost, dry day
6. Cart'd the Trenching Silbury mead | Very cold, damp air
7. SNOW more or less all day, fallow'g RABBIT'S employ'd my Shepherd to destroy doing me a great injury, by no care taken by the Warren man for the last Ten Months in stoping the holes in the Hurdles to Keep them back from my land till this day, find'g I am distroying them.
8. SNOW, heavy before day light, thick fog'y day, SNOW eveng Horses in the Stable, Men winnow'g Pease | HAY, Seed, gave the Ewes this evening, too wet to go on Swedes, Pease Straw morn'g - and Furze D | PORKER, the 1st 86 li fatten'g 6 Weeks from 27th Octr cost 24s 6d (say at 6½d pr li is 46s 7d)
9. Fog'y and SNOW'Y day. Horses in Stable
10. Sunday - Dry day
11. Fine dry day Ploughing
12. WHEAT=RICK, the last OLD fin further Down Piece put into Barn Freezing all day in Shade | Fallowing
13. Horses Thrash'g, and Hurdles removing | Freezing after morn'g in shade | Turnips finish'd Headlands of Swedes
14. Dry day and fallow'g
15. Dry morn'g and fallow'g cold rain evening
16. Very boisterous Cold morn'g, RAIN all afternoon | 21 Sacks & 16 Sacks

Wheat to Calstone Mill for Mr Boman

17. Sunday, RAIN nearly (all day), the land too wet for the Sheep to feed the Swedes.
18. RAIN, driving all day, the Lambs in home mead the Ewes in Silbury mead, Horses in STABLE
19. TERRIFFIC WIND'S. in the night to uncover the Ricks topof Wheat Rick blown off | FLOOD in Silbury mead the first, Fallowing Difficulty in providing Sheep, to wet for Swedes
20. Difficulty in providing the Ewes on Down and Seed hay in Fold, the Lambs in Mead home, and Fold put on home Mead wth hay, Wind like Thunder again, to strip the Barn, mended yesterday. | Dry morn'g, rain wth the Wind after to drive the Horses from Plough
21. Fine dry day - Fallow'g Mr Estcourt, Promised to destroy the Rabbits now destroy'g the Down Turnips
22. Dry morn'g damp after, Fallow'g | The FOX, seen carry'g off the Poultry
23. Very mild dry day, Horses Thrashing
24. Sunday - Fine mild day | Pease haulm the Sheep finish'd, the Dwarf Pease haulm the Sheep will not eat | HAY, consumed by all the Stock since NOVR 20th Old Seed 2 Ton New Field 2 Ton Meadow ½ [Total] 4½ All the Stock eat as many Swedes as can the Lambs 1 Truss Field hay evenings the Ewes Thrash'd hay evenings the Cart Horses Thrash'd hay
25. Xtmas day, fine mild Spring like day The Young field grass, Swedes &c green & growing, | Ram Stag 60li
26. Damp morn'g, mild & the Sun clear after | The Horses Thrash'g
27. Fallow'g, and very mild damp day
28. Fallow'g, and very mild damp day | WHEAT, the last sowing 28th Novr, appearing thick out of ground
29. Fallow'g, Dry morn'g rain in evening
30. Fallow'g mild damp day rain heav'y even'g | Lambs Fold'd home mead again, very dirty at Swedes
31. Sunday, Stanton Church Rev Mr Mog | Foggy morn'g clear & dry after Decr 20th my Nephew Mr Thos Crook, contraced <...> for an establish Grocery Business at Winchester, and entered upon it, and on the 16th Jany Inst 1838 commenced his residing at his new home and his predicessor Mr Robt Cave departed from Winchesr [Diary turned landscape for this entry]

1838

January

Obit 7th January Mr Stephen Hale Aged 83 Yrs May next Buryed at Enfield Midx the honest Miller at Quemerford [Diary turned landscape for this entry and written within a three-sided thick black border]

1. Mr P_ Cook WYETH, commenced Painting our Portraits [329] | Rain heavy

329 This is probably Peyton Cooke Wyeth (1811-1862) who was an American portrait

morn'g mild & dry afternoon too wet for the Sheep to go on Swedes - The Horses fallowing

2. Rain in night dry day | Fold finishd half the middle Down Piece | 5 Horses & 2 Ploughs to KENNETT
3. 5 Horses & 2 Ploughs to KENNETT } Wheat sow'g | Fold put on Nether Down Piece young grass in Wheat Stubbles, no other place for it | Fine mild day
4. Frost and Fine mild day Fallowing
5. Very sharp frost Fine mild day Fallowing
6. Frost & fine day Fallowing
7. Sunday Frost & fine day EWE DIED Murring
8. Very severe Frost, Fallow'g LAMB DIED Giddy
9. Exceeding severe Frost Horse Pump stop d, no one remember colder Winds, Sheep cannot eat the Swedes, frozen like Stones | Cart'g Manure from Inn, to make a Standing Lambing Pen
10. Cart'g Manure from Inn | An other severe Frost, less wind some SNOW
9. FAT PIG the 1st 14 Score
11. Frost continue very severe | Muckle, Carting from the Inn
12. Dung Cart'g from Yard to middle Down Piece for Turnips, - Just Fold'd | Frost continue very severe
13. Frost continue very severe Horses Thrash'g Wheat
14. Sunday Frost continue very severe Mr WYETH, finish'd the five Portraits in 12 day, took him to Cleve y [330]
15. Very piercing cold freezing Winds continue great difficulty, in fixing the HURDLE'S the Fold & Swedes, ground exceed'g hard. | Carting Dung from Yard as the 12th At 10 oClock Wednesday night the 10th Inst a FIRE broke out in the ROYAL EXCHANGE London which totally consumed it.
16. Intense Frost continue, having great plenty of Swedes, the Sheep continue at them and scoope them out of the ground leaving a Dish for each Swede, the ground so hard the Sheep push the Hurdles down | 4 Horses to Devizes for Coal 4„13„0 1-2-2 Waggon Ton 3„10„2 Nt | My Neighbours have stolen 30 Doz of Stakes from my Hurdles in the Skilling in Meadow
17. The Skilling in the lower Fields destroyed by FIRE on BUPTON=FARM, discovered at 12 oClock at noon, (by an Incendiary) | SNOW, heavy at 11 o-Clock & severe Frost | Carting Dung | MR BRIAN RUMBOLD, malitiously shot at in his own Field, returning from Calne Market, and wounded the left Arm broke (SEE Execution the 6th Septr) [331]

---

painter, known to be in Liverpool, England in 1835. *England, United Grand Lodge of England Freemason Membership Registers, 1751-1921 for Peyton Cooke Wyeth* [accessed via Ancestry, 3 July 2020].

330 Clevancy.

331 Although the Devizes & Wiltshire Gazette, Salisbury & Winchester Journal and Wiltshire Independent all reported the injured party as Brian Rumbold, he was

18. Continuation of intense Frost, & some Snow | Carting Dung as yesterday
19. Horses Thrash'g, Frost the same clear day
20. Horses Thrash'g and finish the Rick | Talavera wheat Rick, put into Barn | Freezing in extreme
21. Sunday, Frost continue severe, incline to thaw eve
22. Very gentle THAW, without rain or wind | ASHES, the 1st Load 60 Bushels wth 4 Horses
23. ASHES, the 2nd Load 60 Bushels wth 4 Horses | FROST, sharp freezing again
24. Boisterous winds all night and all day to strip the Ricks & Stables at the Inn, the cold freezing winds, if possible more intense than 20th | Horses Thrashing the Talavera Wheat
25. Ashes - the 3rd - 60 Bushels wth 4 Horses | Very cold sharp Frost continue
26. Very cold sharp Frost continue | Ashes - 120 Bushls - 6 Horses
27. Ashes - 120 Bushls - 6 Horses | Sharp Frost continue, not so cold
28. Sunday Sharp Frost continue, the ROADS as and as DUSTY, as the midst of Summer
29. Ashes 120 Bushls 6 Horses & 45 Sacks old Wheat to Devizes the remainer | THAW in the night and RAIN in the morn'g the SWEDE'S are soft expect are spoiled and lost by the Frost, being the most severe since see 1830 - 1829 - and Decr 28th 1820
30. Mild and thick Fog all day, Thrash'g finishd Talava
Died this morn'g the 30th January my Mother=in=Law, Mrs Mary Pinniger of Tytherton Aged 82 Years 29th Septr last Buryed the 6th Feby at Bremhill with my Fa [332] [This entry written within a three-sided thick black border]
31. ASHES began Sowing, right of Tan hill road 6 Load 60 Bushels each in 6 hours by 4 sowers | Slight Frost in morn'g dry day

February

1. Frost sharp fine day | ASHES, FINISHD sowing, except an other load wanted, to sow the Piece
2. Carting Dung to Down Piece | Severe Frost & fine day
3. ASHES 50 Bushls the last load to finish the Piece 18 Acres with 590 Bushls fm Devizes and of my own 90 [Total] 680 is 38 Bu: pr A | 23½ Sacks

---

actually Bryan Rumboll. In 1841, Bryan Rumboll was a farmer, aged 60 at Lyneham Court, TNA: HO107/1179/7, *Census returns. 1841 census. Wiltshire. Hundred: Kingsbridge including Parish: Lyneham or Lineham, ED 11.* In 1851, Bryan Rumboll was the farmer of 880 acres employing 11 men and 5 boys at Lyneham Court, TNA: HO107/1834. *Census returns. 1851 census. Wiltshire. Registration District: 251, Cricklade. Registration Sub-district: 1 Wootton-Bassett, including Parish: Lyneham or Lineham, ED 2c.*

332 The phrase mother-in-law was used for stepmother: Mary Pinniger was the second wife of Christopher Pinniger, Susannah Pinniger being his first wife and the mother of Thomas Pinniger. Fa - Father.

Talavera Wheat to Devizes | Sharp frost in morn'g & SNOW, slight for several hours

4. Very cold, sharp Frost continue, Sunday

5. Very cold, sharp Frost continue | BARLEY, Rick put into Barn (the 1st this season

6. Sharp Frost, very fine day | 2 Load STONE'S, the first for the BARN

7. 2 Load STONE'S SNOW in the night, Thaw all day with rain in the evening

8. the 2nd PIG 14 Sco 14 li, Rapid Thaw, and RAIN nearly all day | Devizes Wool Fair Ewes 40s Ewe & Teg 42s 42s 6d Teg only 44s pr Tod Newbury Wool Fair 1500 Tod sold Ewes 36s & 37s, mix'd 1/3rd Teg 40s & 42s Merino Teg 50s pr Tod 2 load Stones from Calne

9. 2 load Stones from Calne Thaw & Rain, the ground completely rotton, the Swede ground, like mud, put the Sheep on Down & fold with 3 Truss s HAY, the first time morn'g, Seed hay eves on Swedes every day til now | VERY ANXIOUS TIME NOW, for the fate of the Swedes & Wheat | the green of the Swedes are become white, some are hard & some are soft, the Turnips are supposed to be all rotted & all vegitables in Garden are Kill'd | The Wheat like Fallows | A Months hard FROST see the 9th Jany set in sharp

10. 2 load of Stones from Calne | The wind in the evening, shifting to the East producd an exceeding sharp FROST again, the Sheep return to the Swedes. & HAY, discontinued

11. Sunday, - Sharp FROST and Sun shine day

12. Sharp FROST and Sun shine day | LAMB the 1st, the Ewes time out the 14th 2 Load of Stones from Calne | ASHES sow'd the last load below & adjoining the middle road, the ground as hard as stones by the Frost, no one but myself sow Ashes so early, (observe in Summer if successful)

13. BARLEY=THRASHING=BEGAN, with the Machine for the Sheep in the Lambing Pen | Very sharp Frost, mild'r in the Sun

14. Very sharp Frost, mild'r in the Sun | EWE'S time for Lambing expired, put them in Lambing Pen this evening to have HAY once & STRAW once a day with Swedes the Ewes have but 3 Truss's of Hay this Winter (see the 9th Inst) having had thrash'd seed hay till now, and done better than some Years pass'd having had plenty of Swedes and walk'd from Swedes Tan hill Piece to Fold Nether Down Piece to this day, & Fold'd about 8 Acres since Wheat sowing. The Tegs fold'g home Meadow | HORSES, BEGAN new Field HAY, till now have the old thrash'd seed hay | 2 Load of Stones from Calne | A large Load of Straw from Kennett will weigh 53 & ¾ € at 1s 7d pr €

15. 2 Load Stones from Calne | Oh" exceeding cold boisterous and freezing Winds during the Night and the Day

16. Very cold freezing winds continue, with driving Storms of SNOW | Dung Carted from the Inn, made with it a Standing Pen adjoining the Ewes to

shelter the TEGS, being exceeding cold & bleet on the Swedes & impossible to put the Fold in home mead [333]

17. SNOW in the evening & driving frozen Snow, beating againt the windows all night, thaw and mild evening | Carting Dung from old standing Pen to Gardens
18. Sunday - Frost morn'g, thaw after [Pinniger then recorded a series of notes over the period 12th February to 17th March on sheep and lambing - see Appendix W]
Died this day Feby 18th Mr Richd Pinniger, Aged 57 Years late of Round=robbin Farm, Highworth [This entry written within a three-sided thick black border]
19. Very sharp Frost morn'g Thaw after | HORSE'S in the Stable, Frozen out of work Men winnow'g Barley
20. Very sharp Frost & rapid Thaw again | Horses, Thrashing Barley
21. Horses in the Stable. Men winnowing The THAW continueing covers the land with Water and make it exceeding soft, so as to confine the Sheep to the Standing Pens, on hay Straw & Water, the Swedes being cover'd with Water, the WHEAT looking dead, nearly all
22. Thaw & cold dry day | Horses in Stable - Men winowing Barley
23. Horses Thrashing Barley, SNOW, early in morn'g thin mizly rain remainer of the day | COW, WEIGHD, our Alderny 38 Score & 7 li @ 9s cost with a Calf 18th Feby 1830 - 12 £ then 4 Years old
22. COW & CALF 3 Years old Alderney cost £12 10s 0d bot of Mr Allen
1838 Feby 24th. This is a day that will long be remembered, the RAIN being heavy last evening, with the melted Snow and rain in the day and the Frost not two Inch's out of the Arable ground caused the Waters to decend the sloping lands rapidly carrying the softened earth in its course uprooting the Wheat and burying other parts with the Strand. My lambing Pen by Nine o-Clock in the evening covered with water, continued raining till afternoon causing a great FLOOD | SEE, observation on this day, next Page [Diary turned landscape for this entry]
24. Horses in the Stable, the Men winnowing
25. Sunday, - dry day
26. 2 Load Stones from Calne, damp day.
27. Horses Thrash'g Barley, Rain morn'g dry day, very foggy evening
28. PLOUGH'G, Wheat Stubbles, the first time since the 8th January being 6 WEEK'S FROST, from 8th Jany to 20th Feby with the exception of Thaw, see a day or two | All the Sheep let out of Standing Pen to the SWEDES altho: exceeding dirty to eat them whilst they can, as every Swede appear spoild with the Frost, and expect will not produce the least green, being rotten | Very mild dry day

333 Possibly bleak.

March

1. 14 Qr Barley to Devizes 1m Stales returned for the Barn the first load | The first Barley sold 10S 6li Nt @ 33s pr Qr Mild damp day
2. 2 Load Stones from Calne - Rain nearly all day
3. Dry and very mild day, Fallow'g Stubbles | A=MAD DOG, shot in the yard, after biting our Dog RANGER, which shot also
4. Sunday RAIN heavy morn'g, very dirty for Sheep
5. Fine drying day, finish'd Plough'g Stubbles | My Swedes, 9 out of 10 rotten, some People more fortunate not 1 in an hundred rotten being younger and smaller
6. CHALK for build'g BARN, two Men from Yatesbury began diging | Exceeding boisterous drying winds | Horses finish'd Thrashing 1st Barley Rick
7. Very cold drying Winds | 3 Horses Plough'g for Turnips 2 Horses Plough'g for Oats } Down Pieces | THOS LAD. Died aged 42 Years [334]
8. Cold dry day, Plough'g as yesterday | The Tegs, left the Standing Pen, the land being dryer, are on the Swedes Night & D [335]
9. PIG - 8S 8li, the fourth, the fellow to it, see Porker 4-6 the 8th Decr | Plough'g as yesterday | FROST exceeding hard, very fine mild day | LAMB'S come fast this day 40

   Copy'd from Devizes Gazette Friday eve 23rd Feby at half pass'd 7 oClock, an immense body of Water descended over the frozen ground into the Lambing Pen (the Barn Yard) of Mrs <...> of Watcombe Farm nearly Fawley & Lambourn, filling the lower part of the Yard several feet deep with Water. DROWNING 198 EWES of the 200 in the Yard and 240 LAMBS and injureing in the Barn 60 Qrs of Barley and 30 Qrs of Oats [336] [Diary turned landscape for this entry] [Pencil note in margin] Mr Stratton, narrowly escaped
10. Very cold boisterous winds, Plough'g as yesterday Ray Seed, put in Barn for Machine Distress'g time with the SHEEP, all geting very POOR, expect the perish'd Swedes injure them, the Lambs very thin and weak and die daily 2 [in pencil: 8] already skinned 2 Ewes died to day, heaving, & 2 others same state
11. Sunday, very fine dry day
12. Ray Seed 3 Horses Thrash'g and two Plough'g Beautiful fine mild dry day, the Butterflies, gaily flying

334 Thomas Ladd of Beckhampton buried on 11 March 1838 at Avebury, aged 42, *Wiltshire, England, Church of England Deaths and Burials, 1813-1916 for Thomas Ladd* [accessed via Ancestry, 28 July 2020].

335 Day.

336 In the newspaper report on this event at Watcombe Farm, North Fawley, near Lambourn, the farmer was described as a 'widow having five children', *Devizes and Wiltshire Gazette*, 8 March 1838.

13. Thin RAIN nearly all day, Plough'g for Oats
12th Died Dr Headly <...> of Devizes Aged 82
14th Died Mr M P_ Hitchcock of Avebury near 62 [337] [Pencil note] } 60 - the 13 Novr next [This entry and the previous entry written within a three-sided thick black border]
14. Very mild fine day damp eve Plough'g for Oats
15. 20 Qr Barley to Devizes with 5 Horses - Very fine mild day
16. Barley Rick the small one put into Barn - Plough'g finish'd for Oats | Frost & fine morn'g, boisterous after and very cold hail Storms
17. SNOW in morn'g, quite cover the ground boisterous dry'g winds after | Horses in Stable morn'g, Plough'g for Turnips after
18. Sunday, fine mild day
19. OAT'S, black Tartery 35li pr Bushl @ 24s Qr Sowd the further Down Piece 8A 3R 20P - wth 14 Sacks want 2 Bushls Beautiful grow'g day
20. Horses Thrashing Barley | Very cold boisterous DRIVING winds and storms of hail and rain
21. Thrash'g as yesterday & boisterous winds
17th Died Mr Richd Alexander Chippm Aged 82 Years. [This entry written within a three-sided thick black border]
22. CHALK (began) Cart'g 20 Waggon Load for the Barn | One Man began face'g the Chalk | Very cold Winds and Storms
23. Plough'g began for Barley bottom of Tan hill Piece | Cold SNOW Storms
24. Sharp frost and very fine calm morning - Wind and rain afternoon | WHEAT=RICK (the 2nd) put into Barn, having no Straw for the Coach horses | Plough'g as yesterday
The effect of the SEVERE FROST now felt, in the Wheat, Vetches Winter Beans, Winter Oats, and every thing except Rye, either very much thin'd or whole'y distroyed, MY WHEAT, Ten acres of the 18 Acres not a blade to be seen across many of the Ten stride Lands and the remaining 8 Acres very much thined, and every SWEDE and Turnip perish'd, yet continue the Sheep on them at the risk of doing them an injury, but they do better than expect'd, began giving the LAMB'S, White Pease grittle'd mix'd with Bran[338] [Diary turned landscape for this entry]
1838 (*sic*) My Brother W P of K, have not put his Ewes and Lambs on his rotten Swedes, but Plough'd in for Barley what was an excellent piece of Swedes rather than risk the injureing his Sheep. My Neighbour J W of B, have Plough'd in his Winter Beans and <...> Acres of Wheat [339] [Pencil

337 Michael Pontin Hitchcock of Avebury was buried on 20 March 1838 at Avebury, aged 61, *Wiltshire, England, Church of England Deaths and Burials, 1813-1916 for Michael Hitchcock* [accessed via Ancestry, 28 July 2020].

338 grittle'd – probably griddled.

339 W P of K – William Pinniger of Kennett. J W of B – John Wentworth of Beckhampton.

note] Mr Brown hay

25. Sunday, cold drying day
26. Rain in morn'g, very drying day | Cart'g Dung from Inn, to make a Stand'g Pen for Ewes & Lambs to have Straw & Water when are fold'd on the Fallow for Swedes with Hay, for the purpose of LAMBS RUNING OVER, THE NEW FIELD adjoining having no other GREEN Keep for them till the Water mead is ready which is very backward having no Water, the SPRINGS, being low every precedeing Year, the Springs have risen in Meadow before my house, which is now quite dry
27. Frost & beautiful fine day | Plough'g for Barley
28. Frost & fine warm day Horses Thrashing Wheat
29. Frost & very fine warm day | ROLLING WHEAT, the ground being very light & the Wheat, thin daily
30. COW, BULL'D (Mr Browns) Horses Thrash'g, very fine warm day | POTATOE'S, SEED'S, & PLANT'S, put in the Garden every thing green in Garden Kill'd wth Frost | HORSE of Mr Jno Large, borrow'd to assist in Plough'g
31. Horses Thrash'g & Rolling Wheat | Foggy damp day

April

1. Sunday, Cold dry day
2. 3 Horses to Wroughton wth 24½ Qr Ray Seed | 3 Horses Plough'g, Sharp Frost & fine day | 42 Poles from Wroughton
3. Horses finish'd Thrash'g the Wheat Rick, and Rolling the Oats Down Piece | Dry morn'g, little rain afternoon
4. CHALK, brot 15 Waggon load for Barn | harsh dry winds | LAMB'S, now eat the Pease & Bran, & do well | The rotten Swedes in general are gone mine continue another week, give my Ewes with them 4 Truss's Meadow hay each morn'g and eve'g Barley Straw & Wheat [In pencil: Kerving] mid=day and night do much better than expect'd This Month will be trying to find Sheep Keep
5. 24 Sacks Wheat to Devizes & Plough'g | Boisterous Winds all day
6. Very Boisterous Winds all day | Horses Ploughing Tan hill Piece for Barley
7. Horses Ploughing return d to Stable 11 oClock heavy rain | Very boisterous Winds & Storms all day. | 7 CHERRY Trees planted
8. Sunday, - Dry'g cold winds, boisterous [Pencil note in margin] E.Crook 8-15-2
9. Mild calm Sunshine day | BARLEY=RICK, the last put into Barn The 3 Yatesbury Labourers assist'g got intoxicated and abusive
10. 3 Horses Thrash'g Barley and 3 Plough'g | Cold drying Winds
11. Hot dry winds the Sheep feel it & drank 2 Barrels of Water | 3 Horses Thrash'g & 3 Horses Ploughing | FAT PIG 11 Score @ 9s cost the 22nd Feby 62s Oats appearing sowd 19th March

April 6th Lady day. HAY, half the Mead Rick and half the Field Rick remain, and consumed 5 Tons Seed Hay 9 Tons Old field 15½ Tons New field 7 Tons Meadow [Total] 36½ Tons as Truss'd. 2 Truss's for a Cwt [340]

April the 10th The SWEDES finish'd, were dried up like a Piece of Brown Paper and disappeared, gave the Sheep a Tie of them every day, which they eat without injury to our surprise, and continued in the Lambing Pen to this day being 8 Weeks without injury from the heat of the Dung. [Diary turned landscape for this entry and the previous entry]

[Undated] The Ewes nearly 200, have had from 3 to 4 Truss's of Hay mornings and same evening, with Barley Straw mid day and night with 2 Hogsheads of Water pr Day, with such Keep, I consider it Providential, the Ewes & Lambs do so well as a Compensation for the loss of the Swedes

10. The Ewes & Lambs removed to the Piece for Swedes fold'd wth Hay & go twice a day to a Stand'g Pen to Straw & Water, the Lambs run on new Field & have grittle d PEASE in Standing Pen, doing very well, The Tegs fold'd for Swedes wth HAY & Water, the Sheep never done better this time of the Year, must continue at hay & Water no prospect of grass for them | Mr Brown & Mr Kemm, } in their Meadows for want of other Keep for their Sheep | Mr Sl Wentworths Tegs on Down, no grass

13. Mr Jno Wentworths Ewes & Lambs in Silby Mead

18. Mr W P_ Kennett Ewes & Lambs in his Mead | Robt Philpot Ewes & Lambs in his Mead

12. Dry harsh Winds, 2 Ploughs

13. GOOD FRIDAY 2 Ploughs ICE thick on on the Water Troughs, dry harsh Winds | 2 Ploughs

14. 3 Horses Thrash'g & 3 Horses harrow'g for Drill'g Barley, and Nag removing Hurdles Lamb'g Pen

15. Sunday, very cold Tempestuous, & harsh Winds to carry of the tops of the Straw Ricks

16. Frequent very cold storms of SNOW & hail Finish'd thrash'g the last Barley Rick | Rolling & harrow'g for drilling Barley

17. Exceeding sharp FROST, and SNOW to cover the ground in morn'g, and very cold Snow storms in day | Finish'd Plough'g Swede Ground for Barley

18. BARLEY began Drilling, bottom of Tan hill Piece Wheat Rick the 3rd put into Barn, only one remain out, exceeding cold boisterous winds and Snow Storms

19. VETCHES, winter, Drill'd middle Down Piece where Vetch's quite Kill'd by the Frost | Very cold cuting winds

20. Frequent heavy storms of SNOW and HAIL | Harrow'g & Roll'g for Drill'g Barley tomorrow

21. Barley Drill'd, - about 9 Acres, all the Swede land SNOW in the morn'g

340 2 Truss's for a Cwt – that is, 56 pounds per truss and therefore this is old hay – see Glossary.

to cover the the hills & severe white Frost, mild still morn'g cold blowing afternoon

22. Sunday Cold thin rain at times all day
23. Mild dry morn'g, - cold rain afternoon | The 6 Horses in 3 Ploughs the Pease land for Barley
24. The 6 Horses in 3 Ploughs the Pease land for Barley | Cold harsh dry day
23. W P_ Ploughing at East Kennett where the Wheat Fail'd, being Kill'd by the Frost for sowing Oats, many other People have Plough'd their Wheat Field and sown Spring Corn, half my Wheat Field is nearly distroyed, will wait till the 6th of May KNOWING GOD ALMIGHTY can cause a great change and give me a good Crop in whom I trust, as he did (See the 2nd Kings 19th Ch 29 V) to the good King Hezekiah by giving him a second self sown Crop, spring up from the first.
25. Mr Wentworth, Plough'g his Wheat=failing field to sow Pease | Finish'd Plough'g Tan hill Piece for Barley | Very cold boisterous, dry day
Died at Lowden April 24th Mr Geoe Beames Senr [This entry written between triple lines above and below]
26. Exceeding cold dry harsh winds | Harrowing & Rolling for Drilling Barley Tan hill Pie
27. Cold harsh winds | BARLEY, finish'd Drilling Tan hill Piece 23½ Acres with 26 Sacks
28. CHALK, finish'd diging for building Barn | CHALK, brot home 17 Waggon Loads | Exceeding pierce'g cold harsh winds in the morn'g - mild'r & calmer after | SEED'S sow'g to be roll'd in with with the Barley
29. Sunday, Very cold harsh winds
30. CHALK, 17 loads finish'd Cart'g, about 70 LOADS for the Barn, the air more mild
COPY TO MR WM BROWN OF HORTON Sir By your request I acknowledge the permission from you of useing occasionally a drift road on Horton Down in your occupation, for my conveniance, which I am willing to confirm by paying your Landlord a small annual quit Rent, say Sixpence or a Shilling. I am Sir Your Hble Servt Beckhampton 13th Octr 1836 Thos Pinniger To Mr Willm Brown Horton
N B This day 26th of April 1838, Mr Brown made his first demand on me of Six pence in Devizes Market, which I paid him, having no knowledge of a right of road on his Down. [Rest of page blank]

May

1. Mild morn'g, Cold boisterous Storms after | Tegs began feeding small Turnip greens Whitelands | Thrash'g Grass & rolling Grass | LAMB'S, Tail'd 159 (220 put to R) Cut - 74 Rams [341]
2. Damp blowing morn'g, mild after | Thrash'g Wheat & Roll'g Barley

341 R - Ram.

3. CUCKOO, hear'd 1st even'g Down Firs Plough'g Whitelands for Barley Rolling Grass, Mild Day, little Rain even'g
4. Rolling Barley & Plough'g Whitelands | Beautiful mild morn'g, heavy Shower after
5. THUNDER heavy & lightening fm 5 o-Clock morn till 8_ & heavy rain from 8 to 11 oCl | Horses in Stable, morn'g & Thrash'g Wheat Rick afternoon FINISH'D, no more Thrash'g till have a New Barn - one Wheat Rick in Yard

   Died this morn'g, Mr Broxholms Daughter at Calne Aged [...] [342] [This entry written between double lines above and below]
6. Sunday Cold harsh Winds
7. Cold winds & warm Sun fine Day | 2 Waggons & 6 Horses to Bencroft Hill | FAGGOT'S 1½ Hd & 12 Bdls Rods,
8. Cold Winds & fine day | ROUND=TAIL'G the Sheep 330 by H-Chivers the 2 Shepherd Boys & Geoe Titcomb gave them Bread & Cheese & Beer mid=day, finish'd 3-o-Clock [Pinniger then counted his sheep - see Appendix B]
8. Mr Geoe Brown & Mr Jno Wentworths Sheep finish'd feeding their Meads and began their Field Grass, my Sheep continue at HAY 12 Truss's pr Day & Water with Straw twice a day, the Lambs 1 Bushl grittled Pease with Chaff | Plough'g Whitelands & Rolling Barley
9. Finish'd Plough'g Whitelands and harrowd & Roll'd - Fine warm day
10. BARLEY, Drill'd 5½ Sacks in Whitelands 4¼A Cold boisterous Winds all day | BARN, began uncovering for rebuilding
11. FROST exceeding sharp, People hay their their Cows twice a day | 22 Qr Barley to Devizes with 4 Horses and 1m Slates returned, Harrow'g Wheat 2 Horses
12. Ploughing for Vetches Down Piece Sharp Frost Cold harsh winds
13. Sunday Sharp Frost Cold harsh winds
14. Very sharp FROST, ICE on the Water Troughs | Plough'g for Vetches, Rolling Barley & Seeds sown in Whitelands | People Drilling Barley, where Wheat fail'd | EWE'S & LAMB'S, began feeding Silbury Mead wth 2 Truss's of Hay each night & morn and the Lambs 1 Bushl of grittled PEASE, mornings, continued in Standing Pen till now, to get better Grass, in Mead and keep from the Field Grass, other People finish'd their Meads, & feeding their Field grass The TEG'S continue in Stand'g Pen with HAY, STRAW & Water, having no grass on the Down, a very hard time for the Cattle
15. ICE'Y FROST, warm Sun, frosty cold evening Plough'g & harrow'g for Vetches | Thomas's first day at SCHOOL, Mr Cornwell Avebury [343]

---

342 Inhola Broxholm of Calne buried on 10th May 1838, aged 14, *Wiltshire, England, Church of England Deaths and Burials, 1813-1916 for Broxholme* [accessed via Ancestry, 28 July 2020].

343 William Cornwall, dissenting minister & school master, TNA: HO107/1185/2.

16. Ice'y frost, morn, warm in Sun | VETCHE S, Drill'd 6 Bush of Spring & 4 Bushls of Winter on <...> Ac: home Down P
17. 160 Bushls of Lime, 4 Horses from Calne 2 Horses for Turnips Ploughg | Frosty mornings continue & harsh dry Sun & wind
18. Frosty mornings and exceeding COLD MORN'G, mild'r even'g Ploughing as yesterday | MASON'S 6, began Building the BARN, face'g and sorting the Stone till now.
19. Ploughing finish'd the Turnip land & harrow'g it | BOISTEROUS, in the extreme all the morn, and RAIN, remainer of day, only one DAY RAIN since the 23rd April, the ground very dry and CRACK'D, Vegitation backward.
20. Sunday, rain, thin all day, & mild
21. Carting Dung, Standing Pen 5 Men filling | Fine mild dry & growing Day | HAY, the Tegs DISCONTINUED, & removed from the standing Pen to the Down and fold'd on the grass Nether Down Piece | The Ewes CONTINUE HAY, twice a day, and the Lambs a Bushel of Corn | Thwarting began the Swede ground 20A Piece
22. Thwarting and Carting Dung | Rain in morn'g to stop, Dung Carters & Masons
23. Cold dry rough winds | Plough'g as yesterday & Plough'g for Potatoes
24. Plough'g as yesterday & Plough'g for Potatoes | Cold dry harsh winds | POTATOE'S, began diging in, Tan hill Piece
25. Very cold harsh winds continue | CALF of new Cow weighed 102 li @ 6d pr li | Plough'g for Swedes & harrow'g for Turnips | Harrow g thin Wheat before hoeing
26. HAY, the Ewes this morn'g, and DISCONTINUED would continue to eat hay if given them the Grass in Silbury mead improve, but not half so much as other years, Keep them short of grass to Keep from the Field grass, which grow very little | Very cold morn'gs & evenings & cold harsh winds | Rolling & harrow'g - mid: Down Piece for Turnips 22nd W P_ K & Mr Kemm, finish'd feeding Meads | RAPE=dwarf sow'd 8 li on <...> Acres of Vetches Down Piece
27. Sunday, cold boisterous winds
28. CALF from Corton 15s to graze | Rain in morn'g to stop builders & Labourers | Thwarting for Swedes
29. Rain a little driving in morn'g, harsh boisterous winds after | TURNIP'S began sowing 4¼ Acres, mid:Down P Plough'g as yesterday | WHEAT=HOEING, by 5 Men from Yatesbury being very thin & full of Weeds | SEE the 29th May last Year, altho the Wheat is now very backward, the Trees are fully in leaf except the Ash, which is only in bud
30. Fine mild Day, Plough'g as yesterday | 2 Load of Stone from Calne

---

Census returns. 1841 census. Wiltshire Hundred: Selkley, including Parish: Avebury, ED1 [accessed via Ancestry, 6 July 2020].

31. 2 Load of Lime from Calne 160 Bush 2 Horses Plough'g as yesterday 6 Men hoeing Wheat | Partial growing Showers, afternoon

June

1. PONEY, from Tytherton Mr Clutterbucks 2 Load of Stone from Calne | Warm grow'g Showers afternoon, Plough'g as yes [344]
2. Warm grow'g dry day | BARN, the Window frame's set in 2 Ploughs thwart'g for Swedes
3. Sunday - Some growing Showers in morn'g
4. Fine dry day after the morning | 6 Horses to Honey Street for the first Load of Timber for the Barn on a Carriage
5. 2 Ploughs thwarting for Swedes | Very warm dry day
6. Plough'g as yesterday dry day
7. Plough'g as yesterday dry day cold evening
8. Plough'g finish'd dry day mild evening | POTATOES, plough'g in, having 16 Sacks and no sale for them, and spoiling
9. POTATOES, finish'd Plough'g in. 8 Women droping Kept 2 Ploughs going the 5 lands of East side of upper long Piece 2¾ Acr part Plough'd in & part dug in | POTATOE'S - the Standing Pen. Tanhill Piece | Fine dry Day
10. Sunday, very cold dry morn'g, warm'r, showery eve'g
11. 6 Horses to Honey Street for the Carriage of Timber for the Barn | RAIN in the morn'g, and heavy afternoon, much want'd for the Barley, Grass & Potatoes

8. PEASE the LAMBS discontinued, having consumed <...> Sacks and <...> Sacks of Tail Wheat (Chickens meat in meal) mix'd with the Grittled Pease.

12. RAIN heavy in morn'g, and evening | 3 Load of Stone from Calne
13. 4 Horses to Honey Street for Timber for Barn | 2 Horses harrow'g for Swedes | THUNDER, LIGHTENING & RAIN very heavy at 5 - oClock evening, which Kill'd two of Farmer Philpots Lambs and struck down his Shepherd and Dog, without injury, very much stranded my Potatoe and Swede Land [Pencil note in margin] Cow Bull'd
14. Dry morn'g rain all afternoon, stop the Builders | Ploughing Down Piece Grass fed
15. Dry morn'g rain afternoon | Dung, Carted from the Inn 4 Men fill'g, 6 Horse 4W [345]
16. Dung, Carted from home Yard 4 Men fill'g, 6 Horse 4W | Mild Day, damp at times
17. Sunday, Rain morn'g, dry after
18. SHEEP=WASH'D, at Calstone, by 2 Men Jno Bird the Shepherd & Jno Davis assisting, wash'd 100 pr hour, 2 hours going from Down Piece 16 Score & 6, wash'd by ½ pass 10 'o Clock USAGE, large loaf, Bacon & Cheese in

344 yes - yesterday.

345 W - Waggons.

proportion gave great satisfaction, the Bottle 5 Qts 3 parts Beer & 1 part Ale, wish'd for rather more See usage 1834 | Ploughing Down Piece, Grass fed off - Rain heavy in night & morn'g, Showers after | WHEAT=EARS, at Calstone, the first have seen

19. Rain in morn'g dry after | Plough'g as yesterday
20. 2 Load Stones from Calne | Rain thin driving frequent
21. Grass, Nether Down Piece, the Tegs finish'd the Down only for them | HORSES in the STABLE, no work they can do the land to wet to sow Swedes being full of Charlock, the Men and Boys drawing Charlock from the Wheat
22. Showers in morn'g and winds | SWEDE'S began sow'g uper long Piece on the thwarting, the RAIN the last 11 days prevent'd sowing before | The dry sheep finish'd feeding the Grass 4½ A Nether Down Piece & fold'd, the FOLD began it again on the Fallows for Turnips, having no other land to fold at present.
23. Very fine day | Plough'g and sowing Swedes after | POTATOE'S began hoe'g in the Field, and HARROWING Potatoes, those out of ground and those not out, to check the Charlock | STRAW, discontinued serving the Coach Horses at the Inn, not having enough to thatch the Ricks
24. Sunday Midsummer=day, very sultry warm
23. STRAW to the INN, DISCONTINUED, for the present, not having enough to thatch Ricks
25. Very fine day, RAIN unusually heavy at 8 oClock | Sowing Swedes | MOWING began Field grass Ram stag 60 li
26. Sheep housed very wet, heavy storms morng | DUNG from the INN, and home Yard
27. Sheep on the Down by 3 oClock morng, little rain before 7 and showery after, the Sheep housed and began SHEARING, half past 7 - Wool continue wet, finish d half pass'd 7 o Clock 326 Sheep 6 Men | Ploughing for Swedes
28. Ploughing for Swedes and a load of Stone & lime fm Calne | CORONATION of VICTORIA exceeding fine day | HORSES began Field GRASS in the Stable
29. Swedes, Plough'g & Sowing Very fine till 4 o Clock when commenced violent lightening, Thunder, heavy hail and rain for two hours incessantly
30. Fine day rain evening, Ploughing home Down Piece too wet for Swedes

July

1. WHEAT and BARLEY, Ears, Just appearing | Sunday, Rain in night dry day
2. Plough'g for Swedes, field Grass finish'd mow'g | Fine day | Sheep finish'd, and removed from SILBURY MEAD, having kept 160 Couple's 7 Weeks from the 14th of May, which enables me to mow my Field grass.
3. Very fine warm day, Littlecote | LAMB'S WANED, on the uper little Down

only being very good, the Ewes with the Young Sheep on great Down, all Fold d on home Down Piece | Dry home mead mow'd | SWEDE'S finish'd sowing about 15 Acres with 30li Seed

4. TURNIP'S began sow'g adjoining Swedes Warm fine day | BARN began puting the ROOF on
5. Hay began carry'g, sultry Day | TURNIP HOE'G, began mid=Down Piece
6. Alarming Thunder lightening and rain 2 oClock morn'g to 5 o Clock, dry after | 15 Sacks each Wheat & Barley to Devizes
7. Storms to prevent haymaking, Plough'g for Tu: HORSE'S less CORN, 3 Bushls instead of 5 [346]
8. Sunday, Dry day
9. Harrow'g Turnips, and carry'g Hay again | Very drying day
10. HAY, finish'd carrying, Field & Mead and done for the present
11. Fine day some few drops of rain | 2 Loads of Stone from Calne to finish Barn | The Swedes only shewing rough leaf, the 1st sowing
12. Very warm fine day | TURNIP'S finish'd sowing the 20 Acre Piece adjoining the Swedes & the Little Down
13. Plough'g athwart home Down Piece, Very fine Day
14. Plough'g and 3 Horses to BOX for Barn, RAIN all the afternoon
15. Sunday, Dry morn'g heavy storms after
16. BARN, the Walling finish'd, rain morn'g dry after 25 Sacks Wheat to Devizes 700 Slates returned for Mr Jno Large | HORSE'S CORN, REDUCED, to 2 Bush pr Week for 6 Horses, having little work & hearty grass
17. Plough'g and fine day
18. Plough'g and fine day VETCHE'S, the Lambs go feeding, see, drill'd the 19th April | SLATING, began the Barn
19. Plough'g &c Down Pices for Turnips | MARKET'S, advancing, having shortest supply ever remember, sold Wheat at 36s 6d pr Sack weigh 12 Sco 2 li Net, thrash'd 3 months. See sold Wheat 12 Sco 7 li @ 18s pr Sack 14th Jany 36
20. TURNIP'S, began sowing home Down Piece | Rain morn'g, very fine after
21. Turnip Sow'g as yesterday, very fine after
22. Sunday, very fine after
23. SWEDE'S began hoeing, very small being full of Charlock, in a few days the Swedes could not be seen (the 2nd sowing) N B See the 22nd June sowd, 10 or 12 Days too late, the last=sow'g but just in third leaf, should be hoed | Ploughing & sowing Turnips, very fine day Harrow'g the Turnips that is hoe d
24. Ploughing & sow'g Turnips, very fine day
25. Ploughing & sow'g Turnips, some showers & fine again TURNIP S

346 Tu: - Turnips.

FINISH'D sowing half of home Down Piece FOLD removed from the the home Down Piece to the Vetch ground fed for Wheat mid=Down Piece

26. Fine morn'g rain heavy afternoon, HORSE'S in the STABLE, no land to Plough till have some Vetches fed | The Carters and Boys earthing Potatoes, having but little to employ them whilst the Horses are resting (19¼ € Coal) | VETCH'S geting very good, the Summer ones drill'd the 16 of May as forward as the Winter ones drill'd the 19th of April both shewing blossom, the Ewes follow the Lambs & have plenty [Pinniger then recorded prices at Overton Fair - see Appendix H]
27. Fine day, Men continue at Potatoes, and our two Men only hoe'g all the Turnips & Swedes
28. Dry morn'g, some Showers afternoon | 2 Horses only harrowing Swedes
29. Sunday, heavy Storm of rain & hail, morn'g
30. Frequent slight Showers all day | Waggon to Honey Street for Roof of Machine house
31. Very fine warm day | Horses in the Stable & Brewing

August

1. Horses in the Stable Beautifull fine morn'g, rain all afternoon The Masons building Machine house
2. RAIN till 3 oClock, Dry after Devizes Market ADVANCED 2s pr Sack 41s highest 37s 6d the average pr Sack
3. RAIN morn'g dryer after
4. RAIN morn'g dryer after yet too wet to hoe the Swedes, now growing rapidly, and only half the Piece of <...> Acres hoed the first time | The building part of the Machine House finish'd, which with one side of the Barn to be Slated, and the Walls painted to make a finish
5. Sunday, heavy shower after mid:day EWE, blown on Vetches
6. VETCH'S the Horses began have'g finishd Clover | Many heavy driving Storms, the wetest Tan hill Fair I remember, a very drary prospect from the Hill towards a harvest, Wheat and Barley being of a dark green colour, far from harvest
7. Some Showers, the Barometer now rising the Turnip hoe'rs hindered 4 days by the rain | The Vetches mowed for the Sheep to prevent blowing [Pinniger then recorded the duty on wheat - see Appendix I]
8. Some thin Showers morn'g, fine & dry after | PONEY, returned to Tytherton
9. Fine warm Day, some little rain | Plough'g after the Fold Vetches fed off, for Wheat | SHEEP drew 43 Wether Tegs put wth Lambs for Marlbro fair | BROCOLI & Winter or spring Greens Planted
10. Warm lowering, some rain Plough'g as yesy | GRASS Silbury mead, the Horses began, the Sheep requiring all the Vetches
11. SWEDE'S, finish'd hoe'g the first time only. Harrow'g & Plough'g as

yesterday | Very warm and the finest day for some time

12. Sunday, some little thin Rain
13. MOWING began SILBURY MEAD, very fortunate in the weather, being taken up fine, the hay about before this time spoil'd | Harrowing Turnips next Swedes, want hoe'g | Plough'g Vetches fed off | Beautiful fine day
14. Beautiful fine day Plough'g as yesterday
15. Beautiful fine day 28 Sacks Wheat to Devizes & 1m Slates returned
16. Fine harvest Day, REAPING at Brumham quite general, begin tomorrow at UPAVON &c and to appearance 3 Weeks hence at home [347] | HAY, began carrying Silbury Mead MARKET at Devizes, exceeding large suppose'd to be 6000 Sacks, only 500 sold at a reduction of 2s pr Sack, the late rapid advance to 43s pr Sack, induces the holders of Wheat to bring it to Market, 3 Weeks since the Market only a Tythe as large, see in AUGUST 1835 the best Wheat new 13 Score Net at 18s pr Sack
17. Damp morn'g & even'g | HAY, finish'd carrying Silbury mead 600 Slates fm Devizes to finish Barn
18. Ploughing after Vetches fed | BARN, FINISH'D SLATING 18th July began Slating 18th May began Building 22nd March began Faceing Chalk 6th March began Diging Chalk 4 Carpenters 12th Septr finish'd the Barn [348] | Foggy morn beautiful fine Day | 20 Bu: of Lime from Calne
19. Sunday, RAIN all the morn'g, dry eve'g
20. Very drying all day | Flints, Carting for the Parish
21. Flints, Cart'g for the Parish | Dry morn'g, - Storms afternoon
22. Marlbro fair, heavy Storms, nearly all day Sold 40 - Two tooth Wethers @ 26s cost 12s 3d, Michs 37 [349] | 60 - Uneven Wether Lambs @ 20s | 25 - Cul Lambs of all @ 13s | <...> | Horses in Stable or doing little
23. Horses in Stable or doing little some Storms | Market at Devizes large and still declining
22. CALF (the 2nd) 174 li @ 6½d is £4 14s 3d = cost 15s | CALF from Cleveancey 10 days old 24s
24. Fine drying Day, REAPING, near finishd at Bremhill & neighbourhood | Horses in Stable, Carters cleaning up the Barn
25. Horses in Stable, Carters cleaning up the Barn | RAIN at times all day
26. Sunday, fine dry day
27. TURNIP'S Nether Down Piece began hoeing sowd 20th July Sultry warm day, the most ripening this summer, Horses to Honey Street for Timber a Potatoe Steeming house
28. Ploughing after Vetches & harrowing Down Turnips Rain in night, dry day
29. Ploughing as yesterday, very drying day | SWEDE'S finish'd hoe'g the 2nd

---

347 Brumham - Bromham, 3½ miles north-west of Devizes.

348 The remainder of this page of the diary had been left blank, thereby enabling the later entry of 12th September to be added.

349 Michs 37 - Michaelmas 1837.

time

30. Warm fine day} Messrs Brown, Kemm & Saml Wentworth REAPING
31. Warm fine day
30. TURNIP'S, finish'd hoe'g Down Piece Harrow'g &c after Vetches | OAT'S began REAPING, further Down Piece The Wheat not ripe, the thin Wheat is branch'd out and very green, a part of it is yet in blossom, and all is very uneven ripe
31. Warm ripening day | BARLEY, began mowing Tan hill Piece Gravel Pits. fill'g with road earth | Fine warm harvest day | BARN, the Masons finish'd, by pointing and wash'g inside. SALE EWE'S, drew & put with Chilver Lambs on Vetches & little Down the (& RAM'S) Stock Ewes follow on the Vetches [Pinniger then recorded his sheep numbers - see Appendix B]

September

1. Fine harvest day | Ploughing a Fold drift after Vetches | VETCH'S, the Horses began again | The Potatoe House began building
2. Sunday, very fine day
3. very fine day Plough'g as the 1st
4. Thick fog'y morn'g fine day Plough'g as the 1st | 6 Men DISHEARTENED reaping OATS, further Down Piece and gave up (being heavy crop) and 3 Men & 2 Women continued reaping.
5. Some misly rain at times, & Barometer fallen to N in much rain | OAT'S finish'd reaping, but not tied | Intend begin reaping tomorrow morn'g | GRASS, the 2nd Clover began mowing
6. RAIN, in night very heavey, 1⅝ In of water fallen in a Dish, and Storms in the Day, no reaping | Plough'g a Fold Drift
7. Carting Dung before Fold, frequent Storms | REAPERS 10, returned home, without begin'g the Wheat, waiting till Monday 10th
6. Geoe Maskelyne, convicted of maliciously, shooting at Mr Bryan Rumboll the 17th Jany with intent to murder him was EXECUTED, this forenoon (aged 35 Years
8. Very cold day some small rain | Carting Dung as yesterday
9. Sunday, Partial Storms, fine Day | SERMON by the Revd Mr Young at Lyneham on the Death of Geoe LMaslen 119 Psalm, the 9th Verse
10. Cart'g Chalk from the Rick Yard &c | Barley finish'd mowing Tan hill Piece | Very fine dry Day | REAPING, began 9 Taskers, and only 9A fit to cut, the other 9 Acres, great part very green | HORSE=CORN, none the last month, now allow 2 Bush pr Week for 4 Ho | Exceeding sharp Frost
11. Exceeding sharp Frost & finest harvest day | Horses harrowing
12. [Pinniger recorded prices at Wilton Fair - see Appendix H] CORN, the FIRST began carrying Carried 28 large loads of Oats, from further Down Piece 8A 3R 33P produced 1177 Tythings, is 130 Tythings pr A at 10s 6d pr A | Fine day | 4 CARPENTERS, from Calne the last 6 Days to finish the

BARN, and and the Potatoe Steeming house

13. BARLEY, the first, carried 23 Loads Dble handd Damp morn'g fine day
14. Carrd 28 Loads of Barley by 4 oClock, Dble handed Ten Men and Horse, treading in Barn | Damp day, mizly rain in evening
15. Rain in night to prevent Barley carrying beautifull mild calm afternoon, 4 Men & 7 Women leaze'g Wheat morn'g & haymaking afternoon
16. Sunday Very fine day
17. Fogg'y morn'g, very drying winds after | Harrowing Down Turnips and carried 17 loads of Barley after | Finish'd mowing the 2nd Clover
18. Very fine day Barley dry | WHEAT, the first carried this Harvest 8¾A 554 Tything being 60 pr Acre, the remainer is very green and very bad Crop | BARLEY, carried remainer of Tan hill Piece 23 Acres } 13 Loads 17 18 23 [Total] 81 loads, all in New Barn quite full
19. Dry night, thin rain morn'g, dry after | BARLEY, Whitelands carried 12 loads 4A, in Rick | HAY, the 2nd Clover began carry'g, 6 large loads, bottom of Tan hill P
20. Plough'g and carrying Hay | REAPING, finish'd, rain in night, dry afternoon
21. Thin rain morn'g dry afternoon | Reapers Paid, and our two Men mowing Stubbles no Person at regular harvest Pay | RAM'S, turned the 2 of Mr Browns to Stock Ewes the 2 tooth Ewes with Sale Sheep, and Mr Pumphrys Rams | Plough'g Fold drift & hack'g the Clover Hay
22. Dry night, and as beautiful fine drying day for hay making as any this Summer. | HAY, carried 9 loads, extraordinary weather for 22nd of September
23. Sunday, dry morning rain evening
24. Ploughing, Vetches finish'd feed'g, began, see the 18th July | RAIN, almost incessantly, afternoon
25. Thin rain at times all day, Plough'g | TURNIP'S, began feeding, mid=Down Piece
26. Ploughing Vetches fed off Fine drying Day, the HOT hay Rick opened and a hole cut, expect'g fine to carry hay tomorrow Rainbow very fine evening & little Rain
27. RAIN heavy all night & morn'g, dry'g after Horses in Stable, Men in the Barn
28. Very fine day, Ploughing before Fold.
29. Thick mist all day, Plough'g Oat Stubbles for Vetches
30. Sunday, Very clear fine Sunshine day | OBIT Johnothan Caswell 53 [350]

October

1. Barometer to F in fair, dry, yet thick air the second crop of Field hay spoild

---

350 Jonathan Caswell of Avebury buried on 5 October 1838 at Avebury, aged 53, *Wiltshire, England, Church of England Deaths and Burials, 1813-1916 for Johnothan Caswell* [accessed via Ancestry, 28 July 2020].

by the rain see 24 & 27th Septr which also set the Rick heating, cut two holes through it | Farmer Philpot Cart'g his Wheat from Rick to Field (mowed & carried from Swath the 22 Septr took all the Rain 24th & 27th | Plough'g for Vetches

2. Plough'g for Vetches. Fine drying day | WHEAT=HARVEST finish d by carrying 7 loads from 9 Acres very bad Crop & bad Corn, expect not a Sack pr Acre on a part of it
3. HAY, finish'd carrying, Boisterous & very drying Harrowing [Pencil note] Iron 1-1-14
4. VETCHES sow'd 2 Sacks further Down Piece, Oat Stubble Fine clear drying day | Sheep began feeding Wheat Stubble, and Field Grass after the second Clover Crop of hay for Wheat, too late but no help for it, having had plenty of Vetches & Turnips to feed, to Keep the Sheep from Field grass
5. Cold dry morn'g, very warm afternoon Plough'g the Couch'y Vetch land
6. Plough'g the Couch'y Vetch land Cold morn'g and even'g
7. Sunday - fine day OBIT Thos Ash 66 [351]
8. Fine dry day, Rolling & harrow'g Couch'y land
9. Fine dry day, Plough'g & harrow'g 3rd time
10. Fine dry day, Harrowg & Rolling 3rd time | LAMB'S, dip'd 73
11. Plough'g began for Wheat sowing | middle Down Piece | Windy day damp eve'g | FAT PIG 18 Sco 14 li cost 62s the 22nd Feby fatteng since STORE PIG'S bough 5 @ 29s 6d ea to fatten
12. Exceeding Cold harsh Winds | WILLIAM DAVIS, - Carter & 2 Sons left | DAVID RICHENS Carter & 3 Sons entered | DUNG began Cart'g for Wheat
13. Plough'g Down Piece for Wheat | SNOW, very heavy & cold for an hour mid=day | JOHN DAVIS discharged himself, the last of the Family, Father Mother 6 Sons & Daughter happy release
14. Sunday - Damp day
15. Dry day & mild'r Cart'g Dung
16. Dry day Boisterous Cart'g Dung
17. Plough'g Down Piece for Sow'g (Mr Rumboll Lit: began) Dry day | Sale Sheep drew again 5 - full, mouth Ewes 4 - 6 tooth Ewes 12 - 4 tooth Ewes 27 - 2 tooth Ewes 2 - 2 tooth Wethers [Total] 50
18. Cold dry day Plough'g as yesterday 1T 4C 2Qr Coal Mr W, Waggon
19. Cart'g Dung - fine mild day
20. Ploughing began Tan hill Piece one earth for Whe Damp morn'g, very fine day
21. Sunday dry mild day
22. COLT, a dark gray 2 yr off horse <...> H-Reeves Ploughing as 20th dry

351 Thomas Ash of Beckhampton buried on 12 October 1838 at Avebury, aged 66, *Wiltshire, England, Church of England Deaths and Burials, 1813-1916 for Thomas Ash* [accessed via Ancestry, 28 July 2020].

mild day | POTATOE'S, began diging in Field 36 Sacks filld by 14 Girls & 2 Men

23. Plough'g as yesterday, very mild, Vetch's appe= Sheep continue to have Maggots [352]
24. Harrow'g Down Piece for Drill'g Wheat Rain heavy in night, fine day
25. 40 Sacks of Potatoes, dug. fine day | Harrow'g Tan hill Piece for Drill'g
26. Cart Dung Drying, one Storm mid=day
27. Cart Dung RAIN heavy from 3 o Clock
28. Sunday - Dry day
29. TREMENDOU'S=HURRICANE, commenced this this morning about 1 oClock & continued till day light, blowing down great number of Trees unroofing Buildings & Ricks, our Two best ELM=TREE'S blown down | Rough winds all day Plough'g for Wheat
30. Some rain in night dry day Plough'g for Wheat
31. Rain in night & all morn'g till 9 oClo: dry after Ploughing as yesterday

November

1. X[353] Ploughing as yesterday and Seed Wheat from Devizes | Very heavy Storms
2. Cold Storms and Ploughing
3. Dry morn'g, Cold rain in evening | Plough'g finish'd for Wheat, above road, Rt of Tan hill Road
1. SILBURY=MEAD, the Ewes, began feeding with the old Turnips, Down Piece

   Novr 2nd Died at Monkton Ambrose Ball late of Bishopsgate St London aged 50 Yrs [This entry written between double lines above and below]
4. Sunday Storms
5. HAY, New field, began, the Lambs a little the Grass making them loose (evenings) and the HORSES same HAY, with mead grass being cold and wet do bad with it, geting poor | Cart'g Dung, - dry day, - Giddy Teg Kill'd
6. POTATOE'S, finish'd diging in Field 20 S, by Girls Cart'g Dung in Field finish'd Dry windy Day, RAIN heavy after 5 o Clock eve'g
7. Cart'd the Dung from Yard | CALF 141 li @ 6½d 11 weeks fattening cost 24s 10 days old RAIN all night & morn'g boisterous day
8. Plough'g below road Tan hill Piece | Dry night & fine drying day, rain evening | HAY, consumed from 25th Novr 36, to 16th May 37 Mead 20½ Ton N Field 12½ Ton O Field 10 Ton [Total] 43 | HAY, consumed from 20th Novr 37 to 16th May 38 Mead 14½ Tons N Field 22 Tons N Seed 4¾ Tons O Seed 8¾ Tons [Total] 50
7. FIRE Farmer Philpot, 2 Corn Ricks at Avebury, burnt, discovered at mid=day

---

352 appe= - appear or appearing.

353 [Pencil note in margin] 2Pork s X.

9. Dry night and day | WHEAT=SOWING, began by Drill'g 9 Acres with 6½ Sacks of SALMON on the Nether two Down Pieces
10. Frost & beautiful fine day | WHEAT Drill'd, about 5 Acres, uper part of Tan hill Piece Right hand side | The Down Wheat, trod with the Sheep
11. Sunday, Frost & cold dry day
12. WHEAT, sow'd the 7 Acres above the Road on one earth, 9 Sacks on the 12 Acres Mr Jenners one Ear Wheat 12 S 13 li Nt @ 38s | FROST, very sharp to prevent the Harrows from acting for some time 7 Horses from Kennett assisting | TURN=PIKE=TOLLS, LET for 1452£ LET last year 1308 144 advance
13. FROST, very severe again, Ploughing do badly very cold dry day | 9 Horses with 3 Plough, Press'r 3 Wheels 2 horses follow, on the Clover lay below road Tan hill | POTATOES 242 Sacks fill'd from 3½ Acres 1 Sack from a Peck of Mr Bartletts seed Yorkshire red Kidneys
14. FROST, again very sharp, fine day. Ploughing & Press'g as yesterday | TURNIP'S finish'd feeding, the old ones Down P A 4..1..5, began 25th Septr, sowd 29th May | TURNIPS A 5..0..5 began feeding <...> sowd 20th July small heads, to be fed off for Wheat
15. Damp fogg'y day, Plough'g & Press'g as yesterday | WHEAT sow'd the 6 Acres below the road with 3 Sacks of one ear Wheat 12-13 Nt 38s and 1½ Sack of Salmon Wheat 11..3 Nt 35s | The Wheat=Land being all fold'd & dung'd the Lambs & sale Sheep lie back in Silbury Mead the Stock Ewes on the young Turnips for Wheat
16. The 11 Horses harrow'g off the Wheat sowd Yesterday and Plough'g Down Piece for Wheat Turnips fed Very mild fine day
17. WHEAT=SOWING FINISHD for the present, sow'd the 4¼ Acres Turnip land Down Piece having sown the 18A=Tan hill Piece & 13A of the of the Down, 5 Acres of Turnips to feed & sow to Wheat, making 36 Acres | Very fine mild Day, HORSES=BORROWED of my Brother for Wheat sowing, viz Novr 9th - 3 Horses Novr 10 - 4 Novr 12 - 7 Novr 13 - 6 Novr 14 - 6 Novr 15 - 7 Novr 16 - 6 Novr 17 - 6 Very fortunate, being unusual fine dry weather the eight days sowing my Wheat
18. Sunday RAIN all night & day
19. RAIN all night & day 48 hours Horses in Stable
20. FLOOD, very high to cover Silberry Mead Sheep brot home fm Mead | Dry day Wheat Stubbles, Cartd to Stand'g Pen
21. Cold damp wind'y day, THRASH'G, Barley began wth Machine | BARLEY=STRAW, Ewes began, Stand'g Pen
22. Damp fogg'y day | 10 Qr Barley to Devizes | 1 Young giddy Ewe Killd 40li
23. 40 Young Ewes sold at 19s } 10 Old Ewes sold at 20s } Marlbro Fair | Plough'g Wheat Stubbles began | Dry day
24. Chalk, Cartd to level the Yard - dry day
25. Sunday, exceeding sharp frost all day

26. Thrash'g Barley wth Horses, Frost too hard to Plough - SNOW, all the morn'g, very cold
24. HURDLE'S removed from Horton Down, placed there 22nd March 1836 to Keep the RABBITS from my Wheat, the Rabbits, destroyd and the Hurdles worn out.
27. My 5 HORSES sent to KENNETT to Plough and return'd, being likely to rain, in Stable remainer of day. Dry morn'g, very BOISTEROUS wet day
28. BOISTEROU'S & rain'y, another comfortless day striping the Thatch fm the Skilling | Horses thrash'g Oats
29. BOISTEROU'S & rainey & comfortless day again the fating Pigs exposed to the | rain, the Thatch blown off both sides the Skilling | Horses in Stable, all the Sheep lay in Stand'g Pen | and the Labourers doing but little
30. Some Storms & wind, The Barometer returning from D in very dry | Chalk, Carting to level the Yard

December

1. Dry morn'g rain evening & blow'g | Fallowing Stubbles
2. Sunday, - Rain monr'g, dry afternoon
3. 11 Qr Barley to Devizes Ton of Coal & 3 hd Bricks retd Dry morn'g, heavy Storm after
4. 5 HORSES TO KENNETT Wheat sow'g, fine day
5. 5 HORSES TO KENNETT Wheat sow'g, fine day
6. 5 HORSES TO KENNETT Wheat sow'g, sharp frost
7. 4 Horses to Honey St Wharf for Timber for roof of Skilling | Rain heavy in morning, damp day. | The NUT=MEG, grey HORSE, sent to Wroughton to Mr Hillier Reeves, bot it of him 25th May 1835 for 26£ 3 Yrs old, to be taken to Abingdon Fair the 10th & 11th Inst.
8. Frost and fine day, Horses Thrashg Barley Saturday eve, before closeing the BARN I discovered SECRETED in a Chaff Basket by Geoe Timcomb and Willm Hitchens the two Thrashers my BARLEY, three parts full covered with some rudderings with intention to give the Horses, the CARTER, being accessory to it, admiting it his first offence
9. Sunday, Sharp Frost & thick Fog all day
10. Sharp Frost & thick Fog all day | Fallow'g Stubbles, - Horses began Grains | GRASS, the Horses discontinued since the Flood the 20th Novr about 2 Acres remain to mow, now spoiled the Springs very high, enough water for Meads
11. Fallow'g morn'g, Thrash'g afternoon | Foggy day
12. Fallow'g dry mild day
13. Foggy dul day | Plough'g began the Turnip land fed off, Down
14. 2 Ploughs the Turnip land fed off, Down | WHEAT sowd drill'd the 9th Novr appearing | Fine mild day | TURNIP'S finishd Nether Down Piece began 20 Ac | Ewes Fold'd on Wheat Stubbles with old thrash'd hay

evenings, Straw morning & afternoon & Turnips

15. Married this morning at Mildenhall by the Revd Mr Buxston John Crook of Corton to Elizh Vaisey of Grove Fine clear morning, dry day | Ploughing as yesterday | SLATING, began the Skilling in the Yard
16. Sunday, Frost & fine day
17. Plough'g finishd Turnip land for Wheat Frost & fine day
18. Frost & fine day | WHEAT=SOW'G FINISHD, by sow'g 5 Acres | of Turnip Land Nether Down Piece | Viz 18 Acre Piece Rt of Tan hill Road 18 Acres the two Nether Down Pieces 36 Acres with 27 Sacks is 3 Bu pr A
19. 11 Qr Barley to Devizes & 1000 Slates retd | Sharp Frost & fine day, harrow'g Wheat
20. RAIN all day, Horses in Stable
21. Dry day - Plough'g Stubbles | The 5 Fattening PIG'S began BARLEY=MEAL with Potatoes from the remaining 6 Sacks of old B
22. Rain in Night & morn'g, dry after Plough'g as yesterday
23. RAIN, very heavy in night & morn'g - dry after | Sunday
17. Jno Amors Sale, & left the White Hart N B gave up his Farm 18th Feby 35 to Saml Wentworth[354]
24. RAIN in the night, and nearly all day | The Horses in the Stable
25. Xt day, Very cold, fine clear morning freezing afternoon
26. Exceeding sharp FROST, difficulty in Plough'g Stubbles RAIN heavy evening
27. Frost morn'g dry day | 12 Qr Barley to Devizes 1 Ton Coal & 200 Slates returned
28. Very sharp Frost, fine calm day | Fallow'g Stubbles
29. Frost, Barometer high to FAIR, RAIN all day Horses in Stable
30. Sunday, Frost and dry day
31. Frost and dry day | 12 Doz of Hurdles from Mr Hutchings Marlbro
16th Novr Died in London Willm Chivers, late of Calne Aged 32 Years [This entry written between triple lines above and below]

---

354 This is the White Hart Inn in Devizes. Sale notice, Devizes & Wiltshire Gazette, 13 December 1838.

APPENDICES

LIST OF APPENDICES

A. Fattening pigs: 1823 to 1835 334
B. Stocks of sheep: 1823 to 1838 340
C. Historic cultivation and harvest dates: 1809 to 1826 345
D. Details of the wheat harvest: 1824 347
E. Wool weights: 1823 to 1836 348
F. Sale of Little Bedwyn estate: 1825 356
G. Details of farms: 1825 to 1826 357
H. Market prices: 1825 to 1838 359
I. Corn Laws - duties, prices and yields: 1827 to 1838 366
J. Silbury Hill and the White Horse at Cherhill: 1827 371
K. Three farms sold at Malmesbury: 1827 372
L. Littlecott and Beckhampton Farms: 1828 372
M. Beckhampton enclosure: 1795 373
N. Mr Guy's insolvency: 1829 374
O. Planting up the new grounds at Beckhampton: 1830 to 1834 375
P. A local account of the Swing Riots: 1830 377
Q. The death of Richard Sadler Smith of Bremhill: 1832 378
R. Fattening and selling sheep: 1833 to 1834 380
S. The supply and demand for wheat: 1834 381
T. The price of malt: 1730 to 1833 382
U. The introduction of Imperial Measure: 1835 382
V. Afflicting accident: 1836 383
W. Sheep and lambing: 1838 384
X. Thomas Pinniger the farmer and Thomas Pinniger the timber merchant 384

A. FATTENING PIGS: 1823 TO 1835

1. The economics of fattening pigs in 1823

[Between January and March 1823, Pinniger made a series of extensive comments about the pigs which were killed on the farm, the costs of their keep and the value of the produce. Rather than leaving these entries to interrupt the flow of the diaries, they have been collected together in this Appendix]

January 15

Boar Hog Killd | Put up to fatten 6th Decr 1821 to 12th Jany 1823,- is 57½ Weeks

Say - eat the first	29	Weeks 2B
the last	28½	Weeks 1B
	57½	

	£	s	d
Say cost of Pig when put up	2	-	-
Say 11 Qr Tail Barley @ 14s	7	14	-
Say Grinding Barley	1	2	-
£	10	16	0

	Sc:	li	s	
Weight of Pig	30	15	@ 7s	£10 15 s 3d

	Sco	li	
Head	1	2	
Belly Piece	1	6	
Fleck & Fat[354]	3	4	
Chine[355]		2	13
Feet & Legs		18	
Sweet Bones	1	2	
Griskins[356]		13	

354 Fleck is the white intramuscular fat which forms the marbling effect seen in meat. See X. J. Yang, E. Albrecht, K. Ender, R. Q. Zhao and J. Wegner, 'Computer image analysis of intramuscular adipocytes and marbling in the longissimus muscle of cattle', *Journal of Animal Science*, 84.12 (2006), 3251.

355 The chine bone is sometimes referred to as the back bone. In this context, Chine would have been that joint of pork which included the chine bone.

356 Griskin or griskins – the lean cut of meat from the loin of a pig.

1 Side	9	2
1 Side	8	16
Tongue		5
Spare ribs	1	14
	30	15

January 23

[The majority of this entry is emphasised in the diary with a single line in the margin].

2nd Pig Kill'd – put to fatten for Pork 10th Decr 1821 – small sickly 17 Sco 7 li

February 12

	£	s	d
Fat pig 10 Sco 10 li – put to fatten wth 9			
others 23rd Octr worth then		12	-
to the 10th is 15½ Weeks @ 4s pr W	3	2	-
	3	14	-

S	li	s		£	s	d
10 &	10	@ 7s	Sco is	3	13	6

	£	s	d		£	s	d
Say 3 Pigs @	3	13	6	is	11	1	6
fatten'd in 15½ W & cost					1	16	0

The large Pig

fatten'd in 57½ W & cost	2	0	0
Weigh'd 30S 15li @ 7s	10	15	3
A loss of 40 Weeks Keep			
in the large Pig Note			

did the large Pig eat as much as the 3 small Pigs pr Week

March 3

	S	li
3 Fat Pigs kill'd - own use		
or sell in Bacon	31	17
sold the Offal	4	3
	li	d
Viz Sweet Bones	34 @	4½
Spare ribs & Griskins	49 @	5

March 4

1 Pig	7	13

1 Pig	8	8	
	8	15	
	9	16	
	34	12@ 7s 6d Mr Lavington	
Loose fat		16	
	35	8	

(11th)

1 Pig	5	13	
	6	8	
	6	14	
	8	-	
Fat	2	7	Flak & of the 4
	29	2@ 7s 6d Mr Lavington	

6th

	li	li
1 Porker	33½	28
1 Porker	45	36½
1 Porker	51	45
1 Porker	65	57
	192½	166½
Heads & feet		26
	192½	192½
	li	
1 Porker	46@7s 6d pr Sco Mr Rumbell	

2. The economics of fattening pigs in 1824

[In 1824, Pinniger made further notes on the fattening of pigs]

February 13

Fat Pig – 20 Score, put to fatten, - 1st Novr worth then 2£ now 10s Score, wth 3 other Pigs now weighing about 23 Sco: ea

The make out of 9 Pigs

Viz:

	£	s	d
4 Small broken Belly'd @ 8s ea	1	12	0
1 Small broken Belly'd		14	0
4 Large Pigs @ 40s ea	8	0	0
16 Sacks Tailing Barley @ 8s	6	8	0
26 Sacks Perish'd Vetches @ 8s	10	8	0

Grinding 42 Sacks @ 1s 4d	2	12	0
	29	14	0

	Sco	li
4 Porkers	9	0
1 Pig	7	15
1 Pig	20	0
3 Pig suppose is	66	0
	102	15
	@ 10s £51 7s 6d	

April 2

	Sco	li
Fat Boar Stag, Kill'd	24	1
The Mushroom, Caul & pickings (not weigh'd at Newbury)		14
	24	15

6 Fat Pigs sold Mr Mills, Newbury

1	13 S 16 li
1	15 S 5 li
1	16 S 13 li
1	17 S 5 li
1	18 S 7 li
1	23 S 14 li
	105S @ 9s pr Sco

3. Fattening pigs in 1831

24 March

[The following weights were recorded in the diary.]

	S	li
	13	14
	10	12
	9	14
Sold }	8	8
}	8	6
	7	2
	5	0
Total	62	16

21 October

[The weights of pigs killed between October 1831 and February 1832 were recorded in the diary.]

October 18. 1st Porker Kill'd in House 5 Sco 15 li

October 21. 2nd Porker Mr Harding 5 Sco 11 li
November 4. 1st Bacon Pig Kill'd in House 7 Sco 7 li
November 21. 2nd Bacon Pig Kill'd in House 8 Sco 8 li
December 17. 3rd Bacon Pig Kill'd in House 10 Sco 10 li
February 2 1832. 4th Bacon Pig Kill'd for Sale 12 Sco 18 li
February 2 1832. The Sow Kill'd for Sale 17 Sco 13 li

4. The profit or loss of Fating Pigs - 1833

			£	s	d
10	Pig	at 40s	20	0	0
4	Pigs suckers	@ 12s	2	8	0
100	Sacks Potatoes	@ 3s 6d	17	10	0
3	Sacks Pease	@ 20s	3	0	0
	Grinding Pease	@ 1s 6d		4	6
4	Qr Barley	@ 25s	5	0	0
	Grinding Barley	@ 3s		12	0
A	Ton of Coals		1	2	0
	The labour for the Dung		49	16	6

The make out

			£	s	d
4	Pigs sold	36 Sco: @ 8s 6d	15	6	0
1	Pig sold	23 Sco: @ 8s	9	4	0
3	Porkers sold	145 li @ 6d	3	13	0
4	Pigs for own use	49 Sco: @ 8s 6d	20	16	6
2	Porkers for own use	196 li @ 6d	4	18	0
			53	17	6
			53	17	6
			49	16	6
		Profit	4	1	0

5. The loss made by grazing pigs in 1834.

1833 - The loss by grazing Pigs			£	s	d
Nov 6	7 Store Pigs put to fatten about 4 Sco ea @ 32s		11	4	0
13	Sow farrow'd 6 - say nothing, the Sow kept		0	0	0
	3 Sacks of Offal & grinding say		1	11	6
	10½ Qr Old Barley	@ 27s	14	3	6
	Grinding Barley		1	11	6
	40 Sacks Potatoes	@ 3s 6d	7	0	0
	Coal			12	0
			36	2	6

1834 - The make out				£	s	d
Feby 24	1 Porker	45 li @ 6d		1	2	6
Mar 8	4 Porker to London	@ 4s 4d pr Stone		3	17	0
Mar 21	1 Porker	44 @ 6d		1	2	0
		S li		6	1	6
Novr 26	1st Pig (Porker)	5 16				
Decr 12	2nd Pig	7 9				
Jany 8	3 Pig	10 1				
Jany 8	4 Pig J Chivers	10 5				
Jany 23	5 Pig	12 18				
Jany 23	6 Pig Mr J Large	12 18				
Apl 4	7	18 0				
		77 7	@ 7s	27	1	0
			Loss	3	0	0
				36	2	6

6. The economics of fattening pigs in 1835.

March 6th 1835

3 Fat Pigs, the remainer of 10

		S	li		
	1	11	4	in House	
	1	11	11	James Howell Avebury	
	1	11	14		
Feby 13th	1	10	19	John Chivers	
Feby 13th	1	10	6	In House	
Feby 13th	1	9	11	Mr Kemm	
Jany 23	1	9	9	In House	
Jany 9	1	7	6	In House	
(34) Novr 14	1	5	3	Willm Davis	
Novr 14	1	3	4	In House	
		90	7	@ 6s 6d is	£29 7s 3d
The Sow (Kept and improved) 20½ Sco @ 6s is					£ 6 3s 0d
2 Young Pigs from the Sow, worth					10s 0d
					£36 0s 3d
				Loss in fattening	£ 4 1s 3d
The Cost in Fattening					£40 1s 6d

Novr 17th 1834	£	s	d
10 Pigs put to Fatten, worth 24s ea	12	0	0
5½ Qr Barley @ 32s	8	16	0
1 Qr Barley @ 27s	1	7	0
Grinding Barley		19	6

1 Load of Grains		17	0
100 Sacks of Potatoes @ 2s 6d	12	10	0
1 Ton of Coal	1	2	0
The Sow	2	10	0
(The Dung for Labour &c)	40	1	6

B. STOCKS OF SHEEP: 1823 TO 1838

1 September 1823

579	Stock Ewes & Young Ewes
351	Lambs.
10	Lambs Ram.
10	Rams.
950	
30	Wethers.
70	Wether Lambs.
100	
210	Sale Ewes.
109	Wethers.
319	
319	
100	
950	
1369	The present stock

[There is a pencil note of "Mr Guy" in the margin opposite the wethers and wether lambs.]

14 December 1824

Sheep mark'd to be valued

126	2 Tooths
134	4 Tooths
190	6 Tooths
450	Put to the Ram
223	Ewe Lambs
67	Ewe 2 Tooths
5	Ewe 4 Tooths

1	Ewe 6 Tooth
296	
3	Rams
10	Ram Lambs
9	Ram Stags
2	Ram Wethers
24	Not taken by Mr Pain

11 July 1831

SALE EWES, drew, and put on Vetch's

37	Ewe Tegs	Total	
85	Ewe 6 Tooth	129	- Sale
7	Wether Tegs	230	- Stock
129		1	- Fat Wether
		3	Rams
	Stock Ewes	2	Stags
160	6 Tooths	365	
70	Tegs	205	Lambs
230		570	

July 13	1 Lamb died, murrain
July 17	1 Lamb died, murrain Wether
July 17	1 Ewe died, murrain 6 Tooth
July 20	1 Ewe died, murrain Teg stock
July 24	1 Lamb died, murrain
July 24	1 Lamb died, murrain
July 30	1 Ewe died, Dog bitten & mortified 6 tooth
Aug 1	1 Ewe died, by the Maggots
Aug 6	1 Ewe died, murrain
Aug 6	1 Lamb died, murrain sale
Aug 7	2 Ewes died, murrain Stock
Aug 7	1 Ewe died, murrain Sale
Aug 14	1 Ewe died, murrain
Sept 12	1 Ram Stag Kill'd by fighting
1832	
Feby 10	1 Two tooth Ewe died Murrain

20 October 1831

80	Sale 6 tooth Ewes sold at 30s	£120	0s	0d
35	Sale 2 tooth Ewes sold at 24s	£42	0s	0d
20	Sale Lambs sold at 14s 6d	£14	10s	0d

7	Sale 2 tooth Wethers sold at 23s	£8	6s	0d
142				
70	Sale Lambs sold at 14s 6d	£50	15s	0d
212		£235	15s	0d

1830

80	Lambs @ 13s	£52	0s	0d
20	Lambs @ 10s	£10	0s	0d
	Wool 1830	£49	10s	0d
	Wool (1829)	£32	2s	9d
		£379	7s	9d

Stock Sheep

£316	8s	0d	{154	6 tooth Ewes
			{ 72	2 tooth Ewes
£103	0s	0d	103	Ewe Lambs
	15s	0d	1	Wether Lambs
£6	0s	0d	3	2 tooth Rams
£6	0s	0d	4	Lamb Rams
			337	
£50	0s	0d		Last years Wool
£13	15s	0d		The Balance
£395	18s	0d		

Stock bought in

100	Tegs @ 28s	£140	0s	0d
130	Tegs @ 30s[357]	£188	10s	0d
37	Tegs @ 16s	£29	12s	0d
3	Rams @ 50s	£7	10s	0d
		£365	12s	0d
	Balance	£13	15s	9d
		£379	7s	9d

10 August 1832

The Stock of Sheep

99	2 tooth
70	4 tooth
49	full mouth
218	Ewes - Stock
100	full mouth - Ewes - Sale
91	Wether Lambs
102	Ewe Lambs
3	4 tooth Rams

357 The cost of 130 Tegs at 30s each is actually £195.

4	2 tooth Rams
1	2 tooth Wether
519	

24 July 1833

SALE EWES, drew & put wth Lambs on the 2nd Clover & folded

65 - full mouth last year
17 - 4 Tooth
18 - 2 Tooth
100

STOCK SHEEP, on the Down & follow the Sale Sheep on Clover

76 - 6 Tooths
76 - 4 Tooths
68 - 2 Tooths [Pencil note: 1 Killd 23 Augst]
220
7 - (4 Rams 2 Stags 1 Wether)
227
200 - (98 Wether & 102 Ewe Lambs)
427

The stock of Sheep bought in 1829 - 267 Ewe Tegs out of which have sold 160 & 65 for Sale 225 42 DIED

The Increase is present Stock 227 Lambs 200 } 427 Lambs & Sheep sold 334 761
See for Sale 35 796
Sold Sheep Skins of the Increase 24 820 Sold Sheep Skins of Stock bought in 42

5 December 1833

250	Ewes
131	Tegs
5	Rams
3	Wethers
389	
Viz	
[...]	6 Tooths
[...]	4 Tooths
[...]	2 Tooths
101	Lambs Es[358]
30	Lambs Wethers
2	Stags

358 Ewe Lambs.

1 Wether
5 Rams

25 July 1834

July 25th SALE=SHEEP, drew & put with Rams & Lambs on Clover

80 - full mouth
10 - 6-Tooth
8 - 4-Tooth
2 - 2 Tooth
100

Stock Sheep

77 - 6-Tooth
68 - 4-Tooth
98 - 2-Tooth
243
5 Rams
188 Lambs
29 2 Tooth Wethers
2 Fat Wethers
467
100
567 Total

30 August 1836

For Sale

20 - two-Tooth
20 - four Tooth
10 - six Tooth
81 - full mouth, 17 of which over year & broken
131
24, two, Tooth Wethers
155

The Stock

81 - 6 Tooth
63 - 4 Tooth
86 - 2 Tooth
230
1 - Giddy Ewe
3 - small Culs
6 - Rams

2 - Stags
242

The Earmark
1836. no mark
1835. Right, or off
1834. Left or near
1833. no mark
1832. Right or off
1831. Left or near

8 May 1838

The Sheep Count'd

164	Ewes wth 159 Lambs
40	Ewes wth the Tegs
118	Tegs
1	Wether Sheep
4	Rambs
327	
159	Lambs
486	

31 August 1838

SALE STOCK

7	50	8 Tooth near ear
10	56	6 Tooth off ear
28	54	4 Tooth no mark
35	37	2 Tooth ear notch'd
80	197	
	70	Ewe Lambs
	2	Cripple Lambs
	2	Wethers
	1	Giddy
	4	Rams

C. HISTORIC CULTIVATION AND HARVEST DATES: 1809 TO 1826

1. Bursting and reaping dates for Wheat 1809-1826

[These dates were written in the diary between the entries for 14 June

1824 and 15 June 1824 and have been reordered chronologically]

Year	*Wheat ears bursted*	*Reaping*
1809		17 August
1810	24 June	25 August
1811	15 June	12 August
1812		
1813	25 June	16 August
1814	26 June	17 August
1815	10 June	3 August
1816	21 June	27 August
1817	17 June	9 August
1818	6 June	28 July
1819	28 May	31 July
1820	9 June	8 August
1821	10 June	21 August
1822	29 May	22 July
1823	15 June	29 August
1824	14 June	16 August
1825	31 May	23 July
1826	1 June	18 July

[This table is followed by the following note, written in faint pencil.]
18 - Reaping Hampshire of it at Allington, began <Eve> Netheravon 20th

2. Historic records of the start of hoeing
[These dates were written in the diary within the entry for 16 June 1824 and have been reordered chronologically]

Year	*Began hoe'g*
1813	9 July
1814	20 July
1815	18 July
1816	29 July
1817	3 July
1818	22 August
1819	1 July
1820	21 June
1821	9 July
1822	10 July
1823	22 July

D. DETAILS OF THE WHEAT HARVEST: 1824

3 September 1824.

Particulars, of measure & Crop of Wheat

Nfield		*A*	*R*	*P*	*Tithg*
Potatoe		1	3	16	116
Drill'd		9	0	18	570
Thin brod		12	3	34	720
Drill'd		9	1	16	700
Turnip brod		10	3	8	560
Forebridge} Whitley f }		20	1	8	1350
Hard		7	3	6	422
Red	13	1	10	914	
White	2	3	27	165	
Red	4	3	13	283	
Red	11	0	36	600	
White	3	2	3	183	
Hitchens	9	0	3	626	
Larks lease	8	2	27	476	
Westboro	9	0	10	560	
	134	3	5	8354	

is about 62 Tithing pr Acre at 30 Ti: pr load – is 728. Loads[359] Carr:d in 7 succeed'g days, - is 39 load ea day 1 Pitcher & 2 loaders, - 9 Ricks & 2 Mows filld

20 May 1825.

Wheat, the make out of 1824

A	R	P	Best	Tail
7	3	30	44	8
8	0	0	34	10
13	2	8	55	11
7	3	18	40	6½
12	2	23	70	9
11	3	10	60	10¼
7	3	6	46	6½
12	3	34	52	13¼
22	3	0	117	13
15	0	26	77	6

359 728 should read 278.

14	1	10	74		3½
134	3	5	669		97
		A	97		
	say –	135	766		5½ Sa (&94 Bu over) pr Acre
			675		
			91		
			4		
		135	364	2	
			270		
			94		

E. WOOL WEIGHTS: 1823 TO 1836

1823 and 1824

[One whole page of the diary between the entries for 13th and 14th October 1824 is headed "Wool: weigh'd to Mr Salter 12 Octr".[360]

The basis of accounting for wool is the Tod (28lb): a number of fleeces are weighed at a time, aiming to reach exactly one Tod. The number of fleeces which make up each Tod is recorded along with the number of pounds under (U) or the number of pounds over (O). The following presents an explanation of the first three records for 1823

O	*Fleeces*	*U*	*Explanation*
1	13		13 fleeces in this tod, weighing 1lb over: that is 29lb
	11	1	11 fleeces in this tod, weighing 1lb under: that is, 27lb
	11		11 fleeces in this tod, weighing 28lb

The number of fleeces can then be totalled to ensure that all fleeces have been weighed: the number of pounds over and pounds under are also totalled in order to reach a total weight.

The table of weights for both the years 1823 and 1824 and the series of calculations written in the diary are reproduced below.]

360 Mr Salter is possibly Simon Salter of Kington Langley, who inherited the woollen cloth business of his father, Isaac Salter senior, WSA: P1/1823/70, *Will of Isaac Salter senior of Langley Fitzurse, Kington St Michael, clothier, 1823.*

1823						*1824*					
O	*Fleeces*	*U*	*O*	*Fleeces*	*U*	*O*	*Fleeces*	*U*	*O*	*Fleeces*	*U*
1	13			12			11	1		9	
	11	1		10		1	14			10	1
	11			13	1		12			10	
1	12			9			13			10	
2	11			11		1	12			11	
	12			10			12	1		10	1
	10	1		9			12			11	1
	12			10			13			10	
	11			9			13			11	
	11			10			12			11	
	11			12			11	1		10	
1	11			9		2	9		1	11	
	10			10			12	2		11	1
	10			10			11	1	1	11	
	12			10		1	12			10	1
	10			9		1	10			10	1
	10			11			12			10	1
	11			9	1		12		2	11	
1	11			10		1	12			11	
	12			10		1	11		1	10	
	10			10			11	2	1	11	
	12			9			12		16	7	
	12		1	12			12			810	
1	10		1	11			14	1			
1	11		1	11			12	1			
	11			10			13				
	12		1	10		1	11				
	11		1	11			11				
	10			10			11	1			
	10			10			12				
	11			9			13	1			
	12		1	13			11				
	11			10			13				
1	11			11			13				
	12		1	11			13	1			
	11			11			12				
	12			9	1		13				
	11		8	4			14				
	11			11			13				

1823						*1824*					
O	*Fleeces*	*U*	*O*	*Fleeces*	*U*	*O*	*Fleeces*	*U*	*O*	*Fleeces*	*U*
	11			945		1	14				
	10					19	9				
	10						10				
	11						10	1			
	9	1					10	1			
	11						10				
	10						11				
	10						9	1			
	12						10				
1	11					1	11				
1	11						10				

[Pinniger then makes a rough calculation of the total number of tods of wool on the basis of 11 fleeces to the tod.]

	Fleeces	
	810	
	945	Tods
11	1755	159 is
	11	
	65	11 Fleeces
	55	to the Tod
	105	
	99	
	6	

[Pinniger then summates the number of pounds over & under to determine the adjustment required and then adds it to the total count of tods weighed. The figure 158 in the *Tod* column in the table below should be 161, the actual count of tods recorded in the table above: the subsequent financial calculations are deficient by £5 2s 0d.]

li	
79	Over
28	Under
51	is
Tod	li
1	23

&	158				
	159	23	@ 34s		
			£	s	d
		is	271	14	0
	Belts li 152 @ 3d		1	18	3
	Lambs unwash'd li 89 @ 6d		2	4	0
	Lambs wash'd li 102 @ 10d		4	5	10
			280	2	1

1830

[The majority of one whole page of the diary at the start of the entry for 8 May 1830 is headed "WOOL, weigh'd, Mr Geoe Bailey" with the following table of weights and a series of calculations: these are reproduced below. This would have been the wool from the "11½ Sc Tegs" sheared on 10 June 1829 and therefore was the first wool clip for Pinniger at Beckhampton.[361]]

8	Fleeces 1st Tod	7	
7		8	
8		8	
8		7	
8		8	
7		8	
8		6	
7		8	
8		8	
7	1st Pack	4 17 li	3 Pack
7			
7		30 Tod 17 li	
7		223 Fleeces @ 21s pr Tod is £32 2s 9d	
8			
8			
7		1 Tod 6 Fleeces	
8		12 Tod of 7 Fleeces	
8		17 Tod of 8 Fleeces	
7			

361 George Bailey is probably the proprietor of George Baily and Co., woolstaplers of the Green, Calne , J. Pigot, 'Commercial Directory 1822' in WRS47, p.49.

8
7 2 Pack

11 April 1831
[This would have been the 1830 Clip.[362]]

WOOL weigh'd by Mr Currington of Devizes

9	Fleeces	1 Tod	9	Fleeces	1
10	Fleeces	1	10	Fleeces	1
9	Fleeces	1	10	Fleeces	1
9	Fleeces	1	9	Fleeces	1
10	Fleeces	1	10	Fleeces	1
10	Fleeces	1	10	Fleeces	1
11	Fleeces	1	9	Fleeces	1
9	Fleeces	1	10	Fleeces	1
9	Fleeces	1	10	Fleeces	1
10	Fleeces	1	8	Fleeces	1
9	Fleeces	1	10	Fleeces	1
9	Fleeces	1	10	Fleeces	1
10	Fleeces	1	8	Fleeces	0-25li
10	Fleeces	1	257		26-25li
			At 37s pr Todd		£49 15s 0d
			The Fleeces about 3s 10¼d each		

1831 and 1832

Novr 9th 1832 The 2 Years Wool, weigh'd Mr Connington

1832	1831	1831 is 30 Tod & 6 Fleeces		
11	6	30 |	363	| 12 to Tod
11	12		30	
12	11		63	
13	13		60	
10	12		3	-over
10	12			
11	12			
11	13			
12	11			

362 Mr Currington (also appears below as Connington and Cunnington) is probably William Cunnington, woolstapler of the end of Long Street, Devizes, J. Pigot, 'Commercial Directory 1830' in WRS 47, p.73.

	11	11	
	11	12	1832 is 28 Tod & 7 Fleeces
	11	12	28 | 327 | 11 to Tod
	13	12	28
	12	14	47
	12	14	28
	12	12	19 over
	11	13	
	11	13	Together 59 @ 30s is
	11	12	£88 10s od
	12	13	
	11	13	
	12	12	
	13	11	
	10	11	
	11	11	
	11	11	
	11	11	
	13	11	
	7	10	
		11	
		11	
Fleeces	327	363	

28 August 1833

This Years Wool weighed by Mr John Spackman

11	13	12	13	14	13	14	
11	12	13	13	14	14	14	
11	13	13	14	12	12	329	Fleeces
11	12	12	14	12	12		

1829	223 Teg's	30 Tod 17 li @ 21s	£32 2s 9d
		7 to the Tod	
1830	257 Ewes	26 Tod 25 li @ 37s	£49 15s od
		9½ to the Tod	
{ 1831	363 E[wes]&T[egs]	12 to the Tod	
{ 1832	327 E[wes]&T[egs]	11½ to the Tod	
		59 Tod at 30s	£88 10s od

26 | 329 | 12½ Fleece to a Tod

26
69
54 [*recte* 52]
15

1833	26 Tod @ 46s is	£59 16s 0d
	40 li Belts @ 23s is	£ 1 13s 0d
		£61 9s 0d

1833 WP=K

Tod 60	750 Fleeces	12 ½ to Tod
	60	
	150	
	120	
	30	

25 April 1835 [This would have been the 1834 clip]

WOOL=WEIGHD

Tegs	Ewes	Ewes	
10	15	13	
10	15	13	
10	14	12	
10	14	12	
10	14	12	
9	14	11	is 24li only
9	14	31 Tod & 24 li	
9	14		
9	14	376 Fleeces is near 3s 4¾d ea	
9	14		
9	13		
9	13		
9	13		

Belting Wool at half price 47li, half is 23½ li

31 Tod &	24 li				
	23	Belt'g	£	s	d
32 Tod &	19 li	@ 39s pr Tod is	63	15	0
	Mr Cunnington Devizes				

11 April 1836 [This would have been the 1835 clip]

TEG	EWE		EWE
7	9		11
7	9		11
8	9		11
8	9		11
8	9		11
8	10		11
8	10		11
8	10		11
8	10		11
9	10		11
79	10		11
	10		5
		10	

		£	s	d
321 Fleeces	34 Tod & 18 li			
2 Tod of Belting	1			
	35 & 18 at 47s pr Tod	83	15	0

327 Sheep Shorn, 6 short by some means

Mr Connington Devizes

9 Fleeces to the Tod & 10 Fleeces over

22 July 1836 [This would have been the 1836 clip]

WOOL=WEIGHED

Tegs	Ewes	Ewes	
9	10	10	Fleece's
10	10	11	
9	11	11	
10	10	12	
10	12	11	
10	10	11	
10	11	12	
10	10	7	is 17 li
10	12		
9	12		
10	12		

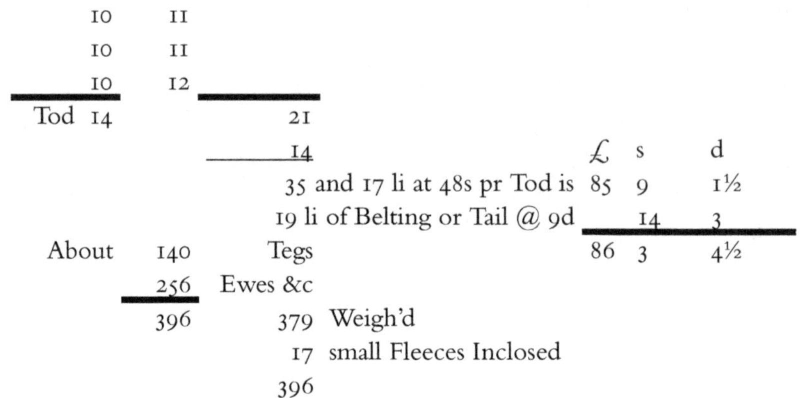

				£	s	d
	10	11				
	10	11				
	10	12				
Tod	14		21			
			14	£	s	d
			35 and 17 li at 48s pr Tod is	85	9	1½
			19 li of Belting or Tail @ 9d		14	3
About		140	Tegs	86	3	4½
		256	Ewes &c			
		396	379 Weigh'd			
			17 small Fleeces Inclosed			
			396			

Mr Connington Devizes

F. SALE OF LITTLE BEDWYN ESTATE: 1825

Saturday 15 January 1825

The Sale of Little Bedwin completed, and Paid for at Marlbro

	£	**s**	**d**
The Estate	32800	0	0
The Timber &c The Timber, cut	4132	0	0
The Live & dead Sk	2766	1	10
The Fixtures	69	1	0
The Makeout	3000	0	0
£	42767	2	10

[The total of £42767 equates to £3.448m at 2018 prices.][363]

363 Measuring Worth, https://www.measuringworth.com/calculators/ukcompare/relativevalue.php [accessed 12 May 2020].

G. DETAILS OF FARMS: 1825 TO 1826

31 January 1825. Rabson Farm

	A	R	P
Arable	355	0	0
Pasture	200	0	0
Down	27	0	0
Roads & Homestead	10	0	0
	592	0	0

£	s	d	
700	0	0	Rent
200	0	0	Tithes
900	0	0	
100	0	0	Poor
1000	0	0	

30 June 1825. Beaversbrook Farm for Sale.

	A	R	P
Pasture	113	3	32
Arable	48	3	23
	162	3	15

About half free of Great Tithes
9000 Guineas offered, and refused
About £58 pr Acre

22 September 1825. Coomb Farm, Enford

425 Acres of Watermeads & Arable
200 Acres of Down
with the Tithes of some adjoining Lands in the Occupation of Mrs Pike
Tithe Free 700£ pr Annum
100 Acres of Wheat in 4 fields

The Sheep on the Farm for Sale

200 full mouths

120	6-Tooths Ewes
140	4-Tooths Ewes
140	2-Tooths Ewes
75	2-Tooths Wethers
300	Chilver & Wether Lambs
5	Rams
980	

1825. Fosbury upper Farm Tithe Free

	A	R	P
Farm House, outbuild Yards, Garden &c.	1	3	16
Old inclosed, meadow & Pasture Land	83	3	33
Down land of superior quality	59	2	0
Coppice Wood	23	0	30
Arable Land	518	3	37
520£ pr Ann	687	1	36

£515 5s 0d is 15s pr Acre

18 April 1826. Monckton Farm

	A	R	P
Arable at Hackpin	228	3	2
Arable, Homefields	276	1	23
Hackpin Down	200	0	0
Windmill Hill Down	55	0	0
The Penings	38	3	15
Pasture	61	2	32
	860	2	27

The Rent, Tithes free, the Poor very low

A		£	s	d
500	Arable	600	0	0
60	Pasture	100	0	0
40	Pasture	40	0	0
255	Down	110	0	0
855		850	0	0

H. MARKET PRICES: 1825 TO 1838

12 September 1826 – Wilton Fair

	1825	1826
Lambs	18s to 25s	8s to 18s
Ewes	27s to 34s	15s to 24s
2 tooth Wethers	25s to 34s	16s to 25s
4 & 6 tooth Wethers	30s to 42s	30s to 36s

1826

Good Lambs 12s, very few @ 25s[364]
Ewes 15s to 20s, very few @ 24s
2 Tooth Wethers 16s to 20s, very few @ 25s
6 Tooth Wethers very few @ 36s

Devizes Market

	1825 (7 Sep)		1826 (8 Sep)	
	Top Price	Average	Top Price	Average
Wheat	37s 6d	33s 6d	32s	28s 1d
	Price range		Price range	
Barley	42s to 48s		34s to 39s	

10 October 1826 - Weyhill Fair

Down Lambs	9s to 12s to 16s up to 21s
Down Ewes	13s to 20s, - prime 28s
Wethers 2 Tooth	15s to 20s to 24s
Wethers 4 Tooth	20s to 32s
Dorset Ewes	18s to 26s
Somerset Ewes	28s to 40s, prime 44s

Hops, the 12-13 & 14th

Country from £5 12s to £7 7s
Farnham fm £6 10s to £8 8s

13,000 Pockets, grounded, the quatity[365] very fine

---

364 This should probably be 18s (the highest price for 1826, as opposed to 1825).
365 Probably quality.

150,000, - Sheep Pen'd

The best Dorset Ewes, which will Lamb 6 weeks before Xtmas was sold at Appleshaw @ 25s. Such Ewes sold last year @ 50s. Such Ewes in 1812 fm 55s to 72s ea

3 April 1827 - Sale of Sheep at Clatford Farm

Ewe Tegs 14s to 21s
Ewe 2 Tooth 15s to 21s mix'd wth warp'd
Ewe 4 Tooth 21s to 24s wth Lambs
Ewe 6 Tooth 21s to 24s wth Lambs
Ram Lambs from 35s to 11£ odd[366]
Old Ram 15s - 25£ have been gave for its use one season

6 August 1827 - Tan Hill Fair

2 Tooth Wethers 26s downwards
Ewes (Mr Paines) 20s and downwards
Lambs 23s to 4s

6 October 1827. Kington Fair

Lambs good 15s to 20s
Wethers 4 tooth 27s
Ewes good 22s

10 October 1827 Weyhill Sheep Fair

150 Thousand Pen'd
Down Ewes 16s to 30s
Down Lambs 10s to 29s
Down 4 Tooth We[367] 20s to 36s
Somerset Ewes 30s to 45s

21 April 1828 - Devizes Fair

Lambs, cost Michs 1826 13s sold at Devis 25s
Lambs, cost Michs 1827 18s sold at Devis 25s

10 September 1828 Weyhill Sheep Fair

---

366 11£ odd - At least £11 but no more than £12.
367 We - Wethers.

Lambs Down 16s to 30s 6d
Ewes Down 18s to 39s
Wethers Down 20s to 40s
Ewes Dorset 40s to 46s
Ewes Summerset 40s to 51s

Note in margin: 5s advance on last Yr

20 October 1828, Devizes Fair

4s to 5s above last Yr for Ewes
Lambs from 8s 6d to 22s
(Messrs Smith & Chivers's Lambs 19s)
Ewes, good Downs from 25s to 30s
Wethers 25s upwards

24 November 1828, Marlbro Fair

Ewes at 39s, Culs sold higher than the Heads, sold at late Fairs
Robt Philpot sold at 38s 6d, offer'd 28s only, at Devizes Fair

10 October 1829, Weyhill Fair

Down Ewes 18s to 28s some 30s
Down Lambs 14s to 20s some 25s
2 Tooth Wethers 18s to 25s
4 Tooth Wethers 26s to 33s

5s to 7s pr head lower than last Years Fair 150 000 Sheep Pen'd

At Apple Shaw
Summerset Ewes 28s to 40s
Dorset Ewes 25s to 35s

Hops, Farnhams, 357 Pockets Pitchd £16 to £19 10s
Hops, Farnhams, last Yrs 180 Pockets Pitchd £10 to £15
Hops, Country Row, 290 Pockets Pitchd New
Hops, Country Row, 860 Pockets Pitchd Old
Hops, In 1828 5000 Pockets Ground'd In 1827 6000 Pockets Ground'd In 1826 10000 Pockets Ground'd

13 September 1830, Wilton St Giles Great Sheep Fair

100,000 Pen'd, 3s under the late Britford Fair
Ewes 16s to 24s
Wethers 17s to 31s
Lambs 12s to 18s - a few choice Pens 4s high'r

3 Choice RAMS of the celebrated Stock of Henry Bays Esqr of Malmains near Dover 200 Miles from Wilton was sold to Mr Stephn Mills of Elston for 100£

9 October 1830, Weyhill Fair

100,000 Sheep Pen'd dul sale
Down Ewes 16s to 28s & 30s
Dorset 22s to 40s
Down Lambs 14s to 24s & 25s
Down Wethers 15s to 23s

1591 Farnham Hops 12£ to 20£ - average 10 to 16£
2752 Country Hops 10 to 16£ average 8£ to 12 Gui[368]
4342 Pockets Pitch'd

12 October 1832, Weyhill Fair

Sheep 20'000, less than last Yr & 2s pr head under Wilton Fair

Wethers 25 to 40s
Ewes 20 to 30s
Lambs 12 to 22s & 26s

Hops 3517 Pockets, Country £12 askd sold £8 8s 0d
Hops 2400 Pockets, Farnham £14 askd sold £13 0s 0d
Hops from Stroud Market, Suffolk 150 Miles

18 July 1833, Great Lamb Fair, Overton

70'000 Ewes, Wethers & Lambs Pen'd prices 3 to 4s head higher

Wethers 25 to 42s
Ewes 21 to 42s
Lambs 14 to 35s

Mr Saunders exhibit'd & Sold 100- Ram Lambs, for one refused 40 Guineas.

368 Guinea.

Mr Twinehams Stock of 1st and 2nd Cross of Cotswolds from the Hampshire Ewe the Wool in Lamb & Teg of the 1st cross, average 8 li his Cotswold Ram produced 10½ li his Ram Lambs sold at 4 & 5 guineas each & his Wether Lambs 35s each

10 October 1833, Weyhill Fair

20'000 Sheep less than last Year Pen'd
10'000 Sheep less than the average of Years.

	Lambs 21s to 27s & 31s
	Ewes 27s to 42s
	Wethers 28s to 36s two tooths
	Wethers 46s four tooths
HOPS	Farnham £10 10s 0d to 12£
	Country £7 15s 0d to 10£

13 October 1834, Weyhill Fair

100,000 Sheep pen'd

Down Ewes from 28 to 30 to 42s
Down Lambs from 20 to 28 to 34s
Wethers from 30 to 35 to 44s

Mr Thos Osmond Lambs realized 40s ea supposed to be the best ever pen'd on the Hill

HOPS nearly 8000 Pocket s ground'd exceed any growth for the last 15 Yrs

3500 Pockets of Farnham from £8 to £9 10s 0d
4340 Pockets Country Rows from £6 to £8 8s 0d
100 Pockets from Kent & Sussex

20 October 1834, Devizes Fair

6 Coops for 80 Ewes full small require 7
2 Coops for 20 Ewes - full large - proper
4 Coops for 60 Lambs, rather small, say proper
1½ Coops for 20 Lambs - full large - proper
1½ Coops for 20 Lambs - full large - proper

Sold 60 Lambs @ 23s and 20 Lambs 20s 6d and 15 Lambs culs @ 15s and 5 Lambs cul Chivers @ 15s avarage'd 21s ea want 10s (see Augst 6th 1831[369] sold the best Lambs 14s 6d)

1830 - Sold best Lambs 13s & seconds 10s
1832 - Sold best Lambs 16s & seconds 13s
1833 - Sold best Lambs 22s & thirds 13s 6d - the 2nds Kept

4 November 1834, Appleshaw Fair

Sheep 3s pr head lower
Wethers 25s to 40s
Ewes 25s to 35s
Lambs 15s to 30s

10 October 1835, Weyhill Fair

Ewes - 18 to 26 & 30s	}	- 28 to 30 & 42s
Lambs - 12 to 20 & 24s	} last year	- 20 to 28 & 34s
Wethers - 20 to 33 & 35s	}	- 30 to 35 & 44s

Hop's Farnham - 1600 Pockets - 8£ to £9 9s 0d - Country - 1700 Pockets - 7£ to £8 8s 0d - Kent & Sussex - 500 Pockets - 4£ to £5 10s 0d

7 January 1836, Depression in Wheat

1834, 20th Septr sold 20 Sa: Wheat at 25s 6d	}
1835, 15 Jany sold 20 Sa Wheat at 22s	} 12 Sco 8 li Nt all of 1833
1836, 7 Jany sold 20 Sa Wheat at 18s	}

The last lot have been in the Warehouse Devizes from the 9th of April 1835, and have not been worth 18s pr Sack since the 27th August 1835 when sold 20 Sacks at 18s to Mr Jefferys of Melksham, and would not then have the other 20 Sa @ 18s which have this day sold at 18s & could not before.

10 October 1836, Weyhill Fair

Sheep from 115 to 120'000
Down Ewes 20s to 32s
Down Lambs 18s to 24s & 30s
Down 2 tooth Wethers 20s to 28s
Down 4 tooth Wethers 28s to 40s
Dorset Ewes 30s to 35s

369 There is no entry in the diary relating to the price of Lambs for 6 August 1831.

Sommerset Ewes 35s to 40s

HOP'S
3026 Pockets Farnhams £8 8s od to 10£
5100 Pockets Country 6£ to 9£

Various sheep fairs in 1837

110000 Sheep Pen'd at Weyhill
Down Ewes - 18s to 30 & 40s
Down Lambs - 14s to 25 & 30s
Down Wethers - 24s to 35 & 45

Devizes Fair nearly the same
(4 Apple=Shaw lower for Ewes & Wethers
23 Marlbro, a large Fair and lower than Devizes

18 July 1838

OVERTON FAIR the 18th very quick sale

Ewes	26s to 42s }	
Wethers	28s to 40s }	90-000 Pen'd
Lambs	18s to 36s }	

12 September 1838

WILTON FAIR

84,000 Sheep Pen'd
Ewes 25s to 35s
Lambs 18s to 32s
Wethers 22s to 32s
Mr Lyne of Brimslade Sale of Sheep
Ewes 40s 6d to 52s
Ewe Lambs 24s to 25s 6d
Ram Lambs 4£ to 13£ 15s each

I. CORN LAWS - DUTIES, PRICES AND YIELDS: 1827 TO 1838

28 February 1827

Propositions, - made for the Corn Laws

Table shewing the operation of the proposed duties on Foreign Wheat, by Mr. Canning.

If the the price per Quarter be	**Duty**	**Price left to the Importer of Foreign Wheat after payment of Duty.**
	s	
70s & upwds		
69 & under 70s	2	
68 & under 69	4	
67 & under 68	6	
66 & under 67	8	
65 & under 66	10	
64 & under 65	12	
63 & under 64	14	
62 & under 63	16	
61 & under 62	18	
Assumed point of protection		
60 & under 61	20	40s up to 41s
59 & under 60	22	37 up to 38
58 & under 59	24	34 up to 35
57 & under 58	26	31 up to 32
56 & under 57	28	28 up to 29
Average price of last 6 Yrs		
55 & under 56	30	25 up to 26
54 & under 55	32	22 up to 23
Average price of 1827 to Feby 15		
53 & under 54	34	19 up to 20

Annual average prices of Wheat for 12 Years previous to 1815 pr Quarter

Year end'd	**s**	**d**	
1803	57	1	
1804	60	5	
1805	87	1	Average of 6 Years 72s 2d
1806	76	9	
1807	73	1	
1808	28	11	
1809	94	5	
1810	103	3	
1811	92	5	Average of 6 Years 98s 6d
1812	122	8	
1813	106	6	
1814	72	1	
	85	4	Average of 12 Years

Year end'd	**s**	**d**	
1815	63	8	
1816	76	2	
1817	94	0	Average of 6 Years 74s 2d
1818	83	8	
1819	72	3	
1820	65	10	
1821	54	5	
1822	43	3	
1823	51	9	Average of 6 Years 55s 9d
1824	62	0	
1825	66	6	
1826	56	11	
	64	11	Average of 12 Years Average of first 6 Weeks of 1827 53s 6d

<u>BARLEY</u>
When average 30s and under 31s pr Qr the duty 10s pr Qr
For every integral (or whole) shilling increase above 30s, the duty decrease 1s 6d pr Q

31s	8s 6d
32s	7s 0d

33s	5s 6d
34s	4s 0d
35s	2s 6d
36s	1s 0d
37s	1s 0d pr Qr

30 and under 29s - duty 11s 0d And for every integral shilling or part of integral shilling under 29s such duty shall increase 1s 6d pr Qr.

OATS
When average 21s and under 22s pr Qr the duty pr Qr 7s
For every integral one shilling above 21s duty decrease 1s pr Qr till such price be 28s
When ever the price be above 28s, the duty to be 1s pr Qr
When ever under 21s and not under 20s, the duty 8s
For every one shilling or part intrical of one shilling by which such price shall be under 20s, - duty increased 1s pr Qr

RYE, PEASE & BEANS
Whenever average shall be 35s & under 36s Qr the duty pr Qr 15s
For every integral one shilling above 35s pr Qr the duty shall be decreased 1s 6d pr Qr until such price be 45s
And at 45s & above for every Qr 1s
When 35s and not under 34s the duty for every Qr 16s 6d
And for every or part of integral one shilling under 34s duty shall be increased 1s 6d pr Qr.

WHEATMEAL & FLOUR
For every Barrel being 196 li a duty equal in amount to the duty payable for 5 Bushl Wheat.

OAT=MEAL
For every quantity of 252 li a duty equal in amount to the duty payable on a Quarter of Oats

MAIZE. INDIAN=CORN = BUCK WHEAT BEER & BIGG
For every Qr a duty equal in amount to the duty payable on a Quarter of Barley.

If the Produce of and imported from any British possessions in North America, or elsewhere out of Europe

WHEAT
For every Quarter 5s 0d

Until the price of British whall [wheat?] be 65s, - when above 65s, - duty to be 6s pr Qr

BARLEY
For every Quarter 2s 6d
Untill average of British Barley shall be 33s 0d Qr, when at or above 33s 0d duty to be 6d pr Qr

OATS
For every Quarter 2s 0d
Until the price of British Oats shall be 24s Qr duty shall be 6d pr Qr

RYE, PEASE & BEANS
For every Quarter 3s 0d
Until the average of British be 40s when at or above 40s Qr duty to be 6d pr Qr

Wheat=meal, and Flour
For every Barrel of 196 li Duty equal to duty of 5 Bushl Wheat.
Oatmeal, for every quantity of 252 li equal to the duty on a Qr Oats
Maize, India Corn, Buck wheat, Beer, and Bigg, duty equal to a Qr Barley

21 November 1828

PORTS OPEN, for Foreign Corn

Average attained the height, for Wheat to be admissible for home consumption upon the payment of the minimum duty of 1s pr Qr, Barley 3s 4d pr Qr, Oats 7s 9d pr Qr
More than 10 Years have now elapsed since the Ports were opened unlimitedly for the Importation of Foreign Wheat
Since 1817, the averages have not been so high as they are now quoted.
Since 1818, This Country with very few exceptions has produced its own supply, and fed its own Population
On 2 or 3 occasions Foreign Wheat spoiling in the Kings Locks has been let into consumption upon payment of Duty, which produced half a Million from the Measure.
The Maximum price from 1817 to 1822 was 80s pr Qr for Wheat,
The Duke of Wellingtons Price is 74s and since 1817, the Averages has not been so high as they have now quoted Viz
The aggregate average prices, for the last 6 Weeks, by which the Duty on Foreign Corn, now in bond is regulated.
Wheat 74s 1d Duty 1s pr Qr
Barley 39s

Oats 26s 4d
Rye 42s 3d
Beans 40s 3d
Pease 42s 7d

28 February 1833

[On this date, Pinniger wrote the following in his diary:]
The Devizes Gazette of this day observes " We scarcely recollects so great a depression in the Corn market as there has been during the last 5 or 6 weeks. We would fain hope that the prices had arrived at their lowest ebb, but we fear they will still tend downwards. The Barley growers suffers the most severely, we have been assured by a most respectable Farmer who sold Barley at our Market on Thursday last for 23s pr Qr that on that very day twelve months he obtained 42s pr Qr for the very same description of Grain, a sinking in price of 20s pr Qr we are persuaded that there is scarcely a Farmer who attends our market but has been a considerable loser during the last 6 months."[370]

11 November 1833

Copied from Bells Paper 21st Octr 1833
A table stateing the comparative produce of the Crop of Corn in Britain of this year to the estimated quantity of an average season, and the yearly consumption; whereby there appears a deficiency of upwards of 11 Millions of quarters of Grain.

	Average Crop in Quartrs	Consumption in Qrs	Crop 1833 in Qrs	Deficiency in Qrs
WHEAT	11 Millions	12 Millis	10 Millis	2 Millis
BARLEY	9 Millis	9 Millis	6½ Millis	2½ Millis
OATS	24 Millis	25 Millis	20 Millis	5 Millis
BEANS	4 Millis	4 Millis	2 Millis	2 Millis
	48	50	38½	11½

7 August 1838

When Wheat is at
69s pr Qr the Duty falls from 16s 8d to 13s 8d
70s pr Qr the Duty falls from 13s 8d to 10s 8d
71s pr Qr the Duty falls from 10s 8d to 6s 8d
72s pr Qr the Duty falls from 6s 8d to 2s 8d
73s pr Qr the Duty falls from 2s 8d to 1s

370 Taken from Devizes and Wiltshire Gazette, Thursday 28 February 1833.

J. SILBURY HILL AND THE WHITE HORSE AT CHERHILL: 1827

April 4, 1827

Silbury Hill, 36 Yards, across the Top

The White Horse, on Cherhill Down Measured, -
the hind leg from the foot to the Body - 20 Yards
The Thigh 5 Yards wide
The small of Leg - 2 Feet
The Feet on a Bank 2½ feet high
The Feet - 2 Yards wide, and a Barrow full of stones on, each foot
From bottom of Foot to Fetlock 3 Yards
12 Yards from hind foot to next
12 Yards from 2nd hind foot to fore
12 Yards from 1 fore foot to other
36 Yards, - full length of Horse
Hind foot being under the point of the Tail, and the fore foot under the Nose.

Standing under the Bank of the Plantation, surveying the Horse, it appear to be about 2½ Poles to its neck, and 1 Pole from Nose to Eye
The Eye is a round hillock 3 feet over
1 Pole from the Nose to the Eye
3 Pole from Eye to the Shoulder
3 Pole from Shoulder to Tail
1 Pole the Tail
Ears 5 feet long, 2 feet wide and 3 feet apart
1 Pole deep the Cheek
4 Pole deep the Body
2 Pole deep the Legs
2 Pole from Nose to Cheek bone end
The head rest on a Bank 6 feet high

K. THREE FARMS SOLD AT MALMESBURY: 1827

Sold by Auction at Malmesbury, July 21 1827

				£	s	d
Farm & Buildings, at Lea & Claverton						
69 Acres Pasture, Tythes 2s 6d pr A				4400	0	0
A	**R**	**P**				
13	2	20	Pasture, Tythes 2s 6d pr A	800	0	0
Farm & Buildings - the Grange Tye[5] free						
102	2	12	Pasture & Arable	7100	0	0
19	1	16	Pasture Tye free	2075	0	0
16	3	32	Pasture Tye free	1500	0	0
9	3	20	Pasture Tye free	900	0	0
4	0	1	Pasture Tye free	400	0	0
4	0	0	Pasture Tye free	480	0	0
Farm & Buildings & Cattle=leaze &						
13	0	0	Pasture Tye free	1750	0	0
2	0	0	Pasture Tye free	225	0	0
6	2	0	Pasture Tye free	550	0	0
4	2	0	Pasture Tye free	410	0	0

Freehold
Within a mile & two of Malmsbury

L. LITTLECOTT AND BECKHAMPTON FARMS: 1828

30 January 1828

Littlecott Farm, Parish of Enford
11 miles to Devizes Market
The value of Littlecott Farm

	A	R	P		£	s	d
Homestall, Garden &	0	3	21				
Orchard				at 50£ pr Acre is	431	11	0
Pasture	7	3	1				
Pasture, the Bank	2	0	16	at 20£ pr Acre is	42	0	0
Watermedow	4	3	39	at 100£ pr Acre is	500	0	0
Arable	152	2	0	at 45£ pr Acre is	6862	10	0
Down	178	1	0	at 11£ pr Acre is	1960	15	0
	346	1	37		9796	16	0

280£ at 30 Years Purchase is 8400£

Outgoings	£	s	d				
Tithes [In pencil - 1833 92£]							
Land Tax under	11	0	0	at 3% pr Ct will			
Chief Rent	2	14	0	take	490	0	0
Chief Rent	1	0	4	from the value of the Estate			
	14	14	4		9306	0	0

at 3% pr Cent is 280£ N.B. the Rent is 300£

27 February 1828

Beckhampton Farm
Compare the value of Beckhampton Farm

	A	R	P		£	s	d
Homestall and Pasture	7	0	21	at 50£ pr Acre is	356	11	0
Watermedow	8	3	37	at 100£ pr Acre is	398	0	0
Arable	111	2	8	at 45£ pr Acre is	5019	18	0
Down	78	3	19	at 11£ pr Acre is	867	13	6
	206	2	5		7142	2	6

215£ at 30 Years Purchase is 6450£

Outgoings	£	s	d
Tithes [In pencil]	48	11	
	5	17	6
	54	8	6

At 3% pr Cent is 215£ N.B. Say the Rent 250£ having many advantages 6½ miles instead of 11 to Devizes Market &c.

M. BECKHAMPTON ENCLOSURE: 1795

Beckhampton Enclosure
Award'd by the Commissioners as follow

Daniel Tanner of Urchfont
John Gale of Sturt, Land Surveyor
Benjn Haynes of Salisbury, Land Surveyor
& Richd Bloxham of West Dean Wilts

Their first meeting held on Tuesday 1st May 1792 at the House of John Rider, Catherine Wheel Beckhampton.
Finished 1st April 1795

	A	R	P
Whitelands	4	0	37
Pasture Waste	0	2	37
Arable & Down	187	0	26
Silbury Mead	8	2	24
	200	3	4
Old meadow & Buildings	6	1	24
	207	0	28

The right of Watering Silbury Mead
Award'd by the Commissioners as under

Peter Holford Esqr 3 days Water exclusively commencing 1st Novr Yearly
Duke of Marlbro, next 3 days
James Sutton Esqr, next 4 days

The expences of Hatches &c. for Watering to be paid by

	£	s	d	
Peter Holford Esqr	0	9	6	in the £
Duke of Marlbro	0	4	6	
James Sutton Esqr	0	6	0	
	1	0	0	

N. MR GUY'S INSOLVENCY: 1829

1829. Questions which Miss Forty of Chippenham propose puting to Mr Guy, on his becoming Insolvent

What sums of Money, have you received of different Persons from the year 1825
What have you done with the Money arising from Property at Beckhampton which you sold very lately to Mr Thos Pinniger

And from Westrip	£9000	0s	0d
And from Little Bedwin		[30000 in pencil]	

And from Cocklebury		[9000 in pencil]	
And from Burgage Houses	£14000	0s	0d
1812 Miss J M Burroughs	£6000	0s	0d
1820 Alton	£5000	0s	0d
1809 Mr Breacher	£20000	0s	0d
1820 Sir John Hawkins	£20000	0s	0d

1794 or 5, Commenced Business with what Property

O. PLANTING UP THE NEW GROUNDS AT BECKHAMPTON, 1830 TO 1834

1830

Front of Garden Wall

3 Morella Cherrys 7s 6d
1 Black Heart[371] 2s 6d
12 Giant or Irish Ivys 6s

Wall, Fruit Trees, Planted

At the end of the House

1st 1 early Nectarine Newington 5s X
2nd 1 Moore Park Apricot 5s
3rd 1 Newington Peach 5s X

X Ripen end of August

Garden Wall, right hand

1st & 2nd Green Gages 5s
3rd 1 Orlean Plumb 2s 6d
4th 1 Magnum Bonum Plumb 2s 6d
5th 1 Swan Egg Pear 2s 6d
6th 1 Charmantelle Pear 2s 6d =
7th 1 Windsor Pear 2s 6d X
8th 1 Autumn Burgamot Pear 2s 6d }

371 Probably Black Heart Cherry tree.

9th 1 Gansells Burgamot Pear 2s 6d }
10th 1 Orange Burgamot Pear 2s 6d }

} Ripen in October & good bearers
X Ripen in August
= Good Winter Table Fruit

Wood house Wall

2 May Dukes,[372] first that ripens 5s
1 Black heart 2s 6d

1831

Trees planted in the Yard, & Climbers

1st	Black Popler	1st	Evergreen Rose	9d
2	Thorn Acacia	2nd	Budlea Globosa	1s 6d
3	Unterio Popler	3rd	Evergreen Clematis	1s 6d
4	Thorn Acacia	4th	Evergreen Honeysuckle	1s
5	Unterio Popler	5th	Lonicera flexiosa	1s
6	Black Popler	6th	Ma-Cartney Rose	9d
	6d each	7th	Sweet scented Clematis	9d

16 February 1833

Flowering Trees, planted in Garden

1st Weeping Cytissus
2nd Evergreen Cytissus
3rd Purpurea Cytissus
4th Nigrecans Cytissus
5th Supinus Cytissus

Pot Flowers Primula Sinensis

21 November 1834

Trees planted by the front gates

Left side   Right side

---

372 Cherry trees.

1st	Chichester Elm	1s	1st	Mountain Ash
2	Mountain Ash	4d	2	Chichester Elm
3	Mountain Ash	4d	3	Thorn Acacia
4	Thorn Acacia	4d	4	Chichester Elm
5	Ontarian Popler	4d	5	Turkey Oak
6	Turkey Oak	4d	6	Mountain Ash
7	Black Popler	4d	7	Black Popler
8	Mountain Ash	4d	8	Mountain Ash
9	Thorn Acacia	4d	9	Ontarian Popler
10	Chichester Elm	4d	10	Thorn Acacia
11	Lime	9d	11	Lime
12	Chichester Elms	4d	12	Chichester Elm

P. A LOCAL ACCOUNT OF THE SWING RIOTS: 1830

November 1830

ALARMING TIMES

Agricultural distress, and great consternation among the Farmers in consequence of Incendiaries traversing the Country, Fireing, Ricks, Barns &c Labourers, at the same time collecting in great numbers 2 & 300, distroying Thrashing Machines, and demanding great Wages (15s pr Week) 7s & 8s now the common wages, dictating to Landlords & Clergymen, to lower their Incomes

The work of distruction commenced in Kent last month, & continue spreading to the adjoining Counties and now Novr 20th, have become general, (see News Papers) having reach'd our neighbourhood

Nov 19 Friday eve 9 O-Clock FIRE, at Mr Fowlers at Oare & other Places

20. Saturday eve 11-O-Clock FIRE Mr Mills at Stanton & other Places

21. Sunday, eve Mr Mills Buildings &c attempted firing again, and Mr Pike, Collingbourne & others fire'd

22. Monday, - no Fires near us this even'g but at distance,

23. Tuesday, Fire at Mr Brown Chisselton and other Places, this eve Rioters taken at Rockly and Burbage by the Cavelry ordered out this day, the Farmers joining

them from Marlbro Fair

24. Wednesday, 400 Horsemen, assembled at Rockly, head'd by Mr Baskerville, the Masgistrate, the Marlbro troop of Cavelry, following went in pursuit of the Ring=leaders of the Rioters, took 20 at Ramsbury with their Captain (Goddard a Tanner) and 8 others, at Chissledon.. | A Troop of the Horse Guards sent to Marlbro, from Trowbridge

25. Thursday, the Horsemen and. Cavelry, scoured the Country Great Bedwin, Shalbourne, Grafton &c many Prisoners taken | Machines, at Corton, Cleveancey &c. broken by the Rioters (200)

29. Monday, Rioters at Bremhill &c. in large numbers, when exacted a promise from the Farmers an advance in Wages to 10s pr Week for Winter and 12s for the Summer

[See also diary entries for November 26 and 27 for Pinniger's personal involvement]

Q. THE DEATH OF RICHARD SADLER SMITH OF BREMHILL: 1832

Saturday 31st March Died aged 64 Years Mr Richard Sadler Smith at Bremhill

Fragments written by his Son B P Smith.

If Heaven permits that there should be
In Man such thing as purity
In <u>him</u> did every virtue blend
of <u>Husband</u> <u>Father</u> <u>Brother</u> <u>Friend</u>

And have I liv'd to see the day
When him I priz'd on earth most dear
My Father snatch'd by death away
And I remain to sojourn here
How much I lov'd him none ere knew
Or Pen can paint or language tell
But all is past, one more sad view
And then a long, a last farewell

On taking a last look at his depart'd Father whis his Coffin was about to be closed.

==

Bremhill when last from thee I parted
My heart was warm, my spirits gay
No tear of grief had ever started
My life had been one summers day.
Since then the winters chilling wind
Has nip'd the bud that promis'd fair
And all the blossoms of the mind
Have wither'd up in cold dispair
And now again I visit thee
Thy beauties still remain the same
But oh how lost they seem to me
For I am changed in all but name.

==

To the memory of Mr R S Smith by the Revd W L Bowles the Minister of the Parish

I think upon that Village scene,
I think of him with tears,
Who there has like a Brother been,
For well nigh thirty years

I see the Garden where we walk'd,
On the calm Sabbath day,
And, looking on the Dial, talk'd
Of this life's shortening way.

When I return the Bells will ring,
The trees in verdure wave,
The Flowers of a new year shall spring,
Whilst He - is in his Grave

Poor Friend! we never more shall kneel
In the same hallow'd pane,
Blest, who like thee, in death shall feel,
He knelt not there in vain

Salisbury April 7th 1832

==

On Saturday March 31-1832- at his residence Bremhill, Wiltshire, in the 64 Year of his age Mr Richard Sadler Smith. The loss of this amiable and philanthropic Gen<antl>tleman, will be long and severely felt by the Poor in the Parish where he resided, having lived nearly the whole of his life amongst them, and to whom he was ever a kind and charitable friend, as a Husband, Father and Master, a Kinder and better never existed; and great as must be the sorrow and affliction of his surviving family and relatives, they have the consolation of knowing that his pity and hopes were of the warmest and best kind and his religion [...] sincere as unostentatious; in him the Church has sustained the loss of one of its most steady and strict adherents, and society one of its best members.
The Purity of his life, and the excellence of his nature, will justly authorize us in saying of this much lamented individual, 'He was, say all that should be, he was that.'

R. FATTENING AND SELLING SHEEP: 1833 TO 1834

23 December 1833

Fat Ram Stag for Labour's at Xtmas

2 Legs	21 li		
2 Shoulders	16¼ li		
2 Loins	19¼ li		
2 Necks	16½ li		
2 Breasts	9 li		
	82 li	say at 6d is	£2 1s 0 d
The Offal			
Skin		3s	
Head & hinge		1s	
Suet (the Caul)	8½ li @ 6d	4s 3d	
Loose Fat	3½ li @ 3d	10½d	
		9s 1½d	£0 9s 1½d
			£2 10s 1½d

24 November 1834

		£	s	d
80 Ewes @ 30s } 20 Ewes @ 28s }	29s 7d	148	0	0

60 Lambs @ 23s }				
20 Lambs @ 20s 6d }	21s	104	10	0
20 Lambs @ 15s }				
29 Wethers @ 27s 6d		39	17	0
		292	7	0
2 Fat Wethers at Xtmas				[...]
Wool				[...]
				[...]

1833	£	s	d
54 Ewes @ 30s	81	0	0
10 Ewes @ 26s	13	0	0
50 Lambs @ 22s	55	0	0
18 Lambs @ 13s 6d	12	3	0
1 Fat Wether at Xtmas	2	10	0
	163	13	0
Wool	61	9	0
	225	2	0

[Pencil notes in margin]
W P_ Ewes 28s 8½d
10 @ 23s
80 @ 26s
120 @ 31s

[The above appears to be a calculation of the average price achieved by his brother William for his ewes: in this case, Thomas achieved more per ewe than his brother]

S. THE SUPPLY AND DEMAND FOR WHEAT: 1834

In the evidence given to the Agricultural Committee, Mr Jacobs stated, that since the year 1827, the Stock of Corn in hand at the time of Harvest, has not exceeded a months consumption, instead of 5 or 6 months consumption as formerly, and that if we were to have as bad a Harvest as that of 1816, the deficiency could not be supplied from all the World.

If we were to diminish the growth of English Wheat by one tenth part of that now produced, we should not be in a safe state in case of a deficient Harvest, for all the World could not make up the deficiency, as all the World could not easily supply at any price. for Wheat is not the Food of Man in any other Country to the same extent as in England.

In France there are only about 17,000,000 of Quarters a Year grown of which near 3,000,000 are wanted for seed, and that for a population of 30,000,000 Persons, while we require nearly as much for half the number of Persons.

The Harvest of 1816 was deficient 3 or 4 Months consumption.
The Harvest of 1828 was deficient 6 Weeks consumption.

The average annual consumption of Wheat in England is about 12 Millions Quarters, of which quantity about nine tenths are the produce of home cultivation and that the remainer is supplied by Importation from abroad, and from Ireland which latter Country has exported of late 600,000 Quarters annually to England.

T. THE PRICE OF MALT: 1730 TO 1833

The manner in which the price of Malt, has been enhanced, as Taxation rose, is thus evinced in the Greenwich Hospital returns, which shew the price paid for Malt. pr Winchester Quarter including the Duty, thus:

	£	s	d		£	s	d
1730	1	0	0	1810	4	4	5
1750	1	4	6	1815	3	9	7
1760	1	4	9	1820	3	8	8
1770	1	8	3	1825	3	2	0
1780	1	11	1	1826	3	5	1
1790	1	15	6	1827	3	4	10
1800	4	4	0	1828	3	1	7
1805	4	5	7	1833	3	0	0

U. THE INTRODUCTION OF IMPERIAL MEASURE: 1835

January 1 1835 IMPERIAL=MEASURE commence this day, - all other measures illegal.
My Bushel enlarged from Winchester to Imperial.
A Sack of Old Wheat Winchester weigh 12 S 7 li Nt. The same Sack made & measured Imperial 12 S 13 li Nt.
In making 38 Sacks (of Old) Winchester measure Wheat to Imperial measure as above, it took one whole sack, consequently 1 Sack of 39 is lost.

It require nearly 3 Quarts which is from 6 li to 6½ li pr Sack to enlarge the Winchester to Imperial.
By puting Wheat at 30s pr Sack & Barley @ 30s pr Qr with somtimes Beans & Pease, my Corn account will shew I loose full 20£ each year, by the change of the measure. Sold 1st my load of Barley 9 S 16 li Nt @ 25s, best Barley selling @ 34s and at Harvest Shavelere sold for 42s.
My Old Wheat weigh now Imperial 12 S 13 li Nt offered this day 21s only pr Sack
The lowest price I have sold such wheat for, was in Octr 1822 - of the Year 1820 - Winchester measure consequently about 12 S 8 li @ 24s pr Sack

V. AFFLICTING ACCIDENT: 1836

[Note in margin] See the account in my Book of the last great wind fall of Timber 31st Octr 1823[373]

On Saturday last the 26th Died Ellen Cordelia aged 20 Daughter of Mr Bartlett of Great Bedwin and this day the 29th of Novr most awfully sudden Henry his youngest Son aged 18, and Miss Susan Bartlett aged 51 Sister to Mr Bartlett

They left Bedwin at 11-o-Clock with his youngest Daughter Elizabeth in Mr Bartletts Phaeton for the purpose of going to Marlbro to order mourning for the Family. as they were proceeding down Knowle hill 2½ miles from Bedwin, the wind which was extremely Violent uprooted one of the large Beech trees that stand upon the edge of the road, at the moment the Carriage was passing: one large limb fell upon the Horse, and two branches diagonally upon the Carriage One of these fell upon Mr Henry Bartlett who was driving and the other upon Miss Susan Bartlett who was sitting in the Seat behind, and their deaths must have been instantaneous. Miss Elizabeth Bartlett who was sitting in front of her Brother, escaped miraculously between the two branches, but was unable to extricate herself on account of a small branch lying over her In this situation they remained for nearly an hour till they were discovered by some Woodmen who were going to their work.

The damage done to the Timber on the Marquis of Ailesbury's Estate during the hurricane is immense, upwards of 300 Trees varying from 2 feet to 12 feet in

373 Entry in diary at the end of October 1823: *The heavy Snow lodging upon the Trees, with the boisterous Winds, blew down innumerable large Trees, and the weight of Snow upon others, splinter'd of the limbs, and injured vast numbers of others, 2 & 3 hundred Trees blown down in many single Parks.*

circumference were entirely uprooted or so much injured as to be necessary to cut them down.

All parts of the Country sustaind injury by the by the ravages committed be the violent Storm

N B Harry Amors Cottage upposite us, by one powerful Gust, took half the Roof Timber & Thatch together clear off the Walls & drop'd it in the Garden adjoining clear from the House, in the same Moment took 6 feet of the Frestone Cress from my new Cottage adjoining the Inn, a short time before 12-oClock noon. A valuable Cow of Mr Jas Smith of Bremhill kill'd by a Tree falling on her

W. SHEEP AND LAMBING: 1838

Feby 12th	1 Lamb first Feby - 1 ewe died dicline
Feby 15	2 Lamb Feb'y 15 1 Teg died murr
Feby 16	2 Lamb
Feby 20	5 Lamb - 1 died
Feby 26	20 Lamb only yet
March 4	40 Lamb - 1 Ewe died heaving
March 5	1 Ewe died heaving
March 10	2 Ewe died heaving
March 11	1 Ewe died heaving
March 12	2 Ewe died heaving
March 13	1 Teg died giddy
March 14	1 Ewe died heaving
March 15	1 Ewe died giddy
March 16	1 Ewe died Evel in the head
March 17	1 Ewe died ill in the head

X. THOMAS PINNIGER THE FARMER AND THOMAS PINNIGER THE TIMBER MERCHANT

On 16 May 1834, Pinniger recorded in his diaries the death of Christopher Pinniger of Potterne who died suddenly, falling in a fit from his horse: this incident had been reported in the Salisbury & Winchester Journal, 12 May 1834, describing Christopher as *butcher of Potterne.*

Christopher Pinniger (then of Pewsey) married Anna Lewis of Calne on 13 November 1826. *Wiltshire, England, Church of England Marriages and*

Banns, 1754-1916, Christopher Pinniger [accessed via Ancestry, 16 April 2020]. Their daughter Martha was baptised at Potterne on 26 October 1827 and their son Thomas was baptised at Potterne on 1 February 1829. *England, Select Births and Christenings, 1538-1975* [accessed via Ancestry, 16 April 2020].

The Pinniger genealogical book contains a number of entries suggesting that Christopher was part of this branch of the Pinniger family, as follows:
(a) Christopher, butcher of Potterne, was the son of William Pinniger, late of Cadenham and Seven Bridges (whose brother Richard was a journeyman butcher): in turn William was the son of Richard Pinniger of Cadenham Farm. p.57;[374]
(b) When she died Miss Elizabeth Gilbert left £500 to Richard Pinniger of Freegrove, which enabled him to go to Cadenham, p.47;
(c) Richard Pinniger of Freegrove was the son of the 1st Christopher Pinniger of Shaw House by his second wife Susannah Broome, p.35;
(d) Memorial in Lyneham churchyard - Richard Pinniger of Freegrove, late of Cadnum [Cadenham] died 16 April 1801, aged 78, p.101.
WSA: 4381/2/1, *Pinniger genealogical memoranda book, c.1820-1943.*

In 1841, Thomas Pinniger, aged 11 was living with his sister Martha (aged 12) at the home of Isaac and Anna Holloway in High Street, Potterne where Isaac was a butcher. TNA: HO107/1184/7, *Census returns. 1841 census. Wiltshire. Hundred: Potterne and Cannings including Parish: Potterne ED2*: it is probable that Isaac took over Christopher's butchery business, in that he married Anna Pinniger on 5 September 1835 at Potterne. *England, Select Marriages, 1538-1973, Isaac Holloway* [accessed via Ancestry, 16 April 2020].

According to the 1861 Census. TNA: RG9/1290, *Census returns. 1861 census. Wiltshire. Registration District: 256, Devizes. Registration Sub-district: 1 Bishops Cannings, including Parish: Stanton St Bernard, ED2.* Thomas Pinniger, aged 32 was a timber merchant living in Honey Street.

In conclusion, Thomas Pinniger the farmer and Thomas Pinniger of Robbins, Lane & Pinniger were related.

374 Cadnum - Cadenham.

INDEX OF PERSONS AND PLACES

n after the page number indicates that this refers to a footnote.
TP – Thomas Pinniger.

Abingdon, Berks 92
  fair 330
Abson, Glos
  Bridgeyate (Bridge yeate) 136
Adelaide, Queen 161
Ailesbury, Marquis of 49, 383
Aldbourne 222
  Dudmore Lodge 271
  Lower Upham 222
Alexander
  family xiii
  Rich, Mr 314
  Susannah xiii
  Wm xiv
All Cannings 244
  Tan Hill 240
    fair (St Ann's Hill) liii, 62, 75, 89, 109, 185, 227, 254, 277, 299, 323, 360
    road *see* Avebury
Allen, Mr 312
Allington *see* Chippenham
Alton livn, 375
Amesbury cxxiv
Amor (Amer)
  Harry 384
  Jn, Mr 102, 105, 226, 235, 241, 331
Andover, Hants 74
  Abbotts Ann xcii
  workhouse xcin
Andrews
  Aaron cix, 103
  J 212
  Jas 245, 252
Appleshaw, Hants
  fair xlvn, liii, 42, 234, 282, 360, 364-5
Ashbury, Berks
  Idstone Farm 230
Ash, Kent lvin
Ash, Thos 327
Ashe, Robt (of Langley House) 119
Atchley, Miss 98
Atherton
  Nath 98
  Nath senior 163
Atworth cxi, cxxii
  Cottles 57
Audry *see* Awdry
Australia cxi
Avebury (Aveby, Avey) xvi, lxviii, lxx-lxxi, lxxiv, cxvin, cxvii, cxix, cxx-cxxiii, cxxxi-cxxxii, cxxxvii, 52-339 *passim*
  Beckhampton (B) ix-x, xiv-xvi, xxxix, 41, 75, 125, 136, 141, 220, 222-3, 243, 264, 275, 285, 287, 298, 313, 314n, 317, 327n
    Beckhampton Farm lxv-cxxxvii *passim*, 82-331 *passim*, 351, 373-7
    Beckhampton Inn lxxi-lxxv, cxvi, cxx, cxxii-cxxiii, 86, 114, 122, 141, 146-7, 149, 150-1, 170, 175, 188-9, 233, 257-262 *and see* Watts, Jn and Treene, Wm
      both inns (Beckhampton Inn & the Waggon & Horses) 264
      building work 238-270 *passim*
      race horse stables 241, 257, 264
    Catherine Wheel inn lxxi, cxvin, 374
    Fields:
      Silbury Mead lxvii-lxix, cxiv, cx, cxiii-cxiv, 83-331 *passim*
      Tan Hill Piece lxxxix, ciii, cxiv, cxxxvi, 101-329 *passim*
      Whitelands lxviii, lxix-lxx, cxv, 99-326 *passim*, 374
    Great & Little Beckhampton Farms lxxn, cxviiin
    Tan Hill Road (St Ann(e)'s Hill) lxx, cxvii, 99-331 *passim*
    Waggon & Horses inn cxxi

church cxxxii-cxxxiii, 129
Cooks cottage 264
roads, measurement cxxxii
Roman road 270
Silbury Hill lxvii, lxix, lxxvi, civ, cxii, 70, 89, 90, 94, 123, 236, 240, 271, 285, 371
turnpike
Beckhampton Turnpike Trust cxxi-cxxii, cxxxii, 282
gates cxvi-cxvii, cxxi-cxxii, 285
road lxvii, lxxi, lxxxvi-lxxxviii, cxxi, cxxviii, 103, 215
tolls cxvi, cxxi-cxxii, 205, 222, 261, 282, 329
West Kennet (Kennet, K, Kent) cxxviii, 16-330 *passim*
Awdry (Audry)
Jane 72n
Peter 60

Badminton, Glos 285n
Bagshot *see* Shalbourne
Bailey
B, Mr 276
Geo (crier of Calne) wool-stapler cvi, 131, 173, 351
Jn, wool-stapler 208
Robt, wool-stapler 125
Thos cxii
Baker
Mr (of Sutton) 277
Thos xviii
Baldown Mills 51
Ball, Ambrose 328
Barbury Farm *see* Ogbourne St Andrew
Barnbridge *see* Chippenham, Tytherton
Barnard, Jn 21
Barry, Revd 226
Bartlett
Eliz, Miss (dau of Mr) 383
Ellen Cordelia (dau of Mr) 383
Hen (son of Mr) 383
Mr 329
Mr (of Great Bedwin) 383
Mrs (of Great Bedwin) 301
Susan, Miss (sister of Mr) 383
Baskerville, Mr 378
Bath lv, lxiv, lxxiin, xciii, cxviii, cxii, cxxiiin, 26, 33, 78-9, 88, 120, 213, 268, 283
Lansdown (Lands down) 136
Landsdowne Monument cxxi
Lyncombe 153n
Paragon Bldgs 87
Quay 184
Spring Gardens 185
Batten (Batton), Ben senior, Ben or Matthew xlv
Bays *see* Boys
Beak, Anne 33n
Beams (Beames)
Geo senior 317
Jn (of Abingdon) 92
Beazells *see* Melksham, Bezell
Beckhampton *see* Avebury
Belcher, Mr (Edw) maltster 129, 291
Berhampore, India xix
Berwick Bassett 186, 222, 239
Common 232
Bethel(l)
Mr (of Chute) 234
Thos (of Trowbridge) 54
Bird
Jn, carter cxii
Jn 21
Jn, shepherd 320
Bishop's Cannings 117, 266
Cannings Hill cxi, 266
Horton 157, 176, 199, 265, 317
Down xc, 253, 317, 330
Old Shepherd's Shore 176
Roundway c, 115, 186
Bitton, Glos 136, 139
Brookham (Brokam) Hill 136
Hole (Hall) Lane colliery 136
Willsbridge (Wellsbridge) 136
Blackman
Abednego (Ebednego, of Charlton) shepherd cvi, 155, 174, 197, 220, 246, 272
Isaac 108, 120-1, 124
Blandy, Mary xxiii
Bloxham, Rich (of West Dean) 373
Boman, Mr (of Calstone) miller 308
Bowen, Dr cxxx
Bowles, Wm Lisle, Revd 76, 379
Bowood 176, 201, 222, 247, 272, 275, 298
Box lxxiii, lxxix, lxxxiv, cxxiii, cxxvii, 61, 99, 109n, 116n, 118, 120-1, 127,

129, 131, 133, 136, 154, 158-9, 178, 258, 278, 322
Haz(s)elbury (Hazleberry) 94, 108, 179, 210-1, 298
Rudloe 93
tunnel cxxiii
Box, Rich 178
Boys (Bays), Mr Hen (of Malmains, Dover, Kent) 182, 362
Brander, Mr 298
Bradford on Avon cxxiiin
Barton Farm 295
Bradford Leigh fair 63, 91
Bramble, Mr (of Devizes) cxx
Breacher, Mr 375
Breadenstoke *see* Lyneham, Bradenstoke
Breadmore, Wm 21
Beversbrook (Beaversbrook) *see* Hilmarton
Bremhill xiiin, lv, lxiii, 32n, 63, 70, 71n, 74, 74n, 76-7, 86-7, 90-1, 102, 116n, 136, 171, 193, 201, 216, 217n, 244, 248, 250, 256, 258, 268, 270, 310, 324, 378-9, 380, 384
Cadenham (Cadnum) 385
Charlcot 271
Farm 199, 204
Bremhill Grove lviii, 50, 54, 173, 242
House 199n
Wick 96
Brewer(s)
Geo (G) cix, 104, 170
Jn 202
Wm 243
Bridgeyate *see* Abson
Brimslade *see* Savernake
Bristol lxxix, xci, cxxiii, 92, 98-9, 109, 164, 184-5, 211
Totterdown 155, 279, 293
Britford 182
fair 362
Broad Chalke cxxivn-cxxv
Broad Hinton 295
Farm cxxviiii
Manor Farm 44
Broad Town 192, 295
Thornhill 237
Broadstock *see* Lyneham, Bradenstoke
Brokam Hill *see* Bitton
Bromham, Jos cx-cxi, 111-2, 134-5
Bromham lxxin, 217, 324
Westbrook 195
Broome
Chris (of Brompton) 124
Rich Pinniger 271-2
Susanna, Mrs 217, 385
Brompton 124
Brown
Algernon (of Broad Hinton) 295
Cath (Mrs Jn of Avebury) 278-9
Eliz 259
Geo (Mr) cxiv-cxv, cxvii, 99, 104, 111-2, 126, 174, 195, 219, 220, 252, 270-1, 279, 283, 294, 318 *and see* Mr (of Avebury)
Jn (of Chiseldon) cxxv, 264, 377
Jn, Mr (of Burderop) 207
Jn (of Lower Upham) 222
Kate, Miss 222
Mr (of Avebury) lxxxvii, xcv, 90, 99, 100, 130, 148, 152, 155-6, 171, 179, 183, 186, 195, 197, 199, 203, 207, 218, 221, 223, 230, 233, 256, 264, 269, 283, 291, 315-6, 325-6 *and see* Geo (Mr)
Mr (of Sutton, Kent) xlv
Thos xcv, cxv
Wm R (of Broad Hinton) lxvi, cxiv-cxv, 44, 100, 220, 223
Wm (of Horton) 265, 317
Broxholm(e)
Inhola 318
Jn, Mr 306, 318
Mary Sophia 306
Miss cxxix
Buckland, Mr [possibly Jn] 253
Budd
Mr cxix, 17-8, 28-9, 262
Mr (of Winterbourne) 104, 111
Bullock, Thos cxii
Bupton *see* Clyffe Pypard
Burbage lxxxiii, 176, 377
Burdett [Fras] 300
Burderop (Burdrupt) *see* Chiseldon
Burgess [Thos] liii
Bishop of Salisbury 64
Burroughs, JM, Miss 375
Burton on Trent, Staffs 252
Bush, Mr (of Trowbridge) cxvii, 118
Bushton *see* Clyffe Pypard
Butler

Mr (of Fowlswick) 86
Jas, parish clerk 272
Robt 142
Buckerfield, Revd Bart 98
Buxton (Buxston)
Cath (Kitty) 234
Mary Dorothea, Miss 204
Mr, Revd (Geo Pococke, of Mildenhall) 76, 78, 331

Cadenham (Cadnum) *see* Bremhill
Calley, Thos, Mr (of Burdrupt Park) 280
Calne lxi, lxiii, lxxvi-lxxvii, xci, cxxv, cxxxin, cxxxvii-cxxxviii, 62-384 *passim*
Blackland Ho xiin, cxxxvii
Butchers Row 80
Chilvester Hill 119, 237
Derry Hill 99, 125, 127, 171
fair 64, 78, 92
Hayle (Hail) Farm cii, 115, 134n, 157, 176, 199, 275
High Penn Farm cxxixn, 82
Linden Grove 187
market xii, 309
Moss's Mill 276
Nuthills Farm 271
Poor Law Union xci
Quarr 113
Quemerford (Quemfd) Mill 118, 191, 308
Silver Street cxxxin, cxxxviii
Stockley Copse 194
Studley 73, 120
brook 120
White Hart Inn 190
Calston(e) Wellington 134, 320-1
Hill cxxi
Mill 308
Cambridge
Robt 184
Wm (of Chippenham) 205
Can[i]ey, Thos 200, 202
Canning
Geo, prime minister 75, 366
Jn, Mr (of Rockley) 133, 196, 209
Robt (of Whitefield, Ogbourne St George) 175
Cannings *see* Bishop's Cannings
Cannon, Jas 21
Castle, Geo, sheep dealer c, 104
Caswell, Jonathan 326
Catcomb *see* Hilmarton
Cave, Robt 308
Chalfield *see* Great Chalfield
Chandler
T 160
Wm 21
Wm cxi
Charlcot *see* Bremhill
Charlton near Pewsey 155, 174n, 246
Charlotte, Princess liv
Cheltenham, Glos 241, 264
Cherhill lxxvi, c, 104, 122, 184, 285
Hill 266
White Horse 70, 371
White House (inn) lxxvi, 122
Chilton Candover, Hants 44
Chilton Foliat 220
Chilvester Hill *see* Calne
Chirton 171
Chippenham (Chippm) ix, xiv-xvi, xxxix, xli-xlii, lii, lv, lvii-lxiii, lxv-lxvi, xci, cii, cxiv, cxvi, cxxiii, 4-314 *passim*
Agricultural Society 265
Allington 60, 89, 346
burgage houses 375
Cocklebury (Cockleberry) lxiii, 56, 58, 65, 72-4, 79, 86, 88, 375
fair 72, 85, 93, 250
Fowlswick 86
Island (of Rea, Mr Guy's) lix, 59, 90, 95
Kellaways 65, 96, 98, 231, 238
Lowden 73, 317
Monkton House 237
Poor Law Union xci
St Andrew's church burial ground 64
Sambourne Farm xli, lix, lx, lxii-lxiii, 51-100 *passim*
Sheldon 61-2
Timber Street 129n, 291n
Tytherton (Tyhtn) xln, 9, 13, 32, 42, 59, 75-8, 95, 97, 117, 127, 130, 144, 156, 159, 192, 208, 216, 221, 225, 246, 274, 284, 306, 310, 320, 323
Barnbridge (Barn Bridge) 60, 192, 268
Tytherton Lucas (West Tytherton) 60n, 102, 190, 192, 225, 234, 242, 265,

268, 268n
White Hart inn 27
Chisbury *see* Little Bedwyn
Chiseldon (Chissledon, Chissleton, Chisselton) cxxv, 264, 377-8
Burderop (Burdrupt) 207, 280
Chivers
Eliz (Mrs Thomas) 217
Hen (H) 293, 304, 318
Jn 179, 212, 241, 253, 339
Jn (father & son John) 173, 184, 195, 218, 245
Jn (of Avebury) 123
Jn (of Calne) 271
Jn junior 230
Lawrence 125, 259
Rich 102
Sarah (Mrs Hart) 125
Wm 112
Wm (late of Calne) 331
Cholsey, Berks 104
Christian Malford (Xn Malford) cxxv, 62
Church
Mr (of Aldbourne) 271
Robt 108
Chute 234
Standen xxiii
Cirencester, Glos 206, 216, 272
Clatford Farm *see* Preshute
Clark, Jacob, Mr (of Devizes) butcher 282, 292
Clarke
& Robbins *see* Robbins, Lane & Pinniger
J lxxxiii, 116
Mr (of Devizes) 292
Cleverton *see* Lea & Cleverton
Clements, Thos 202
Clevancy *see* Hilmarton
Cliffe *see* Clyffe Pypard
Clifton, Mr (Charles) curate 77
Clutterbuck, Mr (of Tytherton) 320
Clyffe Pypard (Cliffe) 69n, 97, 124, 144, 161, 217, 236, 272
Bupton cxxvi, 90, 186, 309
Lower Bupton 214
Bushton 233
Cobbett, Wm, journalist 250
Cobham, Surrey 78
Cocklebury (Cockleberry, Cockelberry) *see* Chippenham
Coke, Mr (Thos of Holkham, Norf) 227
Cole, Thos (of Kington Langley) 221
Coleman, Wm, (W) 140, 160
Coller, Jas cxxii, 283
Collier, Jas, banker cxxxvi
Colerne (Cullern) 159
Collingbourne cxxiv, 86, 377
Colton, Revd (William Collins) 243
Compton, Mr 29
Compton Bassett (Compton) xiii-xiv, xcii, 54, 100-1, 211
Cowage (Cowitch) lxiii-lxiv, lxxix, cxvii, 27, 39, 54, 63-4, 89, 90-1, 93, 97, 99, 120, 147-8, 160, 180, 183, 248, 258, 280
Connington *see* Cunnington
Coombe Bissett 225
Copeland, Wm cxv
Cornwell, Wm, schoolmaster cxxxi, 318
Corsham 191
Pickwick lxxiii, lxxx, cxxvii, 113-4, 126-8, 191-2, 258-9
Westrop (Westrip) 374
Corston *see* Malmesbury
Corton *see* Hilmarton
Cottles *see* Atworth
Coverdale, Miles, Bp of Exeter 257
Cowage *see* Hilmarton
Coward, Mr (of Roundway) c, 115, 123
Cowley, Thomas (of Ramsbury) miller 6
Cowitch *see* Hilmarton, Cowage
Cox cxxii
Cricklade
hiring fair livn
Crook (birth) 65
Ann (sister of Thos P, wife of Hen) cxxviiin, cxxixn, 232, 234, 240n, 242, 268
Cath (Mrs P, wife of John) 211n
E 315
Edw 102
Emma cxxviii
Hen, Mr (of Corton) cxxviiin, cxxixn, 76, 225, 240n
Hen (son of Hen) cxxix
Jn, Mr 260
Jn, Mr (of Corton, nephew of TP) cxxxvi-cxxxvii, 240n, 293, 331
Jn, Mr (of Avebury) 190

Mr cxx, 77, 88, 93, 163
Mr (of Corton) xcvi, 92, 143, 214
Mrs 192
Rich 229
Susannah Mary (of Barn Bridge) 268
T 23
Thos (nephew of TP) grocer cxxix, 308
Walt 73, 266
Mrs 216
Crowdy (Crowder, Wm Morse or Hen Crowdy) cxvi
Crudwell (Crl) 253
Crutwell, Thos M 138
Cullern *see* Colerne
Cunnington (Connington, Currington), Wm, Mr, woolstapler 223, 352, 354-6
Currington see Cunnington

Darley, WB (of Burton upon Trent) 252
Davis
Geo 211
Jacob cxi, 211, 245, 293, 302
Jas cviii, cxi, 176, 199, 253, 293, 302
Jn cxi, 294, 320
Mr 133
Step 176, 199, 211
Wm 166
Wm (W) carter cxi, 253, 296, 339
Days, Jn (of Pickwick) 113
Dean, Wm (of London) 234
Debary, Rich xxii
Derry Hill *see* Calne
Devizes (D) lxvii, lxxii-lxxiv, lxxvi, lxxix-lxxx, lxxii, lxxxiv, lxxxviii, xci, cxiii, cxx, cxxiv, 82-356 *passim*
Assize Courts 246n
Castle Inn cxx
Devizes Green, The Bell 238, 251
fair 70, 78, 84, 93, 101, 115, 185, 210, 232-3, 245, 281-2, 284, 300, 304, 311, 360-1, 363, 365
gaol xl
gas works lxxxii
market [corn] (and crops taken to) cxviii-cxvix, cxxiv, cxxxiv, 10-373 *passim*
New Park 288
Prospect Ho cxxviii
races 257, 279
school cxxxi
White Hart Inn 331
Didmarton, Glos 92, 103
Dobson
Jn cix, 101-2
Wm cix, 101-2
Dorney, Bucks 70
Down Ampney, Glos 80
Downend *see* Mangotsfield
Draycot [Cerne] cxviii, 120
Duck 188
Wm 198
Dudmore Lodge *see* Aldbourne
Durrington 132
Dyke
Dan 111
Robt 104

East Kennet 217, 317
Easton cxxiv, 139, 144
Edmonds (Edmunds, of Beckhampton) 159, 189, 261, 265, 272, 277-9
Ann, Mrs cxv-cxvi
Geo 160
Mr cxvi, 274
Wm (W) lxxin, 213, 287 *and see* Edmonds
Edwards
Ann, Mrs (née Rushly, wife of Jn, of Wilton) xcvi, 143
Ben 208
Dan 116, 306
Harry liii
Jn (of Wilton Manor Farm) xcvi
Josh 76
Jos 228
Mr (Dr) 76
Mr junior 78
Mrs (wid of Dan, of Calne) 306
Mrs (of Devizes Green) 251
Revd Mr 76
T xxv, xxviin, xxx
Eley, Jn 104
Ellman, Mr 227
Elston *see* Orcheston St George
Enfield, Mddx 308
Enford 159, 179, 227, 252
Combe (Coomb) lxiv, 357
Littlecott Farm 81-3, 372-3
Estcourt (Eastcourt), Mr lxix, lxxxix-xc,

252, 265, 272, 308
Etchilhampton (Etchelhampton) 135
Eton college, Berks lxii, 67
Evans, Ezekiel cxxii, 205, 222, 282
Everett, Dame 216
Everleigh (Everley) cxxiv
  petty sessional division 94
Exeter
  Bishop of 257
  mail coach lxxii, 183

Farebrother, Mr 39
Faringdon (Farringdon), Berks lxxivn, 121
Farnham, Surrey liv, 359, 361-5
Fauntleroy, Hen 43
Fawley, Berks
  Watcombe Farm lxxviii, 313
Fell, Mrs (of Tytherton) 216
Fielder, Mr (of Newbury) xliv
Fishlock, David, shepherd cxii
Fittleton ci
Fitzroy, Georgina 285n
Flower (wife of Jn, of Langley) 209
Ford *see* North Wraxall
Fordington 102
Forty, Miss (of Chippenham) xxiin, 83, 374
Fosbury (Forsbury, Forsby) 56, 61, 70, 72, 94, 140, 276
  Upper Farm lxiv, 358
Fowler
  J & Co. (Bath) 184
  Mr (of Oare) 377
Fowlswick *see* Chippenham
Foxham 97, 102n
Freegard, Jn, carter cix 99
Freegrove Farm *see* Lyneham
French, Jn 21
Fromont, Miss cxxiii
Fry, Jas (of Calne) 288

Gaby
  Emma 170
  Hen Headly (of Calne) 204
  Mr 11
  Ralph Hale (of Chippenham) 121
Gale
  Isaac 216
  Jn 260
  Jn (of Savernake Park) 207
  Jn (of Stert (Sturt)) land surveyor 373
  Mr (of Minety) 62
Geo
  King liv
  Prince of Wales livn
Giddings
  G, waterman 247
  Mr (of All Cannings) 244
Gilbert
  Eliz, Miss 385
  Isaac cx, 165, 177-8, 184
Glass, Ben (of Worton) 280
Goatacre *see* Hilmarton
Goddard
  Capt 378
  Jn 287
Godwin
  Jn Hunt (of Holt) 187
  Mr (of Beversbrook) cxviii, 120
Golding, Wm (of London, alias Jn Pin-niger) 185
Goldney, Messrs (of Chippenham) 118
Grafton *see* Great Bedwyn
Grange Farm *see* Malmesbury
Granham *see* Preshute
Gray, Mrs (of Newbury) 235
Green, Mr (of Newbury) xxix
Greenwich *see* Kent
Great Bedwyn (Bedwin) xcvin, 56, 98, 278, 299, 301, 378, 383
  East Grafton Manor Farm xli
  Grafton 166, 185, 378
    Wolfhall xxvii, 12, 209
      Sudden xxvii
  Wilton liii, xcvi, 143
Great Chalfield (Chalfield, Ch) xxvii, lv, lxiv, 54, 57, 60-1, 65, 68, 74-5, 84, 86, 89, 91, 96, 307
Grossett (Jn Rock of Lacock Abbey) lxi
Grove *see* Mildenhall
Gunning, JT 138
Guthrie (Jn) Revd 225
Guy
  Anth, Mr (G, Mr G) xv, xxii-xxiii, xxv, xl-xlii, lvii-lviii, lix-lx, lxv, lxix, lxxi, lxxxv, cxvi-cxviii, cxxxviii, 8, 20, 23, 28-9, 44, 51, 59, 61, 73, 78-9, 82-3, 85-7, 90-1, 114, 118, 120-2, 125, 134, 139, 145, 340, 374-5
  Eleanor 78

Gye, Fred (of London) lxi-lxii, 60, 72

Hacker, Robt liii, 21
Halcombe *see* Malmesbury
Halcomb, Messrs lxxiv, 183, 226, 258, 263
Hale
  Jas (of Calne) 222
  Mrs (wid of Sam) 230
  Sam (of London) 230
  Step, miller 308
Halford 220
Hammond, Jeremiah, plumber lxxxi-lxxxii
Hampshire, Wm (of Con(n)ock) mower 275
Handy, Thos (of Norton, Bristol) 185
Harding
  Mr 338
  Sam 51
Harris
  Jn (of Calne) 287
  Miss 222
Hart
  Robt 125n
  Mrs Sarah (née Chivers, of Seend) 125
Hartnell, Mr (of Bristol) 184
Hastings, E Sussex
  St Leonards 74n
Hatherell, Fred 76
Haz(s)elbury (Hazleberry) *see* Box
Hawkins
  Jn, Sir 375
  Joseph xvii-xviii
Hayle (Hail) Farm *see* Calne
Haynes
  Benjn (of Salisbury) land surveyor 373
  W (of Shalbourne) xviii
Hayward
  Jas Young (of Tufton, Hants) 287
  Mr (of Chirton) 171
Hazle (Hazel), Rich lxxv, cxxxvi
Headly, Dr (of Devizes) 314
Heale, Wm, Mr, nurseryman lxxvii, 249
Hendon, Hen (of Compton) 101
Henl(e)y
  Abram (of Tockenham) 170-1
  Jn (of Goatacre) 33
  Thos (of Cheltenham, Glos) wine merchant 264
Hewett, Jn, Mr (of Mapledurham, Oxon) c, 103-4
Hickley, Rich, Mr (of Avebury) 194
High Penn *see* Calne
Highway 102n, 126, 128, 140n, 275
Highworth cxvin, 79
  Round Rob(b)in Farm 312
Hill
  Rowland, Revd (of London) 194
  Dr 237
Hillier
  Jas, carpenter lxxxv, 142, 145, 157
  Mr (of Granham) 14, 16, 39
  Thos (T) 97, 108, 144, 193
Hilmarton (Hillmarton) xiii, 20n, 95-6n, 125n, 144, 159n, 201, 211n, 225, 237n, 240n, 298
  Beversbrook (Beaversbrook) lxiv, cxviii, 91, 120, 126, 228, 357
  Catcomb 185, 187, 193, 286
  Clevancy (Cleveancy, Cleveancey, Cleve'y) cxxxviii, 57-8, 62, 64, 72, 74, 77, 82-3, 86-7, 89, 93, 95, 107, 109, 115, 143, 162-4, 178, 180, 187, 257, 306, 309, 324, 378
  Corton cxxxvi, 61, 64, 72, 77, 87-9, 92-3, 143-4, 162, 165, 183, 214, 225, 234, 240, 260, 293, 304, 319, 331, 378
  Cowage (Cowitch) lxiii-lxiv, lxxix, cxvii, 27, 39, 54, 63-4, 89-91, 93, 97, 99, 120, 147-8, 160, 180, 183, 201, 211, 248, 258, 280
  Goatacre 33n
Littlecote (Littlecot) lxv, 20, 22, 29, 31, 56, 72, 94, 159, 169, 242, 263, 269-270, 272, 274, 283, 292-4, 321
  Lower Penn Farm cxxixn
  Spillmans Farm 237
  Widcombe Quarry 237
Hitchcock
  Chas (of Mon(c)kton) 209
  Mr cii, 119
  Mich Pontin 314
Hole (Hall) Lane Colliery *see* Bitton
Holford
  Peter, Esq (of Avebury) lxix, 374
  Robt lxx, cxiv-cxv, cxviii
Holloway
  Anna 385
  Isaac 385

Holly 202
Holt 187
Honey Street *see* Woodborough
Hopkins
  Mr (of Cheltenham) 241
  Jn, Mr (of Lyneham) 210
Horn(e), Edw xviii, xx, xxii
Horton *see* Bishop's Cannings
Howell, Jas 189, 339
Hughes
  Dan (of Wick) 96
  Jn (of Tytherton) 75, 79
Hulbert, Mrs C (Seager) 64
Hungerford, Berks lxxvn, cxxviii, 13-4, 21, 42
  sheep fair 104n
Hunt, Hen, orator 242
Huntercombe, Bucks 204n
Hutchings, Mr (of Marlborough) 331

Idstone *see* Ashbury
Isle of Wight lv, lvi, 42
  Rew (Rue) Street 42
Ilsley, Berks
  market c, 104

Jacob, Jacobs
  Mr 381
  Step 211
  Wm Bowman (Boreman) Revd cxxxi
James, Jn 21
Jefferies (Jeffrey, Jeffreys)
  Mr cxiv, 220
  Mr (of Melksham) miller 364
  Step (of Spilman's Farm) 237
Janner *see* Jenner
Jenner (Janner)
  Dan (of Cricklade) 174
  Edw (of Berkeley, Glos) cxxxn
  Mr 329
  Wm (of Bremhill) 71, 272
Jones, Anne (of Stanton St Quinton) 71

Kellaways *see* Chippenham
Kemm (K)
  Thos (of West Kennet) cxxviii
  Wm, Mr (of Avebury) cxvii, 98, 105, 107, 109, 152-3, 171-3, 179, 195, 198, 201, 219, 221, 233, 241, 270-1, 283, 288, 316, 319, 325, 339
Kennet *see* Avebury, West Kennet
Kennet & Avon Canal Navigation lxxxvii, xxix, 207
Kent lx, cxxiii, 182, 363-4, 377 *and see* Avebury, West Kennet
  Ash lvin
  Greenwich 70, 382
  Malmains, near Dover 362
  Trench Farm 70
Kent
  Eliz (wife of Wm) xvii
  Eliz (dau of Wm) xvii, xixn, xxii
  family xv
  Wm xvii, xixn
  Wm Streat xvii-xix
Keynsham (Kainsham), Som 59-60, 65
King
  Geo, carter 104, 114
  Mary, Miss (of Calne) 199
  Step, Mr (of Overton) 190, 207, 231, 272-3
Kingsdown fair, Glos 92
Kington fair, Heref 78, 360
Kington St Michael
  Kington Langley (Langley) cxviii, 120, 221
    Langley Fitzurse 348n
Knight, T 69
Knighton *see* Ramsbury

Lacock (Laycock) 72n
  Abbey 73
  Notton 72, 97
Lad (Ladd)
  Geo 245
  Thos 313
Lambourn(e), Berks 228, 298, 313
  fair xlv
  Upper Lambourn(e) 211, 228
Lane, Ebenezer lxxxiii
Lanfear, Mr (of Sutton) 277
Langley Burrell (Langley) 54-5, 125, 209 *and see* Kington St Michael
  Langley House 119
Lansdown (Lands down) *see* Bath, Lansdown
Lansdowne, Marquis of cxxi, 258
Large (L_e)
  A & R 113
  Abbot xiv, lv, lvi, 57

Abram 115
Ann (AL), Miss (sister-in-law of TP, of Clevancy) 74, 83, 95-96n
Ann, Miss (of Melksham) 96
Eliz 153
family xiii
Jas, Mr 77, 88, 92, 213, 256-7, 260
Jane (mother of Ann (AL)) 95n
Jn, Mr 33, 77, 86, 88, 92, 96n, 107, 115, 186, 232, 269, 315, 322, 339
Jos (of Purton) 13
Jos, carter 266, 289
Martha, Miss (of Tockenham Court) 257
Mary (wife of TP) xiv, lv
Mary (Mrs L, mother of Mary, mother-in-law of TP, née Henly) xiv, lv, cxxxiii, 4, 243
Mary (wife of Thos L Pinniger, of Clevancy) cxxxviii
Mr 115, 118
Mrs Jacob (Lucy, of Lyneham) 145
Rich (father of Ann (AL)) 95n
Robt (of Lyneham) cxxxvi, 137, 141
Latton (Laten) Down c, 104
Lavington 110
Lavington
Mr 336
Thos (of Marlborough) grocer 78
Lawrence, Wm, Mr (of Idstone Farm) 230
Law, W liii
Law, Wm 5, 9, 60, 112
Laycock *see* Lacock
Lea, Wm 21
Lea & Cleverton
Cleverton (Claverton) Farm lxvn, 73, 372
Lewes 69
Mr xxviii
Anth, Mr (of Chilton Candover) 44
C 36
Lewes fair, E Sussex 227
Lewis, Anna 384
Line, Mr ci, 99, 125
Little Bedwyn (Bedwin) ix, xiv-xvi, 28, 56, 227, 278
Bagshot Lane 25
Bedwyn Common 241
Sicily Clump 241
Chisbury 201, 299
Chisbury Farm 50n
Knowle Hill 383
Little Bedwyn Estate xvi-lviii *passim*, 1-51 *passim*, 356, 374
Brick kiln xxviii
Farm (Home Farm) xvii-xviii, xx, xxii, xxvin
Fields:
Burnt Mill (Burntmill) 2, 7, 12, 15, 22, 30-1, 36, 39, 50-1
Dean Heath (Heth) liii, 1, 6, 15, 22, 25, 28-9, 31, 40-1, 44
Forebridge Piece 3, 8-9, 11-7, 32, 40-1, 45, 347
Kite Hill 4, 14-5, 21-2, 26, 33-9, 41
Pedlers piece lii, 1, 9-11, 15, 23-5, 31-3, 36, 43, 46
Sweardown (Swear Down) 6, 9-10, 12, 20, 26-31, 39-40
Wansdyke Furlong 2-4, 14-5, 21, 26, 31-2, 34, 36-9
Westboro 11-3, 15, 17-8, 20, 30, 45, 47, 51, 347
Westboro Copse 4, 8
Whitley Furlong xxxv, 3-4, 10, 13-14, 16-20, 44-5, 347
Jockey Farm xvii-xviii, xxi-xxii, xxvin
Mansion House xviii-xix, xxii, xxvin, xl
Mill xxv-xxviii
Stockwells xxvi, xlviii
Public House (The Jockey) 1, 3, 5, 31
Puthall 101
Workhouse 3
Littlecote (Littlecot) *see* Hilmarton
Littlecott Farm *see* Enford
Lock
Mr 87
Locke (of Rowdeford)
Sarah, Miss 210
Wadham 257
Lockeridge *see* Overton
Long [Walter] 300
London xix, xxii, lx, lxiii, cxxiii, cxxxvii, 70-1, 94, 185, 189, 199, 264, 281, 285, 288, 331, 339
Albany Street 266
Basinghall Street, Masons hall xxii
Bishopgate Street 328
Great Queen Street 74n

Houses of Parliament cxx, 232
Insolvent Debtors court 74n
market 93, 300
the Poultry 230
Royal Exchange 309
Surrey chapel, Blackfriars 194
Wood Street 234n
Looker, Jas (of Monkton) cxxxii, 172
Low
Isaac Selman cxxii, 205, 222
Jas (of Beckhampton) 298
Mr ciii, 141
Lowden *see* Chippenham
Lower Bupton *see* Clyffe Pypard
Lower Upham *see* Aldbourne
Ludgershall cxxiv
Lymington, Hants lv
Lyne, Mr (of Brimslade) 365
Lyneham xiv, lv-lvi, cix, cxxvi, cxxx, cxxxiii, cxxxvi-cxxxvii, 33, 35n, 95-6n, 109, 144-5, 153n, 200n, 210, 243, 272, 325, 385
Bradenstoke (Breadenstoke, Broadstock) lxiv, 62-3
Freegrove Farm 74, 86, 88, 91, 216, 273, 385
Lyneham Court cxxvin, 62, 310n
Tockenham Court Farm lv, 243, 257

M [possibly Moss's] Mill 79 *see* Calne
Maillard, Fred Augustus 68
Maitland, Ebenezer Fuller 60
Maiseyhampton (Maisey Hampton) Glos 272
Malmesbury 74-5, 372
Corston 57
fair 84
Grange Farm lxiv, 73, 372
Halcomb(e) 82
Mangotsfield, Glos 136
Downend 92, 136
Quarry lxxxiv, 136
Mansell, Mr (of Studley) 73, 79
Mapledurham, Hants c, 104
Marden 298
Marlborough, Duke of lxvii, lxix, lxxn, 374
Marlborough (Marlbro) lvii, lxxiv, lxxvi, lxxxii, cxiv, cxxiv, 28, 37, 42, 75, 78, 83, 90, 98, 123, 152, 160, 170, 178, 220, 241, 256, 261, 268, 281, 304, 331, 356, 378, 383
Bridewell cxi-cxii, 135
Castle Inn cxx
fair xlv, liii, 14, 20, 43, 79, 90, 94, 110, 118, 138, 160, 180, 185, 203-4, 210, 235, 254, 323-4, 329, 361, 365, 378
hiring fair livn
Poor Law Union cxix, cxxxii, 259
Quarter Sessions 163
troop of cavalry (cavalry) 378
Marsh, Jn, carter cxii
Marshfield, Glos 136, 159
Martin, Thos (of Bitton) 139
Maskelyne (Maskeline)
Bryan (of Littlecot) 159
Fras, Miss (of Bupton) 186
Geo cxxvi-cxxvii, 325
Jasper 282
Sarah, Mrs (of Bupton) 90
Sarah Rumboll, Miss 69
Maslen lxxix, 105, 108
Geo cxiv, 238
Geo L 325
Isaac cix, 103
Jn 242
Matthews, Jn, brewer lxxv
Meadham
Jn (of Chisbury) 201
Wm 299
Melksham 96, 364
Bezell (Beazells) Farm 79
bridge cxxii
Queenfield Farm 79
turnpike gate cxxii
Merrick, Revd cxvii
Merriman (Merryman, Thomas Baverstock) cxiv-cxv, 220, 223
Methuen [Paul] 300
Michell, Edw, Mr (of Chippenham) 237
Mildenhall 204n, 234n, 331
Grove 130, 263, 269-270, 274, 331
Miller, Mr 45, 59
Millerd, Jn, carter cix, 99
Mills
Mr (of Newbury) 337
Mr (of Savernake) 223
Mr (of Stanton) 377
Step, Mr (of Elston) 362
Thos (of Fittleton) ci

Milton [Lilborne] cxx, 144, 165, 207
Minchin, Jacob 160
Minety 62
Mitchell, Jn, Mr (of Chippenham) 91
Mog(g) Revd 308
Monckton *see* Winterbourne Monkton
Morchard Bishop, Dev lvin
Moore, Mr lxv, 82
Morgan, Revd Mr 253
Moss's Mill *see* Calne
Munday, Mr (of Wedhampton) 142

Nalder, Jn, Mr (of Berwick) 186
Neale 16
Neat(e)
  Geo, Mr (of Monkton) 156
  Wm, Mr 113
  W (of Overton) 127, 150
Netheravon 187, 346
Nettleton Farm xli, lxvi, cix, 99
Newbury, Berks xxix, xliv, 23, 27, 56, 172, 235, 337
  fair 225, 311
Norfolk x
  Duke of 227
North Wraxall
  Ford 136
Norton 82, 84
Norton [Malreward], Som 185
Notton *see* Lacock
Noyes
  Ann, Miss 252
  Rich 211
Nuthills Farm *see* Calne

Oakhanger Farm *see* Shefford Woodlands, Berks
Oare cxxiv, 144, 377
Offer, Jn, hedgelayer 170
Ogbourne 8, 108-9
Ogbourne St Andrew
  Barbury (Barbery) Farm 287
  Rockley (Rockly, Rackly) 134, 144, 196, 377-8
Ogbourne St George
  Whitefield 175
Orcheston St George
  Elston 362
Osmond 16
  Thos, Mr 363
Overton 127, 150, 190, 207-8, 225, 231, 264, 273, 285
  Copse 261
  Lockeridge Farm 231n
Overton fair, Hants 208, 276, 323, 365
Owen, Aneurin cxx
Oxford 168
  coach 81

Pain(e), Mr, Jn xxiv, lvii-lviii, 39, 43-5, 47-8, 51, 341
Palmer
  Jas (of Bath) 185
  Jn, Revd Mr 102
Paradise, Wm lxxxvii, cxxxi, 123, 236
Park Town Farm *see* Ramsbury
Pavy, Mr (of Wroughton) 192-3, 197
Pearce (Pierce) 152, 175, 178
  (of Easton) 139
  Jn (of Standen) xxiii
  Wm (W) 133, 142, 161, 269-270
Peel, Sir Robt cxx
Pegler, Ben (of Foxham) 97
Pepler, Jos, Mr (of Hayle Farm) 134, 275
Pewsey lxxxiii, cxxiv, 144, 384
Philpot
  Betty, Mrs (of Avebury) 229
  Robt (R, Rt Ph_t, Mr, of Beckhampton) farmer lxx, lxxxvi, lxxxix, cxii-cxv, cxviii, cxx, 102-361 *passim*
  Wm (W) (Mr) farmer lxv-lxix, lxxviii-lxxix, lxxxviii, cix, cxv, 83-5, 87, 89-96, 98, 99-101, 103, 105, 108-110, 120, 125, 130-1, 148
  the two Mr (Robt & Wm) 89
Pickwick *see* Corsham
Pierce *see* Pearce
Pike
  Mr (of Collingbourne) 377
  Mr (of Milton) 162, 165
  Mrs 357
  Wm (of Great Bedwyn) 98
Pin *see* Pinniger
Pinckney, Geo (of Wolfhall) 209
Pinniger (Pin, P)
  Abbot Large (son of TP) cxxix-cxxx, 251, 258, 273
  Alex 129
  Ann (dau of TP) lxiii, cxxix-cxxx, cxxxviii, 60, 258

Ann (of Cowage, dau of JP) 211
Ann (sister of TP) *see* Crook
Broome (BP, bro of TP) lawyer xln, cxiv, 76, 87, 95, 192, 231n, 236, 238, 250
C 103
Cath (of Cowage, wife of JP, sister-in-law of TP) 201n, 211n
Cath (of Cowage, niece of TP) lxiv, 54
Cath (of Gt Chalfield, niece of TP) lxiv
Chris cxxxvii
Chris (of Box) 116
Chris (of Hazlebury) 298
Chris (of Langley) 54, 55n, 116, 240n, 270
Chris (of Potterne) 221, 384-5
Chris (father of TP) xiii-xiv, lxiii, 75-6, 310n
Ellen (dau of TP) lv-lvi, 35
Eliz (wife of BP) 95
Emma (née Smith, of Hazlebury) 116, 298
Emma (dau of Emma) 298
family vii, clxi-clxiii, 384-5
Fras 288
HC, draper 304
Jacob (son of WP) xcv, cxxviii 117
Jane 216
Jn (grandfather of TP) xiv
Jn (JP, of Cowage, bro of TP) lxxix, cxvii, 54n, 76, 97-8, 120, 201n, 211n, 248
Jn, (J- P-, of Fosbury) 70
Jn (nephew of TP) 248
Jn alias Wm Golding (of London) cxxv-cxxvi, 185
Martha (dau of TP) cxxix-cxxxi, cxxxviii 187
Mary (Mrs, wife of TP) lv, lxv, cxxix, cxxxvii, cxxxviii, 71, 84, 96n *and see* Large
Mary (step-mother of TP) 310
Mary (dau of TP) lvi, cxxix-cxxxi, 36, 45, 96n, 258, 291
Mary (dau of Jn & Cath) 201
Mary Alice 95
Mr (of Coombe Bissett) 224
Mrs (of Wanborough Leigh) 221
Peter xii-xiii
Rich (of Cadenham Farm) 385
Rich (of Freegrove) 385
Rich (of Highworth) 312
Sarah (sister of TP) 55, 79, 206
Sarah (niece of TP) 294
Susannah 79, 280
Susannah (mother of TP) 310n
Thos (TP)
Beckhampton Turnpike commissioner cxxii, 282
birth, marriage & death xiii-xiv, lv, cxxxvi-cxxxvii
health cxxxi, cxxxv
height & weight cxxix
high constable cxxxii, 153, 172
Kennet & Avon Canal Navigation 207
Marlborough Poor Law Union guardian cxix, 259
road surveyor cxxxii
will cxxxvi-cxxxvii
Thos Large (son of TP) ix, lxxv, xciv, cxx, cxxxv-cxxxviii, 89, 91, 258
Thos (of Down Ampney) 80
Thos (of Honey Street) lxxxiii, 385
Wm (WP_ K, Willm, of Kennet, bro of TP) lxxix, xcii, cii-cv, cxxviii, 73-381 *passim*
Wm (of Calne) millwright 248
Wm (of Cowage, son of Jn & Cath) 148, 201
Wm (of Cadenham) 385
Plain, *see* Salisbury
Pole, Jonathan (of Tytherton) 274
Ponting
Jn (of Bremhill) 136
Nich (of Langley Burrell) 76, 125
Nich Thos Skull 64
Pope
Geo 199
Mr 144, 164-5
Portsmouth lvi, 42
Potter
Edith, Miss 227
Edward, Mr (of Chisbury Farm) 20-2, 33-4, 50
Mr xxxvii
Mrs 30-1
Potterne cxxivn, 221, 385
High St 385
Wick lxxvii, 249n
Worton 280

Poulshot 81
Poulton, Glos 131
Poynder, Mr 248, 258
Preshute
  Clatford Farm 70, 223, 250, 360
  Clatford Mill 274
  Granham 14
Priddy, Som lxxx
Pumphrey (Pumphry) Mr 84, 231, 255-6, 326
Pullen
  Eliz (of Lyneham) 95-6
  Jn (of Lyneham) 153
Purton 13
Puthall *see* Little Bedwyn

Queenfield Farm *see* Melksham
Quemerford (Quemfd) Mill *see* Calne

Rabson Farm *see* Winterbourne Bassett
Ramsbury 6, 25, 42, 378
  Knighton 233
  Park xcii, 103-4
Randall, Rich (of London) xxii
Rawlings, Mr (of Bremhill Grove) lviii, 50
Reading, Berks xxiii, 172
Reeves (Reaves)
  Hillier (H) horse dealer ciii, 193, 220, 245, 248, 269-270, 303, 327, 330
  Mr 62
Rendall, Wm xli
Rich
  Mr 98
  Rich lx
Richardson, Rich (of Devizes) lxvii
Richens
  David, carter cxi-cxii, 327
  Fras, under carter cxii
Richmond, Duke of 227
Rider, Jn 374
Robbins, Lane & Pinniger (of Honey Street) xcii, lxxxiii, 385
  Clarke & Robbins lxxxiii, 116
Robbins
  Jas 107-8, 120-1, 124
  Jn lxxii, 119, 141
  Sam lxxxiii
  Wm (W) lxii, 119, 141
Rockley (Rackley) *see* Ogbourne St Andrew
Rodbourne
  Jn, carter's boy cxxxi
  Step, carter cxii
Rogers
  Jos 100
  Mr 44
Roundway *see* Bishop's Cannings
Rowde 109
  Rowdeford house 210, 257
Rudloe *see* Box
Rugg, Wm (of Chippenham) 287
Rumboll (Rumbold, Rumbell)
  Bryan, Mr (of Littlecot & Lyneham Court) cxxvi, 14, 20, 27, 29, 72, 219, 272, 309, 325, 327, 336
  Chas 294
  Eliz, Miss (of Wootton Bassett) 260
  S, Mrs 91
Rumming
  Mr (of Draycot) cxviii, 120
  Mr (of Notton) 97
  Robt (of Hilmarton) 201
  Wm (W) xcvii, 18, 83, 93, 151-2
Rushly, Ann *see* Edwards, Ann
Rutherford, Thos (of Calne) lxxxn, 113n, 135
Ruttiford *see* Rutherford

St Anns Hill fair *see* All Cannings, Tan Hill, fair
Sainsbury (of Marlborough) 170
Salisbury 60, 88
  assizes 137
  Plain 52
Salter
  Isaac senior 348n
  Simon, Mr (of Kington Langley) cxviii, 120, 348n
Salthrop (Salthrup) *see* Wroughton
Sandell, Mr 238
Saunders, Mr 182-3, 188, 207, 362
Savage, Robt 138
Savernake (Savenake)
  Brimslade 365
  Park 207
Savory, Wm (of Calne) 248
Scott, Thos xxii
Scriven, Mr 33-4
Seager
  Mary, Mrs (of Highway) 140

T (TS) 96, 98
Wm, Mr (of Highway) 140n, 275
Selkley hundred cxxxii, 153n *and see* Pinniger, Thos (TP), high constable
Shalbourne xviii, 378
Bagshot 25n
Shefford Woodlands, Berks
Oakhanger Farm cxxviii
Sheldon *see* Chippenham
Shephard
Chas (of Cobham, Surrey) 78
Rose (dau of Chas) 78
Shepherd (Sheppard)
Giles xviii, xxi, xxii
Short, Revd., Mr, Wm 60, 72, 91, 284
Sidmouth, Visct 288
Silbury hill *see* Avebury
Simkins
(of Littlecote) mole catcher cx, 169
Mr (of Stanton St Bernard) 182-3, 188, 206
Slade
Jn, Mr, glazier 101, 207
Smith, 5
Ab 18
Ann, Miss (of Calne) 238
Anne, Miss (of Bremhill, niece of TP) 244
Broome Pinniger (BPS, son of RSS, nephew of TP) 74, 378
Eliz (wife of RSS, sister of TP) 74n, 268
Eliza (of Thornhill) 237
Emma (née Crook, wife of Sam Hale) cxxixn
Emma (wife of Christopher Pinniger of Box) 116
Jas (of Bremhill) 384
Geo 127, 177, 300
Jn (son of Sarah, nephew of TP) 206
Jos (of Chilvester Hill) 236
Knightly, horse breaker 6
Mr 7-9, 39, 92, 361
Mr (of Beversbrook) 91, 126
Mrs (of Broad Town) 295
Rich Sadler, Mr (RSS, of Bremhill) 70, 74n, 76, 86-7, 90, 171-2, 199n, 378-380
Robt (of Broad Town) 192
Sam Hale (son of RSS, nephew of TP) cxxviii, cxxix
Sarah (of Calne, sister of TP) 185, 206
Tim Caswell (of Avebury) 244
Somerset, Hen, Duke of Beaufort 285n
Sotherton, Mr cxv
Southampton lvi, lxxxii, 42, 225
Southby, Alf, lawyer cxvi
Spackman xiii
Alice (of Tytheron, wid of Jacob) 78
Ann (of Chalfield, wid of Roger) 68
Cath 54
Jacob (of Tytherton) 32, 78
Jn 353
Mary (MS) 96-7
Mr *see* William
Roger (of Chalfield) 68
Susannah (née Pinniger, of Chalfield) 54n, 96
Thos 76
Wm (WS, Mr, of Chalfield) 54n, 75n, 76, 91, 96
Wm (of Chippenham) 250
Spencer, Jn, Mr (of Bradford-on-Avon) 295
Stagg, Chas, Mr (of Netheravon) 187
Standen *see* Chute
Standish, Wm, Mr (of Bowood) 275
Stanton St Bernard cxxiv, 182
Stanton St Quintin (Stanton) 71n, 308, 377
Stedhampton, Oxon 293
Stiles
Robt 76
Step 287
Wm, Mr (W) (of Upper Lambourn, Berks) 211, 283
Stockley Copse *see* Calne
Stone, Jn xl
Stratton
Jn lxvi, 100
Jn (of Milton) cxx
Mr (of Fawley) 313
Mr (of Rabson) 111
Strong, Mr [possibly of Calne] lxxx, 113
Studley *see* Calne
Sudden *see* Great Bedwyn
Sussex 363-4
Sutton
Eleanor (wid of J, late of New Park, Devizes) 288
Jas lxvii-lxix, 374

Sutton (nr Battle, Kent) xlv
Sutton Benger (Sutton) 277
Swindon lxxiv, cxvi, cxxiii, 66
  fair 72, 193, 221, 244, 248

Tan Hill *see* All Cannings
Tanner
  Daniel (of Urchfont) 373
  Jn, Mr (of Yatesbury) 69
  Wm, Mr, (WT, of Kennet) lxxi, cxiii-cxiv, cxxxvii, 103, 155, 185, 194, 196, 215, 236, 238-240, 264, 266, 271-2
Tasker, Robt xcii, 99
Taylor (Tayler)
  Mr (of Baldown Mills) 51
  Mrs (wid, of Chippenham) 192
Thetford fair, Cambs 227
Thomas, Eliz, Mrs (née Chivers) 217
Thornhill *see* Broad Town
Titcomb(e) (Timcombe)
  Geo, carter cxii, 318, 330
  Jn 134, 139
Tockenham 4, 9, 13, 22-3, 54, 62, 75, 78-9, 86, 91, 107, 109, 136, 162, 165, 170, 200-1, 246, 257-8, 265, 272-3, 303-4
Tockenham Court Farm *see* Lyneham
Toms, Wm, brewer lxxv
Townsend, Jas 136
Treen (Treene, Trin), Wm, Mr lxxi-lxxv, cxxxvi, cxxxviii, 239-241, 245, 247, 256-8, 261-3, 268-9, 278, 284
Trin *see* Treen
Trowbridge (Troubridg) 54, 65, 83, 99, 118, 211, 378
Twineham, Mr 363
Tuckey, Hen 45
Tufton, Hants 287
Tytherton (Tythn) Tytherton Lucas *see* Chippenham

Upavon xcv, 225, 273, 304-5, 307, 324
  Widdington Farm 242n
Upton Cheyney, Glos 136
Urchfont cxxxi, 209, 373

Vaisey (Vaizey)
  Eliz 331
  Mr 4, 24
Vincent, Ann, Miss (of Calne) 260
Victoria, Queen 321
Vines, Mr (of Catcomb) 54, 185
Viveash
  Mr (of Berwick Bassett) 239
  Orial, Mr (of Calne) 117, 271, 293
  Sam 76, 276, 293
  Simion 293

Waite, Esau, carter 266n
Walters *see* Waters
Wanborough
  Leigh 221
Wansdyke 176n *and see* Little Bedwyn
Warminster 211
Washbourne
  Jn (of Yatesbury) 207
  Mr 29 104
Ward 198
  Mr (of Marlborough) 83
Warwick, Guy (of Milton) 207
Watcombe Farm *see* Fawley
Waters (Walters, Mr, of Upavon) xcv, 305, 307
Watson, Mrs 261
Watts 103, 134
  Mary, Mrs 234
  Jn, Mr lxxii-lxxiii, cxxii, 238-9, 244-5, 247, 256
Webb 176
Wedhampton 142
Wellington, Duke of 285, 369
Wellsbridge *see* Bitton, Willsbridge
Wentworth (Wth, Wenth, Winkworth)
  Jn lxix, lxxn, lxxxix, cxii-cxvi, cxviii, cxxxii, 238, 281, 294, 316, 318, *and see* Mr
  Mr ciii, cix, 48-331 *passim*
  Sam, Mr (Sl, Sml) cxv-cxvii, 241, 261, 316, 325, 331 *and see* Mr
  Step xxiii
West Kennet *see* Avebury
West Lavington cxxivn
West, Revd, Mr 62, 76
West Tytherton *see* Chippenham, Tytherton Lucas
Westall, Rich 21
Westbrook *see* Bromham
Westrop (Westrip) *see* Corsham
Weyhill fair, Hants liii-liv, 41, 65, 78, 93,

115-6, 142, 183, 208, 233, 258, 281, 359-365
Whales 107
Whorell, Mr cxv
Whyr(e) *see* Winterbourne Bassett
Whites (the two, of Lyneham) cix, 109-111
White Horse *see* Cherhill
White House *see* Cherhill
Whitelock, Jas, Mr xviii, xxii
Wick *see* Bremhill
Wickham, Jas (of Devizes) 201
William, King cxix, 161, 297
Willmot, W, Revd 189
Wilshire, Michael cxv
Wilson & Son (of Bath) dentists lxiv, 87
Wiltshire, Silas 160
Wilton *and see* Great Bedwyn
  fair 64, 115, 182, 325, 359, 362, 365
    St Giles Great Sheep fair 142, 361-2
Winchester, Hants lvi, 42, 308
Windsor, Berks lx, 67, 70
Winkworth *see* Wentworth
Winterbourne c, 104, 110-1, 124, 144, 181
  see Budd, Mr (of Winterbourne)
Winterbourne Basset (Winterbourne) 266n
  Rabson Farm lvii-lvii, lxiv, 46, 111, 154, 357
  Whyr(e) 266, 289
Winterbourne Mon(c)kton (Mon(c)kton) 69, 133, 156, 172, 209, 328
  Down 261
  Farm lxiv, 58, 358
    Hackpin Down 358
    Windmill Hill Down 358
  Monkton Penning 159
Winterslow cxxiv
Wolfhall *see* Great Bedwyn
Woodborough
  Honey Street lxxiii, lxxx, lxxxii-lxxxiv, xcii, 116, 127-8, 175, 258, 262, 277-8, 320, 323-4, 385
    Wharf 114, 117, 258, 330
Wooldridge
  Josiah (of Great Bedwyn) 299
  Mr 44
Woolhampton, Berks xxix
Wootton Bassett 144, 260
  Hay Lane, railway terminus lxxivn, cxxiii
  hiring fair livn
Worton *see* Potterne
Wroughton 99, 190, 192-3, 197, 209, 248, 315, 330
  Salthrop (Salthrup) 78
Wyeth, Peyton Cooke, portrait painter cxxxi, 308-9

Xn Malford *see* Christian Malford

Yatesbury 69n, 110, 180, 207, 214, 313, 315, 319
Yeo, Jas, smith (iron) 107, 157
York, Duke of 67
Young
  Jn (of Marden) 298
  Revd Mr cxxvi, 303, 325

INDEX OF SUBJECTS

n after the page number indicates that this refers to a footnote.

accident cxii, cxxxi, 94, 279, 283
aftergrass 11
arsenic 184, 202, 234, 258

bankrupt lix, lxxv, cxvii–cxviii, 120–1
bankruptcy lix, cxviin, 122, 139, 374-5
bark xxxix, 9, 31
  barking xxxix, liii, xcix, 7–8, 30–1
  de-barking xxxixn
barren xliv, 28, 48, 63, 150, 155, 171, 192, 217–8, 243, 270, 291–2
brewing liv–lv, lxvi, lxxiii, lxxv, lxxix, lxxxii, 208, 263, 323
  ale liv–lv, lxxix, 9, 127, 134, 157, 199, 223, 263, 321
  beer xlii, liv–lv, lxxiii, lxxix, cix, 9, 104, 106, 110, 127, 157, 199, 223, 241, 289, 301, 318, 321
  hogshead lxxix, cxxviii, 19, 127, 208, 210, 289, 316
  hop[s] liv–lv, lxxiii, 70, 359, 361-5
  malt lxxiii, lxxix, 127, 191, 235, 241, 289, 382
buildings
  bakehouse 185
  barn xix-xxi, xxiv, xxvi-xxvii, lviii, lix, lxxviii, lxxx-lxxxi, 1-330 *passim*, 377
    Little Bedwyn
      Cowleaze Barn xxvi, 1, 19, 23, 25-6, 28, 43
      Home Barn xxvi
      Jockey Barn xxvi, 4, 16, 18
      Stockwells Barn xxvi, 3, 16, 22
    mow [part of barn] xxvi, 12-4, 125, 212, 265, 347
  brewhouse lxxix, 100–1, 104, 106, 109–10
  coach house xix
  cowhouse xix, lixn, lxxxii, 140, 277–8
  dairy lxxix, 106
  glasshouse (greenhouse) lxxvii, 249
  henhouse 157, 254
  malthouse xxiii
  outbuildings xx
  outhouse xx
  outshut lxxxi
    catslide lxxxi
  potato steaming (steeming) house lxxxi, 324, 326
  rick house xx
  skilling xxv, lix, lxxix–lxxx, lxxxii, cxxvi, 99, 112, 140, 155–6, 159, 177–8, 233, 306, 309, 330–1
  stable xix-xxi, xxv, xlvi, li, lix, lxxix-lxxx, 99, 101, 106, 108–14, 141, 143, 177–9, 181, 202, 241–2, 255, 257–64, 267, 270, 310
    racing lxxii, 241 *and see* Avebury, Beckhampton Inn, race horse stables
  thatch and thatching xxx, lxxiv, lxxx-lxxxin, xciv, xcvii, 14, 18, 41, 88-90, 101, 108, 110-1, 124, 139, 171, 179-80, 190, 200, 206, 230, 248, 254-5, 280, 321, 330, 384
  well-house (wellshed) lxxix, 106, 279-280
  wood-house lxxx, 155, 376

canal xx, xxix, lxxxvii–lxxxviii, 207
carriage lxxvi, 237, 285, 383
carriage (timber) lxxxiii, 100-1, 116, 128, 258, 320
carriage (water meadow) xxx, cxiii-cxiv, 183-5, 196, 210-1, 213, 239-240, 261, 272, 277, 283n, 293, 307
cavalry cxxiv, 144, 377-8
cheese xlv, liv, lxvi, 9, 143, 157, 164, 187, 199, 223, 265, 318, 320
Christmas (Xtmas, XTmast) 1, 3-4, 67, 120-1, 128, 146, 188, 212, 257, 261-

2, 268, 271, 285-6, 290, 308, 360, 380-1
cider 117, 208, 210, 306
ciderkin 210
coach lxxiin, lxxiv–lxxvi, cxxiii, 51, 81, 94, 120, 122-4, 147, 174–6n, 264, 285, 291, 295
Star lxxiv, 258
White Hart lxxiin, lxxiv, 178, 183, 235, 258
York House lxxiin, lxxiv, lxxvi, 183, 264
coop xlv, 117, 185, 245, 281, 304
corn-kiln xxviii
cottage xix-xx, xxiii, xxviii, lix, lxvi, cxv-cxvi, 19–21, 106, 130, 133, 157, 159–60, 170–1, 217, 232, 246, 261, 264, 269–270, 276–280, 384
crops
barley x–cxxiv *passim*, 1–331 *passim*, 334, 336, 338-9, 359, 367, 369-370, 383
American 265
beer (bigg) 368-9
Chevalier (Shavelear, Shaveleer, Shewclear) xcvi, 225, 246, 267-8
Moldevia xcvi
bean xxxiv, xxxvi, xciv, 4-314 *passim*, 368-370, 383
Heligolands xxxiv, xxxviii
Kidwell xxxvi, xcviii, 150, 192
chert bents xxxv, 195-7
clover xvi, xx, xxvii, xxix-xxx, xxxiv, xlvi, xcn, xcii, civn, cix, 5-329 *passim*, 343-4
hop [clover] xxxv, xxxviii, 4-297 *passim*
mangel (mangel-wurzel) xxxiv, 34, 73
oat xxvi–xcvii *passim*, 1–330 *passim*
Black Siberian xxxiv, 4-6, 37
Georgians xcvi–xcviin, 152–3, 167
Tartery xcvi, 314
pea xxxiv, lii-liii, 3-320 *passim*
Marrowfat xcviii, 171, 292
potato lv, lxii, lxxxi, xcix, cxvi, cxxvi-cxxvii, 3-331 *passim*, 338, 340, 367
Ash leaf (Kidneys) 243, 282
Black xcix, 174, 185, 233, 282
Magpie (Mag Pie) xcix, 142
Prolifficks xcix, 142, 164, 174, 185, 233, 282
Purple xcix, 142, 164, 217, 305
Red 282
Red Apple xcix, 174
White 233, 282
White Apple xcix, 142, 164, 174, 185
(Yorkshire) Red Kidneys 186, 282, 329
ray xxxvii, xlviii, xciv, xcvii, 5–315 *passim*
rye xxxiv, xcvii, cxiii, 143, 180–5, 195–8, 200–3, 205–6, 210, 231, 314, 368-370
Ringwood xxxiv
ryegrass xxxvii, xcivn
sainfoin xxix–xxx, xxxii–xxxiii, xxxv, xxxvii, xlvii–xlviii, xciv, 1–104 *passim*
swede (sweed) xvi–cviii *passim*, 3–324 *passim*
Arial xcviii, cliii, 133
Purple (short top) 214, 218, 223
turnip x–clix *passim*, 1–331 *passim*
Aberdeens xxxiv
Bells xxxiv
Green Rounds (Green Norfolk or Common Green Top White) xxxiii-xxxiv, liii, xcviii, 10-11, 35, 136, 158, 161, 176-9, 202, 222, 249, 284, 297
Norfolk White xxxiiin, xxxiv, xcviii, 174–7, 200, 221–2, 247, 274, 284
Norfolk Red xxxiiin
Red Rounds xxxiv
Tankards xxxiv,xxxvi, 8, 10–2, 106-7, 109, 155, 158
vetch xvi-xvii, xxv, xxxiv, xlvi, xcix, 3-328 *passim, 336, 341*
wheat x–cxxxiv *passim*, 1–331 *passim*
America xxxiv
Golden Drop 304-5, 307
Golden Dun xxxiv
Talavera (Talevara, Talevera) xxxiv, 24–311 *passim*
Taunton xxxiv, 20
cuckoo xv, 7, 28, 49, 58, 70, 84, 101, 130, 153, 173, 195, 219, 224, 246, 251, 270, 273, 294, 301, 318

death ix–cxlix *passim*, 13–331 *passim*, 378-380, 383-4
accidental death lin, 123, 266n
drowning (of humans) 80, 88, 163, 237,

242
drowning (of sheep) 19, 313
suicide 163
disease xlviii, l–li, lxiii, cvii, cxxx
animal
cattle plague li
murrain li, 131, 147, 157–8, 160, 169, 277–8, 341
rinderpest li
strangles lxiii, 61
treatment
bleeding
horse li, 9, 102, 132, 155, 176
salt-petre li, 155, 176
lamb xlix
daisies and salt xlix
tar l, 160, 278, 294
tar stick (stake) 147, 160, 278
human
cholera cxxxiii, 230
influenza cviii, cxxx–cxxxi, 287–8, 290
scarlet fever cxxx, 211
smallpox cxxx
treatment
bleeding 76
vaccination cxxx, 258
disputes
labour cix-cxii, 101-3, 134-5, 253
land and boundary cxii-cxv
tenant cxv-cxvi

emigration cxi
enclosure xvi, lxviii–lxix, lxxi, 83, 373-4

fair xlii, xlv, liii–liv, 14–330 *passim*, 359-365, 378
fence lxxxii, 125, 156–7, 172, 202, 207, 299
fire xxxin, lxii, lxxxiii, xciv, 61-2, 124, 146, 152, 156, 177, 209, 214, 232–4, 264, 267, 281, 309, 328, 377 *and see* law and order, incendiarism
fleece xliii–xliv, cvi, 10, 157, 348-356
funeral xiin, lvi, lxiii, cxxxiii, cxlix, 76–8, 96, 170–2, 194, 225, 243, 257, 259, 268, 272, 279

garden xix-cxxvii *passim*, 3–315 *passim*
American plant 241, 243
asparagus 216
bean [kidney] lv, lxii, cxxvii, 59, 64, 72–3, 84, 102, 104, 130–2, 182, 186, 196–7, 220, 246, 248, 272, 282, 329
bean [scarlet] 71
broccoli lv, cxxvii, 202, 226, 251, 323
cabbage lv, lxii, cxxvii, 24, 59, 100, 132, 143, 150, 161, 175, 193, 208, 232, 246
celery lv
cucumber lxiii, 71
dahlia 193, 197, 217, 233, 258, 270
gooseberry lxii, 59, 92, 240
onion cxxvii, 292
pea lv, lxii, cxxvii
ras(p)berry 92
gardening liv-lv, lxii, cxxvii, 19, 70, 82, 375-7
lawn 19

law and order
arson cxxiv–cxxv
assault 309
assizes cxv, 137, 246n
execution 43, 325
incendiarism cxxiii–cxxv, 156, 209, 214, 309, 377
riot xv, xciii, cxxiii, cxxv, cxxxiii, 144, 164, 377-8
rioter cxxiv, 144, 377-8
robbery, on the turnpike cix, 103
theft xl, cxi, cxxvi, 44, 185, 214, 293, 306, 309
livestock *and see* disease, animal *and* pest, maggot *and* tick
cattle xxv, xxx, xli-xlii, xlv, liv, ci, 52, 61, 63, 66–7, 249, 282, 318
Alderney (Alderny) xlv, ci–cii, 125, 312
calf ci, 125–328 *passim*
cow xiv, ci-cii, 4–320 *passim*
milk xlv, ci, 215, 217
[honey] bee cii, 75, 99, 104
honey 75
skep cii
horse xxv–cxxxiv *passim*, 2–331 *passim*, 383-4
oxen xxxi–xxxii, xxxiv–xxxv, xli–xlii, xlvi–xlvii, l, civ, 2, 10–5, 17–8, 21–4, 29, 31–2, 36–8, 40–3
pig xli, xlv–xlvi, lviii, lxxxi–lxxxii, cii,

cviii–cix, cxxx, 1–331 *passim*, 334-9
boar xlvi, cii, 1, 119, 144, 154, 165, 174, 187, 202, 226, 229, 334, 337
piglet xlvin, cii
porker 1, 23, 33, 45–6, 121, 145, 184, 187, 189, 210, 216–7, 233, 280, 283, 305, 307, 313
poultry 308
sheep x, xv–xvi, xxv, xxx, xxxiii, xxxvii–xlv, xlvii–li, liv, lvii, lxvi–lxvii, lxxv–lxxviii, xc, xcii, c–ci, civ–cviii, cx, cxii, cxxx, cxxxv, 1–330 *passim*, 340-5, 355, 357, 360-5, 380, 384
chilver xliv, 35, 103, 162, 254, 256–7, 278, 301, 307, 325, 358
ewe xvi, xxv, xxxix–xlv, xlvii–li, lxvi, lxxvi, lxxviii, c-ci, civ–cviii, cxviii, cxxx, 2–330 *passim*, 340-5, 353-6, 358-365, 380-1, 384
lamb xvi, xxv, xxxix–xlv, xlviii–l, lxxviii, xcin, xcix–c, cv–cviii, 1-329 *passim*, 340-4, 351, 358-365, 384
lambing xlviii-xlix, lxxviii, ci, cv, cviii, cx, 2–316 *passim*, 384
ram xlii, xliv–xlv, xlvii, l, c–ci, cvi, cviii, 7–326 *passim*, 340-5, 358, 360, 362-3, 365, 380
ruddle l
lodged (crops) liii, cxiii, 14, 29, 89, 176, 178-9, 194, 225, 249-250, 253–4, 271, 276

machines
horse-powered machinery xxvi, xcii–xciii, cxxviii
nine-share xcii
plough (and ploughing) xvi-cxxxiv *passim*, 1–331 *passim*
presser xxxiii, 10, 209
pump lxxxi–lxxxii, 101, 107, 114, 128–9, 146, 189, 238–40, 245, 285, 305, 309
roller xcii, cxvi, 10, 101
sawmill lxxxiii
seed drill xxxv-xxxvi, lxv, xci-xciii, xcvi, xcviii, 4-330 *passim*
seven-share xxxii, xcii
thrashing (threshing) xxv-cxxx *passim*, 6–330 *passim*, 377

makeout xxii, lviii, clvi, 356
marsh 84, 87, 130–1, 141, 151, 153–5, 158, 174, 177, 194, 196–7, 203–4, 273, 275, 298
marter earth lxxiin, 100–1, 106–10, 112, 118, 120, 122, 141, 147–50, 158, 177, 257, 277
mealman 172
Michaelmas xx–xxiii, liv, lvii, lxxii–lxxiii, cxii, 239, 256, 324
mortgage lxv, cxvi, 82
mowing xxxvii, lii, cix, 9-326 *passim*

nitrogen xn, xcvn

occupations
attorney cxivn, cxvin, 220n
banker cxxxvi
brewer lxxvn
butcher 97, 225, 292n, 294
carpenter lxxiv, lxxx, lxxxv, 97–325 *passim*
carter li-liii, cix, cxi-cxii, 17–330 *passim*
under-carter cxii
coroner 266n
dentist lxiv, 87
fellmonger xlviii-xlix
glazier lxxxiin, 207
horse breaker *see* Smith, Knightly
horse dealer ciiin *and see* Reeves (Reaves), Hillier (H)
maltster 129n, 291n
mason xxvii-xxviii, lxix, lxxx, lxxxiv-lxxxxv, cxxvii, 14–325 *passim*
miller 308
mole catcher cx, 169
mower lii, cix, 16, 32, 39, 87, 104-5, 110, 133, 302
nurserymen lxxviin, 154, 249n
ox-driver xxxii
plasterer 136, 142, 261
plumber lxxxi-lxxxiin
sawyer 114, 117–9, 129, 155
school master cxxxi, 318n
sheep dealer xlv, 104
shepherd lii, cx, cxii, 20, 24, 86, 106, 157, 199, 243, 272, 300, 307, 320
shepherd boy lii, 160, 195, 200n, 202, 267, 318
under-shepherd 200

waterman 247
wine merchant 264
wool-stapler 125, 131, 208, 223, 351n, 352n, 354-6
orchard xix–xx, xxiii, 75, 372

pest
black maggot (turnip pest) xcviii–xcix, 254–5, 299
maggot (sheep pest) 302, 328, 341
mice lxxxix, 155, 274, 295–6, 298
rabbit (rabbet) lxxxix–xc, 252, 267, 270, 272, 275–7, 279, 291, 296, 307–8, 330
warren lxxxix, 307
sparrow 296
tick cvii, 184, 202, 209, 234
pew xx-xxi, cxxxiii, 121, 142
portico lxxxiv–lxxxv, 121, 127, 129, 133, 136, 138
prayer, national (form) cxxxiii, 186

quarry lxxxiv, 136, 237
quicklime lxxiin
quicks [hedging] lxxviin

racehorse lxxiii–lxxiv, 257, 261
railway (railroad) lxxiv, cxii, cxxi, cxxiii
rick xxiv, xxviii, xxx, xlvii, lviii, lxvi, xciv, xcvii, cvii, cxxiv-cxxv, 1–328 *passim*, 347, 377
rick staddle xxviii, lxvi, xcii, 15, 100, 108-9, 113, 180, 187, 207, 214, 278, 280, 303
rickyard xxx, xlvii, lxxx, xcix, 11, 101, 109-110, 113, 115, 118, 122-6, 129, 133, 135, 138, 140, 145-7, 150-5, 210, 239, 267, 278, 285-6, 300, 325
road lxxv-lxxvii, 1-331 *passim and see* Avebury, turnpike
chalk carting 25, 46, 99, 113, 122, 140-1, 257
earth removal lv, 1, 45, 108, 113, 117, 123, 128, 132-3, 140, 143, 189, 211, 215, 261,325
road stone carting 24-5, 31, 216, 242, 257-9, 304
surveyor of highways cxxxii, 123

sharefarming xl–xli, cxxxviii
sheaves 15, 110-1, 181, 228, 269
soil, cultivation
drilling [of seed] xxxv-xxxvi, lxi, xci-xciv,xcviii, 4-330 *passim*
fallowing 1–330 *passim*
seven-shearing xxxii, xcii
sowing xvi–xcviii *passim*, 1–329 *passim*
seed broadcast xxxv-xxxvi, xciin
soil, fertilisation and improvement
burn-baking xcvn
burning xvi, 136, 182, 252
chalking lxxxv, lxxxvii, 96–8, 123–4, 233–6, 238
fertiliser xn, xxixn, xxxviii, lxxxviiin, xcin
ashes xxix, lxxxvii–lxxxviii, xci, xcv, 3–311 *passim*
bone manure xci
Daniel's patent manure xc
dung xn, xvi–cxxiii *passim*, 1–329 *passim*
guano xci
hen dung 174, 217
lime xcvi, 139
manure xlviii, lxxxiii, xc–xci, 140n, 309
muckle 8–309 *passim*
pigeon dung xxix
salt lxxxviii, 148, 151-2
soot xxix, lxxxviii, 13-4, 21, 29, 42, 48, 154, 169, 174, 217, 290
urine xcvi, 139
steam, use of
cucumber frame 71
potato steaming lxxxi, 324, 326
steam saw lxxxiii
stone xxviii–cxxvii *passim*, 21–322 *passim*
sarsen lxxiv, 105, 150, 152, 155, 165, 170, 257, 261
staddle xxviii *and see* rick, rick staddle

tar lxxxii, 160, 172, 278
tasker cix, 109, 160, 179–180, 203, 254, 278, 301, 325
tenancy lvi–lvii, cxviii
thatch and thatching *see* buildings
timber
ash 125
Dantzic lxxviii, lxxxiii, 96
deal lxxxiv, 117n, 128, 134, 137, 184-5

Memel lxxxiiin, lxxxiv, clvi, 114, 116, 128
oak 125, 128
treacle 210
tree xxxix, lii, lxxviin, lxxxii, xcix, cxxvii, 9–328 *passim and see* bark
ash lxvi, 31, 59, 222, 296, 319, 377
elm xcix, 86, 145-6, 296, 328
oak 31, 86, 296, 377
willow lxvi, 86
tithe (tythe) xvi, xxxviiin, lviii, lxiv, lxxi, lxxx, lxxxvi, cxiii, cxviii, cxx, cxxiii, 83, 184, 199, 324, 357-8, 372-3

verdigris l

water
flood xxiv-xxv, lx-lxi, lxxxiii, cx, cxiii, 2-330 *passim*
spring 151, 191, 213-5, 239-240, 268-274, 290, 293, 315, 330
well lxxiv, lxxxi-lxxxiii, 34, 96, 125, 186, 209-213, 239, 254, 259-260
water meadow (water mead) x, xx-xxi, xxx-xxxi, xxxiii, lxviii, lxix, lxxi, lxxxiii, civ, cviii, cx, cxii–cxiv, 4–315 *passim and see* carriage (water meadow)
trenching (maintenance of water meadow) 210, 238, 284, 307
weather ix–cxlix *passim*, 12–329 *passim*
barometer xxiv, 101, 303–5, 323, 325–6, 330–1
glass xxiv, 2-302 *passim*
barometric xxiv, 17, 122
lightning lxxvii, 16-322 *passim*
rain xvi-cxiii *passim*, 1-331 *passim*
snow xxiv, lxxv–lxxvi, cxviii, 1–330 *passim*
thunder lxxvii, 9–322 *passim*
thunderstorm lx, 11, 76, 85
whirlwind 85
weeds
charlock lxxxix, 108, 110, 112, 115, 132, 154–7, 175, 198, 225–7, 237, 252, 276, 321–2
docks lxxxix, 101, 127, 140, 198–9
wool xliii-xliv, cvi, 10, 41, 60, 131, 152, 186-7, 195, 199, 203, 205, 214-5, 224, 227, 246, 269, 273-4, 276, 311, 321, 342, 348-356, 363, 381
workhouse xci, cxix, cxxvn, 3

yoke cii, 99

WILTSHIRE RECORD SOCIETY
(AS AT APRIL 2021)

President: DR NEGLEY HARTE
Honorary Treasurer: IVOR M. SLOCOMBE
Honorary Secretary: MISS HELEN TAYLOR
General Editor (Acting): STEVEN HOBBS

Committee:
DR V. BAINBRIDGE
DR D.A. CROWLEY
DR J. HARE
S.D. HOBBS
DR T. PLANT
MRS S. THOMSON
K.H. ROGERS

Honorary Independent Examiner: C.C. DALE

PRIVATE MEMBERS

Note that because of recent legislation the Society no longer publishes members' addresses in its volumes, as it had done since 1953.

Honorary Members
OGBURN, SENR JUDGE R W
SHARMAN-CRAWFORD, MR T

ADAMS, MS S
ANDERSON, MR D M
BAINBRIDGE, DR V
BARNETT, MR B A
BATHE, MR G
BAYLIFFE, MR B G
BENNETT, DR N
BERRETT, MR A M
BERRY, MR C
BLAKE, MR P A
BOX, MR S D
BRAND, DR P A
BROCK, MRS C
BROWN, MR D A
BROWN, MR G R
BROWNING, MR E
BRYSON, DR A
CARTER, MR D
CAWTHORNE, MRS N
CHALMERS, MR D
CHANDLER, DR J H
CLARK, MR G A
COLCOMB, MR D M
COLLINS, MR A T
COOPER, MR S
CRAVEN, DR A
CROOK, MR P H
CROUCH, MR J W
CROWLEY, DR D A
CUNNINGTON, MS J
DAKERS, PROF C
D'ARCY, MR J N
DODD, MR D
DYSON, MRS L
EDE, DR M E
ELLIOTT, DR J
ENGLISH, MS K
FORREST, DR M
GAISFORD, MR J
GALE, MRS J
GHEY, MR J G
GINGER, MR A
GODDARD, MR R G H
GRIFFIN, DR C
GRIST, MR M
HARE, DR J N
HARTE, DR N
HAWKINS, MR D
HEATON, MR R J
HELMHOLZ, PROF R W
HENLY, MR C
HERRON, MRS Pamela M
HICKMAN, MR M R
HICKS, MR I
HICKS, PROF M A
HILLMAN, MR R B
HOBBS, MR S
HOWELLS, DR Jane
INGRAM, DR M J
JOHNSTON, MRS J M
JONES, MS J
KEEN, MR A.G
KENT, MR T A
KITE, MR P J
KNEEBONE, MR W J R
KNOWLES, MRS V A
LANSDOWNE, MARQUIS OF
LAWES, MRS G

Marsh, Rev R
Marshman, Mr M J
Martin, Ms J
Moles, Mrs M I
Morland, Mrs N
Napper, Mr L R
Newbury, Mr C Coles
Newman, Mrs R
Nicolson, Mr A
Nokes, Mr P M A
Noyce, Miss S
Ogbourne, Mr J M V
Ogburn, Mr D A
Parker, Dr P F,
Patience, Mr D C
Perry, Mr W A
Plant, Dr T
Powell, Mrs N
Price, Mr A J R
Railton, Ms A
Raybould, Miss F
Raymond, Mr S
Roberts, Ms M
Rogers, Mr K H
Rolfe, Mr R C
Saunt, Mrs B A
Sheldrake, Mr B
Shewring, Mr P
Skinner, Ms C
Slocombe, Mr I
Smith, Mr P J
Spaeth, Dr D A
Stone, Mr M J
Suter, Mrs C
Sutton, Mr A E
Tatton-Brown, Mr T
Taylor, Miss H
Thomson, Mrs S M
Wadsworth, Mrs S
Williamson, Mr B
Wiltshire, Mr J
Wiltshire, Mrs P E
Woodford, Mr A
Woodward, Mr A S,
Wright, Mr D P
Younger, Mr C

UNITED KINGDOM INSTITUTIONS

Aberystwyth
National Library of Wales
University College of Wales
Birmingham. University Library
Bristol
University of Bristol Library
Cambridge. University Library
Cheltenham. Bristol and Gloucestershire Archaeological Society
Chippenham
Museum & Heritage Centre
Wiltshire and Swindon History Centre
Coventry. University of Warwick Library
Devizes
Wiltshire Archaeological & Natural History Society
Wiltshire Family History Society
Durham. University Library
Edinburgh
University Library
Exeter. University Library
Glasgow. University Library

Liverpool. University Library
London
British Library
College of Arms
Guildhall Library
Inner Temple Library
Institute of Historical Research
London Library
The National Archives
Royal Historical Society
Society of Antiquaries
Society of Genealogists
Manchester. John Rylands Library
Marlborough
Memorial Library, Marlborough College
Merchant's House Trust
Savernake Estate Office
Norwich. University of East Anglia Library
Nottingham. University Library
Oxford
Bodleian Library
Exeter College Library
Reading. University Library
St Andrews. University Library
Salisbury
Bourne Valley Historical Society
Cathedral Library
Salisbury and South Wilts Museum
Swansea. University College Library
Swindon
Historic England
Swindon Borough Council
Taunton. Somerset Archaeological and Natural History Society
Wetherby. British Library Document Supply Centre
York. University Library

INSTITUTIONS OVERSEAS

AUSTRALIA

Adelaide. University Library

Crawley. Reid Library, University of Western Australia

CANADA

Halifax. Killam Library, Dalhousie University

Toronto, Ont
- Pontifical Inst of Medieval Studies
- University of Toronto Library

Victoria, B.C. McPherson Library, University of Victoria

NEW ZEALAND

Wellington. National Library of New Zealand

UNITED STATES OF AMERICA

Ann Arbor, Mich. Hatcher Library, University of Michigan

Athens, Ga. University of Georgia Libraries

Atlanta, Ga. The Robert W Woodruff Library, Emory University

Bloomington, Ind. Indiana University Library

Boston, Mass. New England Historic and Genealogical Society

Boulder, Colo. University of Colorado Library

Cambridge, Mass.
- Harvard College Library
- Harvard Law School Library

Charlottesville, Va. Alderman Library, University of Virginia

Chicago
- Newberry Library
- University of Chicago Library

Dallas, Texas. Public Library

Davis, Calif. University Library

East Lansing, Mich. Michigan State University Library

Evanston, Ill. United Libraries, Garrett/ Evangelical, Seabury

Fort Wayne, Ind. Allen County Public Library

Houston, Texas. M.D. Anderson Library, University of Houston

Iowa City, Iowa. University of Iowa Libraries

Ithaca, NY. Cornell University Library

Los Angeles
- Public Library
- Young Research Library, University of California

Minneapolis, Minn. Wilson Library, University of Minnesota

New York
- Columbia University of the City of New York

Salt Lake City, Utah. Family History Library

San Marino, Calif. Henry E. Huntington Library

Urbana, Ill. University of Illinois Library

Washington. The Folger Shakespeare Library

Winston-Salem, N.C. Z.Smith Reynolds Library, Wake Forest University

LIST OF PUBLICATIONS

The Wiltshire Record Society was founded in 1937, as the Records Branch of the Wiltshire Archaeological and Natural History Society, to promote the publication of the documentary sources for the history of Wiltshire. The annual subscription is £15 for private and institutional members. In return, a member receives a volume each year. Prospective members should apply to the Hon. Secretary, c/o Wiltshire and Swindon History Centre, Cocklebury Road, Chippenham SN15 3QN. Many more members are needed.

The following volumes have been published. Price to members £15, and to non-members £20, postage extra. Most volumes up to 51 are still available from the Wiltshire and Swindon History Centre, Cocklebury Road, Chippenham SN15 3QN. Volumes 52-71 are available from Hobnob Press, c/o 8 Lock Warehouse, Severn Road, Gloucester GL1 2GA. Volumes 1-55 are available online, at www.wiltshirerecordsociety.org.uk.

1. *Abstracts of feet of fines relating to Wiltshire for the reigns of Edward I and Edward II*, ed. R.B. Pugh, 1939
2. *Accounts of the parliamentary garrisons of Great Chalfield and Malmesbury, 1645–1646,* ed. J.H.P. Pafford, 1940
3. *Calendar of Antrobus deeds before 1625,* ed. R.B. Pugh, 1947
4. *Wiltshire county records: minutes of proceedings in sessions, 1563 and 1574 to 1592,* ed. H.C. Johnson, 1949
5. *List of Wiltshire boroughs records earlier in date than 1836,* ed. M.G. Rathbone, 1951
6. *The Trowbridge woollen industry as illustrated by the stock books of John and Thomas Clark, 1804–1824,* ed. R.P. Beckinsale, 1951
7. *Guild stewards' book of the borough of Calne, 1561–1688,* ed. A.W. Mabbs, 1953
8. *Andrews' and Dury's map of Wiltshire, 1773: a reduced facsimile,* ed. Elizabeth Crittall, 1952
9. *Surveys of the manors of Philip, earl of Pembroke and Montgomery, 1631–2,* ed. E. Kerridge, 1953
10. *Two sixteenth century taxations lists, 1545 and 1576,* ed. G.D. Ramsay, 1954
11. *Wiltshire quarter sessions and assizes, 1736,* ed. J.P.M. Fowle, 1955
12. *Collectanea*, ed. N.J. Williams, 1956
13. *Progress notes of Warden Woodward for the Wiltshire estates of New College, Oxford, 1659–1675,* ed. R.L. Rickard, 1957
14. *Accounts and surveys of the Wiltshire lands of Adam de Stratton*, ed. M.W. Farr, 1959
15. *Tradesmen in early-Stuart Wiltshire: a miscellany,* ed. N.J. Williams, 1960
16. *Crown pleas of the Wiltshire eyre, 1249,* ed. C.A.F. Meekings, 1961
17. *Wiltshire apprentices and their masters, 1710–1760,* ed. Christabel Dale, 1961
18. *Hemingby's register,* ed. Helena M. Chew, 1963
19. *Documents illustrating the Wiltshire textile trades in the eighteenth century,* ed. Julia de L. Mann, 1964
20. *The diary of Thomas Naish*, ed. Doreen Slatter, 1965
21–2. *The rolls of Highworth hundred, 1275–1287,* 2 parts, ed. Brenda Farr, 1966, 1968
23. *The earl of Hertford's lieutenancy papers, 1603–1612,* ed. W.P.D. Murphy, 1969
24. *Court rolls of the Wiltshire manors of Adam de Stratton*, ed. R.B. Pugh, 1970
25. *Abstracts of Wiltshire inclosure awards and agreements,* ed. R.E. Sandell, 1971
26. *Civil pleas of the Wiltshire eyre, 1249,* ed. M.T. Clanchy, 1971
27. *Wiltshire returns to the bishop's visitation queries, 1783,* ed. Mary Ransome, 1972
28. *Wiltshire extents for debts, Edward I – Elizabeth I,* ed. Angela Conyers, 1973
29. *Abstracts of feet of fines relating to Wiltshire for the reign of Edward III,* ed. C.R. Elrington, 1974

30. *Abstracts of Wiltshire tithe apportionments*, ed. R.E. Sandell, 1975
31. *Poverty in early-Stuart Salisbury*, ed. Paul Slack, 1975
32. *The subscription book of Bishops Tounson and Davenant, 1620–40*, ed. B. Williams, 1977
33. *Wiltshire gaol delivery and trailbaston trials, 1275–1306*, ed. R.B. Pugh, 1978
34. *Lacock abbey charters*, ed. K.H. Rogers, 1979
35. *The cartulary of Bradenstoke priory*, ed. Vera C.M. London, 1979
36. *Wiltshire coroners' bills, 1752–1796*, ed. R.F. Hunnisett, 1981
37. *The justicing notebook of William Hunt, 1744–1749*, ed. Elizabeth Crittall, 1982
38. *Two Elizabethan women: correspondence of Joan and Maria Thynne, 1575–1611*, ed. Alison D. Wall, 1983
39. *The register of John Chandler, dean of Salisbury, 1404–17*, ed. T.C.B. Timmins, 1984
40. *Wiltshire dissenters' meeting house certificates and registrations, 1689–1852*, ed. J.H. Chandler, 1985
41. *Abstracts of feet of fines relating to Wiltshire, 1377–1509*, ed. J.L. Kirby, 1986
42. *The Edington cartulary*, ed. Janet H. Stevenson, 1987
43. *The commonplace book of Sir Edward Bayntun of Bromham*, ed. Jane Freeman, 1988
44. *The diaries of Jeffery Whitaker, schoolmaster of Bratton, 1739–1741*, ed. Marjorie Reeves and Jean Morrison, 1989
45. *The Wiltshire tax list of 1332*, ed. D.A. Crowley, 1989
46. *Calendar of Bradford-on-Avon settlement examinations and removal orders, 1725–98*, ed. Phyllis Hembry, 1990
47. *Early trade directories of Wiltshire*, ed. K.H. Rogers and indexed by J.H. Chandler, 1992
48. *Star chamber suits of John and Thomas Warneford*, ed. F.E. Warneford, 1993
49. *The Hungerford Cartulary: a calendar of the earl of Radnor's cartulary of the Hungerford family*, ed. J.L. Kirby, 1994
50. *The Letters of John Peniston, Salisbury architect, Catholic, and Yeomanry Officer, 1823–1830*, ed. M. Cowan, 1996
51. *The Apprentice Registers of the Wiltshire Society, 1817– 1922*, ed. H. R. Henly, 1997
52. *Printed Maps of Wiltshire 1787–1844: a selection of topographical, road and canal maps in facsimile*, ed. John Chandler, 1998
53. *Monumental Inscriptions of Wiltshire: an edition, in facsimile, of Monumental Inscriptions in the County of Wilton, by Sir Thomas Phillipps*, ed. Peter Sherlock, 2000
54. *The First General Entry Book of the City of Salisbury, 1387–1452*, ed. David R. Carr, 2001
55. *Devizes Division income tax assessments, 1842–1860*, ed. Robert Colley, 2002
56. *Wiltshire Glebe Terriers, 1588–1827*, ed. Steven Hobbs, 2003
57. *Wiltshire Farming in the Seventeenth Century*, ed. Joseph Bettey, 2005
58. *Early Motor Vehicle Registration in Wiltshire, 1903–1914*, ed. Ian Hicks, 2006
59. *Marlborough Probate Inventories, 1591–1775*, ed. Lorelei Williams and Sally Thomson, 2007
60. *The Hungerford Cartulary, part 2: a calendar of the Hobhouse cartulary of the Hungerford family*, ed. J.L. Kirby, 2007
61. *The Court Records of Brinkworth and Charlton*, ed. Douglas Crowley, 2009
62. *The Diary of William Henry Tucker, 1825–1850*, ed. Helen Rogers, 2009
63. *Gleanings from Wiltshire Parish Registers*, ed. Steven Hobbs, 2010
64. *William Small's Cherished Memories and Associations*, ed. Jane Howells and Ruth Newman, 2011
65. *Crown Pleas of the Wiltshire Eyre, 1268*, ed. Brenda Farr and Christopher Elrington, rev. Henry Summerson, 2012
66. *The Minute Books of Froxfield Almshouse, 1714–1866*, ed. Douglas Crowley, 2013

67. *Wiltshire Quarter Sessions Order Book, 1642–1654,* ed. Ivor Slocombe, 2014
68. *The Register of John Blyth, Bishop of Salisbury, 1493–1499*, ed. David Wright, 2015
69 *The Churchwardens' Accounts of St Mary's, Devizes, 1633–1689*, ed. Alex Craven, 2016
70 *The Account Books and Papers of Everard and Ann Arundell of Ashcombe and Salisbury, 1745–1798*, ed. Barry Williamson, 2017
71 *Letters of Henry Hoare of Stourhead, 1760–81*, ed. Dudley Dodd, 2018
72 *Braydon Forest and the Forest Law*, ed. Douglas Crowley, 2019
73 *The Parish Registers of Thomas Crockford, 1561–1633* ed. John Chandler, 2020

Further details about the Society, its activities and publications, will be found on its website, www.wiltshirerecordsociety.org.uk.